Polymer Physics: From Theory to Experimental Applications

Polymer Physics: From Theory to Experimental Applications

Guest Editors

Célio Pinto Fernandes
Luís Lima Ferrás
Alexandre M. Afonso

Basel • Beijing • Wuhan • Barcelona • Belgrade • Novi Sad • Cluj • Manchester

Guest Editors

Célio Pinto Fernandes
Department of Mechanical
Engineering
University of Porto
Porto
Portugal

Luís Lima Ferrás
Department of Mechanical
Engineering
University of Porto
Porto
Portugal

Alexandre M. Afonso
Department of Mechanical
Engineering
University of Porto
Porto
Portugal

Editorial Office
MDPI AG
Grosspeteranlage 5
4052 Basel, Switzerland

This is a reprint of the Special Issue, published open access by the journal *Polymers* (ISSN 2073-4360), freely accessible at: www.mdpi.com/journal/polymers/special_issues/4763X2G2Z6.

For citation purposes, cite each article independently as indicated on the article page online and using the guide below:

Lastname, A.A.; Lastname, B.B. Article Title. *Journal Name* **Year**, *Volume Number*, Page Range.

ISBN 978-3-7258-2772-5 (Hbk)
ISBN 978-3-7258-2771-8 (PDF)
https://doi.org/10.3390/books978-3-7258-2771-8

© 2024 by the authors. Articles in this book are Open Access and distributed under the Creative Commons Attribution (CC BY) license. The book as a whole is distributed by MDPI under the terms and conditions of the Creative Commons Attribution-NonCommercial-NoDerivs (CC BY-NC-ND) license (https://creativecommons.org/licenses/by-nc-nd/4.0/).

Contents

About the Editors . . . . **vii**

Preface . . . . **ix**

Célio Fernandes, Luís L. Ferrás and Alexandre M. Afonso
Polymer Physics: From Theory to Experimental Applications
Reprinted from: *Polymers* **2024**, *16*, 768, https://doi.org/10.3390/polym16060768 . . . . **1**

Abdallah Barakat, Marc Al Ghazal, Romeo Sephyrin Fono Tamo, Akash Phadatare, John Unser and Joshua Hagan et al.
Development of a Cure Model for Unsaturated Polyester Resin Systems Based on Processing Conditions
Reprinted from: *Polymers* **2024**, *16*, 2391, https://doi.org/10.3390/polym16172391 . . . . **5**

Abdelsalam Alsarkhi and Mustafa Salah
Multiphase Flow Production Enhancement Using Drag Reducing Polymers
Reprinted from: *Polymers* **2023**, *15*, 1108, https://doi.org/10.3390/polym15051108 . . . . **23**

Alexander Semenov and Jörg Baschnagel
General Relations between Stress Fluctuations and Viscoelasticity in Amorphous Polymer and Glass-Forming Systems
Reprinted from: *Polymers* **2024**, *16*, 2336, https://doi.org/10.3390/polym16162336 . . . . **41**

Alya Harichane, Nadhir Toubal Seghir, Paweł Niewiadomski, Łukasz Sadowski and Michał Cisiński
Effectiveness of the Use of Polymers in High-Performance Concrete Containing Silica Fume
Reprinted from: *Polymers* **2023**, *15*, 3730, https://doi.org/10.3390/polym15183730 . . . . **74**

Andrey V. Subbotin, Alexander Ya. Malkin and Valery G. Kulichikhin
The Elasticity of Polymer Melts and Solutions in Shear and Extension Flows
Reprinted from: *Polymers* **2023**, *15*, 1051, https://doi.org/10.3390/polym15041051 . . . . **87**

Evgenii S. Baranovskii
Analytical Solutions to the Unsteady Poiseuille Flow of a Second Grade Fluid with Slip Boundary Conditions
Reprinted from: *Polymers* **2024**, *16*, 179, https://doi.org/10.3390/polym16020179 . . . . **113**

Fernando A. Lugo, Mariya Edeleva, Paul H. M. Van Steenberge and Maarten K. Sabbe
Improved Approach for ab Initio Calculations of Rate Coefficients for Secondary Reactions in Acrylate Free-Radical Polymerization
Reprinted from: *Polymers* **2024**, *16*, 872, https://doi.org/10.3390/polym16070872 . . . . **130**

Furui Shi and P.-Y. Ben Jar
Simulation and Analysis of the Loading, Relaxation, and Recovery Behavior of Polyethylene and Its Pipes
Reprinted from: *Polymers* **2024**, *16*, 3153, https://doi.org/10.3390/polym16223153 . . . . **162**

Jiwoong Ham, Hyeong-U Kim and Nari Jeon
Key Factors in Enhancing Pseudocapacitive Properties of PANI-InO$_x$ Hybrid Thin Films Prepared by Sequential Infiltration Synthesis
Reprinted from: *Polymers* **2023**, *15*, 2616, https://doi.org/10.3390/polym15122616 . . . . **184**

Juliana Bertoco, Antonio Castelo, Luís L. Ferrás and Célio Fernandes
Numerical Simulation of Three-Dimensional Free Surface Flows Using the K–BKZ–PSM Integral Constitutive Equation [†]
Reprinted from: *Polymers* **2023**, *15*, 3705, https://doi.org/10.3390/polym15183705 . . . . **193**

Koh-hei Nitta, Kota Ito and Asae Ito
A Phenomenological Model for Enthalpy Recovery in Polystyrene Using Dynamic Mechanical Spectra
Reprinted from: *Polymers* **2023**, *15*, 3590, https://doi.org/10.3390/polym15173590 . . . . **212**

Michael Roland Larsen, Erik Tomas Holmen Olofsson and Jon Spangenberg
Analyzing Homogeneity of Highly Viscous Polymer Suspensions in Change Can Mixers
Reprinted from: *Polymers* **2024**, *16*, 2675, https://doi.org/10.3390/polym16182675 . . . . **230**

Michael Roland Larsen, Tobias Ottsen, Erik Tomas Holmen Olofsson and Jon Spangenberg
Numerical Modeling of the Mixing of Highly Viscous Polymer Suspensions in Partially Filled Sigma Blade Mixers
Reprinted from: *Polymers* **2023**, *15*, 1938, https://doi.org/10.3390/polym15081938 . . . . **246**

Renáta Rusková and Dušan Račko
Knot Formation on DNA Pushed Inside Chiral Nanochannels
Reprinted from: *Polymers* **2023**, *15*, 4185, https://doi.org/10.3390/polym15204185 . . . . **258**

So Yeon Ahn, Chengbin Yu and Young Seok Song
Cellulose Nanocrystal Embedded Composite Foam and Its Carbonization for Energy Application
Reprinted from: *Polymers* **2023**, *15*, 3454, https://doi.org/10.3390/polym15163454 . . . . **276**

Xiao Li, Rujun Tang, Ding Li, Fengping Li, Leiqing Chen and Dehua Zhu et al.
Investigations of the Laser Ablation Mechanism of PMMA Microchannels Using Single-Pass and Multi-Pass Laser Scans
Reprinted from: *Polymers* **2024**, *16*, 2361, https://doi.org/10.3390/polym16162361 . . . . **291**

Yueqing Xing, Deqiang Sun and Guoliang Chen
Analysis of the Dynamic Cushioning Property of Expanded Polyethylene Based on the Stress–Energy Method
Reprinted from: *Polymers* **2023**, *15*, 3603, https://doi.org/10.3390/polym15173603 . . . . **312**

About the Editors

Célio Pinto Fernandes

C. Fernandes is an assistant professor at the Department of Mechanical Engineering, Faculty of Engineering, University of Porto, (FEUP), Portugal. C. Fernandes obtained his Education of Mathematics bachelor's degree from the University of Minho (Portugal) in 2005. After graduating, C. Fernandes obtained his Applied Mathematics MSc degree at the University of Porto (Portugal) in 2007. During this period, C. Fernandes employed spectral methods to describe the melt flow that occurs in the polymer extrusion process. Afterwards, C. Fernandes joined the Department of Polymer Engineering at the University of Minho (Portugal), where he completed his Ph.D. degree in Science and Engineering of Polymers and Composites in 2012. During this period, C. Fernandes made important contributions to the field of the injection molding process by applying multi-objective evolutionary algorithms to solve the inverse problem of finding the best injection molding parameters to achieve predefined criteria. C. Fernandes was a visiting post-doctoral researcher at MIT, USA, in 2017. C. Fernandes has been working with the open-source computational fluid dynamics software OpenFOAM. He has established new numerical methods for the solution of viscoelastic matrix-based fluids using the finite volume method, such as an immersed boundary method able to fully resolve particle-laden viscoelastic flows and developed a fully implicit log-conformation tensor coupled algorithm for the solution of incompressible non-isothermal viscoelastic flows.

Luís Lima Ferrás

L. L. Ferras is an assistant professor at the Department of Mechanical Engineering, Faculty of Engineering, University of Porto (FEUP), and a researcher at the Centre for Mathematics, University of Minho, Portugal. He received his Ph.D. in Science and Engineering of Polymers and Composites from the University of Minho in 2012, a Ph.D. in Mathematics from the University of Chester in 2019, and was a visiting researcher at MIT in 2016. His current research interests are numerical analysis, applied mathematics, partial and fractional differential equations, mathematical modeling, computational mechanics, computational fluid dynamics, complex viscoelastic flows, rheology, anomalous diffusion, and machine learning.

Alexandre M. Afonso

Afonso graduated with a degree in Chemical Engineering from the Faculty of Engineering at the University of Porto (FEUP) in 2000, with a final-year research project at the Universidad Politécnica de Catalunya graded with Honors (10/10). In 2005, Afonso completed an MSc in Heat and Fluid Mechanics and, in 2010, completed a PhD degree in Biological and Chemical Engineering from FEUP. Currently, Afonso is an assistant professor at the Department of Mechanical Engineering at FEUP.

Preface

The importance of polymer processing techniques in the production of polymer components cannot be overstated. The primary goal is to manufacture parts that meet specific quality standards, typically including mechanical performance, dimensional accuracy, and visual aesthetics. Achieving maximum efficiency in polymer processing relies on advanced modeling tools combined with experimental studies, which will, in turn, help us understand and optimize the underlying processes. This Reprint aims to showcase cutting-edge research papers that advance numerical, theoretical, and experimental knowledge regarding polymer physics.

Célio Pinto Fernandes, Luís Lima Ferrás, and Alexandre M. Afonso
Guest Editors

Editorial

Polymer Physics: From Theory to Experimental Applications

Célio Fernandes [1,2,*], Luís L. Ferrás [1,2] and Alexandre M. Afonso [1,3]

1. Center for Studies of Transport Phenomena (CEFT), Department of Mechanical Engineering, Faculty of Engineering, University of Porto, 4200-465 Porto, Portugal; lferras@fe.up.pt (L.L.F.); aafonso@fe.up.pt (A.M.A.)
2. Center of Mathematics (CMAT), University of Minho, Campus de Azurém, 4800-058 Guimarães, Portugal
3. ALiCE, Associate Laboratory in Chemical Engineering, Faculdade de Engenharia, Universidade do Porto, Rua Dr. Roberto Frias s/n, 4200-465 Porto, Portugal
* Correspondence: cbpf@fe.up.pt

The significance of polymer processing techniques cannot be overstated in the production of polymer components. The primary objective is to create parts that meet specific quality criteria, typically encompassing mechanical performance, dimensional accuracy, and visual aesthetics. Achieving optimal efficiency in polymer processing requires the use of advanced modeling codes in conjunction with experimental work to understand and optimize the underlying processes. In this editorial, we present cutting-edge papers that contribute to the numerical, theoretical, and experimental knowledge of polymer physics.

Regarding numerical methods, Bertoco et al. [1] developed a novel numerical method that was designed to address three-dimensional unsteady free surface flows incorporating integral viscoelastic constitutive equations, specifically the K–BKZ–PSM (Kaye–Bernstein, Kearsley, Zapas–Papanastasiou, Scriven, and Macosko) model. This new proposed methodology employs a second-order finite difference approach along with the deformation fields method to solve the integral constitutive equation and the marker particle method (known as marker-and-cell) to accurately capture the evolution of the fluid's free surface. This newly developed numerical method proved its effectiveness in handling complex fluid flow scenarios, including confined flows and extrudate swell simulations of Boger fluids. Larsen et al. [2] presented a non-isothermal, non-Newtonian Computational Fluid Dynamics model for the mixing of a highly viscous polymer suspension in a partially filled sigma blade mixer. This model accounts for viscous heating and the free surface of the suspension. The rheological model was found by calibration with experimental temperature measurements and was exploited to study the effect of applying heat both before and during mixing on the suspension's mixing quality. Two mixing indexes were used to evaluate the mixing condition, namely, the Ica Manas-Zloczower dispersive index and Kramer's distributive index. The Kramer index results were stable and indicated that the particles in the suspension could be well distributed. Interestingly, the results highlighted that the speed at which the suspension becomes well distributed is almost independent of applying heat both before and during the process. Rusková et al. [3] performed coarse-grained molecular dynamics simulations of DNA polymers pushed inside infinite open chiral and achiral channels. They investigated the behavior of the polymer metrics in terms of span, monomer distributions, and changes in the topological state of the polymer in the channels. These authors also compared the regime of pushing a polymer inside the infinite channel to the case of polymer compression in the finite channels of knot factories investigated in earlier works. It was observed that the compression in the open channels affects the polymer metrics to different extents in the chiral and achiral channels. Also, the chiral channels give rise to the formation of equichiral knots with the same handedness as the chiral channels.On the theoretical side, Nitta et al. [4] studied the effects of annealing time on the specific heat enthalpy of polystyrene above the glass transition temperature. These authors extended the Tool–Narayanaswamy–Moynihan model to describe the endothermic

Citation: Fernandes, C.; Ferrás, L.L.; Afonso, A.M. Polymer Physics: From Theory to Experimental Applications. *Polymers* 2024, *16*, 768. https://doi.org/10.3390/polym16060768

Received: 2 March 2024
Accepted: 7 March 2024
Published: 11 March 2024

Copyright: © 2024 by the authors. Licensee MDPI, Basel, Switzerland. This article is an open access article distributed under the terms and conditions of the Creative Commons Attribution (CC BY) license (https:// creativecommons.org/licenses/by/ 4.0/).

overshoot peaks through the dynamic mechanical spectra. In their work, these authors consider the viewpoint that the enthalpy recovery behavior of glassy polystyrene (PS) has a common structural relaxation mode with linear viscoelastic behavior. As a consequence, the retardation spectrum evaluated from the dynamic mechanical spectra around the primary Tg peak was used as the recovery function of the endothermic overshoot of specific heat. In addition, the sub-Tg shoulder peak around the Tg peak was found to be related to the structural relaxation estimated from light scattering measurements. The enthalpy recovery of annealed PS was quantitatively described using retardation spectra of the primary Tg as well as the kinetic process of the sub-Tg relaxation process. Baranovskii [5] studied the unidirectional pressure-driven flow of a second-grade fluid within a planar channel bounded by impermeable solid walls. These authors examined the well-posed nature of this problem and derived its analytical solution while imposing weak regularity conditions on a function representing the initial velocity distribution. In addition, the notion of a generalized solution, defined as the limit of a uniformly convergent sequence of classical solutions with diminishing perturbations in the initial data, was employed, and the unique solvability of the problem under consideration in the class of generalized solutions was achieved. The conclusion of this work was that the developed analytical solutions facilitate a deeper comprehension of the qualitative characteristics of time-dependent flows involving polymer fluids.

Concerning experimental work, Alsarkhi et al. [6] presented a comprehensive experimental investigation concerning the effect of drag-reducing polymers (DRPs) on enhancing the throughput and reducing the pressure drop for a horizontal pipe carrying a two-phase flow of an air and water mixture. They showed the ability of these polymer entanglements to dampen turbulence waves and change the flow regime, and it was observed that the maximum drag reduction always occurs when the highly fluctuating waves are effectively reduced by DRPs. Furthermore, different empirical correlations have been developed that improve the ability to predict the pressure drop after the addition of DRP. The correlations showed low discrepancies for a wide range of water and air flow rates. Ham et al. [7] investigated the effects of the number of InO_x SIS (sequential infiltration synthesis) cycles on the chemical and electrochemical properties of PANI-InO_x thin films via combined characterization using X-ray photoelectron spectroscopy, ultraviolet–visible spectroscopy, Raman spectroscopy, Fourier-transform infrared spectroscopy, and cyclic voltammetry. The area-specific capacitance values of PANI-InO_x samples prepared with 10, 20, 50, and 100 SIS cycles were 1.1, 0.8, 1.4, and 0.96 mF/cm^2, respectively. These results highlighted that the formation of an enlarged PANI-InO_x mixed region directly exposed to the electrolyte is key to enhancing the pseudocapacitive properties of the composite films. Ahn et al. [8] fabricated a cellulose nanocrystal (CNC)-embedded aerogel-like chitosan foam and carbonized the 3D foam for electrical energy harvesting. The nanocrystal-supported cellulose foam showed a high surface area and porosity, homogeneous size ranging from various microscales, and a high quality of absorbing external additives. In order to prepare the CNC, microcrystalline cellulose (MCC) was chemically treated with sulfuric acid. The CNC incorporates into chitosan, enhancing mechanical properties, crystallization, and the generation of the aerogel-like porous structure. The weight percentage of the CNC was 2 wt% in the chitosan composite. The CNC/chitosan foam was produced using the freeze-drying method, and the CNC-embedded CNC/chitosan foam was carbonized. These authors found that the degree of crystallization of carbon structure increased, including the CNCs. Both CNC and chitosan are degradable materials when CNC includes chitosan, which can form a high surface area with some typical surface-related morphology. The electrical cyclic voltametric results indicated that the vertical composite specimen had superior electrochemical properties compared to the horizontal composite specimen. In addition, the BET measurement indicated that the CNC/chitosan foam possessed a high porosity, especially mesopores with layer structures. At the same time, the carbonized CNC led to a significant increase in the portion of micropore. Xing et al. [9] experimentally studied the dynamic crushing performance of expanded polyethylene (EPE) and analyzed the influence of thickness and dropping height on its mechanical behavior based on the stress–energy method. Hence, a series of impact tests were carried out on EPE foams with different thicknesses and dropping heights. The maximum acceleration, static stress, dynamic stress, and dynamic energy of EPE specimens were obtained through a dynamic impact test. Then, according to the principle of the stress–energy method, the functional relationship between dynamic stress and dynamic energy was obtained through exponential fitting and polynomial fitting, and the cushion material constants a, b, and c were determined. When analyzing the influence of thickness and dropping height on the dynamic cushioning performance curves of EPE, it was found that at the same drop height, with the increase of thickness, the opening of the curve gradually becomes larger. The minimum point on the maximum acceleration–static stress curve also decreased with the increase in thickness. When the dropping height was 400 mm, compared to the foam with a thickness of 60 mm, the tested maximum acceleration value of the lowest point of the specimen with a thickness of 40 mm increased by 45.3%, and the static stress was both 5.5 kPa. When the thickness of the specimen was 50 mm, compared to the dropping height of 300 mm, the tested maximum acceleration value of the lowest point of the specimen with a dropping height of 600 mm increased by 93.3%. Therefore, the dynamic cushioning performance curve of EPE foams can be quickly obtained by the stress–energy method when the precision requirement is not high, which provides a theoretical basis for the design of cushion packaging. Harichane et al. [10] studied the influence of three types of PCEs (polycarboxylate ether superplasticizer), which all have different molecular architectures, on the rheological and mechanical behavior of high-performance concretes containing 10% SF (silica fume) as a partial replacement of cement. Their results revealed that the carboxylic density of PCE has an influence on its compatibility with SF.

Finally, Subbotin et al. [11] presented a review devoted to understanding the role of elasticity in the main flow modes of polymeric viscoelastic liquids—shearing and extension. The flow through short capillaries is the central topic for discussing the input of elasticity to the effects, which are especially interesting for shear. An analysis of the experimental data made it possible to show that the energy losses in such flows are determined by the Deborah and Weissenberg numbers. These criteria are responsible for abnormally high entrance effects as well as for mechanical losses in short capillaries. In addition, the Weissenberg number determines the threshold of the flow instability due to the liquid-to-solid transition. In extension, this criterion shows whether deformation takes place as flow or as elastic strain. However, the stability of a free jet in extension not only depends on the viscoelastic properties of a polymeric substance but also on the driving forces: gravity, surface tension, etc.

The editors express confidence that this editorial will facilitate researchers in comprehending the fundamental principles of polymer physics from both numerical and experimental viewpoints. Moreover, this editorial serves as a valuable reference for those keen on staying abreast of cutting-edge technologies.

Author Contributions: Conceptualization, C.F., L.L.F. and A.M.A.; writing, C.F., L.L.F. and A.M.A.; review and editing, C.F., L.L.F. and A.M.A. All authors have read and agreed to the published version of the manuscript.

Funding: This research was funded by national funds through the FCT/MCTES (PIDDAC), LA/P/0045/2020 (ALiCE), UIDB/00532/2020, and UIDP/00532/2020 (CEFT) projects. It was also funded by the FCT through the CMAT (Centre of Mathematics of the University of Minho) projects UIDB/00013/2020 and UIDP/00013/2020. This work was also financially supported by national funds through the FCT/MCTES (PIDDAC) under the project 2022.06672.PTDC—iMAD—Improving the Modeling of Anomalous Diffusion and Viscoelasticity: Solutions to Industrial Problems. This work was also financially supported by national funds through the FCT under the project 2022.00753.CEECIND.

Conflicts of Interest: The authors declare no conflicts of interest.

References

1. Bertoco, J.; Castelo, A.; Ferrás, L.; Fernandes, C. Numerical Simulation of Three-Dimensional Free Surface Flows Using the K-BKZ-PSM Integral Constitutive Equation. *Polymers* **2023**, *15*, 3705. [CrossRef] [PubMed]
2. Larsen, M.; Ottsen, T.; Holmen Olofsson, E.; Spangenberg, J. Numerical Modeling of the Mixing of Highly Viscous Polymer Suspensions in Partially Filled Sigma Blade Mixers. *Polymers* **2023**, *15*, 1938. [CrossRef] [PubMed]
3. Rusková, R.; Račko, D. Knot Formation on DNA Pushed Inside Chiral Nanochannels. *Polymers* **2023**, *15*, 4185. [CrossRef] [PubMed]
4. Nitta, K.; Ito, K.; Ito, A. A Phenomenological Model for Enthalpy Recovery in Polystyrene Using Dynamic Mechanical Spectra. *Polymers* **2023**, *15*, 3590. [CrossRef] [PubMed]
5. Baranovskii, E.S. Analytical Solutions to the Unsteady Poiseuille Flow of a Second Grade Fluid with Slip Boundary Conditions. *Polymers* **2024**, *16*, 179. [CrossRef] [PubMed]
6. Alsarkhi, A.; Salah, M. Multiphase Flow Production Enhancement Using Drag Reducing Polymers. *Polymers* **2023**, *15*, 1108. [CrossRef] [PubMed]
7. Ham, J.; Kim, H.; Jeon, N. Key Factors in Enhancing Pseudocapacitive Properties of PANI-InOx Hybrid Thin Films Prepared by Sequential Infiltration Synthesis. *Polymers* **2023**, *15*, 2616. [CrossRef]
8. Ahn, S.; Yu, C.; Song, Y. Cellulose Nanocrystal Embedded Composite Foam and Its Carbonization for Energy Application. *Polymers* **2023**, *15*, 3454. [CrossRef] [PubMed]
9. Xing, Y.; Sun, D.; Chen, G. Analysis of the Dynamic Cushioning Property of Expanded Polyethylene Based on the Stress-Energy Method. *Polymers* **2023**, *15*, 3603. [CrossRef]
10. Harichane, A.; Seghir, N.; Niewiadomski, P.; Sadowski, Ł.; Cisiński, M. Effectiveness of the Use of Polymers in High-Performance Concrete Containing Silica Fume. *Polymers* **2023**, *15*, 3730. [CrossRef] [PubMed]
11. Subbotin, A.; Malkin, A.; Kulichikhin, V. The Elasticity of Polymer Melts and Solutions in Shear and Extension Flows. *Polymers* **2023**, *15*, 105. [CrossRef] [PubMed]

Disclaimer/Publisher's Note: The statements, opinions and data contained in all publications are solely those of the individual author(s) and contributor(s) and not of MDPI and/or the editor(s). MDPI and/or the editor(s) disclaim responsibility for any injury to people or property resulting from any ideas, methods, instructions or products referred to in the content.

Article

Development of a Cure Model for Unsaturated Polyester Resin Systems Based on Processing Conditions

Abdallah Barakat [1], Marc Al Ghazal [1], Romeo Sephyrin Fono Tamo [1], Akash Phadatare [1], John Unser [2], Joshua Hagan [3] and Uday Vaidya [1,4,5,*]

[1] Department of Mechanical, Aerospace, and Biomedical Engineering, University of Tennessee, Knoxville, TN 37996, USA; aragab@vols.utk.edu (A.B.); malghaza@vols.utk.edu (M.A.G.); rfonotam@utk.edu (R.S.F.T.); aphadata@vols.utk.edu (A.P.)
[2] Composite Applications Group (CAG), 3137 Waterfront Dr, Chattanooga, TN 37419, USA; john@compositeapplicationsgroup.com
[3] Research and Development Department, Wabash National Corporation, 3550 Veterans Memorial Pkwy S, Lafayette, IN 47909, USA; joshua.hagan@onewabash.com
[4] Manufacturing Sciences Division (MSD), Oak Ridge National Laboratory (ORNL), 2350 Cherahala Blvd, Knoxville, TN 37932, USA
[5] The Institute for Advanced Composites Manufacturing Innovation, 2370 Cherahala Blvd, Knoxville, TN 37932, USA
* Correspondence: uvaidya@utk.edu

Citation: Barakat, A.; Al Ghazal, M.; Fono Tamo, R.S.; Phadatare, A.; Unser, J.; Hagan, J.; Vaidya, U. Development of a Cure Model for Unsaturated Polyester Resin Systems Based on Processing Conditions. *Polymers* **2024**, *16*, 2391. https://doi.org/10.3390/polym16172391

Academic Editor: Chengji Zhao

Received: 11 July 2024
Revised: 19 August 2024
Accepted: 20 August 2024
Published: 23 August 2024

Copyright: © 2024 by the authors. Licensee MDPI, Basel, Switzerland. This article is an open access article distributed under the terms and conditions of the Creative Commons Attribution (CC BY) license (https://creativecommons.org/licenses/by/4.0/).

Abstract: Unsaturated polyester resin (UPR) systems are extensively used in composite materials for applications in the transportation, marine, and infrastructure sectors. There are continually evolving formulations of UPRs that need to be evaluated and optimized for processing. Differential Scanning Calorimetry (DSC) provides valuable insight into the non-isothermal and isothermal behavior of UPRs within a prescribed temperature range. In the present work, non-isothermal DSC tests were carried out between temperatures of 0.0 °C and 250 °C, through different heating and cooling ramp rates. The isothermal DSC tests were carried out between 0.0 and 170 °C. The instantaneous rate of cure of the tested temperatures were measured. The application of an autocatalytic model in a calculator was used to simulate curing behaviors under different processing conditions. As the temperature increased from 10 °C up to 170 °C, the rate of cure reduced, and the heat of reaction increased. The simulated cure behavior from the DSC data showed that the degree of cure (α) maximum value of 71.25% was achieved at the highest heating temperature of 85 °C. For the low heating temperature, i.e., 5 °C, the maximum degree of cure (α) did not exceed 12% because there was not enough heat to activate the catalyst to crosslink further.

Keywords: unsaturated polyester resin; DSC; autocatalytic model; cure kinetics; cure behavior; cure simulation

1. Introduction

Due to the synthetic nature and liquid phase of polyester, its processing conditions require specific parameters and steps to realize the final product [1–3]. These parameters affect the final mechanical and thermal properties of the polyester. Polymerization begins with the raw materials, typically terephthalic acid or its esters, along with ethylene glycol [4,5].

Unsaturated polyester resins (UPRs) are widely used in various sectors such as marine, automotive, and construction infrastructure, to name a few. The chemical process of UPRs includes creating UPRs through poly-esterification or gradual ionic copolymerization. The resulting UPR is dissolved in an unsaturated monomer and subsequently crosslinked using radical polymerization [6,7].

The cure behavior of UPRs is influenced by temperature, catalysts, catalyst concentration, promoters and inhibitors, stoichiometry, resin formulation, atmosphere, curing time, and post-cure processes. These parameters control the transition rate of resin from liquid to solid state through crosslinking. The selection of these parameters is highly dependent on the end-use application and the target properties of the cured material [8,9]. It is also highly dependent on the physical characteristics of the resin system, i.e., resin viscosity, gel point, and vitrification [10].

The resin formulation of UPR influences its curing behavior, which is based on the type of monomers and their functionality [10]. Also, the presence of additives, i.e., reinforcements and fillers, would impact the overall curing process [11,12]. The type of catalysts that initiate the curing reaction, i.e., peroxides, amines, and metal salts, as well as their concentration and amount, significantly influence the curing behavior. These factors affect the desired rate of the curing reaction and the specific properties of the final cured material too [13,14].

Using the proper exact ratio of resin (monomers) and catalyst (crosslinking agents), the optimal curing behavior would be achieved. Moreover, promoters and inhibitors are additives used to control the curing behavior to achieve the desired properties. Promoters and inhibitors are added, if desired, to enhance the activity of catalysts and slow down the curing reaction, respectively.

External parameters also impact the curing behavior, e.g., temperature, pressure, volume, and processing time [15,16]. As the curing reactions are temperature-dependent, heat input provides the necessary activation energy for the reaction [10]. Increasing temperature accelerates curing, while excessive heat degrades the material. Applying pressure to a certain volume of the resin system has a direct effect on its cure behavior. For example, ultra-high curing pressure, around 100–400 MPa, shortens the curing time and reduces the degree of cure and glass transition temperature [17,18]. A significant increase in the curing pressure also might cause a weak glue and fiber distortion, as the resin flows faster in the fiber medium [19,20]. However, decreasing the curing pressure (8–10 MPa, depending on the resin and thickness) allows void accumulation in the matrix [21–24]. Allowing sufficient time for the curing process achieves the desired mechanical and chemical properties.

Some resin systems require a post-cure stage to enhance the final properties. Environmental conditions also have a critical influence on the curing process. In some cases, the humidity level, the presence of oxygen, and the surrounding temperature influence the curing reactions.

Different studies have optimized the cure profile [25–31] to minimize the residual stresses and shape distortion, with a reduction in the residual stresses up to 30% [32–37]. As the part geometry and thickness influence the cure behavior, it is desired to control the excessive exothermic heat of the thick thermosetting composites. The cure control reduces the process duration for thick parts by 30% [26–28], and 50% in the ultra-thick parts [29–31]. Several studies have simulated the cure behavior of different resin systems to optimize the cure profile [38–40] and to obtain the desired part performance of different processing conditions [41–46].

Based on their structure, UPRs are classified as ortho-resins, iso-resins, biphenol-A fumarates, Chlorendics, and vinyl ester resins. Several studies tackled the use of promoters and co-promoters to modify the cure profile of UPRs [47–49]. Naderi et al. (2015) [50] added 1–5% nano clay in a UPR containing Na-Montmorillonite to investigate the cure behavior and the cure kinetic parameters. Their investigation showed a decrease in the gel time and exothermic peak, while there was an increase in the cure rate. The inclusion of carbon nanotubes in a polyester resin system (D-EP/CNTs-H20) significantly shifted the cure temperature to a lower temperature while accelerating the cure [51]. This shows the cure reaction can be accelerated at a lower temperature provided the relevant catalyst is used. Calabrese et al. (2023) [52] demonstrated that commercial unsaturated polyester imide resins possessed excellent thermal resistance of up to 320 °C.

An examination of the effect of cure temperature on the mechanical properties of a polyester resin was carried out by Silva et al. (2020) [15]. Their results indicated a significant improvement of the bending strength when the resin was post-cured at 40 °C and 60 °C as compared to room temperature. Nacher et al. (2007) [53] examined how the curing rate influences the mechanical properties of a polyester resin immersed in saline water for 500 h. They realized that the highly crosslinked internal structure of the resin presents a smaller capacity to absorb water than a structure with a low degree of crosslinking. Moreover, examination of the influence of curing rate on the mechanical properties of polyester resin under saline water revealed that the highly crosslinked resin structure withstands more compared to the resin system with a low degree of crosslinking due to lower water absorption [53]. Furthermore, an important increase in tensile strength was observed with the crosslinking grade for each cured temperature. For instance, at the cured temperature of 60 °C, tensile strength increased from 14.6 MPa for 60% of crosslinking to 34.1 MPa for 90% in reference samples. But in the degradation samples, this increase was from 11.3 to 28.2 MPa.

The glass transition phase is an important factor in the cure condition of unsaturated polyester resins. There has been keen interest among researchers to increase the glass transition temperature without altering the crosslink of the UPRs. By raising the styrene content from 35 to 50 wt% in their newly developed thermal curing profile without thermal initiators, Stuck et al. (2020) [54] saw a drastic increase in the glass transition temperature (195–215 °C) at 10 Hz. Delaite et al. (2020) [55] enhanced the glass transition temperature (T_g) from 197 °C to 130 °C by replacing 35 wt% of styrene with butyl methacrylate.

The degradation of unsaturated polyester resin depends mostly on the environmental conditions, and it can take various courses, as evaluated by Pączkowski et al. (2020) [56]. UPR is subjected to accelerated aging, immersion in different solvents, and high temperatures for the purpose of comparing their degradation.

There are many processing parameters involved during resin curing; therefore, extensive testing and optimization are needed for new UPR formulations. The main objective of this work was to create a tool to determine the optimal process conditions for the UPR system. Material characterizations were conducted to evaluate the conversion rate of the resin system with different processing parameters. A model was developed to simulate the curing process based on the desired input processing parameters. The study aims to employ an approach to comprehend the behavior of resin systems under any given processing condition through a systematic solution to optimize processing conditions. The objectives of the present work are summarized as follows:

Conducting gel tests and TGA and DSC runs to characterize and study the cure kinetics of two resin systems.

Simulating the cure kinetics behaviors under different temperatures for the two resin systems using the autocatalytic model.

2. Experimental and Numerical Methodology

A material characterization study was carried out on two UPR systems through various techniques including the gelation test, thermographic analysis (TGA), and DSC. These two resin systems are referred to as Resin system 1 (COR61-AA-248S) and Resin system 2 (COR61-AA-270LF).

2.1. Polymer Characterization

Three gel tests were conducted on each UPR system 1 and 2. Temperature was recorded throughout the curing process by the HH374-Omega data logger thermometer and Omega thermocouples. For each test, 100 g of resin was mixed with 1.0 g (1.0 wt%) of Norox methyl ethyl ketone peroxide (MEKP)-925H initiator. The mixture was placed in an open polypropylene container made by FibreGlast, with a thermal conductivity coefficient of 0.2 W/(m°C).

The TGA Q50 (V20.13, Build 39) instrument was used to understand the mass loss of both polyester resins. Around 15.0 mg of resin was placed in a platinum pan. The sample chamber was flushed with nitrogen. The ramp rate was 10 °C/min up to a maximum temperature of 500 °C. The resin degradation temperature was recorded.

TA Instruments DSC Q-2000 was used for the measurement of evolved heat during the cure reaction of UPR with 1.0% MEKP. All the DSC experiments were carried out using hermetic aluminum sample pans with sample weights of 20 mg (±4 mg). The isothermal reaction rate versus time profiles were measured at 10–120 °C and 170 °C. The isothermal runs were conducted as per Table 1. Isothermal DSC runs with no initiator were performed for 10 min to evaluate the impact of heat on the resin.

Table 1. DSC run time with respect to isothermal temperatures.

Temperature (°C)	DSC Run Time (min)
10–20	180
30–50	120
60–120	30
170	30

The total heat of the cure reaction (HU) was investigated using the measurement of heat flow with heating ramps for 5 °C/min, 10 °C/min, and 20 °C/min using dynamic DSC experiments carried out from temperature range 0 to 250 °C. For non-isothermal DSC runs, Tert-butyl peroxybenzoate (Trigonox C) was used as an initiator due to its thermal stability at high temperature. The following equations were used for the heat of reaction calculation [57–60].

The total heat of the reaction (HU) can be computed by the non-isothermal DSC runs. Equation (1) is used for HU calculation:

$$HU = \int_0^{td} \left(\frac{dQ}{dt}\right) d \, dt \tag{1}$$

where $\left(\frac{dQ}{dt}\right)d$ represents the instantaneous rate of heat produced, while td corresponds to the time required for the completion of the reaction in the non-isothermal DSC experiment.

Equation (2) gives the amount of energy (heat generation) in an isothermal experiment form the start of the cure up to the time t:

$$H(t) = \int_0^t \left(\frac{dQ}{dt}\right) T \, dt \tag{2}$$

where $\left(\frac{dQ}{dt}\right)T$ represents the instantaneous rate of energy generated at temperature T.

The heat generation $H(t)$ is directly proportional to the degree of reaction, $\alpha(t)$, characterized by Equation (3):

$$\alpha(t) = \frac{H(t)}{HU} \tag{3}$$

The total isothermal heat of reaction (HT) from isothermal scanning experiments can be expressed by Equation (4):

$$HT = \int_0^{ti} \left(\frac{dQ}{dt}\right) i \, dt \tag{4}$$

Similarly, in case of isothermal scanning runs, $\left(\frac{dQ}{dt}\right)i$ represents the instantaneous rate of heat generated and ti is the amount of time required to complete the reaction.

The actual rate of cure as function of time $\left(\frac{d\alpha}{dt}\right)$ and the isothermal rate of cure $\left(\frac{d\beta}{dt}\right)$ can be defined by Equations (5) and (6) [60]:

$$\frac{d\alpha}{dt} = \frac{1}{HU}\left(\frac{dQ}{dt}\right)T \tag{5}$$

$$\frac{d\beta}{dt} = \frac{1}{HT}\left(\frac{dQ}{dt}\right)T \tag{6}$$

Furthermore, the actual rate of cure $\left(\frac{d\alpha}{dt}\right)$ can be related to the isothermal rate of cure $\left(\frac{d\beta}{dt}\right)$, as per Equation (7):

$$\frac{d\alpha}{dt} = \frac{HT}{HU}\frac{1}{HT}\left(\frac{dQ}{dt}\right)T = \frac{HT}{HU}\frac{d\beta}{dt} \qquad (7)$$

$\beta(t)$ is defined as the extent of reaction in the isothermal scan at time t which can be expressed by Equation (8):

$$\beta(t) = \frac{H(t)}{HT} \qquad (8)$$

The ratio between HT and HU expresses the degree of incomplete reaction and can be thought of as a piecewise function, as shown in Equation (9):

$$\frac{HT}{HU} = \begin{cases} f(T) & T < Tc \\ 1 & T \geq Tc \end{cases} \qquad (9)$$

For a given heating time, and above some critical temperature T_c, the isothermal cure will reach complete reaction.

The process of studying cure kinetics using DSC involves a crucial step where the obtained reaction rate profile from experiments is fitted to a kinetic model. These models can be categorized as either phenomenological or mechanistic [61]. In phenomenological models, a relatively simple equation is utilized, disregarding the intricate details of the specifics of reactive species involvement in the reaction. On the other hand, mechanistic models are developed from balances of the reactive species participating in the reaction. Although mechanistic models provide more accurate predictions, deriving them is often not feasible due to the complex nature of cure reactions. In this study, the resin system exhibited the autocatalytic effects defined by Equation (10) [62].

$$\frac{d\alpha}{dt} = k\alpha^m(1-\alpha)^n \qquad (10)$$

where k represents the reaction constant defined by the Arrhenius law ($k[T] = Ae^{\frac{-E}{RT}}$), T is the absolute temperature, R is the universal gas constant, A is the pre-exponential factor, E is the activation energy, and m and n are the reaction orders. The rate coefficients and reaction order values are calculated and used as proprietary information.

2.2. Cure Simulation

A numerical simulation was used to simulate the cure behavior of different processing conditions. The temperature variation of the heat input and the surrounding area are critical to control the cure behavior. Predicting correct cure behavior at each temperature through simulations is essential. However, testing all the possible temperatures is not feasible through DSC.

In Figure 1, a schematic shows the analysis steps and the simulation sequence. The material characterization process generates the data to evaluate the curing behavior of the UPR system. The evaluation of those data were conducted through the autocatalytic model, as mentioned previously. There is a relation between the rate of cure, degree of cure, curing temperature, and curing time [63–66], as shown in Equation (1) to Equation (9). This relation accurately predicts the cure behavior [65]. Also, it could be used to capture the effect of the reinforcing materials on the curing kinetics [67–72]. An iterative process fits the DSC data and evaluates the corresponding activation energy of the resin system. The activation energy governs the reaction behavior, speed, and the exothermic heat. The process also evaluates some main values based on the autocatalytic model equations.

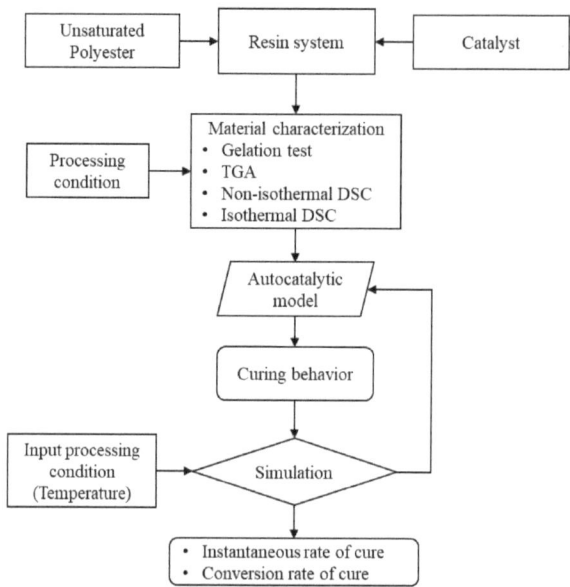

Figure 1. A flow diagram illustrates the simulation process of the cure behavior. The autocatalytic model utilizes the heat of reaction of the resin system and yields the rate of cure and conversion rate. These values vary based on the input processing temperature, which in turn influences the heat of reaction.

The input parameters were introduced, and an interpolation process calculates the results corresponding to the desired value. For example, the tested DSC for a certain temperature is already existing; however, a fraction of the temperature up or down is not available. Therefore, the simulation finds that results and evaluates the cure behavior accordingly. The model interpolates between each time step to find the degree of conversion corresponding to the exact temperature. The unknown degree of conversion was evaluated though an interpolation process.

As the used catalyst type is heat activated, mainly the heat input represented in the heating temperature is essential. The used temperature would be used as a constant temperature, or even a variable temperature. In the case of a variable temperature, the controlled temperature is divided into three main zones: first the heating zone with a ramp rate, dwell time with a constant temperature, then a cooling temperature with a ramp rate.

A MATLAB code was developed with a graphical user interface (GUI) to process these inputs through the autocatalytic model using MATLAB R2023b. The code uses a limited number of DSC data to predict the cure behavior of the resin system. Moreover, the model evaluates the heat of reaction, heat output, and degree of cure (α).

3. Results and Discussion

3.1. Characterization Results

No visible separation between the samples and respective containers was observed during the gel tests. The average temperature against time plots were concluded on both resin systems and are shown in Figure 2. The initial temperature of the resins was dependent on the room temperature. COR61-AA-248S started at 26 °C, while the initial temperature of COR61-AA-270LF was 22 °C. There are significant differences between the two curves. COR61-AA-248S started transitioning from a liquid to gel state after 13 min. For COR61-AA-270LF, the reaction started after 60 min, a 361.5% longer time. The peak exotherms were also different; COR61-AA-248S reached its highest temperature of 162 °C after 34 min, while COR61-AA-270LF exhibited a 20% lower peak (126 °C) after 70 min, a 105.9% longer time. The difference in behaviors between the two polyester resins was caused by their distinct chemistries and crosslinking networks. The gelation results were compared to the available literature [73,74].

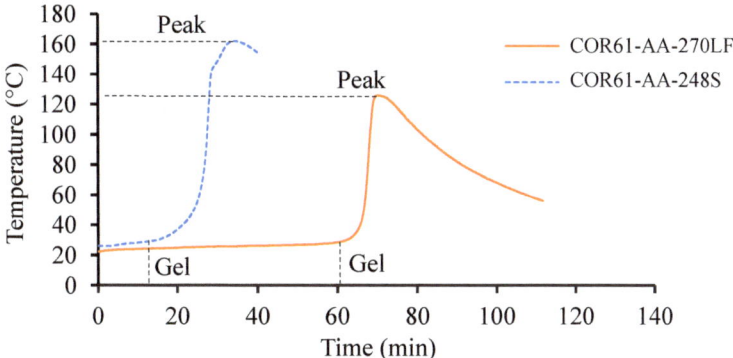

Figure 2. Gel test results for Resin 1 and Resin 2. The test was conducted for a sufficient period until the peak temperature was achieved and no further heat release was recorded. Resin 2, as a fast-cure system, exhibits a higher release rate and provides a greater exothermic heat.

The peak exotherms obtained in this research (126 °C and 162 °C) compared well with the literature values (120 °C and 150 °C). The higher values (162 °C and 350 °C) had only a 7.3% difference and the lower ones (126 °C and 120 °C) had an even smaller difference of 4.4%. The gel times were harder to compare because of the effects of accelerators mixed with the UPR/MEKP system, but they served as indicators for the starting times of the cure reactions.

Based on the TGA results, the weight change for the corresponding temperature plots for Resin 1 and Resin 2 are shown in Figure 3. Both resins experienced a single stage decomposition. The temperatures at 10 wt% weight loss (T_{10}) and the maximum temperatures (T_{max}) are given in Table 2 for both UPRs. Similar results were observed for both resins; Resin 1 had a T_5 of 243.03 °C, T_{10} of 304.64 °C, and a T_{max} of 416.88 °C, while Resin 2 gave higher values of T_5 of 296.44 °C (22% higher), T_{10} of 324.87 °C (6.6% higher), and a T_{max} of 423.84 °C (1.7% higher). To compare the results of this study with other research, the literature was reviewed.

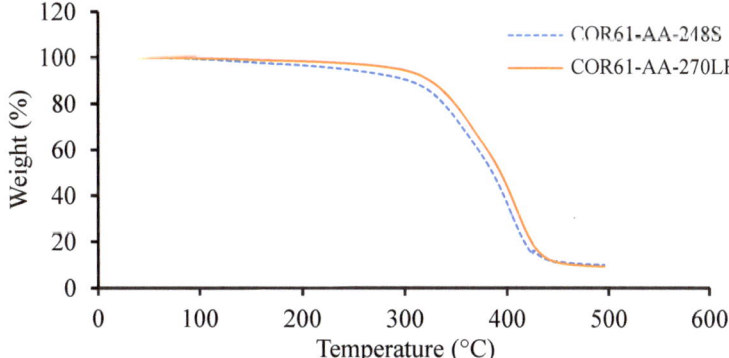

Figure 3. TGA results for Resin 1 and Resin 2. Both resin systems exhibited similar trends; however, COR61-AA-248S begins to lose weight more rapidly than COR61-AA-270LF. The decomposition of Resin 1 and Resin 2 at T_5 occurs at 243 °C and 296 °C, respectively.

Table 2. TGA data for Resin 1 and Resin 2 at different heating temperatures.

Resin Type	T_5 (°C)	T_{10} (°C)	T_{max} (°C)
Resin 1	243.05	304.64	416.88
Resin 2	296.44	324.87	423.84

Dai et al. [75] experimented on a UPR and obtained a T_{10} of 348 °C and T_{max} of 419 °C. Pączkowski et al. [56] studied the properties of an orthophthalic-based UPR and observed a T_{10} of 336 °C and T_{max} of 394 °C. Similarly, Tibiletti et al. [76] obtained a T_5 of 287 °C and T_{max} of 427 °C for their orthophthalic-based UPR. Finally, Bai et al. [77] also performed TGA on an orthophthalic-based UPR and obtained resulting temperatures of T_5 equal to 307 °C and T_{max} equal to 433 °C. All the literature results are given in Table 3. The results of Resin 1 and Resin 2 are comparable to the literature values. The TGA results were used to ensure that no temperature values exceeded the initial degradation temperatures (T_5) while conducting DSC.

Table 3. Review of TGA for different UPR systems.

Literature Study	T_5 (°C)	T_{10} (°C)	T_{max} (°C)
Dai et al. [75]	-	348	419
Pączkowski et al. [56]	-	336	394
Tibiletti et al. [76]	287	-	427
Bai et al. [77]	307	-	433
Average value	297	342	418

Figure 4 shows the non-isothermal DSC curves of the UPR for the two resin systems at three different heating ramps of 5, 10, and 20 °C. It is evident in these graphs that the cure reactions took place in one stage, regardless of the heating rate for both resin systems. Both the initiation temperature and final temperature increased with an increase in scan rates, similar to the observation reported by Sultania et al. (2012) [78]. An estimation of the area under the curve of the high heating rate is higher than in the other two cases. This behavior could be explained by a complete curing at a higher heating rate. In Figure 4a, the heat flow rate is substantially higher (6.8 mW/mg) as compared to the same in Figure 4b. The peak exothermic temperature for the highest heat flow rate is approximately the same in both cases and situated around 145 °C.

The total heat of reaction (HU) was computed around 341.62 J/g with 13.88 J/g standard deviation (S.D.) for Resin 1. Similarly, for Resin 2, the HU was measured about 365.8 J/g with an S.D. of 11.37 J/g. Furthermore, the analysis of non-isothermal and isothermal DSC data resulted in the parametric constants of the autocatalytic model using a weighted least squares non-linear regression method with a 95% confidence interval [79]. The obtained parameters could not be revealed due to sensitivity of the project outcome. Unlike the work of Gohn et al. (2019) [80], increasing cooling rate tends to suppress the cure temperature to lower temperatures. Meanwhile, Figure 4, displaying the non-isothermal cure curves, shows that increasing the ramp rate pushes the cure temperatures to higher values.

At the lowest rate, 5 °C/min, Resin system 1 has a peak cure temperature (T_c) of 110 °C and Resin 2 has a T_c of 115.9 °C, resulting in a 5.3 °C difference. At the maximum rate, 20 °C/min, Resin 1 has a T_c of 146.3 °C, while Resin 2 has a T_c of 148.1 °C, resulting in a 2.0 °C difference. It is apparent that the Resin 2 additive has a strong effect on non-isothermal cure at low cooling rates but is more pronounced at higher rates.

Figure 5 shows the isothermal reaction of the two resin systems at low temperature, from 10 °C to 50 °C. Both resin systems behave differently without following a particular pattern. For instance, Figure 5a shows that at 10 °C, there is no significant impact on the curing behavior of the resin. It takes up to 79 min to record a low heat flow (0.027 mW/mg). In Figure 5b, the initiation process is faster as it occurs in less than a minute at the start of the test and the heat flow remains low (0.062 mW/mg), although this value is higher than that recorded in Resin 1. At the isothermal temperature of 50 °C, both resin systems achieve relatively similar heat flow patterns of 0.195 and 0.190 mW/mg, respectively, but at a shorter time for Resin 1 (3 min) as compared to Resin 2 (16 min). Furthermore, Figure 5b shows heat flow peaks at 30 and 40 °C only. The curves are flat throughout at 10, 20, and 50 °C. Such instability could be explained by the nature of the initiators used in both resin systems, whereby the crosslinking of the polymeric chains does not follow a pattern. Due to the large area under the curves in Figure 5b, it can be determined that the total heat released during the isothermal cycle is considerable compared to the same area in Figure 5a.

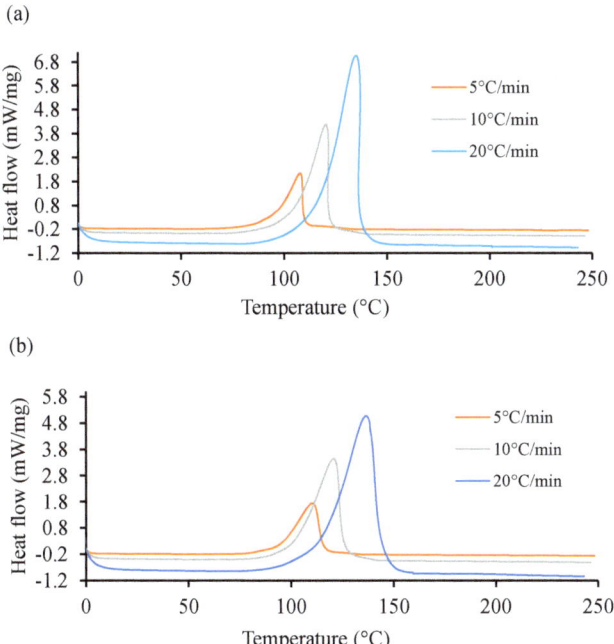

Figure 4. Heat flow of resin systems corresponding to three heating ramp rates, 5, 10, and 20 °C. (a) Non-isothermal DSC runs for Resin 1 and (b) non-isothermal DSC runs for Resin 2.

To further investigate the discrepancies observed in the heat flow of the two resin systems at low temperatures with initiators, more DSC experiments were conducted at higher temperatures between 60 and 170 °C. The acquired curves are shown in Figure 6. The curves exhibit different induction times. Unlike in the previous case at low temperature, the heat flow here increases as the temperature increases. The highest heat flow corresponds to the highest temperature in both cases. Resin 2, as shown Figure 6b, can achieve a higher temperature and higher heat flow rate as compared to Resin 1. The optimal interval is not easily identifiable as the induction of reactions is around 1 min. One distinctive observation here is that at higher temperatures, the curing time is lower in both resin systems. This same trend was observed by Salla and Ramis (1996) [59] while using different procedures to study the cure kinetics of a UPR.

In Figure 6, both resin systems were subjected to a high temperature isothermal cure ranging between 10 and 170 °C. It is shown that the isothermal cure of Resin 1 could only be quantified up to temperatures of 120 °C. At higher temperatures, the cure rate was too slow to produce a detectable calorimetric signal. However, Resin 2 was able to crosslink at temperatures up to 170 °C. A similar trend was deduced by Gohn et al. (2019) [80] in their study on understanding the cure kinetics in two different composites.

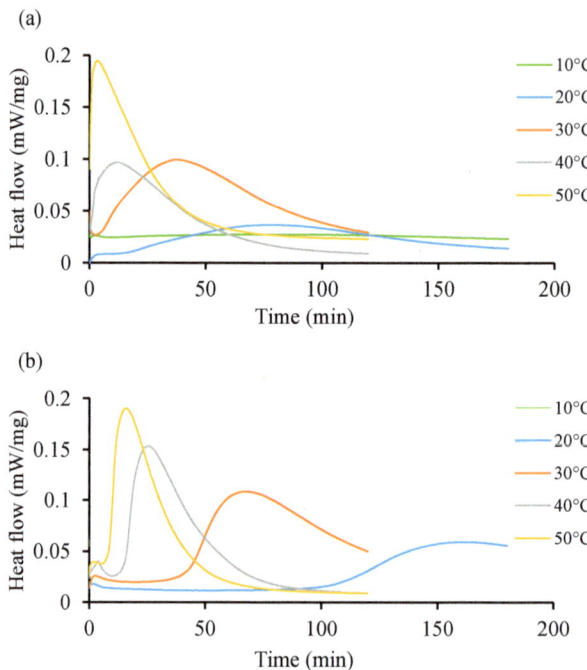

Figure 5. Heat flow of resin systems corresponding to temperature range from 10 °C to 50 °C without initiator. (**a**) Isothermal DSC runs for Resin 1 and (**b**) isothermal DSC runs for Resin 2.

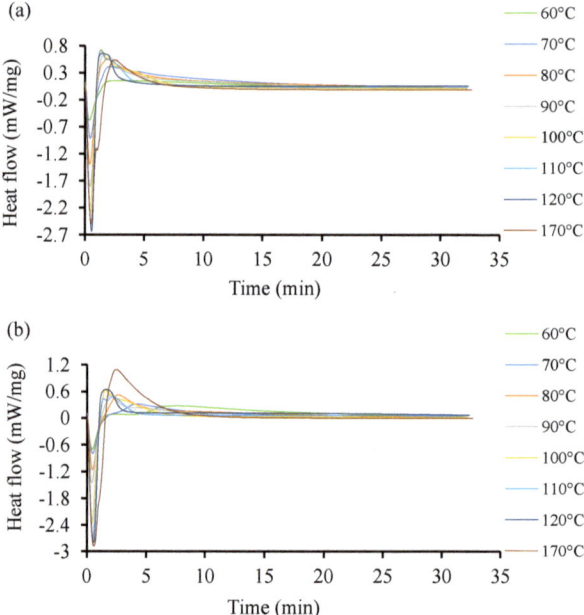

Figure 6. Heat flow of resin systems corresponding to temperature range from 60 °C to 170 °C with initiator. (**a**) Isothermal DSC runs for Resin 1 and (**b**) isothermal DSC runs for Resin 2.

The molecular-level information determines the behavior of the systems. The chemical properties of the cured resin system at 70 °C of both Resin 1 and Resin 2, as characterized by Fourier-transform infrared spectroscopy (FTIR), are shown in Figure 7. It highlights the main functional groups of UPR, i.e., the polyester linkages of carbon–oxygen bonds (C-O-C, C=O and C-O), aromatic hydrocarbons of styrene (C-H), and the hydroxyl (OH) groups. These results match with data from the literature [81].

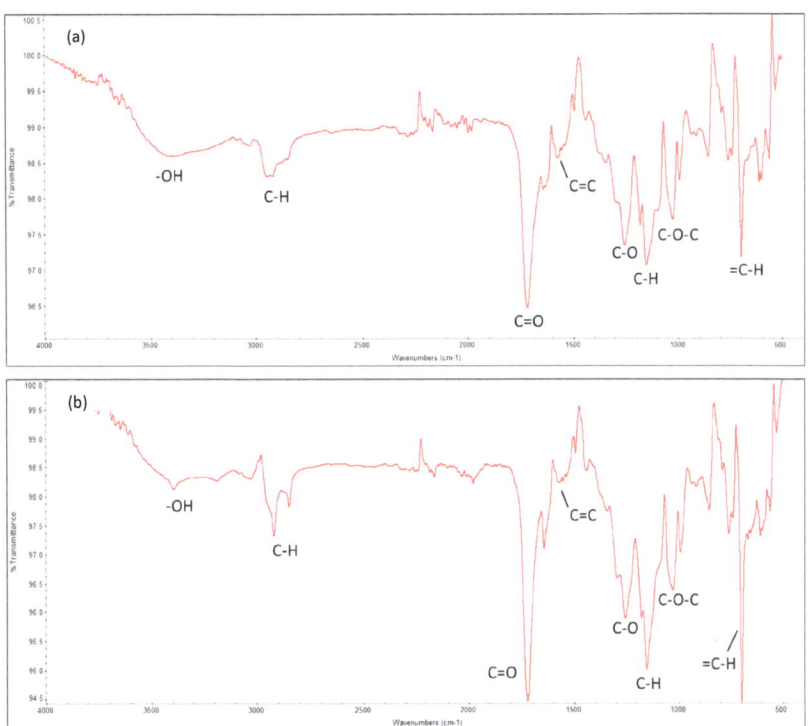

Figure 7. Fourier-transform infrared (FTIR) spectrum of the cured UPRs for both resin systems. (**a**) Resin 1 and (**b**) Resin 2.

3.2. Cure Behavior

The heat of reaction from the DSC represents the endothermic and exothermic behavior of the reaction but does not represent the conversion rate. The autocatalytic curing model was used to evaluate the degree of cure (α). The non-isothermal DSC data were used to evaluate the total heat of reaction (HU), which was used to represent the heat behavior of the resin system under different conditions. The glass transition temperature (T_g) was specified from the peak value of the reaction, as shown in Figure 4.

The non-isothermal tests were carried out between temperatures of 0.0 °C and 250 °C, through different heating and cooling ramp rates. It ensures the full cure of the samples, which determines the exact heat of reaction of the resin system. It was noticed that there was no peak in the second heat cycle, which indicates that it was difficult to further break the crosslinking, as the polymeric chains could not vibrate due to heat. The heat of reaction of the resin system is given by the heat capacity value.

As shown in Figure 6, the DSC results are limited to the tested temperatures, e.g., 60.0 °C, 120.0 °C, etc. But, as the applied temperature varies due to internal or external factors, the cure behavior changes accordingly. Implementing all the possible cases of DSC testing could be an extensive process. Therefore, the simulation of the cure behavior solves this problem. To obtain the cure behavior correctly, the rate of cure was captured to identify the reaction speed and intensity. The instantaneous rate of cure was measured based on Equation (5).

The heat input for the resin system is the main contributor to activate the catalyst and the crosslinking reaction. The amount of added catalyst defines the total degree of cure, which is based on the total activation energy. Hence, the pressure applied, volume of the system, and the temperature applied through the PVT method dictate the cure behavior, cure rate, and cure amount [82]. Controlling that temperature is essential, and mainly the temperature was used as a constant temperature or a variable temperature. The definition of the temperature profile depends on the resin system, the cycle time, and the end use application. The simulation functionality finds the degree of cure of any given constant or variable temperatures in between the tested values to optimize the heating cycle. The variable heating temperature was used in several end-use applications, e.g., autoclave, pultrusion, etc.

As the temperature increases from 10 °C up to 170 °C, the rate of cure reduces, and the heat of reaction increases. The amount of heat released from the different samples is close to each other. However, the external heating temperature defines the distribution and heat release from the exothermic reaction. Also, the amount of catalyst in the end-use application affects the heat release distribution, as it impacts the crosslinking process and, consequently, the amount of heat due to the reaction. Figure 8a shows an example of the instantaneous rate of cure $\left(\frac{dQ}{dt}\right)$ calculated based on constant processing temperatures of 15 °C, 25 °C, 35 °C, 45 °C, 55 °C, 65 °C, 75 °C, and 85 °C.

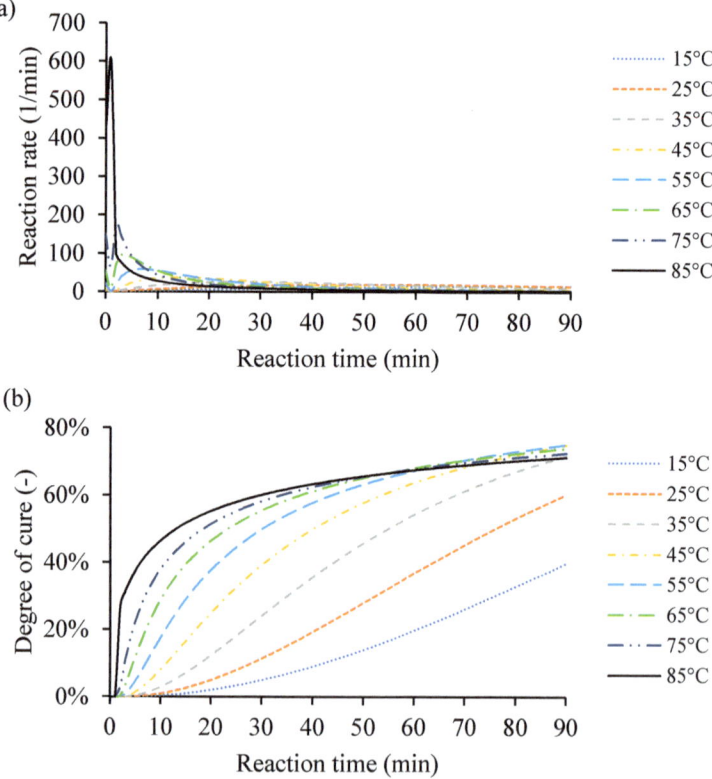

Figure 8. An example of the cure kinetics behavior of Resin system 1 simulated by the MATLAB GUI using constant processing temperatures of 15 °C, 25 °C, 35 °C, 45 °C, 55 °C, 65 °C, 75 °C, and 85 °C. (**a**) Instantaneous rate of cure corresponding to reaction time. (**b**) Degree of cure over time.

Based on the raw DSC data, the cure behavior was simulated from the same temperatures used in that case study (10–170 °C). For the simulated low heating temperature, i.e., 5 °C, the maximum degree of cure (α) did not exceed 12% because there was not enough heat to activate the catalyst to crosslink more. The high heating temperature evaporates the remaining monomers in the resin system, which limits the resin system to have a higher conversion value. Finding the proper heating input for the resin system is essential to obtain the desired conversion rate in time. Thus, the highest heating temperature was limited to 85 °C and achieved degree of cure (α) value of 71.25%, to avoid material vitrification. The type and amount of catalyst dictates the cure behaviors. Any excess heat input leads to vitrification of the resin system, and then degradation of the system. Figure 8b shows the corresponding degrees of cure (α) of the constant heating temperatures 15 °C, 25 °C, 35 °C, 45 °C, 55 °C, 65 °C, 75 °C, and 85 °C during 90 min. The degree of cure was increased by increasing the processing temperature to the maximum of 85 °C. Any given processing temperature within that range for any of the processed resin systems would be simulated through the MATLAB GUI. The cure behavior at this point is assumed to simulate a very thin plate. The kinetic equations are necessary; however, the thermal conductivity coefficient and heat transfer equations are neglected for simplicity.

For the variable heating temperature, the curing cycle would be controlled as desired. Usually, the heat cycle consists of three main heating zones, i.e., the heating up, dwelling, and cooling down. Each heat zone has a function that affects the curing behavior of the resin system. The heating and the cooling cycles use a ramp rate to avoid any thermal stresses on the cooling and the rapid curing, which causes undesired shrinkage of the part. However, the dwell zone provides sufficient heat to cure the resin volume. Figure 9 shows three temperature profiles and the yielded instantaneous rate of cure and degree of cure. The variable heating temperatures consists of three main cycles, heating up with a ramp rate, dwell time for consistent temperature, and cooling down with a ramp rate, as shown in Table 4.

Table 4. Parameters of the heating profile for the variable processing temperatures and degree of cure.

Temperature Profile	Ramp Up (°C/min) for 30 min	Dwell Time (min)	Ramp Down (°C/min) for 30 min	Maximum Calculated DOC (%)
1	0.15	30	0.10	63.19
2	0.25	30	0.25	66.05
3	0.75	30	0.50	73.02

In Figure 9a, the instantaneous rate of cure of Resin 1 based on variable heating temperatures shows the heat distribution over time. The high heating temperature, as in case 3, increases the reaction rate rapidly unlike the lower heating temperatures. However, the conversion rate behaves in a different fashion than the constant heating temperatures, as shown in Figure 9b. The increased temperature did not increase the degree of cure significantly, which could save energy during manufacturing. The calculated degrees of cure for 90 min were 63.19%, 66.05%, and 73.02% for cases 1, 2, and 3, respectively (see Table 4). The instantaneous rate of cure controls the cure cycle, which depends on the variable thermal cycle to reach the desired amount of cure. Thus, the curing behavior of the resin system could be controlled and adapted for different manufacturing processes under varying processing conditions.

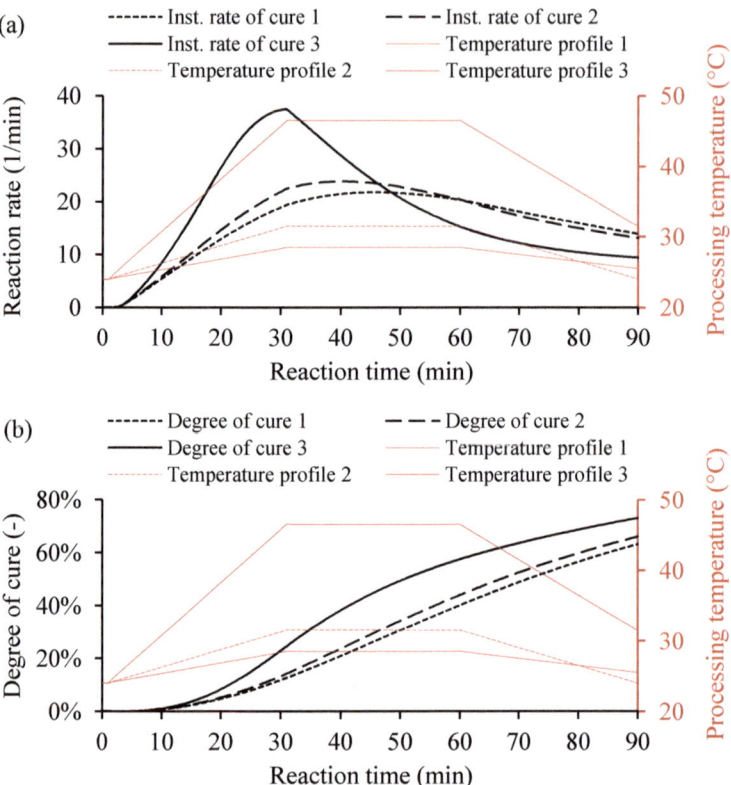

Figure 9. Examples of cure kinetics behaviors of Resin system 1 simulated by MATLAB GUI using variable heating temperature. (**a**) Corresponding instantaneous rate of cure. (**b**) Degree of cure for a variable processing temperature, including heating up with a ramp rate, dwell time at a constant temperature, and cooling down with a ramp rate.

4. Summary and Conclusions

In this work, an autocatalytic model was used to develop a calculator that can accurately provide cure profiles for different UPRs considering the processing conditions. The initial characterization of the UPRs involved the gel time, the TGA, the non-isothermal, and the isothermal DSC tests. The following were observed:

- For the gel time, COR61-AA-248S started transitioning from a liquid to gel state after 13 min, while the COR61-AA-270LF reaction started after 60 min, making it a 361.5% longer time. COR61-AA-248S reached its highest temperature of 162 °C after 34 min, while COR61-AA-270LF exhibited a 20% lower peak (126 °C) after 70 min, a 105.9% longer time. The difference in behaviors between the two polyester resins was caused by their distinct chemistries and crosslinking networks.
- The TGA showed that Resin 1 had a T_5 of 243.03 °C, T_{10} of 304.64 °C, and a T_{max} of 416.88 °C, while Resin 2 gave higher values of T_5 of 296.44 °C (22% higher), T_{10} of 324.87 °C (6.6% higher), and a T_{max} of 423.84 °C (1.7% higher).
- At the lowest rate of 5 °C/min, Resin system 1 had a peak cure temperature (T_c) of 110 °C, and Resin 2 had a T_c of 115.9 °C, resulting in a 5.3 °C difference. At the maximum rate of 20 °C/min, Resin 1 had a T_c of 146.3 °C, while Resin 2 had a T_c of 148.1 °C, resulting in a 2.0 °C difference. The Resin 2 additive has a strong effect on non-isothermal cure at low cooling rates but is more pronounced at higher rates.

- The instantaneous rate of cure of the tested temperatures were measured. It was observed that as the temperature increased from 10 °C up to 170 °C, the rate of cure reduced, and the heat of reaction increased. The amount of heat released from the different samples were close to each other.
- For a constant applied temperature, the conversion rate behaved in a constant fashion and depended on the instantaneous rate of cure. The amount of added catalyst defined the total degree of cure and was based on the total activation energy. For the variable heating temperature, the curing cycle was controlled as desired and provided the desired amount of cure with a proper peak during the thermal cycle.

The simulation of the degree of cure not only matched the experimental data but it also helped in predicting the cure at other temperatures not determined experimentally. This GUI calculator will be a useful tool to determine the cure behaviors of resin systems and will save time and materials. The calculator is useful for different manufacturing processes to understand resin curing behavior and to optimize the processing conditions. Further study would analyze the localized rate of conversion and degree of cure of engineering parts, including thick composite plates with further modification of the numerical model.

Author Contributions: Conceptualization, A.B. and J.H.; Methodology, A.B., M.A.G., R.S.F.T. and A.P.; Software, A.B., M.A.G. and R.S.F.T.; Formal analysis, A.B., M.A.G., R.S.F.T. and A.P.; Investigation, A.B.; Resources, J.U. and J.H.; Data curation, M.A.G. and R.S.F.T.; Writing—original draft, A.B., M.A.G., R.S.F.T. and A.P.; Writing—review & editing, R.S.F.T. and U.V.; Supervision, U.V.; Project administration, J.H. and U.V.; Funding acquisition, J.H. and U.V. All authors have read and agreed to the published version of the manuscript.

Funding: The authors would like to acknowledge Wabash National Corporation, 3550 Veterans Memorial Pkwy S, Lafayette, IN 47909, USA, for the financial support of this work. This material is based upon work supported by the U.S. Department of Energy's Office of Energy Efficiency and Renewable Energy (EERE) under the Advanced Materials and Manufacturing Technologies Office (AMMTO), Award Number [DE-EE0010659], through IACMI-The Composites Institute, Knoxville, Tennessee. Funding for open access to this research was provided by University of Tennessee's Open Publishing Support Fund under funding number [006301].

Data Availability Statement: The original contributions presented in the study are included in the article, further inquiries can be directed to the corresponding author/s.

Conflicts of Interest: Author John Unser was employed by the company Composite Applications Group (CAG). Author Joshua Hagan was employed by the company Wabash National Corporation. The remaining authors declare that the research was conducted in the absence of any commercial or financial relationships that could be construed as a potential conflict of interest.

References

1. Nitin, M.S.; Suresh Kumar, S. Ballistic performance of synergistically toughened Kevlar/epoxy composite targets reinforced with multiwalled carbon nanotubes/graphene nanofillers. *Polym. Compos.* **2022**, *43*, 782–797. [CrossRef]
2. Gieparda, W.; Rojewski, S.; Wüstenhagen, S.; Kicinska-Jakubowska, A.; Krombholz, A. Chemical modification of natural fibres to epoxy laminate for lightweight constructions. *Compos. Part A Appl. Sci. Manuf.* **2021**, *140*, 106171. [CrossRef]
3. Li, Q.; Li, Y.; Chen, Y.; Wu, Q.; Wang, S. An Effective Method for Preparation of Liquid Phosphoric Anhydride and Its Application in Flame Retardant Epoxy Resin. *Materials* **2021**, *14*, 2205. [CrossRef]
4. Zhang, M.; Singh, R.P. Mechanical reinforcement of unsaturated polyester by Al_2O_3 nanoparticles. *Mater. Lett.* **2004**, *58*, 408–412. [CrossRef]
5. Boyard, N.; Vayer, M.; Sinturel, C.; Seifert, S.; Erre, R. Investigation of phase separation mechanisms of thermoset polymer blends by time-resolved SAXS. *Eur. Polym. J.* **2005**, *41*, 1333–1341. [CrossRef]
6. Penczek, P.; Czub, P.; Pielichowski, J. Unsaturated polyester resins: Chemistry and technology. In *Crosslinking in Materials Sciencel*; Springer: Berlin/Heidelberg, Germany, 2005; pp. 1–95.
7. Saleh, H.E.-D.M. *Polyester*; BoD–Books on Demand: Norderstedt, Germany, 2012.
8. Leroy, E.; Dupuy, J.; Maazouz, A.; Seytre, G. Evolution of the coefficient of thermal expansion of a thermosetting polymer during cure reaction. *Polymer* **2005**, *46*, 9919–9927. [CrossRef]
9. Chatham, C.A.; Washington, A.L. A framework for forming thermoset polymer networks during laser powder bed fusion additive manufacturing. *Addit. Manuf.* **2023**, *72*, 103620. [CrossRef]

10. Liang, Q.; Feng, X.-P.; Zhang, K.; Hui, X.-M.; Hou, X.; Ye, J.-R. Effect of curing pressure on the curing behavior of an epoxy system: Curing kinetics and simulation verification. *Polymer* **2022**, *256*, 125162. [CrossRef]
11. Hsu, C.P.; Kinkelaar, M.; Hu, P.; Lee, L.J. Effects of thermoplastic additives on the cure of unsaturated polyester resins. *Polym. Eng. Sci.* **1991**, *31*, 1450–1460. [CrossRef]
12. Li, W.; Lee, L.J. Low temperature cure of unsaturated polyester resins with thermoplastic additives. II. Structure formation and shrinkage control mechanism. *Polymer* **2000**, *41*, 697–710. [CrossRef]
13. de la Caba, K.; Guerrero, P.; Eceiza, A.; Mondragon, I. Kinetic and rheological studies of an unsaturated polyester cured with different catalyst amounts. *Polymer* **1996**, *37*, 275–280. [CrossRef]
14. Sachin, W.; Babu, B.J.C.; Amit, R. Curing studies of unsaturated polyester resin used in FRP products. *Indian J. Eng. Mater. Sci.* **2011**, *18*, 31–39.
15. Silva, M.P.; Santos, P.; Parente, J.M.; Valvez, S.; Reis, P.N.B.; Piedade, A.P. Effect of Post-Cure on the Static and Viscoelastic Properties of a Polyester Resin. *Polymers* **2020**, *12*, 1927. [CrossRef]
16. Akbari, S.; Root, A.; Skrifvars, M.; Ramamoorthy, S.K.; Åkesson, D. Novel Bio-based Branched Unsaturated Polyester Resins for High-Temperature Applications. *J. Polym. Environ.* **2023**, *32*, 2031–2044. [CrossRef]
17. Tarnacka, M.; Madejczyk, O.; Dulski, M.; Wikarek, M.; Pawlus, S.; Adrjanowicz, K.; Kaminski, K.; Paluch, M. Kinetics and Dynamics of the Curing System. High Pressure Studies. *Macromolecules* **2014**, *47*, 4288–4297. [CrossRef]
18. Tarnacka, M.; Wikarek, M.; Pawlus, S.; Kaminski, K.; Paluch, M. Impact of high pressure on the progress of polymerization of DGEBA cured with different amine hardeners: Dielectric and DSC studies. *RSC Adv.* **2015**, *5*, 105934–105942. [CrossRef]
19. Li, S.J.; Zhan, L.H.; Chen, R.; Peng, W.F.; Zhang, Y.A.; Zhou, Y.Q.; Zeng, L.R. The influence of cure pressure on microstructure, temperature field and mechanical properties of advanced polymer-matrix composite laminates. *Fibers Polym.* **2014**, *15*, 2404–2409. [CrossRef]
20. Xie, F.; Wang, X.; Li, M.; Zhang, Z. Experimental research on pressure distribution and resin flow of T-stiffened skins in autoclave process. *Fuhe Cailiao Xuebao/Acta Mater. Compos. Sin.* **2009**, *26*, 66–71.
21. Muric-Nesic, J.; Compston, P.; Stachurski, Z.H. On the void reduction mechanisms in vibration assisted consolidation of fibre reinforced polymer composites. *Compos. Part A Appl. Sci. Manuf.* **2011**, *42*, 320–327. [CrossRef]
22. Hernández, S.; Sket, F.; Molina-Aldareguía, J.M.; González, C.; Llorca, J. Effect of curing cycle on void distribution and interlaminar shear strength in polymer-matrix composites. *Compos. Sci. Technol.* **2011**, *71*, 1331–1341. [CrossRef]
23. Hernández, S.; Sket, F.; González, C.; Llorca, J. Optimization of curing cycle in carbon fiber-reinforced laminates: Void distribution and mechanical properties. *Compos. Sci. Technol.* **2013**, *85*, 73–82. [CrossRef]
24. Yang, X.; Zhan, L.; Jiang, C.; Zhao, X.; Guan, C.; Chang, T. Evaluating random vibration assisted vacuum processing of carbon/epoxy composites in terms of interlaminar shear strength and porosity. *J. Compos. Mater.* **2019**, *53*, 2367–2376. [CrossRef]
25. Struzziero, G.; Skordos, A.A. Multi-objective optimisation of the cure of thick components. *Compos. Part A Appl. Sci. Manuf.* **2017**, *93*, 126–136. [CrossRef]
26. Pillai, V.; Beris, A.N.; Dhurjati, P. Heuristics guided optimization of a batch autoclave curing process. *Comput. Chem. Eng.* **1996**, *20*, 275–294. [CrossRef]
27. Rai, N.; Pitchumani, R. Optimal cure cycles for the fabrication of thermosetting-matrix composites. *Polym. Compos.* **1997**, *18*, 566–581. [CrossRef]
28. Skordos, A.A.; Partridge, I.K. Inverse heat transfer for optimization and on-line thermal properties estimation in composites curing. *Inverse Probl. Sci. Eng.* **2004**, *12*, 157–172. [CrossRef]
29. Li, M.; Zhu, Q.; Geubelle, P.H.; Tucker Iii, C.L. Optimal curing for thermoset matrix composites: Thermochemical considerations. *Polym. Compos.* **2001**, *22*, 118–131. [CrossRef]
30. Yang, Z.L.; Lee, S. Optimized curing of thick section composite laminates. *Mater. Manuf. Process.* **2001**, *16*, 541–560. [CrossRef]
31. Carlone, P.; Palazzo, G.S. A Simulation Based Metaheuristic Optimization of the Thermal Cure Cycle of Carbon-Epoxy Composite Laminates. *AIP Conf. Proc.* **2011**, *1353*, 5–10. [CrossRef]
32. White, S.R.; Hahn, H.T. Cure Cycle Optimization for the Reduction of Processing-Induced Residual Stresses in Composite Materials. *J. Compos. Mater.* **1993**, *27*, 1352–1378. [CrossRef]
33. Olivier, P.; Cottu, J.P. Optimisation of the co-curing of two different composites with the aim of minimising residual curing stress levels. *Compos. Sci. Technol.* **1998**, *58*, 645–651. [CrossRef]
34. Gopal, A.K.; Adali, S.; Verijenko, V.E. Optimal temperature profiles for minimum residual stress in the cure process of polymer composites. *Compos. Struct.* **2000**, *48*, 99–106. [CrossRef]
35. Bailleul, J.L.; Sobotka, V.; Delaunay, D.; Jarny, Y. Inverse algorithm for optimal processing of composite materials. *Compos. Part A Appl. Sci. Manuf.* **2003**, *34*, 695–708. [CrossRef]
36. Zhu, Q.; Geubelle, P.H. Dimensional accuracy of thermoset composites: Shape optimization. *J. Compos. Mater.* **2002**, *36*, 647–672. [CrossRef]
37. Khorsand, A.; Raghavan, J.; Wang, G. Tool-shape optimization to minimize warpage in autoclave processed L-shaped composite part. In Proceedings of the International SAMPE Technical Conference, Memphis, TN, USA, 8–11 September 2008; pp. 8–11.
38. Yuan, Z.; Tong, X.; Yang, G.; Yang, Z.; Song, D.; Li, S.; Li, Y. Curing Cycle Optimization for Thick Composite Laminates Using the Multi-Physics Coupling Model. *Appl. Compos. Mater.* **2020**, *27*, 839–860. [CrossRef]

39. Tziamtzi, C.K.; Chrissafis, K. Optimization of a commercial epoxy curing cycle via DSC data kinetics modelling and TTT plot construction. *Polymer* **2021**, *230*, 124091. [CrossRef]
40. Hu, J.; Xie, H.; Zhu, Z.; Yang, W.; Tan, W.; Zeng, K.; Yang, G. Reducing the melting point and curing temperature of aromatic cyano-based resins simultaneously through a Brønsted acid-base synergistic strategy. *Polymer* **2022**, *246*, 124745. [CrossRef]
41. Ding, A.; Li, S.; Wang, J.; Ni, A.; Zu, L. A new path-dependent constitutive model predicting cure-induced distortions in composite structures. *Compos. Part A Appl. Sci. Manuf.* **2017**, *95*, 183–196. [CrossRef]
42. Svanberg, J.M.; Holmberg, J.A. Prediction of shape distortions Part I. FE-implementation of a path dependent constitutive model. *Compos. Part A Appl. Sci. Manuf.* **2004**, *35*, 711–721. [CrossRef]
43. Sorrentino, L.; Polini, W.; Bellini, C. To design the cure process of thick composite parts: Experimental and numerical results. *Adv. Compos. Mater.* **2014**, *23*, 225–238. [CrossRef]
44. Sorrentino, L.; Tersigni, L. A Method for Cure Process Design of Thick Composite Components Manufactured by Closed Die Technology. *Appl. Compos. Mater.* **2012**, *19*, 31–45. [CrossRef]
45. Bogetti, T.A.; Gillespie, J.W. Two-Dimensional Cure Simulation of Thick Thermosetting Composites. *J. Compos. Mater.* **1991**, *25*, 239–273. [CrossRef]
46. Johnston, A.A. *An Integrated Model of the Development of Process-Induced Deformation in Autoclave Processing of Composite Structures*; University of British Columbia: Vancouver, BC, Canada, 1997.
47. Malik, M.; Choudhary, V.; Varma, I.K. Current Status of Unsaturated Polyester Resins. *J. Macromol. Sci. Part C* **2000**, *40*, 139–165. [CrossRef]
48. Özeroğlu, C. The Effect of Tertiary Amines on the Free Radical Copolymerization of Unsaturated Polyester and Styrene. *Polym. Plast. Technol. Eng.* **2004**, *43*, 661–670. [CrossRef]
49. Matušková, E.; Vinklárek, J.; Honzíček, J. Effect of Accelerators on the Curing of Unsaturated Polyester Resins: Kinetic Model for Room Temperature Curing. *Ind. Eng. Chem. Res.* **2021**, *60*, 14143–14153. [CrossRef]
50. Naderi, N.; Mazinani, S.; Beheshty, M.H.; Rajab, M.M. Cure kinetics of hot cured unsaturated polyester (UP)/nanoclay nanocomposite including dual initiators. *Plast. Rubber Compos.* **2015**, *44*, 19–25. [CrossRef]
51. Li, L.; Liao, X.; Hao, Z.; Sheng, X.; Zhang, Y.; Liu, P. Investigation on cure kinetics of epoxy resin containing carbon nanotubes modified with hyper-branched polyester. *RSC Adv. R. Soc. Chem.* **2018**, *8*, 29830–29839. [CrossRef]
52. Calabrese, E.; Raimondo, M.; Catauro, M.; Vertuccio, L.; Lamberti, P.; Raimo, R.; Tucci, V.; Guadagno, L. Thermal and Electrical Characterization of Polyester Resins Suitable for Electric Motor Insulation. *Polymers* **2023**, *15*, 1374. [CrossRef]
53. Nacher, L.S.; Amoros, J.E.C.; Moya, M.D.S.; Martinez, J.L. Mechanical Properties of Polyester Resins in Saline Water Environments. *Int. J. Polym. Anal. Charact.* **2007**, *12*, 373–390. [CrossRef]
54. Stuck, M.; Krenz, I.; Kökelsum, B.S.; Boye, S.; Voit, B.; Lorenz, R. Improving glass transition temperature of unsaturated polyester thermosets: Conventional unsaturated polyester resins. *J. Appl. Polym. Sci.* **2021**, *138*, 49825. [CrossRef]
55. Delaite, C.; Bistac, S.; Dreyer, E.; Schuller, A.-S. Influence of glass transition temperature of crosslinked unsaturated polyester resin/styrene formulations on the final conversion after an isothermal curing at 100 °C. *Polym. Adv. Technol.* **2020**, *31*, 2031–2037. [CrossRef]
56. Pączkowski, P.; Puszka, A.; Gawdzik, B. Investigation of Degradation of Composites Based on Unsaturated Polyester Resin and Vinyl Ester Resin. *Materials* **2022**, *15*, 1286. [CrossRef]
57. Aktas, A.; Krishnan, L.; Kandola, B.; Boyd, S.; Shenoi, R. A Cure Modelling Study of an Unsaturated Polyester Resin System for the Simulation of Curing of Fibre-Reinforced Composites during the Vacuum Infusion Process. *J. Compos. Mater.* **2015**, *49*, 2529–2540. [CrossRef]
58. Ton-That, M.-T.; Cole, K.C.; Jen, C.-K.; França, D.R. Polyester Cure Monitoring by Means of Different Techniques. *Polym. Compos.* **2000**, *21*, 605–618. [CrossRef]
59. Salla, J.M.; Ramis, X. Comparative Study of the Cure Kinetics of an Unsaturated Polyester Resin Using Different Procedures. *Polym. Eng. Sci.* **1996**, *36*, 835–851. [CrossRef]
60. Dusi, M.R.; Lee, W.I.; Ciriscioli, P.R.; Springer, G.S. Cure Kinetics and Viscosity of Fiberite 976 Resin. *J. Compos. Mater.* **1987**, *21*, 243–261. [CrossRef]
61. Lang, M.; Hirner, S.; Wiesbrock, F.; Fuchs, P. A Review on Modeling Cure Kinetics and Mechanisms of Photopolymerization. *Polymers* **2022**, *14*, 2074. [CrossRef]
62. Cure, T. Determination of autocatalytic kinetic model parameters describing thermoset cure. *J. Appl. Polym. Sci.* **1994**, *51*, 761–764.
63. Voto, G.; Sequeira, L.; Skordos, A.A. Formulation based predictive cure kinetics modelling of epoxy resins. *Polymer* **2021**, *236*, 124304. [CrossRef]
64. Toyota, K. Products and kinetics of the cationic ring-opening polymerization of 3-glycidoxypropylmethyldimethoxysilane by lithium perchlorate. *Polymer* **2021**, *218*, 123490. [CrossRef]
65. Sourour, S.; Kamal, M.R. Differential scanning calorimetry of epoxy cure: Isothermal cure kinetics. *Thermochim. Acta* **1976**, *14*, 41–59. [CrossRef]
66. Vilas, J.L.; Laza, J.M.; Garay, M.T.; Rodríguez, M.; León, L.M. Unsaturated polyester resins cure: Kinetic, rheologic, and mechanical-dynamical analysis. I. Cure kinetics by DSC and TSR. *J. Appl. Polym. Sci.* **2001**, *79*, 447–457. [CrossRef]
67. Pattanaik, A.; Mukherjee, M.; Mishra, S.B. Influence of curing condition on thermo-mechanical properties of fly ash reinforced epoxy composite. *Compos. Part B Eng.* **2019**, *176*, 107301. [CrossRef]

68. Putzien, S.; Louis, E.; Nuyken, O.; Crivello, J.V.; Kühn, F.E. UV curing of epoxy functional hybrid silicones. *J. Appl. Polym. Sci.* **2012**, *126*, 1188–1197. [CrossRef]
69. Johnston, K.; Pavuluri, S.; Leonard, M.; Desmulliez, M.; Arrighi, V. Microwave and thermal curing of an epoxy resin for microelectronic applications. *Thermochim. Acta* **2015**, *616*, 100–109. [CrossRef]
70. Kim, Y.C.; Min, H.; Yu, J.; Suhr, J.; Lee, Y.K.; Kim, K.J.; Kim, S.H.; Nam, J.-D. Nonlinear and complex cure kinetics of ultra-thin glass fiber epoxy prepreg with highly-loaded silica bead under isothermal and dynamic-heating conditions. *Thermochim. Acta* **2016**, *644*, 28–32. [CrossRef]
71. Mphahlele, K.; Ray, S.S.; Kolesnikov, A. Cure kinetics, morphology development, and rheology of a high-performance carbon-fiber-reinforced epoxy composite. *Compos. Part B Eng.* **2019**, *176*, 107300. [CrossRef]
72. Baghad, A.; El Mabrouk, K. The isothermal curing kinetics of a new carbon fiber/epoxy resin and the physical properties of its autoclaved composite laminates. *Mater. Today Proc.* **2022**, *57*, 922–929. [CrossRef]
73. Cook, W.D.; Lau, M.; Mehrabi, M.; Dean, K.; Zipper, M. Control of gel time and exotherm behaviour during cure of unsaturated polyester resins. *Polym. Int.* **2001**, *50*, 129–134. [CrossRef]
74. Pandiyan Kuppusamy, R.R.; Neogi, S. Influence of curing agents on gelation and exotherm behaviour of an unsaturated polyester resin. *Bull. Mater. Sci.* **2013**, *36*, 1217–1224. [CrossRef]
75. Dai, K.; Song, L.; Jiang, S.; Yu, B.; Yang, W.; Yuen, R.K.; Hu, Y. Unsaturated polyester resins modified with phosphorus-containing groups: Effects on thermal properties and flammability. *Polym. Degrad. Stab.* **2013**, *98*, 2033–2040. [CrossRef]
76. Tibiletti, L.; Longuet, C.; Ferry, L.; Coutelen, P.; Mas, A.; Robin, J.-J.; Lopez-Cuesta, J.-M. Thermal degradation and fire behaviour of unsaturated polyesters filled with metallic oxides. *Polym. Degrad. Stab.* **2011**, *96*, 67–75. [CrossRef]
77. Bai, Z.; Song, L.; Hu, Y.; Yuen, R.K. Preparation, flame retardancy, and thermal degradation of unsaturated polyester resin modified with a novel phosphorus containing acrylate. *Ind. Eng. Chem. Res.* **2013**, *52*, 12855–12864. [CrossRef]
78. Sultania, M.; Rai, J.S.P.; Srivastava, D. Modeling and simulation of curing kinetics for the cardanol-based vinyl ester resin by means of non-isothermal DSC measurements. *Mater. Chem. Phys.* **2012**, *132*, 180–186. [CrossRef]
79. Hwang, S.S.; Park, S.Y.; Kwon, G.C.; Choi, W.J. Cure Kinetics and Viscosity Modeling for the Optimization of Cure Cycles in a Vacuum-Bag-Only Prepreg Process. *Int. J. Adv. Manuf. Technol.* **2018**, *99*, 2743–2753. [CrossRef]
80. Gohn, A.M.; Seo, J.; Ferris, T.; Venkatraman, P.; Foster, E.J.; Rhoades, A.M. Quiescent and flow-induced crystallization in polyamide 12/cellulose nanocrystal composites. *Thermochim. Acta* **2019**, *677*, 99–108. [CrossRef]
81. Halim, Z.A.A.; Yajid, M.A.M.; Idris, M.H.; Hamdan, H. Effects of Rice Husk Derived Amorphous Silica on the Thermal-Mechanical Properties of Unsaturated Polyester Composites. *J. Macromol. Sci. Part B* **2018**, *57*, 479–496. [CrossRef]
82. Ramos, J.A.; Pagani, N.; Riccardi, C.C.; Borrajo, J.; Goyanes, S.N.; Mondragon, I. Cure kinetics and shrinkage model for epoxy-amine systems. *Polymer* **2005**, *46*, 3323–3328. [CrossRef]

Disclaimer/Publisher's Note: The statements, opinions and data contained in all publications are solely those of the individual author(s) and contributor(s) and not of MDPI and/or the editor(s). MDPI and/or the editor(s) disclaim responsibility for any injury to people or property resulting from any ideas, methods, instructions or products referred to in the content.

Article

Multiphase Flow Production Enhancement Using Drag Reducing Polymers

Abdelsalam Alsarkhi *[ ] and Mustafa Salah

Department of Mechanical Engineering, Center for Integrative Petroleum Research, King Fahd University of Petroleum and Minerals, Dhahran 31261, Saudi Arabia
* Correspondence: alsarkhi@kfupm.edu.sa

Abstract: This paper presents a comprehensive experimental investigation concerning the effect of drag reducing polymers (DRP) on enhancing the throughput and reducing the pressure drop for a horizontal pipe carrying two-phase flow of air and water mixture. Moreover, the ability of these polymer entanglements to damp turbulence waves and changing the flow regime has been tested at various conditions, and a clear observation showed that the maximum drag reduction always occurs when the highly fluctuated waves were reduced effectively by DRP (and that, accordingly, phase transition (flow regime changed) appeared. This may also help in improving the separation process and enhancing the separator performance. The present experimental set-up has been constructed using a test section of 1.016-cm ID; an acrylic tube section was used to enable visual observations of the flow patterns. A new injection technique has been utilized and, with the use of different injection rates of DRP, the results have shown that the reduction in pressure drop occurred in all flow configurations. Furthermore, different empirical correlations have been developed which improve the ability to predict the pressure drop after the addition of DRP. The correlations showed low discrepancy for a wide range of water and air flow rates.

Keywords: drag reducing polymers; multiphase flow; production enhancement; flow pattern transition; disturbance waves

Citation: Alsarkhi, A.; Salah, M. Multiphase Flow Production Enhancement Using Drag Reducing Polymers. *Polymers* 2023, 15, 1108. https://doi.org/10.3390/polym15051108

Academic Editors: Célio Pinto Fernandes, Luís L. Ferrás and Alexandre M. Afonso

Received: 31 December 2022
Revised: 14 February 2023
Accepted: 18 February 2023
Published: 23 February 2023

Copyright: © 2023 by the authors. Licensee MDPI, Basel, Switzerland. This article is an open access article distributed under the terms and conditions of the Creative Commons Attribution (CC BY) license (https:// creativecommons.org/licenses/by/ 4.0/).

1. Introduction and Literature Review

Drag reducer chemicals are high molecular weight polymers (greater than 2×10^6). Typical species include polyacrylamides and both natural and Xanthan gums. Their mode of action is believed to be by reducing turbulent eddies and extending the laminar boundary layer at the pipe wall and they are considered to be effective under turbulent conditions.

During the transportation of the multiphase (gas-liquid) in the pipelines industry, several flow regimes might form, leading to a large pressure gradient. To reduce the frictional pressure drop, different techniques have been proposed in the literature. Addition of a few parts per million (ppm) of polymers liquid in the pipe is one way to achieve good drag reduction and reduce the frictional pressure losses.

The concept of adding high molecular weight long-chain polymers into a single-phase liquid flow was first published by Tom [1] and is known as Tom's phenomenon. In Tom's work, high reduction was observed on the frictional resistance at the pipe wall which finally leads to the possibility of increasing the pipeline capacities and flow rates.

For the gas-liquid flow in pipes, the effect of the drag reducing polymers on the existing system have been investigated experimentally and several scientific researches have been published. Oliver et al. [2] were the first to investigate the effect of drag reducing polymers in gas-liquid flows using 1.3% polyethylene (PEO) aqueous solution and air. They reported that the liquid in the slug flow where the wave was absorbed gave smooth liquid film.

Al-Sarkhi et al. [3] studied the drag reducing polymers on air-water flow in horizontal pipes, they found that the DRA destroys the turbulent waves which affect the flow rates and the pressure of the system. The maximum drag reduction obtained was about 48% for annular flow configuration. The discussion has been carried out about the effectiveness of the drag reduction agent which is depend on the way of DRA been introduced into the regime, they suggested an injection of a well mixed master solution to the film in order to have good distribution along the pipe circumference.

Soleimani et al. [4] examined the influence of adding polymers on the pseudo slug flow and the transition to slug flow patterns for the air-water two phase configuration using pipe diameter of 2.54 cm. They studied the effect of the polymer concentration on the pressure gradient for different superficial liquid velocities (U_{sl}) and they also noted that the decrease in the pressure gradient is not monotonic with the polymer ppm since the polymers will enlarge the liquid holdup while decreasing the interfacial friction which have an opposite effect on the pressure drop. Therefore an increase or decrease could be realized.

Stratified flow configuration and its transition to slug flow in small pipes may exhibits some complicity than the one flows in large diameters since the interface between gas and the liquid is hidden by the large waves which touch the pipe top wall. Baik et al. [5] investigated the effect of the drag reducing polymers on these waves at high and low superficial gas velocity, they reported that the wave amplitude decreased dramatically using the polymer solution and a drag reduction of about 42% was noted.

Fernandes et al. [6] conducted an experimental study using high molecular weight poly-alpha-olefin polymers on two phase flow(gas-condensate flow) that operate in annular flow regime, they developed a mechanistic model and comparative study by applying the DRA on similar experimental loop of Al-Sarkhi et al. [3] to show the applicability of their model and its limitation. The error between the model and the experimental data was 5%, finally they concluded that as the pipe diameter increases, the drag reduction increases due to the reduction of the entrainment.

Many researchers in the literature continued delivering numerous empirical correlations that help in evaluating the pressure drop occurs within multiphase flows, with respect to various operational conditions those models have been developed. Recently, Al-sarkhi et al. [7] developed two correlations for the friction factor (based on the asymptotic value of drag reduction) for a wide range of pipe diameter from 0.019 to 0.0953 m and using the results of the published data of air–liquid annular flows and liquid–liquid flows to realize the capability of the prediction for any flow pattern with the presence of DRP in pipes.

Al-sarkhi [8] investigated the influence of mixing technique of a drag reducing polymer and the way it is introduced to the gas-liquid annular flow on the percentage of drag reduction by DRP. Effect of different master solution (the injected liquid polymers) concentrations were studied. Al-sarkhi [9] published a very extensive literature review of drag reduction by polymers in gas-liquid and liquid-liquid flows in pipes. In this work, the mechanisms of drag reduction proposals were discussed and the need for further research in this area were identified.

Wang et al. [10] used direct numerical simulation method to investigate the gas liquid drag reducing cavity flow using the volume of fluid and level set method. It was concluded that a high concentration of polymers enhances the drag reduction.

The novelty of the present work is that the experiments were conducted in a 1.016-cm ID stainless steel tube in which the experimental data is rarely exited in open literatures. Moreover, different flow regimes were tested and new empirical correlations are developed which enables predicting the pressure drop and friction factor after the addition of DRP compared to gas-liquid without DRP to give an estimate of the amount of drag reduction.

2. Description of Experimental Setup and Procedure

The experimental flow loop depicted in Figure 1 below is designed to investigate the influence of the DRP additives on the flow behavior of liquid and air mixture. The loop comprises two 200 liter barrels for water and an instrument air connection for the air supply. The flow rate of the feed streams is measured and can be adjusted using regulating valves. The additive is added to the flow system via a nozzle into the mixed fluid stream via diaphragm pump.

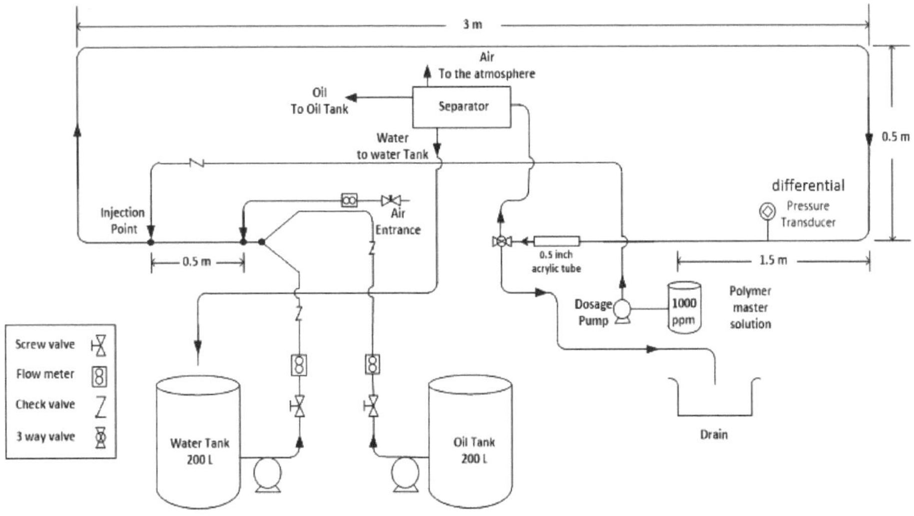

Figure 1. Sketch of the flow facility.

The feed pumps for the liquids (water) are rotary pumps equipped with axial face sealings. Water, and air can be separated in the separator or using cyclone and separator which are connected to the outlet of the test section. However, in the present paper only air and water are used as the two phases but the loop has the capability of having three phases air, water, and oil.

The test section is made of stainless-steel tube with an outer diameter of 1.27 cm and an inner diameter of 1.016 cm. its total length is approximately 5 m divided into two straight horizontal sections separated by elbows (90-degree elbow). The horizontal sections are equipped with differential pressure transducer to measure the pressure drop inside the test section along a distance of 1.5 m. At the end of the test section an acrylic section of 20 cm long allows the visible inspection of the flow behavior. After having passed the test section, the fluid can be directed to the phase separator where water and air can be separated by gravity or alternatively to the cyclone whose outlet which is connected to the phase separator.

2.1. Preparation of the Polymer Solution

A polymer in a powder format is mixed with water in rotating magnetic mixer at low speed in order to avoid polymer shear degradation. Then rotation is stopped when the mixture completely dissolved in the water and having a conglomerated consistency. The mixing process may take several hours and sometimes a heat addition up to 50 °C was used to accelerate the solubility. Water and polymer specifications are given in Tables 1 and 2. Table 2 shows data from the manufacturer product sheet indicating the viscosity of the DRP at the specific concentration. It is worth to be mentioned here that at 100 PPM DRP in water which is the maximum concentration used in the present experiments neither the viscosity of the water with 100 ppm DRP nor the density or surfactant will be affected by the DRP presence in such a small amount.

Table 1. Fluids standard properties.

Water Density	1000 $\frac{kg}{m^3}$
Water viscosity	0.000891 Pa s
Ph	7–8
Gas Density	1.28 $\frac{kg}{m^3}$
Gas viscosity	0.0000185 Pa·s

Table 2. Polymer technical properties from manufacturer.

Product Name	Coopolymer of Acrylamide and Quaternized Cationic Monomer		
Product Type	Powder		
physical form	off-white granular solid		
cationic charge	Medium-high		
Molecular weight	very high		
specific gravity	0.75		
Bulk density	749.66 kg/m³		
Ph 1% solution	4–6		
Apparent Viscosity/(cP) 25 °C			
Concentration	0.0025	0.005	0.01
Viscosity	650	1200	3000

2.2. System Operation

The generated air-water two phase flow is circulated through the flow loop using a vertical centrifugal pump that can provide a maximum flow rate of 40 L/min of water. On the other hand, the air is introduced to the system (from the laboratory main source) using a pressure regulator (with a maximum 16 bar inlet pressure and maximum 10 bar outlet pressure) connected at the inlet of the compressed air. A thermal mass flow rate measures the air flow in range from (0–150) L/min. The flow rate of the water is measured using an electromagnetic flow meter for flow range up to 40 L/min, a check valve is connected after the flow meter to prevent back flow of water.

3. Results and Discussions

3.1. Effect of DRP on Frictional Pressure Drop

In this study, the effect of adding the DRP on frictional pressure gradient for air-water mixture has been tested for a wide range of liquid and gas flow rates, the liquid flow rate starts from 3 to 25 L/min while gas flow is up to 70 L/min. The corresponding pressure drop has been recorded for the entire range with and without the DRP.

Figure 2 provides a clear comparison for the frictional pressure drop reported with different liquid superficial velocities (the points symbols stand for pressure drop after the injection of 0.6 L/min DRP and the lines represent pressure drop without adding the DRP). It is indicated that from this figure as the gas superficial velocity increases from 2.6 to 4.11 m/s the pressure drop increases accordingly, one possible justification for this behavior is that once the gas velocity increases an additional pressure loss in the mixture of the two-phase flow appear due to the disturbance in the liquid flow caused by the gas.

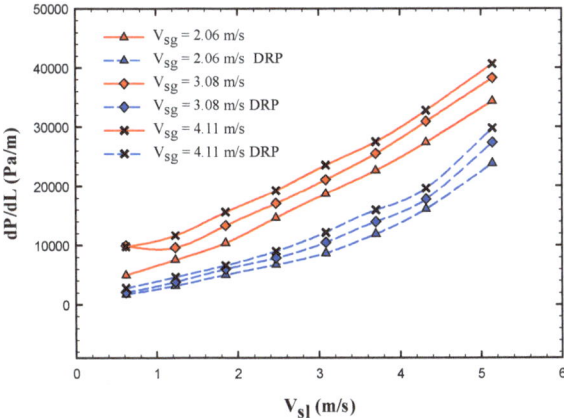

Figure 2. Comparison of the frictional pressure drop variation with respect to liquid superficial velocity at different gas superficial velocities of 2.06, 3.08 and 4.11 m/s.

3.2. Effect of DRP in Two-Phase Flow Pattern Transition

The observed flow patterns gas-liquid two phase with and without the addition of 40 ppm Drag Reducing polymer (DRP) results are illustrated in Appendix A. As can be seen, most of the flow regimes were changed with DRP except the smooth stratified. The minimum percentage of drag reduction was in the stratified flow regime and the maximum was for the slug flow when it is changed to wavy stratified after the addition the DRP.

3.2.1. Stratified and Stratified Wavy Flow Regimes

The reductions in the role waves and ripples have been realized and the flow has become more stable. Furthermore, the range of the smooth stratified flow pattern increased primarily at the transition region between slug and stratified wavy flows.

As it can be seen from Figure 3 that the frictional pressure gradient increases significantly as the dimensionless superficial velocity increases, which is mainly due to the increase in gas superficial velocity that adds more disturbance to the gas liquid interface.

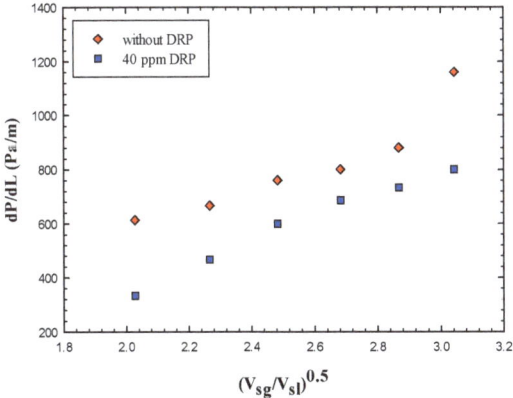

Figure 3. Effect of drag reducing polymer on the stratified flow regime.

A minimal effect of DRP has been noted for the stratified regime due to uniform and quite stable interface between gas and liquid (air-water). Also, there is no clear transition effects from wavy stratified to stratified flow pattern. However, the role of the waves and their intensity have been damped further with the presence of DRP. Also, as emphasized by Baik et al. [5], the DRP effectively reduced the wave amplitude and delayed transition to slug flow regime. Figures 4 and 5 depict the stratified and wavy stratified flows before and after adding the DRP.

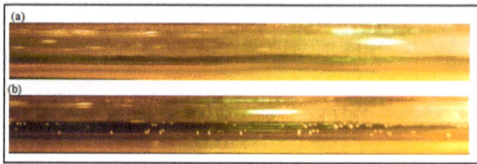

Figure 4. (**a**) Stratified Wavy flow without DRP (V_{sl} = 0.1 m/s, V_{sg} = 0.41 m/s); (**b**) Stratified Wavy flow with 40 ppm DRP.

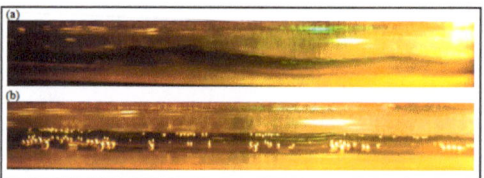

Figure 5. (**a**) Stratified Wavy flow without DRP (V_{sl} = 0.1 m/s, V_{sg} = 2.88 m/s); (**b**) Stratified Wavy flow with 40 ppm DRP.

3.2.2. Annular and Wavy Annular Flow Patterns

Annular and wavy annular flow patterns have been studied to show the effectiveness of adding a small concentration of drag reducing polymer. Figure 6 illustrates how the DRP can reduce the pressure drop at various gas superficial velocities. It can be seen clearly that the DRP was able to suppress the waves at the bottom of annular film for all gas flow rates been studied. Thus, a drag reduction has been observed.

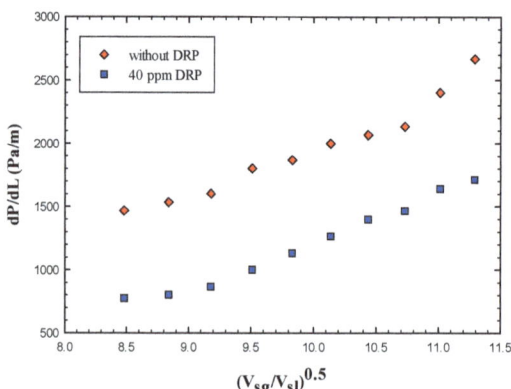

Figure 6. Effect of drag reducing polymer on annular flow regime.

Moreover, the transition from wavy annular regime to stratified wavy occurred with the addition of only 40 ppm of DRP. Appendix A summarizes the ranges at which these transitions have been observed, and the maximum drag reduction obtained for the annular and wavy annular region was 48%, this effectiveness decreases as more waves propagate at the annular liquid film.

Taylor et al. [11] divided the annular flow regime into three distinct regions according to liquid film disturbance, first region in which the wave starts to form then augments more in region two, and finally the wave oscillations go down in third region. The overall frequency of the interfacial waves decreases as far as it move downstream Zhao et al. [12]. The energy associated with forming these waves always results in reduction in the total pressure, thus using DRP to damp and delay these oscillations lead to a reduction in pressure drop. Figures 7 and 8 represent typical features of annular and wavy annular flow regimes.

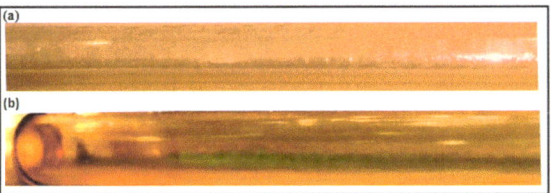

Figure 7. (a) Annular flow regime without DRP (V_{sl} = 0.1 m/s, V_{sg} = 9.05 m/s); (b) Annular flow regime with 40 ppm of DRP.

Figure 8. Annular wavy flow regime (V_{sl} = 0.1 m/s, V_{sg} = 12.75 m/s).

The effectiveness of the drag reducing polymers is very sensitive to the way that the DRP been introduced to the system (Al-Sarkhi et al. [3]), in the present study a diaphragm pump has been utilized to inject the polymer into liquid film of the annular flow to avoid polymer molecules breakup. Figures 7 and 8 indicated that the wavy annular flow shifted slightly to stratified wavy regime as DRP injected.

It should be noted that drag reducing polymers acting to stabilize the liquid film by damping disturbance waves at the gas liquid interface, thus a reduction in pressure drop occurs and also an increase in the mean liquid thickness could be observed, this realization in a good agreement with Spedding et al. [13] and Thwaites et al. [14] findings for the annular flow regime.

3.2.3. Dispersed Bubbly Flow Regime

As the liquid superficial velocity further increases the dispersed bubbly regime would be a possible candidate and generally this flow pattern characterized by small bubbles introduced as a discrete particle in the liquid continuous phase. Figure 9a shows the typical behavior of the bubbly flow. The performance of the DRP has been examined for this type of flow; Figure 10 exhibits the variation of the pressure drop with the dimensionless superficial velocity with and without the DRP. The results are reported in Appendix B, it can be seen that the maximum drag reduction percentage occurred was about 55% and the flow changed slightly to pseudo slug flow regime; these changes were limited up to 5 m/s of gas superficial velocity.

The onset of transition to pseudo slug flow is clearly indicated in Figure 9b with the presence of 40 ppm DRP. The mechanism of the transition is that; with these polymers the separated bubbles tends to coalesce together forming gas pseudo slugs, due to the decrease in the level of turbulence which contribute in keeping the air bubble dispersed in the liquid.

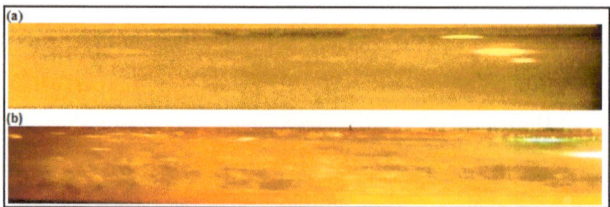

Figure 9. (a) Typical feature of a Dispersed Bubbly flow regime (V_{sl} = 3.08 m/s, V_{sg} = 1.03 m/s) (b) Transition from Dispersed Bubbly to Pseudo slug flow regime with 40 ppm DRP.

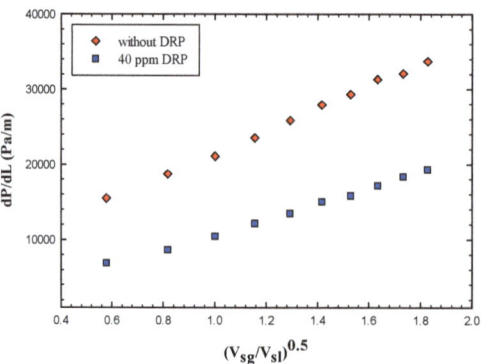

Figure 10. Effect of drag reducing polymer on dispersed bubbly flow regime.

3.2.4. Slug and Pseudo Slug Flow Regimes

A distinctive study has been carried out to examine the effect of the DRP on the characteristics of slug and pseudo slug flows utilizing two polymer concentrations namely 40 and 100 ppm; to show the effectiveness of the DRP in changing the flow patterns at low and relatively high concentrations.

Appendices B and C articulate the frictional pressure gradient with and without adding DRP of 40 and 100 ppm, respectively. It is noted that adding 40 ppm DRP could results in a decrease of turbulence wave's intensity and slug frequency with no clear transition from the slug to the stratified wavy regime. The inception of this transition is illustrated with the presence of adding 100 ppm (Figure 11).

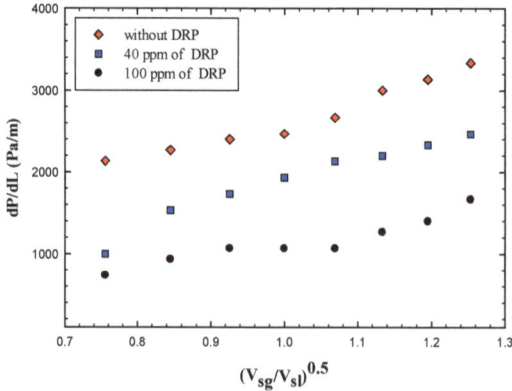

Figure 11. Effect of drag reducing polymer on slug flow regime using a concentration of 40 and 100 ppm.

As seen from Figure 11 that the pressure drop reduced more in the case of 100 ppm and the maximum Drag Reduction effectiveness reported in the case of 40 ppm was 53%, and 66% for a situation where 100 ppm added to the flow.

It should be noted that the transition from Slug to Stratified wavy flow has been observed for all of the range studied with addition of 100 ppm concentration unlike the case where no transition noted with utilizing only 40 ppm.

The effectiveness of DRP on the pseudo slug regime is exhibited in Appendices D and E. Here, the possible transition to wavy annular flow started earlier when 40 ppm has been added. Also it has been realized that more disturbance appears as the gas flow rate increases, and there is no changes in the characteristics of pseudo slug regime is observed, though the DRP only acts to decrease the turbulence intensity at the gas liquid interface which is totally support the claims of More et al. [15].

The maximum effectiveness reported was about 41% and 64% for the case of 40 and 100 ppm, respectively. Figure 12 shows the variation of pressure drop with the dimensionless superficial velocity, it is clearly indicated that the pressure drop has been reduced further more in case of adding 100 ppm DRP concentration for both slug and pseudo slug regimes. Increasing the DRP concentration even more up to 100 ppm enhanced the transition to Wavy annular for the whole superficial gas velocity range (1.03–6.17 m/s) and this could justify why the drag reduction has been increased. Figure 13 depicts this transition clearly.

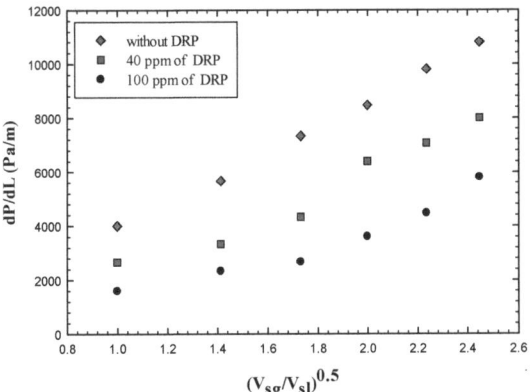

Figure 12. Effect of drag reducing polymer on pseudo slug flow regime using a concentration of 40 and 100 ppm.

The formulation of slug flow pattern is always accompanied by a formation of two components (gas pocket and liquid film). As it can be seen from Figure 13a that the gas pocket (at gas liquid interface) penetrates in the stratified liquid film causing an increase in turbulence intensity. Adding the drag reducing polymer is believed to reduce these penetrations, suppresses turbulence patches and also enlarges the stratified liquid film region (Figure 13b). Daas et al. [16] pointed out similar explanation for the drag reduction mechanism in slug flow regimes. Figure 14 is also showing the transition at higher superficial liquid and gas velocities.

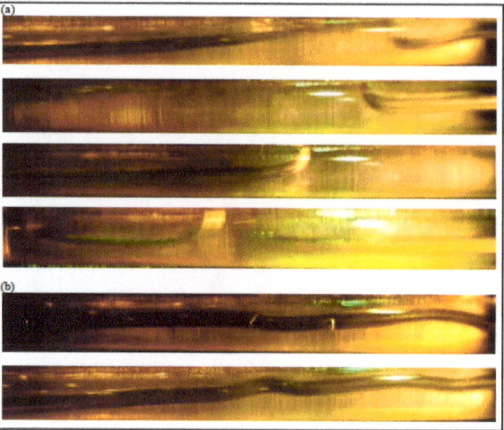

Figure 13. (a) Slug flow regime without DRP (V_{sl} = 0.72 m/s, V_{sg} = 0.41 m/s); (b) Transition from Slug to Stratified Wavy flow regime using 100 ppm DRP.

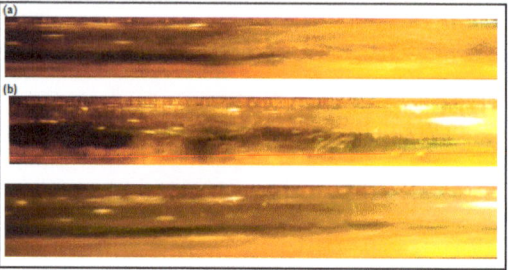

Figure 14. (a) Pseudo Slug flow without DRP (V_{sl} = 1.03 m/s, V_{sg} = 3.08 m/s); (b) Transition from Pseudo Slug to Wavy Annular flow regime using 100 ppm.

3.3. Correlations for Gas –Liquid Flow with Addition of DRP

An experimental study has been carried out for a horizontal pipeline to examine the addition of water-soluble polymer on the two-phase water-air flow. The experimental data has been generated based on the two phase (air-water without DRP) map in Figure 15.

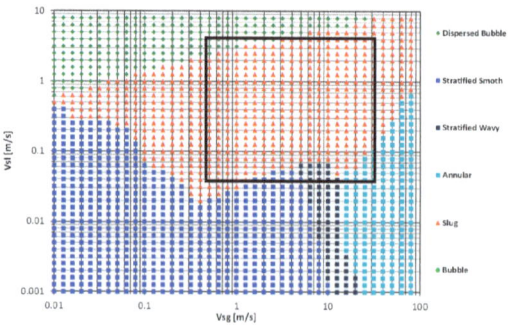

Figure 15. Air-water (without DRP) flow pattern map using Unified [17] Model in a 1.01-cm pipe. (Dashed box is the present work flow conditions) Where: **DB:** dispersed bubble, **SL:** Slug, **IN:** Intermittent, **SS:** Smooth Stratified, **SW:** Stratified Wavy, **AN:** Annular.

In this study, correlations have been developed to allow further understanding of the drag reducing polymers in reducing the frictional pressure gradient and the parameters that could be affected by these additions also explained.

The mixture friction factor f_M and the mixture Reynolds number Re_M for the two-phase water-air flow are playing a key role in developing good relations that predict and represent the experimental data more properly. The definitions of the mixture friction factor and mixture Reynolds number has been illustrated in the studies of García et al. [18].

3.3.1. Correlation Development

Usually the two-phase water-air is very complex in nature and this complexity is more obviously when the detecting of a flow pattern that could be existed before and after adding the DRP is required. Therefore, the need for developing a correlation that appropriately relate different flow parameters and characteristics receives high attention especially in predicting the pressure drop in pipelines without knowing the flow regime. Possible dimensionless parameter could be the one includes various parameters such as the mixture Reynolds number Re_M which comprises pipe diameter, density, viscosity and mixture velocity in one dimensionless number.

Several studies has been performed to correlate the two phase using dimensionless groups, for example García et al. [18,19] developed a correlation of friction factor that covered a wide range of laminar and turbulent flow of gas-liquid regimes. The correlation that been produced was based on liquid holdup ranges to differentiate between the experimental data used in their analysis. However, these correlations have been carried out without the addition of the drag reducing polymers, and hence different trends and correlations could be realized as the DRP added to the system.

Alsarkhi et al. [7] Studied the effectiveness of two correlations for predicting the effect of the drag reducing polymers on the mixture friction factor using published experimental data of air-liquid and oil-water flows in literature. This was the only attempt been found in the open literature at least for predicting the drag reduction in different pipe diameters namely from 0.019 to 0.0953 m.

In the present work, experiments on water-air flow were conducted and several correlation has been developed based on various water superficial velocities using different dimensionless groups and parameters that used in Al-sarkhi et al. [7] and García et al. [18].

3.3.2. Dimensionless Parameters

The mixture friction factor for water–air mixture without the addition of DRP ($f_{Mwithout-DRP}$) is expressed as follows:

$$f_{Mwithout-DRP} = \frac{2 \times D \times \frac{dP}{dL}\bigg|_{withoutDRP}}{\rho_M \times V_M^2} \quad (1)$$

where D is the diameter of the pipe, and V_M is the mixture velocity which is defined as the summation of liquid and gas superficial velocities ($V_M = V_{sl} + V_{sg}$).

The superficial liquid and gas velocities are calculated using the below equations:

$$V_{sl} = \frac{4Q_l}{\pi D^2} \quad (2)$$

$$V_{sg} = \frac{4Q_g}{\pi D^2} \quad (3)$$

where Q_l, Q_g are the flow rate for the liquid and the gas, respectively.

The mixture density (ρ_M) is defined as:

$$\rho_M = \rho_l \lambda_l + \rho_g (1 - \lambda_l) \quad (4)$$

where: $\lambda_l = \frac{Q_l}{Q_l+Q_g}$ the volumetric flow rate fraction, and the ρ_l, ρ_g are the densities of liquid and the gas, respectively.

The mixture friction factor with the drag reducing polymer being added is formulated using the same parameters on Equation (1) above with only changing the pressure drop to $\frac{dP}{dL}\big|_{DRP}$ which is the one with DRP added to the system.

$$f_{M\text{-}DRP} = \frac{2 \times D \times \frac{dP}{dL}\big|_{DRP}}{\rho_M \times V_M^2} \quad (5)$$

Reynolds number on this analysis is based on the liquid kinematic viscosity v_L

$$Re_M = \frac{V_M \times D}{v_L} \quad (6)$$

The regression analysis is conducted based on the experimental data obtained at different liquid superficial velocities ranged from 1.85 to 4.317 m/s forming around 100 data set points. As it can be seen from Figure 16, that all of the data points are following the same trend of the fitted curve (Equation (7) presents the correlation).

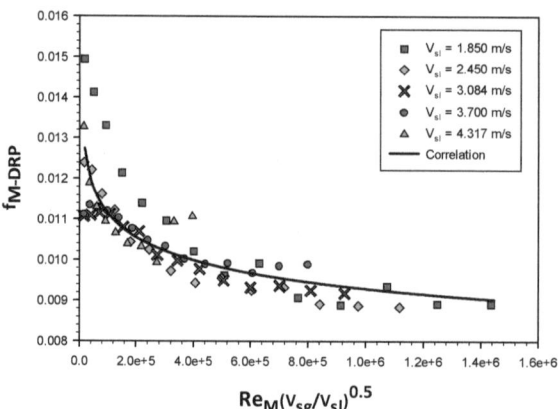

Figure 16. Friction factor variation with the mixture Reynolds number times the square root of the superficial velocities ratio for different liquid superficial velocities (1.85, 2.45, 3.08, 3.7 and 4.32 m/s).

The scatter data conclude a wide range of flow types and regimes of slug (pseudo slug), annular and dispersed bubbly flow regimes. This could enable better prediction of the correlation under the study.

Figure 17 exhibits the comparison between the measured values of friction factor and the predicted one, however as shown in the plot all scatter data has been predicted within band of ± 15%, and the correlation for the friction factor can be represented as:

$$f_{(M\text{-}DRP)} = 0.0276 \left(Re_M \left(\frac{V_{sg}}{V_{sl}} \right)^{(0.5)} \right)^{(-0.079)} \quad (7)$$

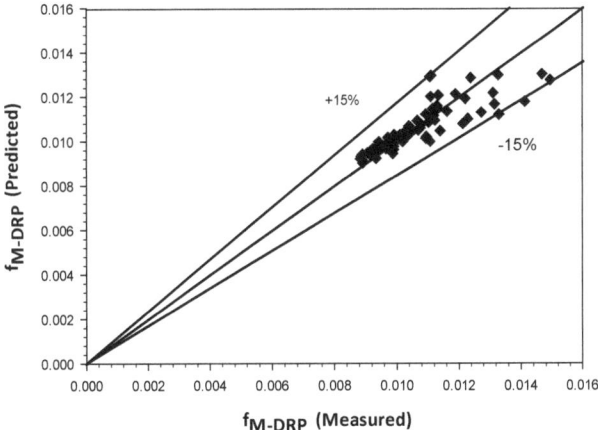

Figure 17. Comparison between measured friction factor and predicted by Equation (7).

Using the same experimental data, we were able to generate another correlation that fits the data points exponentially. Since the frictional pressure drop will increase as more liquid flow rate added to the flow; then it could be more interesting to describe such dimensionless pressure drop that includes the frictional pressure gradient when the flow assumed to be liquid only (P_{sl}) and the pressure drop with the addition of the drag reducing polymer (P_{DRP}).

Where P_{sl} defined as:

$$P_{sl} = \frac{f \times \rho_l \times V_{sl}^2}{2D} \quad (8)$$

The friction factor (f) for single phase liquid shown in Equation (8) is calculated using the well-known Blasius equation (Equation (9)):

$$f = 0.184 \text{Re}_{sl}^{-0.2} \quad (9)$$

And Re_{sl} is Reynolds number that can be expressed as:

$$\text{Re}_{sl} = \frac{\rho_l \times V_{sl} \times D}{\mu_l} \quad (10)$$

Figure 18 presents the relation between the dimensionless pressure drop ratio $\frac{P_{DRP}}{P_{sl}}$ and the normalized superficial velocity $\frac{V_{sg}}{V_{sl}}$. As it can be seen that the superficial gas and liquid velocity are in great impact on the pressure drop ratio and could confirm that it is controlling the drag reduction and that makes the correlation behaves better.

The importance of such correlation appears when the Drag Reducing polymers are added to the two-phase system in order to broaden the understanding of the reduction mechanism by using general descriptive model. Using this correlation, we can predict the pressure drop after adding the DRP without knowing the flow pattern and this is quite useful for industrial applications.

The regression analysis is performed for the data points and it can be seen from Figure 19 that all of the scattered data are now in between ±10% spread with correlation goodness of fit ($R^2 = 0.97$). The correlation can be expressed as:

$$\frac{P_{DRP}}{P_{sl}} = (0.5648) \exp\left(0.6456 \left(\frac{V_{sg}}{V_{sl}}\right)^{0.5}\right) \quad (11)$$

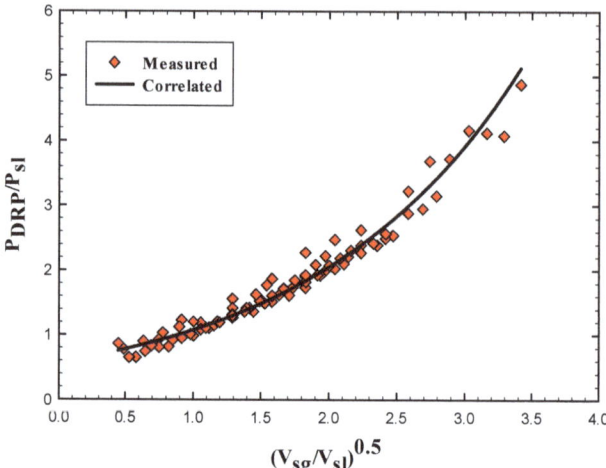

Figure 18. Dimensionless pressure drop ratio versus square root of the normalized superficial velocities (correlation is Equation (11)).

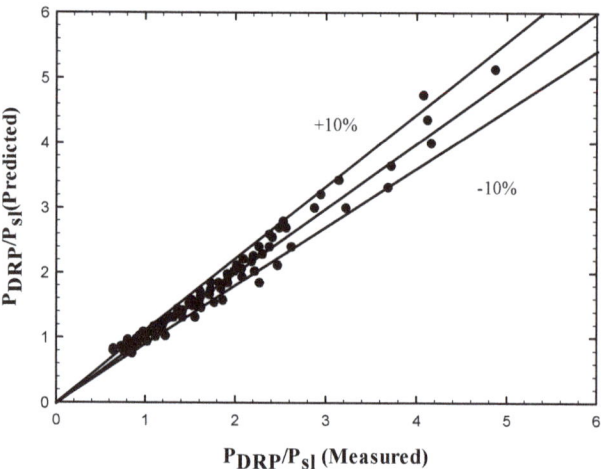

Figure 19. Comparison between measured dimensionless pressure drop ratio and the predicted by Equation (11).

4. Conclusions

It can be concluded that the drag reduction in two phase of air-water mixture occurs at all liquid and gas flow rates. The mechanism of drag reduction can be explained as that the DRP leads to suppress the interfacial friction; the polymer solution stretched along the interface to increase laminar sub-layer thickness and result in more unidirectional free of eddies flow.

Utilizing the DRP with low concentrations attributed in pressure drop reduction for all cases been studied, the effectiveness of the DRP varied tremendously from 14% to 80% as reported at the stratified and intermittent flows, respectively.

The DRP can affect flow behavior and found to be more efficient in suppression of highly disturbed waves and found to be able to shift the flow from slug and pseudo slug to stratified wavy and wavy annular regimes, respectively.

In order to produce high drag reduction, the DRP should be able to damp turbulence intensity and fluctuations. The maximum drag reduction always occurs when the highly fluctuated waves were reduced effectively by DRP and accordingly phase transition appeared. For example, DR% of 63% is reported when the slug flow altered to stratified wavy regime with the presence of only 40 ppm DRP.

As discussed previously, the DRP injection mechanism is very sensitive. Therefore, in this study using a diaphragm pump (gives DRP solution in dosages) offered an effective technique by introducing the DRP without causing any polymer shear degradation. Moreover, with an easy adjustable flow rate controller attached to the pump this method of injection can give a reliable way for industrial applications, since it can provide a wide range of flow rates with changeable speeds and torques by which the exact needed amount of polymer solution would be controlled more accurately.

Friction factor correlation as a function of mixture Reynolds number obtained in this study is evaluated and it has been successfully covered a wide range of liquid and gas flow rates (including different flow regimes) and predicted the data effectively within ±15% when it is compared with the measured values.

The pressure drop after the addition of DRP (P_{DRP}) has been predicted using the experimental data, the correlation has been presented as a function of the superficial frictional pressure drop and normalized superficial velocity. Regression analysis is conducted over a wide range of the experimental data, the correlation effectively predicted the data within ±10% spread with correlation goodness of fit ($R^2 = 0.97$).

Author Contributions: Data curation, M.S.; Methodology, A.A.; Validation, A.A.; Writing—original draft, M.S. and A.A. All authors have read and agreed to the published version of the manuscript.

Funding: This research received no external funding.

Data Availability Statement: All the data is presented in the article.

Acknowledgments: The authors would like to thank King Fahd University of Petroleum & Minerals for supporting this work.

Conflicts of Interest: The authors declare no conflict of interest.

Appendix A

Table A1. Frictional Pressure Drop and Flow Pattern with and without 40 ppm DRP.

run	$V_{sl}\left(\frac{m}{s}\right)$	$V_{sg}\left(\frac{m}{s}\right)$	$\frac{dP}{dL}\left(\frac{Pa}{m}\right)$	Flow pattern	$\frac{dP}{dL}\left(\frac{Pa}{m}\right)$	Flow pattern	DR%
			without DRP		40 ppm DRP		
1	0.1	0.41	613	St	400	St	35
2	0.1	0.51	667	St	467	St	30
3	0.1	0.62	760	St	600	St	21
4	0.1	0.72	800	St	687	St	14
5	0.1	0.82	880	St	733	St	17
6	0.1	0.93	1160	St—Slug	800	St Wavy	31
7	0.1	1.03	1200	St—Slug	933	St Wavy	22
8	0.1	1.64	1800	Slug	667	St Wavy	63
9	0.1	2.06	1933	Slug	800	St Wavy	59
10	0.1	2.47	2467	Slug	1000	St Wavy	59
11	0.1	2.88	3000	St Wavy	1067	St Wavy	64
12	0.1	3.70	1467	St Wavy	667	St Wavy	55
13	0.1	4.11	1667	St Wavy	733	St Wavy	56
14	0.1	4.52	1800	An	867	St Wavy	52
15	0.1	4.93	1867	An	867	St Wavy	54
16	0.1	5.34	2000	An	1000	St Wavy	50
17	0.1	7.20	1467	An	773	St Wavy	47

Table A1. Cont.

run	$V_{sl}\left(\frac{m}{s}\right)$	$V_{sg}\left(\frac{m}{s}\right)$	$\frac{dP}{dL}\left(\frac{Pa}{m}\right)$	Flow pattern	$\frac{dP}{dL}\left(\frac{Pa}{m}\right)$	Flow pattern	DR%
			without DRP		40 ppm DRP		
18	0.1	7.81	1533	An	800	St Wavy	48
19	0.1	8.43	1600	An	867	St Wavy	46
20	0.1	9.05	1800	An	1000	St Wavy	44
21	0.1	9.66	1867	W An	1133	St Wavy	39
22	0.1	10.28	2000	W An	1267	St Wavy	37
23	0.1	10.90	2067	W An	1400	St Wavy	32
24	0.1	11.51	2133	W An	1467	St Wavy	31
25	0.1	12.13	2400	W An	1640	W An	32
26	0.1	12.75	2667	W An	1713	W An	36
27	3.08	1.03	15,480	DB	6905	Pseudo Slug	55
28	3.08	2.06	18,713	DB	8659	Pseudo Slug	54
29	3.08	3.08	21,100	DB	10,456	Pseudo Slug	50
30	3.08	4.11	23,553	DB	12,171	Pseudo Slug	48
31	3.08	5.14	25,847	DB	13,500	Pseudo Slug	48
32	3.08	6.17	27,920	DB	15,050	DB	46
33	3.08	7.20	29,287	DB	15,823	DB	46
34	3.08	8.22	31,280	DB	17,172	DB	45
35	3.08	9.25	32,067	DB	18,359	DB	43
36	3.08	10.28	33,680	DB	19,323	DB	43

Appendix B

Table A2. Frictional Pressure Drop Associated with the Slug Flow Regime and Drag Reduction Effectiveness Using 40 ppm DRP.

run	$V_{sl}\left(\frac{m}{s}\right)$	$V_{sg}\left(\frac{m}{s}\right)$	$\frac{dP}{dL}\left(\frac{Pa}{m}\right)$	Flow pattern	$\frac{dP}{dL}\left(\frac{Pa}{m}\right)$	Flow pattern	DR%
			without DRP		40 ppm DRP		
1	0.72	0.41	2133	Slug	1000	Slug	53
2	0.72	0.51	2267	Slug	1533	Slug	32
3	0.72	0.62	2400	Slug	1733	Slug	28
4	0.72	0.72	2467	Slug	1933	Slug	22
5	0.72	0.82	2667	Slug	2133	Slug	20
6	0.72	0.93	3000	Slug	2200	Slug	27
7	0.72	1.03	3133	Slug	2333	Slug	26
8	0.72	1.13	3333	Slug	2467	Slug	26

Appendix C

Table A3. Frictional Pressure Drop and Transition of the Slug Flow Regime Using 100 ppm DRP.

run	$V_{sl}\left(\frac{m}{s}\right)$	$V_{sg}\left(\frac{m}{s}\right)$	$\frac{dP}{dL}\left(\frac{Pa}{m}\right)$	Flow pattern	$\frac{dP}{dL}\left(\frac{Pa}{m}\right)$	Flow pattern	DR%
			without DRP		100 ppm DRP		
1	0.72	0.41	2133	Slug	733	St Wavy	66
2	0.72	0.51	2267	Slug	933	St Wavy	59
3	0.72	0.62	2400	Slug	1067	St Wavy	56
4	0.72	0.72	2467	Slug	1067	St Wavy	57
5	0.72	0.82	2667	Slug	1067	St Wavy	60
6	0.72	0.93	3000	Slug	1267	St Wavy	58
7	0.72	1.03	3133	Slug	1400	St Wavy	55
8	0.72	1.13	3333	Slug	1667	St Wavy	50

Appendix D

Table A4. Frictional pressure gradient and transition of Pseudo Slug flow regime using 40 ppm DRP.

run	$V_{sl}\left(\frac{m}{s}\right)$	$V_{sg}\left(\frac{m}{s}\right)$	$\frac{dP}{dL}\left(\frac{Pa}{m}\right)$	Flow pattern	$\frac{dP}{dL}\left(\frac{Pa}{m}\right)$	Flow pattern	DR%
			without DRP		40 ppm DRP		
1	1.03	1.03	4000	Pseudo Slug	2667	W An	33
2	1.03	2.06	5667	Pseudo Slug	3333	W An	41
3	1.03	3.08	7333	Pseudo Slug	4333	W An	41
4	1.03	4.11	8467	Pseudo Slug	6400	Pseudo Slug	24
5	1.03	5.14	9800	Pseudo Slug	7067	Pseudo Slug	28
6	1.03	6.17	10800	Pseudo Slug	8000	Pseudo Slug	26

Appendix E

Table A5. Frictional Pressure Gradient and Transition of Pseudo Slug Flow Regime Using 100 ppm DRP.

run	$V_{sl}\left(\frac{m}{s}\right)$	$V_{sg}\left(\frac{m}{s}\right)$	$\frac{dP}{dL}\left(\frac{Pa}{m}\right)$	Flow pattern	$\frac{dP}{dL}\left(\frac{Pa}{m}\right)$	Flow pattern	DR%
			without DRP		100 ppm DRP		
1	1.03	1.03	4000	Pseudo Slug	1600	W An	60
2	1.03	2.06	5667	Pseudo Slug	2333	W An	59
3	1.03	3.08	7333	Pseudo Slug	2667	W An	64
4	1.03	4.11	8467	Pseudo Slug	3600	W An	57
5	1.03	5.14	9800	Pseudo Slug	4467	W An	54
6	1.03	6.17	10800	Pseudo Slug	5800	W An	46

References

1. Toms, B. Some observations on the flow of linear polymer solutions through straight tubes at large Reynolds numbers, Proceeding. In *Proceedings of the 1st International Congress on Rheology*; North-Holland: Amsterdam, The Netherlands, 1948; Volume 2, pp. 135–141.
2. Oliver, D.R.; Hoon, A.Y. Two-phase non-Newtonian flow. *Trans. Inst. Chem. Eng.* **1968**, 26, T106.
3. Al-Sarkhi, T.; Hanratty, A. Effect of drag-reducing polymers on annular gas-liquid flow in a horizontal pipe. *Int. J. Multiph. Flow* **2001**, 27, 1151–1162. [CrossRef]
4. Soleimani, A.; Al-sarkhi, A.; Hanratty, T. Effect of drag reducing polymers on Pseudo-Slugs-Interfacial drag and transition to slug flow. *Int. J. Multiph. Flow* **2002**, 28, 1911–1927. [CrossRef]
5. Baik, S.; Hanratty, T. Effects of a drag reducing polymer on stratified gas–liquid flow in a large diameter horizontal pipe. *Int. J. Multiph. Flow* **2003**, 29, 1749–1757. [CrossRef]
6. Fernandes, R.; Jutte, B.; Rodriguez, M. Drag reduction in horizontal annular two-phase flow. *Int. J. Multiph. Flow* **2004**, 30, 1051–1069. [CrossRef]
7. Al-sarkhi, A.; El Nakla, M.; Ahmed, W. Friction factor correlations for gas—Liquid/liquid—Liquid flows with drag-reducing polymers in horizontal pipes. *Int. J. Multiph. Flow* **2011**, 37, 501–506. [CrossRef]
8. Al-Sarkhi, A. Effect of mixing on frictional loss reduction by drag reducing polymer in annular horizontal two-phase flows. *Int. J. Multiph. Flow* **2012**, 39, 186–192. [CrossRef]
9. Al-Sarkhi, A. Drag reduction with polymers in gas-liquid/liquid-liquid flows in pipes: A literature review. *J. Nat. Gas Sci. Eng.* **2010**, 2, 41–48. [CrossRef]
10. Wang, Y.; Yan, W.; Cheng, Z. Direct Numerical Simulation of Gas-Liquid Drag-Reducing Cavity Flow by the VOSET Method. *Polymers* **2019**, 11, 596. [CrossRef]
11. Taylor, N.H.; Nedderman, R. The coalescence of disturbance waves in annular two phase flow. *Chem. Eng. Sci.* **1968**, 23, 551–564. [CrossRef]
12. Zhao, Y.; Markides, C.; Matar, O.; Hewitt, G. Disturbance wave development in two-phase gas-liquid upwards vertical annular flow. *Int. J. Multiph. Flow* **2013**, 55, 111–129. [CrossRef]
13. Spedding, P.; Hand, N. A Revised Analysis of the Effect of Surfactants on Two-Phase Phenomena in Horizontal Air-Water Pipe Flow. *Dev. Chem. Eng. Miner. Process.* **1997**, 5, 267–279. [CrossRef]
14. Thwaites, G.; Kulov, N.; Nedderman, R. Liquid film properties in two-phase annular flow. *Chem. Eng. Sci.* **1976**, 31, 481–486. [CrossRef]
15. More, P.; Kang, C.; Magalhães, A. The Performance of Drag Reducing Agents in Multiphase Flow Conditions at High Pressure; Positive and Negative Effects. In Proceedings of the 2008 7th International Pipeline Conference, Calgary, AB, Canada, 29 September–3 October 2008; pp. 1–9.

16. Daas, M.; Bleyle, D. Computational and experimental investigation of the drag reduction and the components of pressure drop in horizontal slug flow using liquids of different viscosities. *Exp. Therm. Fluid Sci.* **2006**, *30*, 307–317. [CrossRef]
17. Zhang, H.-Q.; Wang, Q.; Sarica, C.; Brill, J. Unified model for gas-liquid pipe flow via slug dynamics—Part 1: Model development. *J. Energy Resour. Technol.* **2003**, *125*, 266–273. [CrossRef]
18. Garcı, J.M. Friction factor improved correlations for laminar and turbulent gas—Liquid flow in horizontal pipelines. *Int. J. Multiph. Flow* **2007**, *33*, 1320–1336. [CrossRef]
19. Garcı, R. Composite power law holdup correlations in horizontal pipes. *Int. J. Multiph. Flow* **2005**, *31*, 1276–1303. [CrossRef]

Disclaimer/Publisher's Note: The statements, opinions and data contained in all publications are solely those of the individual author(s) and contributor(s) and not of MDPI and/or the editor(s). MDPI and/or the editor(s) disclaim responsibility for any injury to people or property resulting from any ideas, methods, instructions or products referred to in the content.

Article

General Relations between Stress Fluctuations and Viscoelasticity in Amorphous Polymer and Glass-Forming Systems

Alexander Semenov * and Jörg Baschnagel

Institut Charles Sadron, CNRS–UPR 22, University of Strasbourg, 67034 Strasbourg, France
* Correspondence: semenov@unistra.fr

Abstract: Mechanical stress governs the dynamics of viscoelastic polymer systems and supercooled glass-forming fluids. It was recently established that liquids with long terminal relaxation times are characterized by transiently frozen stress fields, which, moreover, exhibit long-range correlations contributing to the dynamically heterogeneous nature of such systems. Recent studies show that stress correlations and relaxation elastic moduli are intimately related in isotropic viscoelastic systems. However, the origin of these relations (involving spatially resolved material relaxation functions) is non-trivial: some relations are based on the fluctuation-dissipation theorem (FDT), while others involve approximations. Generalizing our recent results on 2D systems, we here rigorously derive three exact FDT relations (already established in our recent investigations and, partially, in classical studies) between spatio-temporal stress correlations and generalized relaxation moduli, and a couple of new exact relations. We also derive several new approximate relations valid in the hydrodynamic regime, taking into account the effects of thermal conductivity and composition fluctuations for arbitrary space dimension. One approximate relation was heuristically obtained in our previous studies and verified using our extended simulation data on two-dimensional (2D) glass-forming systems. As a result, we provide the means to obtain, in any spatial dimension, all stress-correlation functions in terms of relaxation moduli and vice versa. The new approximate relations are tested using simulation data on 2D systems of polydisperse Lennard–Jones particles.

Keywords: supercooled liquids; polymers; viscoelasticity; amorphous solids

Citation: Semenov, A.; Baschnagel, J. General Relations between Stress Fluctuations and Viscoelasticity in Amorphous Polymer and Glass-Forming Systems. *Polymers* **2024**, *16*, 2336. https://doi.org/10.3390/polym16162336

Academic Editors: Célio Pinto Fernandes, Luís L. Jorge Lima Ferraz and Alexandre M. Afonso

Received: 28 June 2024
Revised: 10 August 2024
Accepted: 14 August 2024
Published: 18 August 2024

Copyright: © 2024 by the authors. Licensee MDPI, Basel, Switzerland. This article is an open access article distributed under the terms and conditions of the Creative Commons Attribution (CC BY) license (https://creativecommons.org/licenses/by/4.0/).

1. Introduction

Viscoelastic liquids and amorphous materials are characterized by long-lasting memory effects often involving a wide spectrum of relaxation times correlating the flow to the prior external forces and strains [1–4]. Examples of such materials include complex fluids like viscoelastic polymer melts and solutions, molten metallic alloys, glass-forming (supercooled) liquids and soft-matter systems [3–19]. The central physical quantity governing the dynamics of such materials is the mechanical stress [20]. Stress-correlation functions can provide important information on the rheological properties of amorphous systems, including the most important rheological functions like shear and longitudinal relaxation moduli (and the corresponding dynamical moduli) [3,4,21]. Moreover, glass-forming liquids are known to be highly heterogeneous (near or below the glass transition temperature T_g) [22–29], leading to a significant wave-vector dependence of their shear viscosity and relaxation moduli [30–32]. A similar behavior was observed [33] and predicted [34] for polymer liquids, and it is expected to be even more important for high-molecular-weight polymers.

Useful relations between the spatio-temporal stress correlation functions and the generalized (length-scale dependent) relaxation moduli (GRMs) have recently been obtained using the Zwanzig–Mori projection operator formalism [35,36] and the fluctuation-dissipation theorem (FDT) [32,37]. Based on these theoretical relations, it was established that liquids with long terminal relaxation times are characterized by transiently frozen stress fields, which, moreover, exhibit long-range correlations supporting the dynamically heterogeneous nature of glass-forming systems [32,35–39]. These theoretical predictions reinforce the conclusions of the prior extensive and pioneering simulation studies on stress-correlations in supercooled liquids [40–43] and also agree with more recent simulation results [32,44,45].

While the recent theoretical studies show that stress correlations and viscoelastic relaxation moduli are intimately related in complex fluids like polymer and supercooled liquids [32,35–37], the origin of these relations (involving spatially resolved relaxation functions) appears to be non-trivial: a number of relations are exact and follow from the fluctuation-dissipation theorem (FDT), while alternative physical arguments are required to derive other relations [32,37]. Generalizing our recent results on two-dimensional (2D) systems [32,37], we here obtain and discuss the full set of such stress–fluctuation relations valid for arbitrary space dimension.

In the next two sections, we reprise the relevant classical results on the bulk elastic and viscoelastic properties of amorphous systems, presenting the fully tensorial relations between fluctuations of volume-averaged stress and elastic (relaxation) moduli. The bulk equations are then generalized in Section 4 to deal with wave-vector (q)-dependent stress correlations (characterized by the tensorial correlation function $C = \bar{C}(q, t)$) and spatially resolved relaxation moduli (elasticity tensor $E = E(q, t)$). The methodologically new point here is that we first present a detailed derivation of the general *tensorial* equation linking C- and E-tensor fields, which then yields three basic relations between the generalized shear, longitudinal and transverse (mixed) relaxation moduli ($G(q,t)$, $L(q,t)$ and $M(q,t)$) on the one hand, and the invariant correlation functions on the other hand. The recently discovered M-relation (Equation (64)) then follows from the general tensorial equation in exactly the same way as other relations (Equations (62) and (63)). Different aspects concerning the definition of the q-dependent elasticity tensor E are discussed in Section 5. In particular, it is highlighted there that not all the components of E can be unambiguously defined for $q \neq 0$ based on a stress-to-strain response. In Section 6, we introduce the concept of stress noise σ^n and propose a new definition of all components of the elasticity tensor $E(q, t)$ in terms of σ^n. It is also demonstrated there that the new definition is consistent with all the known properties of this tensor. On this basis, we derive the full set of exact relations between the correlation and elasticity tensors, $C(q,t)$ and $E(q,t)$, and establish two approximate relations allowing to obtain the full correlation tensor $C(q,t)$ in terms of only three material functions, $G(q,t)$, $L(q,t)$ and $M(q,t)$, also known as viscoelastic memory functions (VMFs). We also discuss how to improve the accuracy of an approximate relation for two-dimensional systems. The theoretical predictions are then compared with simulation results on 2D polydisperse systems of Lennard–Jones (LJ) particles. Such 2D systems have been recently studied experimentally [46–48] and have received a lot of attention in simulation studies [32,44]. The main results of the paper are summarized in the last Section 7. In particular, the most important novel results are highlighted in the last point 11 of this section.

2. Classical Elasticity

For tutorial purposes, we start with the linear (affine) elasticity of a macroscopically uniform isotropic solid body. Its deformation is defined by the (coarse-grained) field of displacements $\underline{u} = \underline{u}(\underline{r})$ of its material elements (here, $\underline{r}$ is the initial position of an element). At equilibrium ($\underline{u} = 0$), the mean mechanical stress $\underline{\underline{\sigma}} = \underline{\underline{\sigma}}^{(0)}$ is isotropic, $\sigma^{(0)}_{\alpha\beta} = -p_0 \delta_{\alpha\beta}$, where p_0 is the external pressure [49]. (Note that in computer simulations the mean stress tensor for a given configuration can be slightly anisotropic. In this case, $\underline{\underline{\sigma}}^{(0)}$ should be considered as the mean stress tensor averaged over a sufficiently large ensemble of configurations). Let us consider a weak affine deformation

$$\gamma_{\alpha\beta} \equiv \partial u_\alpha / \partial r_\beta \tag{1}$$

where $\gamma_{\alpha\beta}$ is the strain tensor which does not change in time, and α, β refer to Cartesian components. It leads to a macroscopically homogeneous stress as a linear response, $\Delta\sigma_{\alpha\beta} = \sigma_{\alpha\beta} - \sigma_{\alpha\beta}^{(0)}$, where $\sigma_{\alpha\beta}$ is the mean (time-averaged) stress in the deformed state. According to Hooke's law

$$\Delta\sigma_{\alpha\beta} = E_{\alpha\beta\alpha'\beta'}\gamma_{\alpha'\beta'} \quad (2)$$

where $E_{\alpha\beta\alpha'\beta'}$ is the tensor of static elastic moduli. (Here and below, the Einstein convention for summation over repeated indices is assumed). For systems (liquids) with no orientational order, the stress tensor is always symmetric, leading to

$$E_{\alpha\beta\alpha'\beta'} = E_{\beta\alpha\alpha'\beta'} \quad (3)$$

To exclude a rotation of the system as a whole (which does not cost any energy), we can consider the symmetric part of $\underline{\underline{\gamma}}$,

$$\epsilon_{\alpha\beta} = \frac{1}{2}(\gamma_{\alpha\beta} + \gamma_{\beta\alpha}) \quad (4)$$

The symmetric $\epsilon_{\alpha\beta}$ will be referred to as the classical strain. Then, Equation (2) transforms into the classical relation

$$\Delta\sigma_{\alpha\beta} = E_{\alpha\beta\alpha'\beta'}\epsilon_{\alpha'\beta'} \quad (5)$$

The reason for the equivalence of the two equations, (2) and (5), is the minor symmetry of the E-tensor:

$$E_{\alpha\beta\alpha'\beta'} = E_{\alpha\beta\beta'\alpha'} \quad (6)$$

coming from the assumed isotropy of the system demanding that E must be an isotropic tensor [44], which, together with the symmetry relation (3), leads to a well-known equation

$$E_{\alpha\beta\alpha'\beta'} = \lambda\delta_{\alpha\beta}\delta_{\alpha'\beta'} + \mu\left(\delta_{\alpha\alpha'}\delta_{\beta\beta'} + \delta_{\alpha\beta'}\delta_{\alpha'\beta}\right) \quad (7)$$

with λ and μ being the Lamé coefficients [20,44,49]. (Note that Equation (6) also comes directly from Equation (2) since rotations of the body as a whole (corresponding to an anti-symmetric $\underline{\underline{\gamma}}$) must not lead to any change in the mean (ensemble-averaged) stress. The usefulness (convenience) of Equation (2) is also clarified in Section 5.3 in relation to the wave-vector- dependent elasticity).

The free energy increment ΔF associated with a small strain $\underline{\underline{\epsilon}}$ is

$$\Delta F \simeq \frac{V}{2} E_{\alpha\beta\alpha'\beta'}\epsilon_{\alpha\beta}\epsilon_{\alpha'\beta'} \quad (8)$$

where V is the system volume and the higher-order terms in $\underline{\underline{\epsilon}}$ are omitted. (Note that Equation (8) remains valid also if $\epsilon_{\alpha\beta}$ is replaced with $\gamma_{\alpha\beta}$). Suppose the elastic body has a free surface, so it can be deformed by a thermal fluctuation. Then, by virtue of the Boltzmann equipartition principle Equation (8), it leads to the following correlation properties for thermal fluctuations of the classical strain, $\underline{\underline{\epsilon}}$:

$$\langle\epsilon_{\alpha\beta}\epsilon_{\alpha'\beta'}\rangle = \frac{T}{4\mu V}\left[\delta_{\alpha\alpha'}\delta_{\beta\beta'} + \delta_{\alpha\beta'}\delta_{\alpha'\beta} - \frac{2\lambda}{2\mu + \lambda d}\delta_{\alpha\beta}\delta_{\alpha'\beta'}\right] \quad (9)$$

where $\epsilon_{\alpha\beta}$ and $\epsilon_{\alpha'\beta'}$ are strain components taken at the same time, the brackets $\langle..\rangle$ mean the complete ensemble- and time-averaging, d is the space dimension, and T is the temperature in energy units.

It is also possible to relate stress fluctuations with the E-tensor. Here, it is useful to recall that the latter tensor reflects the long-time (static) stress response (cf Equation (2)); hence, the relevant stress correlation function

$$C_{\alpha\beta\alpha'\beta'} \equiv \frac{V}{T}\langle \delta\sigma_{\alpha\beta}\delta\sigma_{\alpha'\beta'}\rangle, \quad \delta\sigma_{\alpha\beta} \equiv \sigma_{\alpha\beta} - \langle\sigma_{\alpha\beta}\rangle$$

must involve a *quasi-static* (rather than instantaneous) stress $\sigma_{\alpha\beta}$ (note that $\langle\sigma_{\alpha\beta}\rangle = \sigma_{\alpha\beta}^{(0)}$). To define it, one has to assume that the strain fluctuations are sufficiently slow (cf refs. [50,51]) as compared to the internal structural relaxation of the system (with terminal relaxation time τ); the characteristic time of the strain fluctuations, τ_{strain}, must be much longer than τ [50]. Therefore, $\sigma_{\alpha\beta}$ (involved in the definition of the C-tensor) must be considered as the stress component coarse-grained over a time interval Δt such that $\tau \ll \Delta t \ll \tau_{strain}$. In this case,

$$\delta\sigma_{\alpha\beta} \simeq E_{\alpha\beta\alpha'\beta'}\epsilon_{\alpha'\beta'}$$

(note that ϵ is a slow strain fluctuation, hence $\delta\sigma$, being a stress response to ϵ, cf Equation (5), is a fluctuation as well; that is why we use the notation $\delta\sigma$ instead of $\Delta\sigma$ here) so that using Equations (7) and (9), we obtain

$$C_{\alpha\beta\alpha'\beta'} \simeq E_{\alpha\beta\alpha'\beta'} \quad (10)$$

Equation (9) can be obtained in a different way using the classical fluctuation-dissipation theorem (FDT) [52,53]. As before, one can define the mean strain, $\gamma_{\alpha\beta} = (1/V)\int \gamma_{\alpha\beta}(\underline{r})d^d r$ (cf Equation (1)) based on the coarse-grained displacement field. The strain $\gamma_{\alpha\beta}$ can be considered as a tensorial variable conjugate to the external 'force', $\sigma_{\alpha\beta}^{ex}$, such that the external potential energy (the contribution of the 'force' to the free energy of the system) is

$$U_{ext} = -V\gamma_{\alpha\beta}\sigma_{\alpha\beta}^{ex} \quad (11)$$

Obviously, the 'force' tensor $\sigma_{\alpha\beta}^{ex}$ is the external stress applied to the system [20]. The symmetric part of the deformation, $\epsilon_{\alpha\beta}$, induced by a *weak* external stress can be obtained by minimization of the total free energy $F_{tot} = \Delta F + U_{ext}$, where ΔF is defined in Equation (8). As a result, we obtain a relation between the mean deformation, $\langle\epsilon_{\alpha\beta}\rangle$, and $\sigma_{\alpha\beta}^{ex}$ (cf Equation (2)):

$$E_{\alpha\beta\alpha'\beta'}\langle\epsilon_{\alpha'\beta'}\rangle = \sigma_{\alpha\beta}^{ex} \quad (12)$$

According to the classical FDT [52,53], the correlation tensor $\langle\epsilon_{\alpha\beta}\epsilon_{\alpha'\beta'}\rangle$ must be proportional to the susceptibility, $\partial\langle\epsilon_{\alpha'\beta'}\rangle/\partial\sigma_{\alpha\beta}^{ex}$:

$$\langle\epsilon_{\alpha\beta}\epsilon_{\alpha'\beta'}\rangle = \frac{T}{V}\partial\langle\epsilon_{\alpha'\beta'}\rangle/\partial\sigma_{\alpha\beta}^{ex} \quad (13)$$

Equation (9) can then be deduced from Equations (7), (12) and (13).

3. Classical Viscoelasticity

Let us turn to viscoelastic systems, including polymer melts and solutions and glass-forming supercooled liquids, but also, in principle, glassy amorphous solids below the glass transition temperature T_g. Such systems are characterized by time-dependent relaxation moduli, like the shear relaxation modulus $G(t)$. The goal is to find relations between the stress-correlation functions and the relaxation moduli. Obviously, the argument leading to Equation (10) is not applicable in this case due to its static nature. By contrast, it is well-known that the classical FDT can be applied to relaxation processes [52,53]. There is, however, a fundamental problem with its application to flows of liquids using the deformation tensor as a variable, which is based on the concept of a necessarily continuous (coarse-grained) displacement field $\underline{u}(\underline{r})$, cf Equation (1). The point is that during long relaxation times characteristic of most viscoelastic liquids, the initially neighboring particles can go far away from each other (by self-diffusion), which means that $\underline{u}(\underline{r})$ becomes ill-defined (virtually discontinuous). In other words, here, we arrive at a contradiction between continuum-field and corpuscular views on the fluid dynamics [20]. To avoid such problems, another version of the FDT [32,34,37] should be employed here. It is outlined below. (Note that this approach is applicable more generally also for networks with transient or permanent bonds and solid amorphous systems where $\underline{u}(\underline{r})$ is well defined).

(i) We use a more precise definition of the elastic moduli. The tensor of relaxation moduli $E_{\alpha\beta\alpha'\beta'}(t)$ is defined via the stress tensor response, $\Delta\sigma_{\alpha\beta}(t)$, to a small but instantaneous deformation of the system, $\gamma_{\alpha\beta}$, at $t = 0$:

$$\Delta\sigma_{\alpha\beta}(t) = E_{\alpha\beta\alpha'\beta'}(t)\gamma_{\alpha'\beta'} \tag{14}$$

(cf Equation (2); note that $\Delta\sigma_{\alpha\beta}(t)$ is the mean, ensemble-averaged, stress increment at time t). This deformation must be *affine-canonical* [54–56]. (The transformation is necessarily canonical since we assume that the system dynamics remains Hamiltonian also with the perturbation), which implies changes of both the coordinates ($\underline{r}$) and the velocities ($\underline{v}$) of all particles:

$$r_\alpha \to r_\alpha + \gamma_{\alpha\beta}r_\beta, \quad v_\alpha \to v_\alpha - \gamma_{\beta\alpha}v_\beta \quad \text{at } t = 0 \tag{15}$$

(ii) We assume that before the perturbation (at $t < 0$), the system was at equilibrium, being characterized by an isothermal-isobaric distribution in the phase space:

$$\mathcal{P}(\Gamma) = \mathcal{P}_0(\Gamma) = \text{const}\, e^{-H(\Gamma)/T} \tag{16}$$

where $\mathcal{P}$ is the probability density, and Γ stands for the microstate in the phase space (coordinates and velocities of all particles), $H(\Gamma) = H_0(\Gamma) + p_0 V$, $H_0(\Gamma)$ is the system Hamiltonian, $V = V(\Gamma)$, its volume, and p_0, the imposed pressure. (Note the normalization condition: $\int_\Gamma \mathcal{P}_0(\Gamma) = 1$.) Importantly, we consider a liquid or a *fully equilibrated* amorphous system here, so that their shape variations, which are allowed, do not bring about any correction to the Hamiltonian; the external stress corresponds solely to an *isotropic* pressure p_0.

Right after the instantaneous deformation (at $t = 0^+$), the distribution changes to

$$\mathcal{P}(\Gamma) = \mathcal{P}_0(\Gamma) + \Delta\mathcal{P}(\Gamma) \tag{17}$$

The microscopic definition of the stress tensor reads [57]

$$\sigma_{\alpha\beta}(\Gamma) = \frac{1}{V}\sum_{i>j} u'_{ij}(r)\frac{r_\alpha r_\beta}{r} - \frac{1}{V}\sum_i m_i v_{i\alpha}v_{i\beta} \tag{18}$$

where $u_{ij}(r)$ is the interaction energy of a pair of interacting particles (i,j), r is the distance between them, $\underline{r}$ is the corresponding displacement vector (from i to j), m_i is the mass of particle i, and $v_{i\alpha}$ is the α-component of its velocity. Using Equation (18), we find that the change in $H(\Gamma)$ generated by the transformation of Equation (15) (for a system initially in microstate Γ) is

$$\Delta H = V\delta\sigma_{\alpha\beta}(\Gamma)\gamma_{\alpha\beta} \tag{19}$$

where $\delta\sigma_{\alpha\beta}(\Gamma) = \sigma_{\alpha\beta}(\Gamma) - \langle\sigma_{\alpha\beta}\rangle = \sigma_{\alpha\beta}(\Gamma) + p_0\delta_{\alpha\beta}$ is the stress fluctuation, and $\langle\sigma_{\alpha\beta}\rangle$ is the equilibrium ensemble-averaged stress, $\langle\sigma_{\alpha\beta}\rangle = -p_0\delta_{\alpha\beta}$. Taking also into account that the transformation conserves the phase space, we obtain

$$\Delta\mathcal{P}(\Gamma)/\mathcal{P}_0(\Gamma) = \frac{\Delta H}{T} = \frac{V}{T}\delta\sigma_{\alpha\beta}(\Gamma)\gamma_{\alpha\beta} \tag{20}$$

(iii) From the above equations, it immediately follows that

$$\Delta\sigma_{\alpha\beta}(t) = \int_\Gamma \sigma_{\alpha\beta}(t|\Gamma)\Delta\mathcal{P}(\Gamma) = \gamma_{\alpha'\beta'}\int_\Gamma \frac{V}{T}\sigma_{\alpha\beta}(t|\Gamma)\mathcal{P}_0(\Gamma)\delta\sigma_{\alpha'\beta'}(\Gamma) \qquad (21)$$

where $\sigma_{\alpha\beta}(t|\Gamma)$ is the stress at time t under the condition that at $t = 0$ the system was in the microstate Γ, and $\Delta\sigma_{\alpha\beta}(t) = \langle\sigma_{\alpha\beta}(t) - \sigma_{\alpha\beta}(0^-)\rangle$ is the ensemble-averaged stress increment due to the deformation (recall that at $t < 0$, the system was assumed to be fully equilibrated; hence, $\langle\sigma_{\alpha\beta}(0^-)\rangle = -p_0\delta_{\alpha\beta}$). Note that after the deformation, at $t > 0$, the volume and shape of a system are not allowed to vary. Thus, e.g., the volume varies only within the ensemble, but not in time.

The last integral in Equation (21) is obviously equal to the stress correlation function

$$C_{\alpha\beta\alpha'\beta'}(t) \equiv \frac{V}{T}\left\langle\delta\sigma_{\alpha\beta}(t)\delta\sigma_{\alpha'\beta'}(0)\right\rangle \qquad (22)$$

where '0' means $t = 0$. On using the above equation and Equations (14) and (21), we find the FDT relation:

$$E_{\alpha\beta\alpha'\beta'}(t) = C_{\alpha\beta\alpha'\beta'}(t) \qquad (23)$$

It is important to note that, strictly speaking, Equation (23) is valid for a perfectly equilibrated isothermal-isobaric ensemble (as reflected in its probability distribution in the phase space at any instant before the perturbation) of either liquid or fully equilibrated amorphous solid systems whose *equilibrium* shear modulus G_e is vanishing, $G_e = 0$. (Note that $G_e = 0$ does not exclude that the static modulus μ is positive since, in the case of an amorphous glassy system, μ corresponds to the long-time glassy plateau of $G(t)$, which, however, eventually relaxes to 0 at $t \to \infty$.)

It is also noteworthy that the condition of perfect equilibration still allows for dynamical fluctuations of energy for each *individual system*, including the case of no such fluctuations, ie an energy-conserving and isochoric dynamics for each system (perhaps also involving periodic boundary conditions useful for simulations) [58,59]. In the general case, including isotropic liquids in the canonical or microcanonical isochoric ensembles, systems with canonical (Nosé–Hoover) or non-canonical (Gaussian isokinetic) thermostats, and amorphous systems equilibrated in a glassy state (a metabasin), a constant (time-independent) tensor must be added on the rhs of Equation (23) [51,58,59].

For isotropic systems, the general structure of $E_{\alpha\beta\alpha'\beta'}(t)$ is analogous to the static Equation (7):

$$E_{\alpha\beta\alpha'\beta'}(t) = M(t)\delta_{\alpha\beta}\delta_{\alpha'\beta'} + G(t)\left(\delta_{\alpha\alpha'}\delta_{\beta\beta'} + \delta_{\alpha\beta'}\delta_{\alpha'\beta}\right) \qquad (24)$$

where $M(t)$ and $G(t)$ are the generalized Lamé coefficients, λ and μ, respectively. At long times, in the quasi-static regime, $t > \tau_s$, where the relaxation moduli change weakly or vanish (note that at $T < T_g$, this regime corresponds to the glassy plateau), Equation (24) provides the static response in agreement with Equation (7): $M(t > \tau_s) \approx \lambda$, $G(t > \tau_s) \approx \mu$.

Note that in the general case, the stress response depends on whether the deformation was isothermic, adiabatic or, else, an imperfect control of temperature is involved (in numerical studies, it may correspond to an isokinetic thermostat, energy-conserving microcanonical simulation, and Nose–Hoover thermostatting, respectively). The thermostatting issues do not affect the shear modulus $G(t)$, but can be important for $M(t)$ [32,60]. Note also that the instantaneous response reflected in the affine moduli, $G(0)$ and $M(0)$, is always adiabatic (unless a perfect thermostatting of the system is provided) [58].

The above approach can be used to find an increment of any variable $X = X(\Gamma)$ upon the affine-canonical deformation. The result is

$$\Delta X(t) = \frac{1}{T}\langle X(t)\Delta H\rangle = \frac{V}{T}\gamma_{\alpha\beta}\langle X(t)\delta\sigma_{\alpha\beta}(0)\rangle, \; t > 0. \qquad (25)$$

The variable $X(t)$ in the rhs of the above equation can be replaced with $\delta X(t)$ since $\langle\delta\sigma_{\alpha\beta}(0)\rangle=0$ by definition ($\delta\sigma_{\alpha\beta}=\sigma_{\alpha\beta}-\langle\sigma_{\alpha\beta}\rangle$, see text below Equation (19)).

4. Space-Resolved Viscoelasticity

In the previous section, we considered classical relaxation moduli related to a response of the volume-averaged stress to a perturbative affine deformation. Let us turn to the space-resolved viscoelasticity providing, in particular, a position-dependent response to a (possibly) localized perturbation. Inspired by the Boltzmann superposition principle [61], one can regard a weak continuous deformation of the system (e.g., in the course of its slow flow) as a superposition of small strains $d\gamma_{\alpha\beta}=\dot\gamma_{\alpha\beta}dt$, where (cf Equation (1))

$$\dot\gamma_{\alpha\beta}=\frac{\partial v_\alpha}{\partial r_\beta} \qquad (26)$$

is the strain rate and $\underline{v}$ is the flow velocity. Thus, generalizing Equation (14) (considering the stress field $\sigma_{\alpha\beta}(\underline{r},t)$ [20,62], instead of the volume-averaged stress, $\sigma_{\alpha\beta}(t)$), we can write (based on the superposition principle due to the adopted linear response approximation and recalling the space-time uniformity of the equilibrium macroscopically homogeneous systems we consider)

$$\Delta\sigma_{\alpha\beta}(\underline{r},t)=\int_{0^-}^{t}\tilde{E}_{\alpha\beta\alpha'\beta'}(\underline{r}-\underline{r}',t-t')\dot\gamma_{\alpha'\beta'}(\underline{r}',t')dt'd^dr' \qquad (27)$$

where $\Delta\sigma_{\alpha\beta}(\underline{r},t)$ is the ensemble-averaged stress response to the flow (which was absent at $t<0$). (A canonical ensemble is assumed by default in this section focused on space-resolved viscoelasticity for an arbitrary but finite wave-vector, $\underline{q}\neq 0$. By contrast to $\underline{q}=0$, which is sensitive to the thermodynamic boundary conditions (fixed volume or fixed pressure), all fluctuations at a finite $\underline{q}$ are disentangled from fluctuations of global variables like total energy or volume.) Note that the response relation, Equation (27), is different in nature from Equation (21) since the condition of no deformation (no flow) at $t>0$ was assumed in the relevant part of Section 3. Note that the response relation, Equation (27), is also different in nature from Equation (72) since in the latter case (of Equation (72)), the flow at $t>0$ is not prescribed in contrast to Equation (27), where $\dot\gamma_{\alpha'\beta'}(\underline{r}',t')$ is considered as a known field.

It is now instructive to provide a microscopic definition of the velocity field:

$$\underline{v}(\underline{r},t)=\rho_0^{-1}\sum_{i=1}^{N}m_i\underline{v}_i\delta(\underline{r}-\underline{r}_i) \qquad (28)$$

where $\rho_0=\mathcal{M}/V$ is the mean mass density ($\mathcal{M}$ is the total mass of the system and N is the total number of particles), and $\underline{r}_i=\underline{r}_i(t)$, $\underline{v}_i=\underline{v}_i(t)$ are the position and velocity of particle i. Equation (27) can be rewritten in terms of Fourier transforms (FT) of the position-dependent functions. For example, the FT of $\tilde{E}$ is the wave-vector ($\underline{q}$)-dependent tensor of relaxation moduli:

$$E_{\alpha\beta\alpha'\beta'}(\underline{q},t)\equiv\int\tilde{E}_{\alpha\beta\alpha'\beta'}(\underline{r},t)\exp(-i\underline{q}\cdot\underline{r})d^dr \qquad (29)$$

The FT of any other relevant function, $f(\underline{r})$ (where f may stay for a component of $\underline{\underline{\sigma}}$ or $\dot{\underline{\underline{\gamma}}}$ tensors) is generically defined as

$$f(\underline{q})=\frac{1}{V}\int f(\underline{r})\exp(-i\underline{q}\cdot\underline{r})d^dr \qquad (30)$$

This way, $f(\underline{q})$ and $f(\underline{r})$ have the same physical dimension (note that here and below, we distinguish the original function from its FT by the argument only, $\underline{q}$ or $\underline{r}$). The inverse Fourier transform is given by

$$f(\underline{r}) = \sum_{\underline{q}} f(\underline{q}) \exp\left(i\underline{q} \cdot \underline{r}\right) \tag{31}$$

where the sum runs over all $\underline{q}$-modes defined by the system size. In particular, the strain rate in Equation (27) is

$$\dot{\gamma}_{\alpha\beta}(\underline{r},t) = \sum_{\underline{q}} \dot{\gamma}_{\alpha\beta}(\underline{q},t) \exp\left(i\underline{q} \cdot \underline{r}\right)$$

where (cf Equation (26))

$$\dot{\gamma}_{\alpha\beta}(\underline{q},t) = iv_{\alpha}(\underline{q},t) q_{\beta} \tag{32}$$

Thus, Equation (27) leads to

$$\Delta\sigma_{\alpha\beta}(\underline{q},t) = \int_{0^-}^{t} E_{\alpha\beta\alpha'\beta'}(\underline{q},t-t') \dot{\gamma}_{\alpha'\beta'}(\underline{q},t') dt' \tag{33}$$

The kernel $E_{\alpha\beta\alpha'\beta'}$ here is the tensor of the generalized ($\underline{q}$-dependent) relaxation moduli. Equation (33) can be considered as a $(\underline{q},t)$-dependent generalization of Equation (2). As to why we use here the γ-strain instead of the classical symmetrized ϵ-strain, see Section 5.3.

Since physical variables like $\underline{r}_a$, $\underline{v}_a$, $\sigma_{\alpha\beta}(\underline{r})$ are necessarily real, changing $\underline{q}$ to $-\underline{q}$ always leads to complex conjugation of a $\underline{q}$-dependent variable. Hence, for example, $\sigma_{\alpha\beta}(-\underline{q},t) = \sigma^*_{\alpha\beta}(\underline{q},t)$, where star ($*$) means complex conjugate (cf Equation (30)). Also, obviously, the tensor $\tilde{E}_{\alpha\beta\alpha'\beta'}(\underline{r},t)$ must be real and for all isotropic achiral systems, it is an isotropic tensor field [44], which (being a 4th-rank tensor) must be even in $\underline{r}$:

$$\tilde{E}_{\alpha\beta\alpha'\beta'}(\underline{r},t) = \tilde{E}_{\alpha\beta\alpha'\beta'}(-\underline{r},t)$$

Hence, by virtue of Equation (29), the same must be true for its Fourier transform:

$$E_{\alpha\beta\alpha'\beta'}(\underline{q},t) = E_{\alpha\beta\alpha'\beta'}(-\underline{q},t) = E^*_{\alpha\beta\alpha'\beta'}(\underline{q},t) \tag{34}$$

Thus, the tensor $E_{\alpha\beta\alpha'\beta'}(\underline{q},t)$ is real; it is also obviously symmetric with respect to permutation of α and β (cf Equation (3)).

Note that for $q = 0$ and $\dot{\gamma}_{\alpha\beta}(0,t') = \gamma_{\alpha\beta}\delta(t')$ corresponding to instantaneous affine deformation $\gamma_{\alpha\beta}$, Equation (33) becomes identical to Equation (14). It is also obvious that the velocity $\underline{v}(\underline{q},t)$ is proportional to the 'current' (momentum density) $\underline{J}$:

$$\underline{v}(\underline{q},t) = \underline{J}(\underline{q},t)/\rho_0 \tag{35}$$

where

$$\underline{J}(\underline{q},t) = V^{-1} \sum_{i=1}^{N} m_i \underline{v}_i(t) \exp\left(-i\underline{q} \cdot \underline{r}_i(t)\right) \tag{36}$$

The current in the real space is

$$\underline{J}(\underline{r},t) = \sum_i m_i \underline{v}_i(t) \delta(\underline{r} - \underline{r}_i(t)) \tag{37}$$

Our next step will be to accept Equation (33) (which is equivalent to Equation (27)) and to try and find a relation of its kernel E with the stress correlation functions. To this end, consider a system which was at equilibrium at $t < 0$ (with no flow on the average)

$$\underline{v} = 0 \text{ at } t < 0$$

but where the flow was generated at $t \geq 0$ by a perturbative external force field such that the force on particle i is
$$\underline{F}_i(t) = m_i \underline{A}(\underline{r}_i(t), t)$$
where $\underline{A} = \underline{A}(\underline{r}, t)$ is a continuous vector field. Such forces F_i provide a coherent acceleration $\underline{A}(\underline{r}, t)$ of all particles. The external force density, therefore, is
$$\underline{f}(\underline{r}, t) = \underline{A}(\underline{r}, t) \rho(\underline{r}, t) \tag{38}$$

where
$$\rho(\underline{r}, t) = \sum_i m_i \delta(\underline{r} - \underline{r}_i(t)), \quad \rho(\underline{q}, t) = \frac{1}{V} \sum_{i=1}^{N} m_i \exp\left(-i\underline{q} \cdot \underline{r}_i(t)\right) \tag{39}$$

is the microscopic mass density.

To simplify the argument, let us consider a very short perturbation,
$$\underline{A}(\underline{r}, t) = \underline{V}(\underline{r}) \delta(t) \tag{40}$$

where $\underline{V}(\underline{r})$ is the coherent velocity increment at position $\underline{r}$. Just like the rapid deformation considered in the previous section, this perturbation, $\underline{v} \to \underline{v} + \underline{V}(\underline{r})$, must lead to a change in the distribution $\mathcal{P}(\Gamma)$ in the phase space, from the canonical $\mathcal{P}_0(\Gamma)$ to $\mathcal{P}_0(\Gamma) + \Delta \mathcal{P}(\Gamma)$. As before, this transformation conserves the phase-space measure and leads (for a given initial microstate Γ) to the energy (H) increment

$$\Delta H = \sum_i m_i \underline{v}_i \cdot \underline{V}(\underline{r}_i) = \int \underline{J}(\underline{r}, 0) \cdot \underline{V}(\underline{r}) d^d r \tag{41}$$

Hence, the transformation leads to the following increment of a variable X at $t > 0$ (cf Equation (25)):

$$\Delta X(t) = \frac{1}{T} \langle X(t) \Delta H \rangle = \frac{1}{T} \int \langle X(t) \underline{J}(\underline{r}, 0) \rangle \cdot \underline{V}(\underline{r}) d^d r \tag{42}$$

where $\underline{J}(\underline{r}, t)$ is defined in Equation (37). At this point, it is convenient to focus on just a single wave-vector $\underline{q}$ setting
$$\underline{V}(\underline{r}) = \underline{V}(\underline{q}) \exp(i\underline{q} \cdot \underline{r}) \tag{43}$$
and choosing $X(t) = \sigma_{\alpha\beta}(\underline{q}, t)$. (Note that $\underline{V}(\underline{q})$ here is a constant vector.) Then, Equation (41) transforms to
$$\Delta H = V \underline{J}(-\underline{q}, 0) \cdot \underline{V}(\underline{q}) \tag{44}$$

while Equation (42) reduces to
$$\Delta \sigma_{\alpha\beta}(\underline{q}, t) = C^{\sigma J}_{\alpha\beta\alpha'}(\underline{q}, t) \mathcal{V}_{\alpha'}(\underline{q}) \tag{45}$$

where
$$C^{\sigma J}_{\alpha\beta\alpha'}(\underline{q}, t) \equiv \frac{V}{T} \langle \sigma_{\alpha\beta}(\underline{q}, t+t') J_{\alpha'}(-\underline{q}, t') \rangle \tag{46}$$

is the cross-correlation function of the stress and current, which does not depend on t' since time is uniform (and we consider a stationary well-equilibrated system). Note that
$$C^{\sigma J}_{\alpha\beta\alpha'}(\underline{q}, 0) = 0, \quad \Delta \sigma_{\alpha\beta}(\underline{q}, 0) = 0 \tag{47}$$

due to time reversibility.

The function $C^{\sigma J}_{\alpha\beta\alpha'}(\underline{q}, t)$ is related to the generalized stress-correlation function (cf Equation (22))

$$C_{\alpha\beta\alpha'\beta'}(\underline{q}, t) \equiv \frac{V}{T} \langle \delta \sigma_{\alpha\beta}(\underline{q}, t+t') \delta \sigma_{\alpha'\beta'}(-\underline{q}, t') \rangle \tag{48}$$

To establish this relation, we employ the fundamental momentum equation [21,37]

$$\frac{\partial J_\alpha}{\partial t} = \frac{\partial \sigma_{\alpha\beta}}{\partial r_\beta}$$

which reads in Fourier space:

$$\frac{\partial J_\alpha(\underline{q},t)}{\partial t} = iq_\beta \sigma_{\alpha\beta}(\underline{q},t) \qquad (49)$$

Note that Equation (49) does not assume an ensemble averaging; it is valid microscopically for each system. On using Equations (46) and (48), it leads to

$$\frac{\partial}{\partial t} C^{\sigma J}_{\alpha\beta\alpha'}(\underline{q},t) = i C_{\alpha\beta\alpha'\beta'}(\underline{q},t) q_{\beta'} \qquad (50)$$

It is now convenient to deal with time-dependent functions, $f(t)$, in terms of their modified Laplace transform (s-transform) [32,44]

$$f(s) \equiv s \int_{0^-}^{\infty} f(t) e^{-st} dt \qquad (51)$$

The transformed Equation (50) reads

$$C^{\sigma J}_{\alpha\beta\alpha'}(\underline{q},s) = \frac{i}{s} C_{\alpha\beta\alpha'\beta'}(\underline{q},s) q_{\beta'} \qquad (52)$$

so that Equation (45) leads to

$$\Delta \sigma_{\alpha\beta}(\underline{q},s) = \frac{i}{s} C_{\alpha\beta\alpha'\beta'}(\underline{q},s) q_{\beta'} V_{\alpha'}(\underline{q}) \qquad (53)$$

(Note that Equation (47) was taken into account here). Setting $X(t) = J_\alpha(\underline{q},t)$ in the general Equation, we obtain the response

$$\Delta J_\alpha(\underline{q},s) = C^{JJ}_{\alpha\alpha'}(\underline{q},s) V_{\alpha'}(\underline{q}) \qquad (54)$$

where

$$C^{JJ}_{\alpha\alpha'}(\underline{q},t) = \frac{V}{T} \left\langle J_\alpha(\underline{q},t+t') J_{\alpha'}(-\underline{q},t') \right\rangle \qquad (55)$$

is the current correlation function [21,57] whose s-transform, $C^{JJ}_{\alpha\alpha'}(\underline{q},s)$, is related to the stress correlation function [37]:

$$C^{JJ}_{\alpha\alpha'}(\underline{q},s) = \rho_0 \delta_{\alpha\alpha'} - C_{\alpha\beta\alpha'\beta'}(\underline{q},s) q_\beta q_{\beta'}/s^2 \qquad (56)$$

The above equation can be derived using the momentum equation just like Equation (52) (with the only difference that $C^{JJ}_{\alpha\alpha'}(\underline{q},t=0) = \rho_0 \delta_{\alpha\alpha'}$ is nonzero since, as follows from Equations (36), (40) and (43), $\Delta \underline{J}(\underline{q},t=0^+) = \rho_0 \underline{V}(\underline{q})$).
The s-transform of Equation (33) reads:

$$\Delta \sigma_{\alpha\beta}(\underline{q},s) = E_{\alpha\beta\alpha'\beta'}(\underline{q},s) \gamma_{\alpha'\beta'}(\underline{q},s) \qquad (57)$$

where

$$\gamma_{\alpha'\beta'}(\underline{q},s) = \frac{1}{s} \dot{\gamma}_{\alpha'\beta'}(\underline{q},s) = \frac{i}{s} \frac{1}{\rho_0} \Delta J_{\alpha'}(\underline{q},s) q_{\beta'} \qquad (58)$$

since $\dot{\gamma}_{\alpha'\beta'}(\underline{q},t) = (i/\rho_0) J_{\alpha'}(\underline{q},t) q_{\beta'}$, as follows from Equations (32) and (35). Using Equations (53), (54) and (56)–(58) and taking into account that $\underline{\mathcal{V}}(\underline{q})$ is an arbitrary vector, we obtain

$$C_{\alpha\beta\alpha'\beta'}(\underline{q},s) q_{\beta'} = E_{\alpha\beta\alpha'\beta'}(\underline{q},s) q_{\beta'} - \frac{E_{\alpha\beta\mu\beta'}(\underline{q},s) q_{\beta'}}{\rho_0 s^2} C_{\mu\delta\alpha'\delta'}(\underline{q},s) q_\delta q_{\delta'} \tag{59}$$

The above equation provides the general FDT relation between the stress correlation function C and the tensor E of the viscoelastic relaxation moduli. It can be simplified using the naturally rotated coordinate frame (NRC) [37,44] with axis 1 parallel to wave-vector $\underline{q}$:

$$C_{\alpha\beta\gamma 1}(\underline{q},s) = E_{\alpha\beta\gamma 1}(\underline{q},s) - \frac{q^2}{\rho_0 s^2} E_{\alpha\beta\mu 1}(\underline{q},s) C_{\mu 1\gamma 1}(\underline{q},s) \tag{60}$$

Equation (60) can be solved for $C_{\alpha\beta\gamma 1}(\underline{q},s)$ with a given $E_{\alpha\beta\gamma 1}(\underline{q},s)$ by first setting $\beta = 1$ and then treating it as a standard matrix equation. Once $C_{\alpha 1\gamma 1}(\underline{q},s)$ is known, $C_{\alpha\beta\gamma 1}(\underline{q},s)$ can be obtained directly from Equation (60). As follows from the $(\alpha\beta)$ symmetry of the tensor $E_{\alpha\beta\gamma 1}$ (which is obviously also applicable to the tensor $C_{\alpha\beta\gamma 1}$ [37]) and their invariance with respect to rotations around the main axis 1, there are only three independent components (involved in Equation (60)) in each tensor, $E_{\alpha\beta\gamma 1}(\underline{q},s)$ and $C_{\alpha\beta\gamma 1}(\underline{q},s)$. For the elasticity tensor, these components are

$$L(q,s) \equiv E_{1111}, \; G(q,s) \equiv E_{2121} \; \text{and} \; M(q,s) \equiv E_{2211} \tag{61}$$

known as the longitudinal, shear and mixed (transverse) modulus, respectively (cf ref. [37]). (Another way to identify the three material functions, L, G and M, is expressed by Equation (79) in the next section.) Note that the functions L, G and M are all real and do not depend on the orientation of $\underline{q}$ (cf Equation (34) and ref. [37]). The relevant three general relations derived from Equation (60) (for $q \neq 0$) in the NRC are:

$$C_G(q,s) \equiv C_{2121}(\underline{q},s) = \frac{G(q,s)}{1 + G(q,s) q^2/(\rho_0 s^2)} \tag{62}$$

$$C_L(q,s) \equiv C_{1111}(\underline{q},s) = \frac{L(q,s)}{1 + L(q,s) q^2/(\rho_0 s^2)} \tag{63}$$

$$C_M(q,s) \equiv C_{2211}(\underline{q},s) = \frac{M(q,s)}{1 + L(q,s) q^2/(\rho_0 s^2)} \tag{64}$$

These relations have already been stated in refs. [32,34,37,59]. The first Equation (62) is rather well known [57,63,64]. Noteworthily, the above relations are valid both for liquid systems (above the glass transition) and for amorphous solids (vitrified liquids), provided that they are completely equilibrated thermodynamically. Still, they are also valid for metastable glassy systems (trapped in a metabasin), provided that the lifetime of the metastable state is much longer than $1/s$ [32].

The three basic relations (62)–(64) allow to obtain all the three GRMs, $G(q,t)$, $L(q,t)$ and $M(q,t)$, based on the stress-correlation functions. However, the reverse (to obtain all components of $C_{\alpha\beta\alpha'\beta'}(\underline{q},t)$ based on the moduli) is, strictly speaking, impossible. Nevertheless, it is still possible to approximately find the undefined components of $C_{\alpha\beta\alpha'\beta'}$ for small q using the three material functions. This is performed in Section 6.4. Furthermore, the basic Equation (59) is rederived and generalized in Section 6 using a different method (the mesoscopic approach involving the concept of stress noise). This way, we both demonstrate the consistency of our approaches and provide a framework for the approximate hydrodynamic theory.

The main results obtained above are discussed in the next section.

5. Preliminary Discussion
5.1. Applied Strain vs. External Force

Equation (33) can be converted to a rather standard definition of $E_{\alpha\beta\gamma\delta}(\underline{q},t)$ as a stress response, $\Delta\underline{\sigma} = \Delta\underline{\sigma}^{(r)}$, to a small instantaneous q-dependent canonical strain $\underline{\gamma}$ applied to the system at $t = 0$ under the condition of no flow at $t > 0$ (ie no further strain is allowed, $\underline{\dot{\gamma}} = 0$ at $t > 0$) indicated with the superscript '(r)': [34,37]

$$\Delta\sigma^{(r)}_{\alpha\beta}(\underline{q},t) = E_{\alpha\beta\gamma\delta}(\underline{q},t)\gamma_{\gamma\delta}(\underline{q}) \tag{65}$$

Here, $\underline{\gamma}(\underline{r}) = \underline{\gamma}(\underline{q})e^{i\underline{q}\cdot\underline{r}}$ is defined by the particle displacement field (a particle located at $\underline{r}$ instantly moves to position $\underline{r} + \underline{u}(\underline{r})$ at $t = 0$)

$$\underline{u}(\underline{r}) = \underline{u}(\underline{q})e^{i\underline{q}\cdot\underline{r}} \tag{66}$$

Note that $\underline{q}$ is fixed here, so $\underline{u}(\underline{q})$ and $\underline{\gamma}(\underline{q})$ are just a constant vector and tensor, respectively. Obviously (cf Equation (1)),

$$\underline{\gamma}(\underline{q}) \equiv i\underline{u}(\underline{q})\underline{q} \tag{67}$$

To make the whole transformation canonical, a particle velocity $\underline{v}$ must also be changed as [37]

$$\underline{v} \to \underline{v} - \underline{v}\cdot\underline{\gamma} \tag{68}$$

The whole transformation, being canonical, conserves the phase space measure $d\Gamma$, and leads to the energy increment (cf Equation (19))

$$\Delta H(\Gamma) = V\sigma_{\alpha\beta}(-\underline{q})\gamma_{\alpha\beta}(\underline{q}) \tag{69}$$

Note that Equation (69) applies to each microstate Γ (the argument Γ being omitted in the rhs). The above equation can be verified, for example, using the microscopic definition of the q-dependent stress [57]

$$\sigma_{\alpha\beta}(\underline{q}) = \frac{1}{V}\int \sigma_{\alpha\beta}(\underline{r})e^{-i\underline{q}\cdot\underline{r}}d^d r =$$

$$= \frac{1}{V}\sum_{i>j} u'_{ij}(r)\frac{r_\alpha r_\beta}{i\underline{q}\cdot\underline{r}}\frac{1}{r}\left(e^{-i\underline{q}\cdot\underline{r}_i} - e^{-i\underline{q}\cdot\underline{r}_j}\right) - \frac{1}{V}\sum_i m_i v_{i\alpha}v_{i\beta}e^{-i\underline{q}\cdot\underline{r}_i} \tag{70}$$

where the first sum includes all disordered pairs i,j of interacting particles (with positions $\underline{r}_i, \underline{r}_j$), $u_{ij}(r)$ is their interaction energy, $u'_{ij}(r) = \frac{du_{ij}(r)}{dr}$, $\underline{r} = \underline{r}_j - \underline{r}_i$, and $v_{i\alpha}$ is component α of the velocity of particle i. In the limit, $q \to 0$, Equation (70) agrees with Equation (18).

The first Equation (20), which is generally valid for transformations conserving $d\Gamma$, reads

$$\Delta\mathcal{P}(\Gamma)/\mathcal{P}_0(\Gamma) = \frac{\Delta H(\Gamma)}{T} \tag{71}$$

Using it with Equation (69), we find that the perturbation of the system distribution in the phase space (due to the deformation, Equation (66)) is proportional to $\sigma_{\alpha\beta}(-\underline{q})$.

Furthermore, from Equations (69) and (71), we deduce that if no external force is applied to the system at $t > 0$ (ie the internal flow is allowed at $t > 0$), the stress response $\Delta\underline{\sigma}$ (to a weak instantaneous strain $\underline{\gamma}$ at $t = 0$) is provided by the stress correlation function defined in Equation (48):

$$\Delta\sigma_{\alpha\beta}(\underline{q},t) = C_{\alpha\beta\gamma\delta}(\underline{q},t)\gamma_{\gamma\delta}(\underline{q}), \quad t > 0 \tag{72}$$

(cf Equation (65)). The above equation was obtained in ref. [37] using slightly different notations (see Equations (6) and (7) there).

Importantly, exactly the same perturbation of $\mathcal{P}(\Gamma)$ can be achieved by application of an external force field, Equation (38), corresponding to the coherent acceleration field (cf Equations (40) and (43)) [32]

$$A(\underline{r},t) = A(t)e^{i\underline{q}\cdot\underline{r}}, \quad A(t) = \underline{u}(\underline{q})\dot{\delta}(t) \tag{73}$$

where $\dot{\delta}(t) = d\delta(t)/dt$ is the first derivative of Dirac's δ:

$$\dot{\delta}(t) = \lim_{\Delta t \to 0} \frac{\delta(t) - \delta(t - \Delta t)}{\Delta t}$$

It is instructive to consider a finite Δt. Then, the effect of the external field can be viewed as a combination of a push (on each particle) at $t = 0$ and the opposite push at $t = \Delta t$:

$$A(\underline{r},t) = \underline{V}(\underline{q})(\delta(t) - \delta(t-\Delta t))e^{i\underline{q}\cdot\underline{r}}, \quad \underline{V}(\underline{q}) \equiv \underline{u}(\underline{q})/\Delta t \tag{74}$$

The velocity increment due to the first push is $\Delta\underline{v} = \left(\underline{u}(\underline{q})/\Delta t\right)e^{i\underline{q}\cdot\underline{r}}$, which leads to additional displacement $\Delta\underline{r} = (\Delta\underline{v})\Delta t = \underline{u}(\underline{q})e^{i\underline{q}\cdot\underline{r}}$ before braking, which agrees with Equation (66). Since without external forces, the dynamics is energy conserving, the energy change, $\Delta H = \Delta H(\Gamma)$, comes solely from the two pushes, Equation (74). By virtue of Equation (44), the effect of the first push is

$$\Delta H_1 = V\underline{J}\left(-\underline{q},0\right)\cdot\underline{V}(\underline{q})$$

while the second (negative) push must lead to

$$\Delta H_2 = -V\underline{J}\left(-\underline{q},\Delta t\right)\cdot\underline{V}(\underline{q})$$

Taking into account both pushes, using Equation (49) and taking the limit $\Delta t \to 0$, we obtain

$$\Delta H = \Delta H_1 + \Delta H_2 = Vu_\alpha(\underline{q})iq_\beta\sigma_{\alpha\beta}(-\underline{q}) \tag{75}$$

which coincides with Equation (69). Hence, the effects of an instantaneous deformation (Equations (66) and (68)) and of an appropriate external acceleration (Equation (73)) for any ensemble-averaged quantity are exactly the same. (Note, however, that the particle velocity perturbations, $\Delta\underline{v}$, are different in the two cases: Equation (68) implies that $\Delta v_\alpha = -iv_\beta u_\beta(\underline{q})q_\alpha e^{i\underline{q}\cdot\underline{r}}$, while from Equation (73), it follows that $\Delta v_\alpha = -iv_\beta u_\alpha(\underline{q})q_\beta e^{i\underline{q}\cdot\underline{r}}$). This applies in particular to the stress response. The definitions of the elastic moduli in terms of the instantaneous canonical deformation and via the external force field (Equation (38)) are, therefore, totally equivalent. The latter definition can be applied also for an arbitrary time-dependence of the external acceleration field, $A(\underline{r},t)$ (defining the external force, $\underline{f}(\underline{r},t)$, cf Equation (38)). Therefore, using the external force field as a perturbation appears to be a more versatile approach than that of Section 4.

Noteworthily, the above results are consistent with Equations (53) and (54) of ref. [37], providing another illustration of a close relationship between responses to a canonical deformation and to an external force of the type defined in Equation (38).

5.2. The Reduced Elasticity Tensor

Equation (59) looks like a general relationship between the C- and E-tensors of the forth rank. However, in fact it yields only three relations between the components of the two tensors. The main reason for this is that there are only three independent components of the elasticity tensor E for $q \neq 0$, while some of its components simply cannot be determined based on the standard definition (see Equation (65) or Equation (33)): the structure of the deformation tensor $\gamma_{\alpha\beta}(\underline{q}) = u_\alpha(\underline{q})q_\beta$ (cf Equation (67)) is such that $E_{\alpha\beta\gamma\delta}(\underline{q},t)$ always comes in combination with q_δ, ie as

$$\mathcal{E}_{\alpha\beta\gamma}(\underline{q},t) = E_{\alpha\beta\gamma\delta}(\underline{q},t)q_\delta \qquad (76)$$

which is a third-rank tensor. That is why, for example, the component E_{2222} (in the NRC) cannot be obtained. This is in contrast with the bulk case, $q = 0$, where *all* components of the elasticity tensor $E_{\alpha\beta\gamma\delta}$ have physical significance and can be measured. It is natural to expect that the tensor $E_{\alpha\beta\gamma\delta}(\underline{q},t)$ at $\underline{q} \to 0$ must coincide with the classical adiabatic moduli of the whole system (because the diffusive transport of heat, whose rate scales as q^2, vanishes for $\underline{q} \to 0$) [32,58]:

$$E_{\alpha\beta\gamma\delta}(\underline{q} \to 0, t) = E^{(A)}_{\alpha\beta\gamma\delta}(t)$$

On using Equations (24) and (61), we, therefore, find:

$$M(q,t) \to M_A(t), \quad G(q,t) \to G(t), \quad L(q,t) \to M_A(t) + 2G(t) \text{ as } q \to 0 \qquad (77)$$

where the subscript 'A' (and the superscript '(A)') stand for 'adiabatic' (which is irrelevant for G, see text below Equation (24)).

Equation (65) can be written in terms of the tensor $\mathcal{E}$ as

$$\Delta\sigma_{\alpha\beta}(\underline{q},t) = i\mathcal{E}_{\alpha\beta\gamma}(\underline{q},t)u_\gamma(\underline{q}) \qquad (78)$$

Since the displacement u_γ can take any value (independently of $\underline{q}$), the above equation unambiguously define *all* components of the reduced elasticity tensor $\mathcal{E}_{\alpha\beta\gamma}(\underline{q},t)$. Recalling the definitions of the material functions G, L, M (cf Equation (61)) and taking into account that $\mathcal{E}_{\alpha\beta\gamma}(\underline{q},t) = \mathcal{E}_{\beta\alpha\gamma}(\underline{q},t)$ is an isotropic tensor field, we find its unique expression in terms of the GRMs (cp Equation (11) of ref. [37]):

$$\mathcal{E}_{\alpha\beta\gamma}(\underline{q},s) = G(q,s)(q_\alpha\delta_{\beta\gamma} + q_\beta\delta_{\alpha\gamma}) + M(q,s)\delta_{\alpha\beta}q_\gamma$$

$$+ (L(q,s) - M(q,s) - 2G(q,s))q_\alpha q_\beta q_\gamma/q^2 \qquad (79)$$

Furthermore, Equation (59) can be rewritten in terms of the third-rank tensors as

$$C_{\alpha\beta\gamma}(\underline{q},s) = \mathcal{E}_{\alpha\beta\gamma}(\underline{q},s) - \frac{q_{\beta'}}{\rho_0 s^2}\mathcal{E}_{\alpha\beta\alpha'}(\underline{q},s)C_{\alpha'\beta'\gamma}(\underline{q},s) \qquad (80)$$

where

$$C_{\alpha\beta\gamma}(\underline{q},s) \equiv C_{\alpha\beta\gamma\delta}(\underline{q},s)q_\delta \qquad (81)$$

Equation (80) allows to obtain all components of $C_{\alpha\beta\gamma}$ in terms of s-transforms of three VMFs (relaxation moduli): longitudinal, $L(q,t)$, mixed/transverse, $M(q,t)$, and shear, $G(q,t)$.

To sum up, the elasticity tensor $E_{\alpha\beta\gamma\delta}(\underline{q},t)$ at $\underline{q} \neq 0$ shows a sort of gauge invariance; its components can be varied without changing the material properties of the system, provided that the related (reduced) tensor $\mathcal{E}_{\alpha\beta\gamma}(\underline{q},t)$ (cf Equation (76)) remains unchanged.

The inverse Fourier transform of $i\mathcal{E}_{\alpha\beta\gamma}(\underline{q},t)$ is

$$\tilde{\mathcal{E}}_{\alpha\beta\gamma}(\underline{r},t) = \frac{\partial}{\partial r_\delta}\tilde{E}_{\alpha\beta\gamma\delta}(\underline{r},t) \qquad (82)$$

Indeed, Equation (29) implies that

$$i\mathcal{E}_{\alpha\beta\gamma}(\underline{q},t) = \int \tilde{\mathcal{E}}_{\alpha\beta\gamma}(\underline{r},t)\exp\left(-i\underline{q}\cdot\underline{r}\right)d^d r$$

and, hence, Equation (78) in real space becomes

$$\Delta\sigma_{\alpha\beta}(\underline{r},t) = \int \tilde{\mathcal{E}}_{\alpha\beta\gamma}(\underline{r}-\underline{r}',t)u_\gamma(\underline{r}')d^d r' \quad (83)$$

For systems with short-range interactions, we consider (with interaction range $\sim b$, the molecular size) the function $\tilde{\mathcal{E}}_{\alpha\beta\gamma}(\underline{r},t)$ must be localized within $r \lesssim b$ at $t = 0$. For any $t > 0$, the localization size increases in time, but remains finite. Hence, any integral over the real space involving $\tilde{\mathcal{E}}_{\alpha\beta\gamma}(\underline{r},t)$ must converge. In particular, we obtain (recalling that a translation of the system as a whole does not lead to any stress):

$$\int \tilde{\mathcal{E}}_{\alpha\beta\gamma}(\underline{r},t)d^d r = 0 \quad (84)$$

Moreover, taking into account that for affine deformations, $\underline{u}$ is a linear function of $\underline{r}$, we find

$$\int \tilde{\mathcal{E}}_{\alpha\beta\gamma}(\underline{r},t)r_\delta d^d r = -E_{\alpha\beta\gamma\delta}(\underline{q}=0,t) = -E^{(A)}_{\alpha\beta\gamma\delta}(t) \quad (85)$$

ie the linear moments of $\tilde{\mathcal{E}}_{\alpha\beta\gamma}(\underline{r},t)$ are related to the bulk relaxation moduli.

5.3. Why Asymmetric Strain?

Suppose the minor $(\gamma\delta)$ symmetry of the elastic tensor holds: $E_{\alpha\beta\gamma\delta}(\underline{q},t) = E_{\alpha\beta\delta\gamma}(\underline{q},t)$. Then, the classical form of response (to the generalized q-dependent canonical deformation) involving symmetrized strain $\epsilon_{\alpha\beta} = (\gamma_{\alpha\beta} + \gamma_{\beta\alpha})/2$,

$$\Delta\sigma_{\alpha\beta}(\underline{q},t) = E_{\alpha\beta\gamma\delta}(\underline{q},t)\epsilon_{\gamma\delta} \quad (86)$$

must give exactly the same stress response, $\Delta\sigma_{\alpha\beta}(\underline{q},t)$, as that given in Equation (65). However, while the $(\gamma\delta)$ symmetry of E is guaranteed for $q = 0$ (affine deformations), this generally may not be the case for nonzero q. Noteworthily, if the elasticity tensor shows a physically meaningful asymmetry, $E_{\alpha\beta\gamma\delta} - E_{\alpha\beta\delta\gamma} \neq 0$, then Equation (65) would capture the effect of this asymmetry, while Equation (86) would certainly miss it.

In fact, in terms of the classical affine deformation, a small shear along x with gradient along y is equivalent to a similar shear along y with gradient along x simply because the difference of these two shears is equivalent to a rotation of the system as a whole. The tensor $E_{\alpha\beta\gamma\delta}$ must, therefore, be symmetric with respect to permutation of γ and δ. However, such symmetry is not guaranteed in the case of q-dependent deformations, where (for $\underline{q}$, say, parallel to the x-axis) a shear along y is possible, while a shear along x is not. Hence, the xy and yx shears at $q \neq 0$ can not be physically equivalent any more, and the strain symmetrization does not make sense.

The definition of the E-tensor with Equation (65) (and strain with Equation (1)) is, therefore, more general, and that is why it is used in the present paper.

6. Alternative Derivation of the C-E Relations Using the Concept of Stress Noise

6.1. Stress Noise and Flow-Induced Deterministic Stress

Following ref. [37], we split the instantaneous stress into two parts:

$$\sigma_{\alpha\beta}(\underline{q},t) = \sigma^n_{\alpha\beta}(\underline{q},t) + \sigma^d_{\alpha\beta}(\underline{q},t) \quad (87)$$

where σ^d is the flow-induced deterministic stress defined by the strain history (as in Equation (33)), while σ^n is the 'stress noise' collecting all contributions to stress other than those caused by the strain history. The noise term σ^n is omnipresent; it can be considered as a genuine stress fluctuation inherent in a system (or a system element) kept at zero strain. Such 'random' stress (σ^n) can be present even in an ideal gas due to temperature fluctuations, but can also stem from (possibly frozen) structural fluctuations (of local molecular packing) in the case of liquids and amorphous solids.

6.2. Deterministic and Noise Stresses with No Flow

Let us consider an ensemble of systems where the flow is arrested at *all times*. In what follows, we focus on a given (arbitrarily selected) wave-vector $\underline{q}$. So, the condition to be satisfied is 'no flow' at wave-vector $\underline{q}$:

$$\underline{J}(\underline{q},t) = 0 \tag{88}$$

How to impose this condition given that the stress noise always tends to generate a fluctuative flow? The natural solution is to apply an appropriate external force field to the system (cf Equation (38)). The relevant external force must be harmonic in space:

$$\underline{F}_i(t) = m_i \underline{A}(t) e^{i\underline{q}\cdot\underline{r}_i(t)} + c.c. \tag{89}$$

where $\underline{F}_i$ is applied to particle i and *c.c.* stands for complex conjugate. (Note that the *c.c.* term is generally needed to keep the force real. It was omitted in Sections 4 and 5 dealing with linear response since linearity implies additivity allowing to consider complex perturbations). The function $\underline{A}(t)$ provides a coherent 'external' acceleration of all particles. It must be chosen in such a way as to suppress the current $\underline{J}(\underline{q},t)$ in order to satisfy the condition (88) (cf Equation (36)) and thus suppressing also the strain rate (cf Equations (32) and (35)). Importantly, in this case, the total energy stays conserved in spite of the external force field, Equation (89).

The no-flow condition (88) does not imply that the deterministic stress σ^d is absent, rather it ensures that σ^d is time-independent for each system in the ensemble. In fact, the condition $\underline{J}(\underline{q},t) = 0$ means (by virtue of mass conservation) that the mass density field at wave-vector $\underline{q}$ is frozen: $\rho(\underline{q},t) = \rho(\underline{q})$ (cf Equation (39)). This frozen density fluctuation may be considered as having been created long ago as a result of a longitudinal deformation $\gamma_{\alpha\beta}(\underline{q}) = -\rho(\underline{q})\hat{q}_\alpha\hat{q}_\beta/\rho_0$ at $t \to -\infty$, where $\hat{q}_\alpha = q_\alpha/q$ is a unit vector in the direction of $\underline{q}$. (Naturally, one has to demand that just before the constraint, Equation (88), was imposed at $t = -\infty$, the ensemble of systems was fully equilibrated to reach the canonical distribution in the phase space. Hence, the distribution of $\rho(\underline{q})$, being frozen, must remain canonical at all times). Such an initial deformation leads to a time-independent deterministic stress (for a given system)

$$\sigma^d_{\alpha\beta}(\underline{q},t) = \sigma^d_{\alpha\beta}(\underline{q}) = -E^e_{\alpha\beta\gamma\delta}\hat{q}_\gamma\hat{q}_\delta\rho(\underline{q})/\rho_0 \tag{90}$$

where E^e is the tensor of perfectly static (equilibrium) elastic moduli. As we consider a conceptually liquid regime (where all relaxation times are finite, albeit some of them may be extremely long), the equilibrium shear modulus (G_e) is zero, and (cf Equation (79))

$$E^e_{\alpha\beta\gamma\delta}(q)\hat{q}_\delta = (L_e(q) - M_e(q))\hat{q}_\alpha\hat{q}_\beta\hat{q}_\gamma + M_e(q)\delta_{\alpha\beta}\hat{q}_\gamma \tag{91}$$

where $L_e(q)$ is the equilibrium longitudinal modulus (which is close to the bulk compression modulus in the liquid regime at low q), and $M_e(q)$ is the analogous mixed (transverse) modulus. The above equations lead to

$$\sigma^d_{\alpha\beta}(\underline{q}) = -\kappa_{\alpha\beta}\rho(\underline{q})/\rho_0 \tag{92}$$

where

$$\kappa_{\alpha\beta} \equiv (L_e(q) - M_e(q))\hat{q}_\alpha\hat{q}_\beta + M_e(q)\delta_{\alpha\beta} \tag{93}$$

Note that since the time-averaged stress noise, $\overline{\sigma^n}$, is always zero, $\sigma^d_{\alpha\beta}(\underline{q})$ can also be interpreted as the time-averaged total stress, $\overline{\sigma_{\alpha\beta}(\underline{q},t)}$, for a given system with restricted dynamics, where Equation (88) is imposed for any t. In the above equation, we allow for a q-dependence of the equilibrium elastic moduli, although it is weak at low q, so that $L_e(q) \simeq L_e(0)$, and $M_e(q) \simeq M_e(0)$ (note also that $M_e(0) = L_e(0)$ and they both are equal to the equilibrium (static) bulk compression modulus since $G_e = 0$ [37]). Therefore, taking also into account that all the material functions are even in q,

$$L_e(q) - M_e(q) \propto q^2 \tag{94}$$

The time-independent correlation function of σ^d,

$$C^{d(r)}_{\alpha\beta\gamma\delta}(\underline{q}) \equiv \frac{V}{T}\left\langle \sigma^d_{\alpha\beta}(\underline{q})\sigma^d_{\gamma\delta}(-\underline{q}) \right\rangle_r \tag{95}$$

is then simply defined by the density fluctuation and, hence, eventually by the equilibrium elastic moduli (the superscript '(r)' and subscript 'r' mean with restricted dynamics). Using the generalized compressibility equation [57]

$$\left\langle \left|\rho(\underline{q})\right|^2 \right\rangle / \rho_0^2 = T/(VL_e(q))$$

we then find that

$$C^{d(r)}_{\alpha\beta\gamma\delta}(\underline{q}) = \kappa_{\alpha\beta}\kappa_{\gamma\delta}/L_e(q) \tag{96}$$

because $\left\langle \rho(\underline{q})\rho(-\underline{q}) \right\rangle_r = \left\langle \rho(\underline{q})\rho(-\underline{q}) \right\rangle$ since density fluctuations (with restricted dynamics) are frozen in at their value at $t \to -\infty$ when the system was fully equilibrated before the no-flow condition (88) was turned on.

By the concept introduced in ref. [37], the total stress is always a sum of the deterministic stress and noise, Equation (87), including the case of restricted dynamics, Equation (88). In the latter case σ^d is constant (independent of time, cf Equation (90)), while the time-averaged stress noise, $\overline{\sigma^n_{\alpha\beta}(\underline{q},t)}$, must vanish for each system (by virtue of its stochastic nature):

$$\overline{\sigma^n_{\alpha\beta}(\underline{q},t)} = 0$$

This means that σ^n and σ^d are never correlated:

$$\left\langle \sigma^d_{\alpha\beta}(\underline{q})\sigma^n_{\gamma\delta}(-\underline{q},t) \right\rangle \equiv 0 \tag{97}$$

and that the stress–noise correlation function

$$C^n_{\alpha\beta\gamma\delta}(\underline{q},t) \equiv \frac{V}{T}\left\langle \sigma^n_{\alpha\beta}(\underline{q},t+t')\sigma^n_{\gamma\delta}(-\underline{q},t') \right\rangle \tag{98}$$

must vanish at $t \to \infty$. As a result, the correlation function of the total stress, $\sigma_{\alpha\beta}(\underline{q},t) = \sigma^d_{\alpha\beta}(\underline{q},t) + \sigma^n_{\alpha\beta}(\underline{q},t)$,

$$C^{(r)}_{\alpha\beta\gamma\delta}(\underline{q},t) \equiv \frac{V}{T}\left\langle \sigma_{\alpha\beta}(\underline{q},t+t')\sigma_{\gamma\delta}(-\underline{q},t') \right\rangle_r \tag{99}$$

where 'r' indicates the restricted dynamics, becomes a sum of the deterministic (cf Equation (96)) and noise terms:

$$C^{(r)}_{\alpha\beta\gamma\delta}(\underline{q},t) = C^{d(r)}_{\alpha\beta\gamma\delta}(\underline{q}) + C^n_{\alpha\beta\gamma\delta}(\underline{q},t)$$

On the other hand, the total stress correlation function $C^{(r)}$ can be related to the stress-to-strain response using FDT just as is done in Section 5.1, so that Equation (72) becomes

$$\Delta\sigma_{\alpha\beta}\left(\underline{q},t\right) = C^{(r)}_{\alpha\beta\gamma\delta}\left(\underline{q},t\right)\gamma_{\gamma\delta}(\underline{q}), \quad t > 0 \tag{100}$$

Moreover, as the restricted dynamics do not allow for any further deformation (at wavevector $\underline{q}$) for $t > 0$, Equation (65) must be valid together with Equation (100):

$$\Delta\sigma_{\alpha\beta}\left(\underline{q},t\right) = E_{\alpha\beta\gamma\delta}\left(\underline{q},t\right)\gamma_{\gamma\delta}(\underline{q}), \quad t > 0 \tag{101}$$

The above two equations lead to

$$C^{(r)}_{\alpha\beta\gamma\delta}\left(\underline{q},t\right)q_\delta = E_{\alpha\beta\gamma\delta}\left(\underline{q},t\right)q_\delta, \quad t > 0 \tag{102}$$

Equation (102), together with Equations (76) and (79), allow to obtain all the three material functions, $G(q,t)$, $L(q,t)$, $M(q,t)$ from the stress correlation function $C^{(r)}$. Moreover, using simple properties of the stress-correlation functions stated above, we also obtain from Equation (102):

$$C^{d(r)}_{\alpha\beta\gamma\delta}\left(\underline{q}\right)q_\delta = E^e_{\alpha\beta\gamma\delta}\left(\underline{q}\right)q_\delta \tag{103}$$

(as follows from Equations (91), (93) and (96)) and

$$C^n_{\alpha\beta\gamma\delta}\left(\underline{q},t\right)q_\delta = \left[E_{\alpha\beta\gamma\delta}\left(\underline{q},|t|\right) - E^e_{\alpha\beta\gamma\delta}\left(\underline{q}\right)\right]q_\delta \tag{104}$$

Note that $|t|$ in the rhs renders the above equation valid also for $t < 0$ due to time-reversibility of the restricted dynamics. A relation equivalent to Equation (104) was established (in a different form and using a different argument unrelated to the restricted dynamics) in ref. [37] (see Equation (51) there).

It is important that Equation (104) is general—the correlation properties of stress noise are the same no matter if external force is applied or not. Using Equation (104), together with the momentum equation for the classical (unconstrained) dynamics, one can (following the approach developed in ref. [37]) derive again Equation (59) for the stress correlation tensor and the three exact relations for its components, Equations (62)–(64). It shows that the concept of stress noise is consistent with the FDT-based approach developed in Section 4.

In the general case, Equation (59) can serve as a basis of the method to obtain all elastic moduli E in terms of stress correlation functions. Furthermore, due to Equation (102), this task become trivial with the constrained dynamics once $C^{(r)}$ is known. How about the reverse task: to obtain all components of C based on material functions? It may seem impossible since C generally involves five unknown invariant scalar functions (cf Equation (A16) in Appendix B), while E is characterized by only three well-defined functions, which naturally define just three independent components of the C-tensor (cf Equations (62)–(64)). Nevertheless, below (in Section 6.4), we consider an approximate way to obtain the remaining two equations in order to completely define the stress correlation tensor field C.

The fact that only three independent components of the E-tensor can be defined for $q > 0$ based on the stress-to-strain response, Equation (65) (cf Section 5.2), means that we have some freedom in defining the other components of the E-tensor. The situation here is similar to the classical electrodynamics, where the electric and magnetic fields are measurable and are, therefore, unambiguously defined for a given system, while the scalar and vector ($\mathcal{A}$) potentials (whose derivatives do define the physical fields) are not uniquely defined themselves, so there is a certain freedom of choosing $\mathcal{A}$, which is known as gauge invariance. In this regard, an appropriate definition of the whole E-tensor can be based on a straightforward generalization of Equations (102)–(104) by postulating that

$$E_{\alpha\beta\gamma\delta}(\underline{q},t) \equiv C^{(r)}_{\alpha\beta\gamma\delta}(\underline{q},t) = C^{n}_{\alpha\beta\gamma\delta}(\underline{q},t) + E^{e}_{\alpha\beta\gamma\delta}(\underline{q}) \tag{105}$$

where

$$E^{e}_{\alpha\beta\gamma\delta}(\underline{q}) \equiv C^{d(r)}_{\alpha\beta\gamma\delta}(\underline{q}) = \kappa_{\alpha\beta}\kappa_{\gamma\delta}/L_e(q) \tag{106}$$

(cf Equation (96)) and $\kappa_{\alpha\beta}$ is defined in Equation (93). Obviously, this definition is consistent with Equations (102)–(104), and, therefore, it provides correct values of already defined components (like $E_{2121} = G(q,t)$). Moreover, $E_{\alpha\beta\gamma\delta}$ according to Equation (105) correctly tends to an isotropic tensor in the limit $q \to 0$ (cf Equation (7)) since (i) $C^{n}_{\alpha\beta\gamma\delta}$ always shows a finite correlation range and, therefore, must tend to an isotropic tensor for $q \to 0$ in Fourier space, and (ii) the factor $\kappa_{\alpha\beta}$ becomes isotropic as well, $\kappa_{\alpha\beta} \to M_e \delta_{\alpha\beta}$ (cf Equation (93)) as $L_e(q) - M_e(q) \to 0$ at $q \to 0$ (cf Equation (94)). Recall that all components of the E-tensor become measurable at $q \to 0$ and its general definition given above is necessarily correct in this limit (cf Equation (23) and note that $C = C^{(r)}$ at $q = 0$). It is also obvious that $E_{\alpha\beta\gamma\delta}$ defined in Equation (105) must show all minor and major symmetries with respect to index permutations just like the bulk elasticity tensor (cf Equation (7)) because Equation (105) identifies $E_{\alpha\beta\gamma\delta}$ with the stress correlation function $C^{(r)}$ that does show all these symmetries for equilibrium isotropic systems (cf ref. [37]) since the restricted dynamics are time-reversible just like the classical dynamics.

6.3. No Flow at $T < 0$: A Route to New Relations

For the purpose of argument, it is now convenient to consider the system where the flow at wave-vector $\underline{q}$ is arrested at $t < 0$ (cf Equation (88)), but this constraint is released at $t > 0$. In this case, $\sigma^d = $ const at $t < 0$. After the constraint is released (external force suppressed at $t > 0$), the system rapidly equilibrates to arrive at the genuine equilibrium distribution (with proper fluctuations of the current J). It may seem, therefore, that at short times, $t > 0$, the stress correlation function defined as

$$C^{r0}_{\alpha\beta\alpha'\beta'}(\underline{q},t) \equiv \frac{V}{T}\left\langle\sigma_{\alpha\beta}(\underline{q},t)\sigma_{\alpha'\beta'}(-\underline{q},0)\right\rangle_{r0} \tag{107}$$

with the constraint, $J(\underline{q},t) = 0$ at $t < 0$ (indicated by the superscript 'r0') may be different from the genuine *equilibrium* correlation function of the total stress σ (obtained with the classical unconstrained dynamics)

$$C_{\alpha\beta\alpha'\beta'}(\underline{q},t) \equiv \frac{V}{T}\left\langle\sigma_{\alpha\beta}(\underline{q},t+t')\sigma_{\alpha'\beta'}(-\underline{q},t')\right\rangle \tag{108}$$

The latter function does not depend on t' since the equilibrium state is obviously stationary. We show, however, that in fact

$$C^{r0}_{\alpha\beta\alpha'\beta'}(\underline{q},t) = C_{\alpha\beta\alpha'\beta'}(\underline{q},t), \quad t > 0 \tag{109}$$

(see Appendix A).

The total stress is always a sum of the deterministic stress σ^d and the noise, σ^n, Equation (87). The deterministic stress at $t = 0$, $\sigma^d_{\alpha\beta}(\underline{q},0)$, is never correlated with σ^n:

$$\left\langle\sigma^d_{\alpha\beta}(\underline{q},0)\sigma^n_{\gamma\delta}(-\underline{q},t)\right\rangle = 0 \tag{110}$$

Indeed, for $t \leq 0$, the above equation simply follows from the results of the previous section (Equation (97)) since the release of the constraint (Equation (88)) at $t = 0$ does not affect the system dynamics at $t < 0$. On the other hand, the validity of Equation (110) at $t > 0$ is hinged on the stochastic nature of the noise, $\sigma^n_{\alpha\beta}(\underline{q},t)$ (its independence of a weak flow in the linear regime). Recall now that the stress noise is omnipresent (both at $t < 0$ and $t > 0$) and its autocorrelation function is always related to the elasticity tensor, cf Equations (105) and (106):

$$C^n_{\alpha\beta\gamma\delta}(\underline{q}, t) = E_{\alpha\beta\gamma\delta}(\underline{q}, |t|) - E^e_{\alpha\beta\gamma\delta}(\underline{q}) \tag{111}$$

The autocorrelation function of the deterministic stress at $t = 0$, $\sigma^d_{\alpha\beta}(\underline{q}, 0)$, is defined by the rhs of Equation (96) (cf also Equations (92) and (93)) leading to (on recalling Equation (106)):

$$(V/T)\langle \sigma^d_{\alpha\beta}(\underline{q}, 0)\sigma^d_{\gamma\delta}(-\underline{q}, 0)\rangle = E^e_{\alpha\beta\gamma\delta}(\underline{q}) \tag{112}$$

Turning to $\sigma^d_{\alpha\beta}(\underline{q}, t)$ at $t > 0$, based on the Boltzmann superposition principle, it must be a sum of $\sigma^d_{\alpha\beta}(\underline{q}, t = 0)$ and the contributions due to the flow at $t > 0$. Thus, the total stress at $t > 0$ is

$$\sigma_{\alpha\beta}(\underline{q}, t) = \sigma^n_{\alpha\beta}(\underline{q}, t) + \sigma^d_{\alpha\beta}(\underline{q}, 0) + \int_0^t E_{\alpha\beta\gamma\delta}(\underline{q}, t - t')iv_\gamma(\underline{q}, t')q_\delta dt' \tag{113}$$

where $\underline{v} = \underline{J}/\rho_0$ is the flow velocity ($\underline{J}$ is the mass current), and $iv_\gamma(\underline{q}, t')q_\delta$ is the relevant strain rate at t' corresponding to $\partial v_\gamma / \partial r_\delta$ (cf Equation (33)). Note that the last term in Equation (113) is the time-dependent part of the deterministic stress, which is defined by the flow field history and the generalized elasticity tensor (cf Equations (27) and (33)).

Doing the s-transform (cf Equation (51)) of the above relation and taking into account the equation of motion (cf Equation (49))

$$\frac{\partial v_\gamma}{\partial t} = \frac{iq_{\gamma'}}{\rho_0}\sigma_{\gamma\gamma'}$$

we find

$$\sigma_{\alpha\beta}(\underline{q}, s) = \sigma^n_{\alpha\beta}(\underline{q}, s) + \sigma^d_{\alpha\beta}(\underline{q}, t = 0) - \frac{q_\delta q_{\gamma'}}{\rho_0 s^2} E_{\alpha\beta\gamma\delta}(\underline{q}, s)\sigma_{\gamma\gamma'}(\underline{q}, s) \tag{114}$$

Note that the term

$$\frac{iq_\delta}{s} E_{\alpha\beta\gamma\delta}(\underline{q}, s)v_\gamma(\underline{q}, t = 0)$$

is omitted in the above equation since $v_\gamma(\underline{q}, t = 0) = 0$ by preparation of the system. Next, multiplying Equation (114) with $\sigma_{\alpha'\beta'}(-\underline{q}, t = 0) = \sigma^d_{\alpha'\beta'}(-\underline{q}, t = 0) + \sigma^n_{\alpha'\beta'}(-\underline{q}, t = 0)$, we obtain using Equations (109)–(112):

$$C_{\alpha\beta\alpha'\beta'}(\underline{q}, s) = E_{\alpha\beta\alpha'\beta'}(\underline{q}, s) - \frac{q_\delta q_{\gamma'}}{\rho_0 s^2} E_{\alpha\beta\gamma\delta}(\underline{q}, s)C_{\gamma\gamma'\alpha'\beta'}(\underline{q}, s) \tag{115}$$

The above equation can be compared with Equation (45) of ref. [65]. The latter paper deals with overdamped systems of identical particles (with concentration n), where the friction forces overwhelm the inertia. In the overdamped regime, the term $\rho_0 s^2$ must be replaced with $\zeta_0 ns$. With this and other trivial reductions, Equation (115) becomes similar in structure to Equation (45) of ref. [65], suggesting that the irreducible memory kernel $M_{\alpha\beta\gamma\delta}$ (defined in Equation (22) of ref. [65]) is likely to correspond to the elasticity tensor $E_{\alpha\beta\gamma\delta}$. However, any rigorous proof of such a correspondence is missing at present (not to mention a significant effect of the overdamped dynamics on the relaxation moduli). This issue could be an interesting point for further study.

Noteworthily, the FDT relation, Equation (59), simply comes from Equation (115) after multiplying it by $q_{\beta'}$. We tend to view Equation (115) as exact, just like Equation (59). Equation (115) allows to predict all components of the correlation tensor C based on the elasticity tensor E. First, obviously, Equation (115) leads to the already stated relations (62)–(64) defining three independent components of the C-tensor (C_G, C_L, C_M) in terms of measurable material functions (G, L, M) since these relations follow from Equation (59), which, in turn, follows from Equation (115). Second, Equation (115) also defines the remaining two independent components C_N and C_P (see Appendix B):

$$C_N(q,s) = N(q,s) - \frac{M(q,s)^2}{\rho_0 s^2/q^2 + L(q,s)}, \quad C_P(q,s) = P(q,s) - \frac{M(q,s)^2}{\rho_0 s^2/q^2 + L(q,s)} \quad (116)$$

where $N(q,s) = E_{2222}(q,s)$, $P(q,s) = E_{2233}(q,s)$ using NRC.

We emphasize that the new functions $N(q,s)$ and $P(q,s)$ cannot be 'measured' directly based on the stress-to-strain response. Of course, they can be obtained (for example, in simulations) using their relation to the correlation function of stress noise, Equation (105). However, it would be better to try and obtain as much as possible from the measurable quantities (in particular, the G, L and M functions). This is performed in the next section, where we derive approximate versions of Equation (116) involving only the latter three material functions.

Noteworthily, in ergodic systems, all the stress correlation functions must vanish at $t \to \infty$; hence, in particular, $C_N(q, s \to 0) = 0$, $C_P(q, s \to 0) = 0$, leading to

$$N(q,s)L(q,s) = M(q,s)^2, \quad N(q,s) = P(q,s), \quad s \to 0$$

The above equations are valid in the liquid state ($T > T_g$). However, they are not valid any more in the glassy regime ($T < T_g$), unless we treat the condition $s \to 0$ literally (ie allowing for astronomical times t). In the glassy case, the transiently frozen stresses, $\sigma_{22}(\underline{q})$, $\sigma_{33}(\underline{q})$, $\sigma_{23}(\underline{q})$, are generally present in the system for the experimentally accessible time-scales (cf ref. [32]). It is noteworthy, however, that other stress components, like σ_{11} and $\sigma_{12} = \sigma_{21}$, never include any frozen part (see Discussion point 9 in ref. [32]), which is in line with the fact that the correlation functions C_G, C_L and C_M are all tending to 0 at low s even in glassy systems (cf Equations (62)–(64)). By contrast, the correlation functions involving only the 'transverse' stress components ($\sigma_{22}(\underline{q})$, $\sigma_{33}(\underline{q})$, $\sigma_{23}(\underline{q})$) are generally nonzero in vitrified (amorphous) systems:

$$C_N(q,s) > 0, \quad C_P(q,s) \neq 0, \quad C_{2323}(q,s) > 0, \quad s \ll 1/\tau_s$$

where the '2323' component refers to the NRC, and $s \ll 1/\tau_s$ actually means the glassy plateau regime. The last condition also implies that $C_N(q,s) > C_P(q,s)$, as follows from the relation $2C_{2323}(q,s) = C_N(q,s) - C_P(q,s) = N(q,s) - P(q,s)$, which, in turn, comes from the general 'isotropy' relation, Equation (A16). The above inequalities reflect the presence of frozen stress fields in amorphous systems [44].

6.4. Approximate Relations between C and E Tensors

Let us consider the low-q regime, $qb \ll 1$. At the end of Section 6.2, we already argued that $E_{\alpha\beta\gamma\delta}(\underline{q}, t)$ must become isotropic at $q \to 0$. To expand on this point now, let us turn to the last term, E^e in Equation (105) defining $E_{\alpha\beta\gamma\delta}(\underline{q},t)$. According to Equation (106)

$$E^e_{\alpha\beta\gamma\delta}(\underline{q}) = e_0(q)\delta_{\alpha\beta}\delta_{\gamma\delta} + e_2(q)\left(q_\alpha q_\beta \delta_{\gamma\delta} + q_\gamma q_\delta \delta_{\alpha\beta}\right) + e_4(q) q_\alpha q_\beta q_\gamma q_\delta \quad (117)$$

where

$$e_0(q) = \frac{M_e(q)^2}{L_e(q)}, \quad e_2(q) = \frac{L_e(q) - M_e(q)}{q^2}\frac{M_e(q)}{L_e(q)}, \quad e_4(q) = \frac{(L_e(q) - M_e(q))^2}{q^4 L_e(q)} \quad (118)$$

The equilibrium (perfectly static) material functions $L_e(q)$, $M_e(q)$ must be continuous and analytical (at least near $q = 0$) and, moreover, equal at $q = 0$: $L_e(0) = M_e(0)$ since a liquid does not show any shear elasticity in the static regime ($G_e(q) = 0$); in other words, its static stress response must be isotropic. As the characteristic length-scale of static elasticity is expected to be structural in nature (cf the paragraph below Equation (123)), it should be typically defined by the molecular size $\sim b$ and/or the interaction range (assumed to be similar). Taking also into account that the material functions are even in q, we, therefore, expect that (cf Equation (94))

$$L_e(q) - M_e(q) \sim L_e(0)(qb)^2, \quad qb \ll 1$$

The above relation shows that all the prefactor functions (e_0, e_2, e_4) in Equation (117) are continuous and finite at $q = 0$ and, hence, $E^e_{\alpha\beta\gamma\delta}(\underline{q})$ is analytical as a function of *vector* $\underline{q}$. The main isotropic e_0-term in Equation (117) is $\sim L_e(0)$, while the other two terms provide small corrections (for $qb \ll 1$): e_2-term is $\propto (qb)^2$, e_4-term is $\propto (qb)^4$.

A similar argument works for the stress noise correlation tensor $C^n_{\alpha\beta\gamma\delta}(\underline{q}, t)$ involved in Equation (105). Considering the stress noise σ^n, it is convenient to assume that the flow is arrested (cf Equation (88)), so that the elements of the system are not deformed. Then, variations of σ^n are primarily due to structural (molecular packing) fluctuations in the system, which are expected to be short-range (with correlation length $\sim b$). In this case, the correlation tensor $C^n_{\alpha\beta\gamma\delta}(\underline{q}, t)$ would be nearly q-independent and, therefore, isotropic for $qb \ll 1$. There is, however, one complication—apart from structural fluctuations, σ^n is also affected by fluctuations of conserved fields, like energy density and composition (in the widely encountered case of multi-component polydisperse systems) [32,66]. Importantly, these fields are scalar and, therefore, they mainly contribute to the isotropic part of $\sigma^n_{\alpha\beta}$ corresponding to pressure: $\delta\sigma^n_{\alpha\beta} = -(\delta p)\delta_{\alpha\beta}$, where δp is the pressure fluctuation due to temperature or composition variations. The eventual contribution of scalar fields to $C^n_{\alpha\beta\gamma\delta}(\underline{q}, t)$ is, therefore, also mainly isotropic (proportional to $\delta_{\alpha\beta}\delta_{\gamma\delta}$). Anisotropic contributions come from the gradients of the conserved fields, which are small for $qb \ll 1$. The resultant structure of the correlation tensor $C^n_{\alpha\beta\gamma\delta}(\underline{q}, t)$ (and of the elasticity tensor $E_{\alpha\beta\gamma\delta}(\underline{q}, t)$ in view of Equation (105)) must, therefore, be similar to the rhs of Equation (117) involving the main isotropic term plus some quadratic and quartic terms of order of $(qb)^2$ and $(qb)^4$, respectively. Neglecting the latter terms (depending on the wave-vector orientation $\hat{q}$), we arrive to the main approximation for the E-tensor at small qb:

$$E_{\alpha\beta\gamma\delta}(\underline{q}, t) \simeq a_E(q, t)\delta_{\alpha\beta}\delta_{\gamma\delta} + b_E(q, t)(\delta_{\alpha\gamma}\delta_{\beta\delta} + \delta_{\alpha\delta}\delta_{\beta\gamma}), \quad qb \ll 1 \qquad (119)$$

(cf Equation (7)), where a_E and b_E must be identified as

$$a_E(q, t) = L(q, t) - 2G(q, t), \quad b_E(q, t) = G(q, t) \qquad (120)$$

in order to provide exact results for the shear and longitudinal moduli, $E_{2121}(q, t) = G(q, t)$ (note that $G(q, t) \simeq G(t)$, the bulk shear relaxation modulus) and $E_{1111}(q, t) = L(q, t)$ using the NRC, cf Equation (61). Generally Equation (119) is valid up to a correction of $\mathcal{O}(q^2b^2)$. In particular, for the transverse modulus M, it gives:

$$M(q, t) = L(q, t) - 2G(q, t) + \mathcal{O}(q^2b^2) \qquad (121)$$

as already noted in ref. [32] (see also ref. [37]). Importantly, Equation (119) also yields the remaining two independent functions N and P (see text below Equation (116)):

$$N(q, t) = L(q, t) + \mathcal{O}(q^2b^2), \quad P(q, t) = M(q, t) + \mathcal{O}(q^2b^2) \qquad (122)$$

All the components of the correlation tensor C can then be obtained based on $G(q, t)$ and $L(q, t)$ using Equations (62)–(64) and Equations (116). In particular, the above approximation, $N(q, t) \simeq L(q, t)$, leads to

$$C_N(q, s) \simeq L(q, s) - \frac{M(q, s)^2}{\rho_0 s^2/q^2 + L(q, s)} \qquad (123)$$

which agrees with Equation (23) of ref. [37] (an equivalent equation was also stated as Equation (75) in ref. [32]). Note that Equation (23) was derived in ref. [37] for monodisperse systems with infinite thermal conductivity. However, as demonstrated above, it remains valid also for *polydisperse* systems with any thermal conductivity. Equation (123) was verified at $qb \lesssim 0.5$ in simulation studies on two-dimensional (2D) systems of polydisperse LJ particles [32,44]. In particular, Figures 3 and 4 of ref. [44] show that well below the glass transition temperature (T_g), the stress correlation function $C_N(q,t)$ is nearly constant for $qb \lesssim 0.5$ at long times t. In this regime, it is independent of q and t, $C_N \approx L - M^2/L \approx 4G(1 - G/L)$, and is close to the Young's modulus $e_Y = 4G(1 - G/L)$ for glassy 2D systems, cf refs. [37,44] (here, G, L and M are the long-time plateau values of the bulk moduli).

In a similar way, we obtain

$$C_P(q,s) \simeq M(q,s) - \frac{M(q,s)^2}{\rho_0 s^2/q^2 + L(q,s)} \tag{124}$$

Note that Equation (124) is useful for three-dimensional systems, but is irrelevant in two dimensions, where $C_P(q,s) = C_{2233}(q,s)$ is not defined (cf Equation (A19)).

Noteworthily, at $q \to 0$ and $t \to 0$, the function $C_N(q,t)$ is related to the affine moduli, $G(0)$ and $M(0)$:

$$C_N(0,0) = M(0) + 2G(0)$$

as follows from Equations (121), (122) and (116), which also lead to $C_N(0,0) = N(0,0) = L(0,0) = C_L(0,0)$.

At long t, the response of conserved fields to a local strain must relax; hence, the q-dependence of a_E and b_E at $t \to \infty$ should come solely from structural correlations (of molecular packing), which are short-range in liquids and glasses. As a result, the q-dependence of the elastic response (cf Equation (119)) can be neglected for $qb \ll 1$, ie the elastic response is expected to be essentially local at $t \to \infty$. This conclusion supports the argument presented below Equation (118).

Is it possible to improve on Equation (119) for the generalized elasticity tensor by obtaining, for example, the quadratic correction $\sim \mathcal{O}(q^2 b^2)$? This problem is tackled below, but only for 2D systems ($d = 2$). In this case, the general expression for $E_{\alpha\beta\gamma\delta}(q,t)$ is given in Equation (A21) (see Appendix B). As discussed above, the terms depending on the wave-vector orientation $\hat{q}$ must be small there:

$$c_E(q,t) \propto (qb)^2, \; e_E(q,t) \propto (qb)^4$$

Therefore, in the quadratic approximation, we can neglect the last e_E-term:

$$E_{\alpha\beta\gamma\delta}(q,t) \simeq a_E(q,t)\delta_{\alpha\beta}\delta_{\gamma\delta} + b_E(q,t)(\delta_{\alpha\gamma}\delta_{\beta\delta} + \delta_{\alpha\delta}\delta_{\beta\gamma})$$
$$+ c_E(q,t)(\hat{q}_\alpha \hat{q}_\beta \delta_{\gamma\delta} + \hat{q}_\gamma \hat{q}_\delta \delta_{\alpha\beta})$$

Next, we can express the unknown functions a_E, b_E and c_E in terms of the measurable material functions G, L and M using Equations (61):

$$a_E(q,t) = M(q,t) - \Delta L(q,t), \; b_E(q,t) - G(q,t), \; c_E(q,t) - \Delta L(q,t) \tag{125}$$

where

$$\Delta L(q,t) \equiv L(q,t) - M(q,t) - 2G(q,t) \tag{126}$$

Therefore, we obtain

$$N(q,t) \simeq L(q,t) - 2\Delta L(q,t) \tag{127}$$

which must be valid up to the 'quartic' correction $\sim \mathcal{O}(q^4 b^4)$. Equation (127) is in agreement with the first Equation (122) (improving it) since $\Delta L(q,t) \propto (qb)^2$, as follows from Equations (121) and (126). Interestingly, Equation (127) is also in agreement with equation

$$N_e(q) = \frac{M_e(q)^2}{L_e(q)} \qquad (128)$$

coming from Equations (117) and (118). Indeed, using Equations (126) and (127), taking into account that $G_e(q) = G(q, t \to \infty) = 0$ and neglecting the $\mathcal{O}(q^4 b^4)$ correction, we obtain $N_e(q) = N(q, t \to \infty) = L_e(q) - 2(L_e(q) - M_e(q)) = 2M_e(q) - L_e(q)$, which coincides (to $\mathcal{O}(q^4 b^4)$) with Equation (128) since $L_e(q) - M_e(q) \propto (qb)^2$ (cf Equation (121)).

Noteworthily, Equation (128) can be written as

$$N_e(q) = L_e(q) - 2\Delta L_e(q) + (\Delta L_e(q))^2 / L_e(q) \qquad (129)$$

where $\Delta L_e(q)$ is defined by Equation (126) for $t \to \infty$. The last term in the above equation comes from the quartic term in Equation (117), which is proportional to q^4 since generally $\Delta L(q,t) \propto q^2$. Generalizing Equation (129), we propose the following heuristic approximation

$$N(q,s) \simeq L(q,s) - 2\Delta L(q,s) + \frac{(\Delta L(q,s))^2}{L(q,s)} \equiv \frac{[M(q,s) + 2G(q,s)]^2}{L(q,s)} \qquad (130)$$

which is supposed to include not only the quadratic (q^2) but also the quartic correction ($\propto q^4$) to the main approximation, Equation (122). The correlation function $C_N(q,t)$ can then be obtained more precisely using the first Equation (116), with N defined either in Equation (127) or Equation (130).

To verify the above predictions for $N(q,t)$, we performed a simple test for $t = 0$ (corresponding to $s \to \infty$) using simulation data for a 2D system of polydisperse LJ particles. This glass-forming system involving a weak polydispersity of the particle size (rather than mass) has been described elsewhere [32,58,60,66]. For $s \to \infty$, the basic Equation (115) gives:

$$C_{\alpha\beta\alpha'\beta'}(\underline{q}, t = 0) = E_{\alpha\beta\alpha'\beta'}(\underline{q}, t = 0)$$

leading to (the time argument $t = 0$ is omitted):

$$G(q) = C_G(q), \; L(q) = C_L(q), \; M(q) = C_M(q), \; N(q) = C_N(q),$$

(Note that G, L, M here are the adiabatic moduli corresponding to the instantaneous stress response to an appropriate strain.) Thus, all the elastic material functions ($G(q), \ldots, N(q)$) can be obtained directly based on the stress-correlation data from simulations. Figure 1 shows a comparison of the simulated $N(q)$ with its approximations: 0th, $N_{a0}(q)$ based on $N(q) = L(q)$, cf Equation (122); 1st, $N_{a1}(q)$ based on Equation (127); and second, $N_{a2}(q)$, from Equation (130). One can observe that the 0th approximation, N_{a0}, works well at $qb < 0.5$, while $N_{a1}(q)$ and $N_{a2}(q)$ show excellent agreement with the simulated $N(q)$ for $qb < 1.5$ and $qb < 3$, respectively (see Figure 1a). Thus, the second approximation allows to widen the q-region of applicability of the theory by a factor of 6. Moreover, the second approximation is also reasonable for larger q ($qb > 3$), where it reproduces the peak of $N(q)$ at $qb \sim 6$ in a qualitatively correct fashion, while the other two approximations show a qualitatively incorrect behavior at $qb \gtrsim 6$ with $N_{a0}(q)$, showing a minimum instead of a peak there (cf Figure 1b).

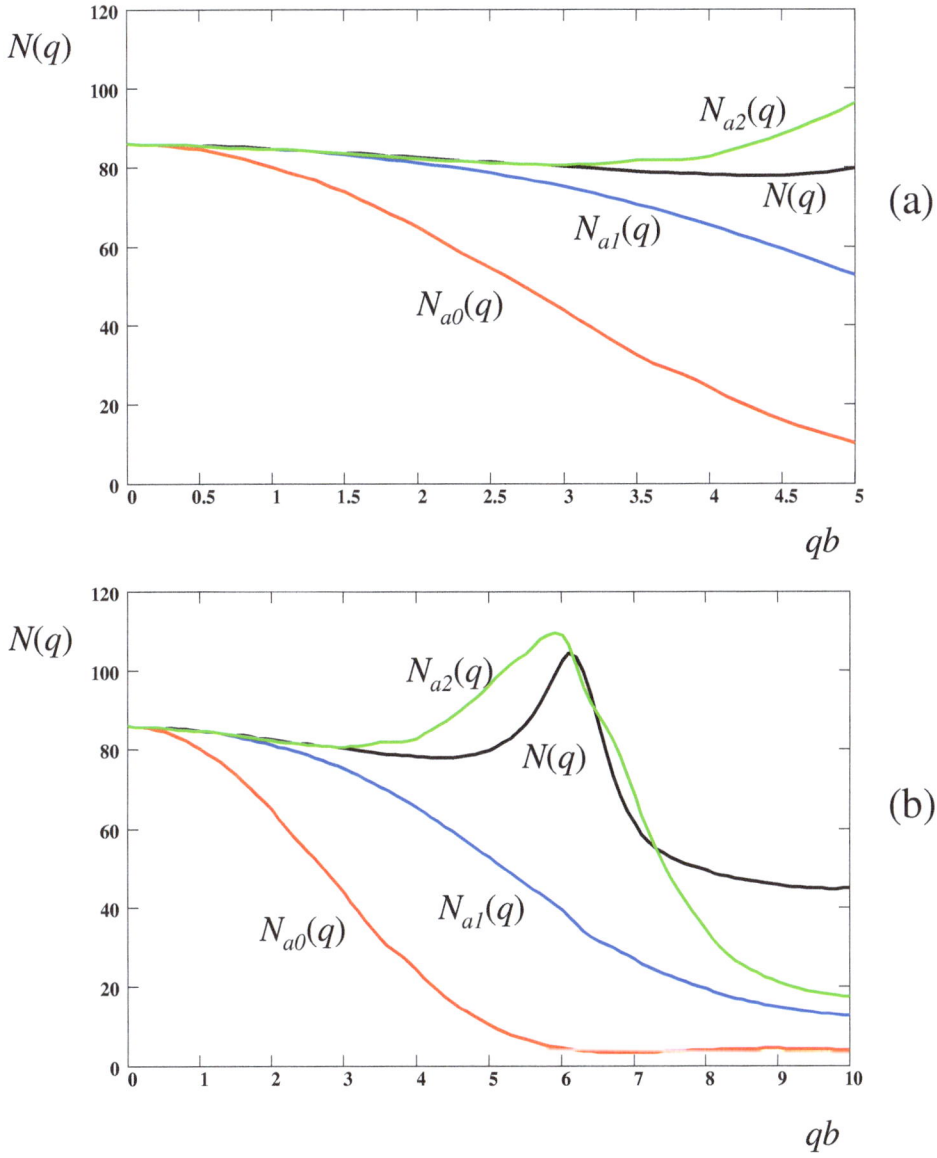

Figure 1. The wave-number (q) dependence of the instantaneous (adiabatic) elastic modulus $N(q) \equiv N(q, t = 0)$, black curve, and its three approximations based on the shear ($G(q)$), longitudinal ($L(q)$) and transverse ($M(q)$) elastic moduli: $N_{a0}(q)$, Equation (122) (red curve), $N_{a1}(q)$, Equation (127) (blue curve), and $N_{a2}(q)$, Equation (130) (green curve). All the moduli (N, G, L, M) are based on the stress correlation functions $C_N(q)$, $C_G(q)$, ... obtained by MD simulations for a polydisperse system of LJ particles [32,58,60,66]. Panel (**a**) highlights the range $0 < q < 5$, while panel (**b**) shows a wider range, $0 < q < 10$, including the main structural peak at $q \approx 6.3$. Temperature $T = 0.4$ (in LJ energy units) and the mean particle size $b = 1$.

7. Summary

1. In the present paper, we established and discussed a number of general relations between the 4th-rank tensor fields of stress correlations, $C_{\alpha\beta\alpha'\beta'}(q,t)$, cf Equation (48), and the tensor of generalized (viscoelastic) relaxation moduli, $E_{\alpha\beta\alpha'\beta'}(q,t)$, cf Equation (33). The C-tensor field is generally characterized by, at most, five independent components (invariant correlation functions, see Equation (A16) in Appendix B) as long as the minor and major symmetries of $C_{\alpha\beta\alpha'\beta'}$ are taken into account (cf Equations (A17)) [37]. By contrast, the E-tensor involves only three material functions, the generalized relaxation moduli (GRMs), $G(q,t)$, $L(q,t)$ and $M(q,t)$ (cf Equations (61) and (79)), that can be measured according to their definition via stress response to a weak strain as given in Equation (33). It is, therefore, not surprising that there exist only three exact relationships (Equations (62)–(64)) linking the independent components of the C-tensor with the three material functions (since obviously five independent correlation functions cannot be expressed using only three material functions). Noteworthily, all the three GRMs can be obtained based on the correlation tensor using Equations (62)–(64) (which follow from the exact tensorial relation, Equation (59)). These three equations are rigorously derived in Section 4 based on the FDT. Equations (62) and (63) have been established before (cf refs. [34,37,57,59,63,64]). The last relation, Equation (64), was presented in ref. [37] and verified numerically in ref. [32]. It is also noteworthy that Equations (62) and (63) have been recently derived using the Zwanzig–Mori formalism [35].

2. In the case of affine deformations, the strain tensor is normally defined as the symmetric part of the tensor $\gamma_{\alpha\beta}$ of particle displacement gradients (cf Equations (1) and (4)). However, in the more general case of inhomogeneous deformations (which can be considered as a superposition of harmonic waves), the nonsymmetrized definition of strain, Equation (1), is more appropriate, as argued in Section 5.3.

3. We considered two definitions of the viscoelastic memory functions (VMFs): in terms of the stress response to a harmonic canonical strain (Equations (66)–(68)) and as a response to a coherent external acceleration field (Equations (73) and (38)). Importantly, it is demonstrated (see Section 5.1) that the two definitions lead to exactly the same response functions ($G(q,t)$, $L(q,t)$, $M(q,t)$). Remarkably, the approach involving the external force, Equation (38), appears to be more general than imposing a q-dependent canonical deformation: the latter can be reproduced with a singularity time-dependence of the external field, Equation (73).

4. It is also remarkable that the stress response to an arbitrary prescribed deformation of an amorphous system can be *completely* defined in terms of the reduced elasticity tensor, $\mathcal{E}_{\alpha\beta\gamma}(q,t)$, introduced in Section 5.2 (cf Equation (76)). All components of this tensor can be obtained based on just three GRMs ($=$ VMFs), $G(q,t)$, $L(q,t)$, $M(q,t)$, cf Equation (79). The isotropic nature of the system dictates that these material functions are real and do not depend on the orientation of $\underline{q}$ (cf Equations (34), (76) and (79)). Moreover, as we argue in Section 6.4, these functions must be generally continuous and, moreover, analytical functions of q^2. At $t=0$, the elastic response is local. It is also likely that the same is virtually true at $t \to \infty$ (cf Section 6.4), so that, for example, $L(q, t \to \infty) \simeq L(0, t \to \infty) = M_e + 2G_e$ at $qb \ll 1$ (with relative error $\sim q^2 b^2$, where b is the particle interaction range). Importantly, at low q ($qb \ll 1$), the three GRMs are related for any time t (cf Equation (121) and refs. [32,37]).

5. As mentioned above, in this study, we consider the elasticity tensor in terms of the stress response to a prescribed small strain or to an external force perturbation (in the latter case, the force generally depends on the particle position). Noteworthily, considering another type of perturbation by changing the system Hamiltonian from H_0 to $H = H_0 + \Delta H$ with $\Delta H = -\epsilon_{\alpha\beta}(t)\sigma_{\alpha\beta}(q,t)$, involving a prescribed weak 'deformation' function $\epsilon_{\alpha\beta}(t)$, does not make much sense: On the one hand, it allows to employ the classical FDT [52], but on the other hand, it is unclear how the prescribed $\epsilon_{\alpha\beta}(t)$ can be possibly linked with the physical strain in the system given that the introduction of ΔH changes the classical

relations between particle velocities and momenta leading to an anisotropic and position-dependent particle mass.

6. To uncover new relationships between the stress correlations and the elasticity tensor (cf Section 6), we employ the concept of the stress noise, σ^n, proposed in our previous paper [37]. It is defined as $\sigma^n(q,t) = \sigma(q,t) - \sigma^d(q,t)$, where σ^d is the deterministic stress due to the flow history in the system (cf Equation (87)). The stress noise σ^n can, thus, be considered as a genuine stress fluctuation unrelated to deformation and flow. This concept opened up the possibility to define *all components* of the generalized (q-dependent) elasticity tensor, $E_{\alpha\beta\gamma\delta}(q,t)$, in terms of the stress noise correlation function (cf Equations (98) and (105)). It is important that the new definition is totally consistent with the classical linear response way to introduce the elasticity tensor, Equation (33), and, therefore, leads to exactly the same GRMs, $G(q,t)$, $L(q,t)$, $M(q,t)$. The latter statement is valid since Equations (59) and (62)–(64) trivially follow from Equation (115). On the other hand, the new definition, Equation (105), implies both minor and major symmetries of $E_{\alpha\beta\gamma\delta}(q,t)$, which are inherent in the classical *bulk* elasticity tensor. Moreover, the bulk tensor coincides with $E_{\alpha\beta\gamma\delta}(q,t)$ at $q = 0$ since the latter tensor field is continuous and analytical as a function of q (see end of Sections 6.2 and 6.4). The definition of the generalized elasticity tensor, Equation (105), therefore, combines the best of both worlds (of affine strains, $q = 0$, and harmonic deformations, $q \neq 0$).

7. One may wonder how to obtain the correlation function of stress noise, $C^n_{\alpha\beta\gamma\delta}(q,t)$. The answer is given in Section 6.2: it can be done using simulations with arrested flow at wave-vector q implying the condition, Equation (88). This condition can be imposed using an external force field (cf Equation (89)) leading to an appropriate coherent harmonic acceleration of particles. With the constrained dynamics, the deterministic stress is always constant (time-independent); it is defined by the 'quenched' concentration fluctuation at q. Then,

$$C^n_{\alpha\beta\gamma\delta}(q,t) = C^{(r)}_{\alpha\beta\gamma\delta}(q,t) - \text{Const}$$

where $C^{(r)}_{\alpha\beta\gamma\delta}(q,t)$ is the total stress correlation function with restricted dynamics (cf Equation (99)) and $\text{Const} = E^e_{\alpha\beta\gamma\delta}(q) = C^{d(r)}_{\alpha\beta\gamma\delta}(q)$ is a time-independent tensor, which, however, generally depends on q (cf Equations (96) and (106)). This tensor (Const) simply equals to $C^{(r)}_{\alpha\beta\gamma\delta}(q, t \to \infty)$; it is related to the equilibrium elastic moduli (at $t \to \infty$), cf Equations (92) and (95).

8. In Section 6.2, we introduced the equilibrium elasticity tensor $E^e_{\alpha\beta\gamma\delta}(q)$ defined in Equation (106). In the liquid regime, $E^e_{\alpha\beta\gamma\delta}(q)$ coincides with the static elasticity tensor, $E^e_{\alpha\beta\gamma\delta}(q) = E_{\alpha\beta\gamma\delta}(q, t \gg \tau_s)$, so that $E^e_{\alpha\beta\gamma\delta}(q)$ can be considered as a generalization of the classical static elasticity tensor (cf Section ?) for nonzero q. However, in the glassy (amorphous solid) state, the two tensors, equilibrium and static, are different since even a very long waiting time, $t \gg \tau_s$, may not ensure a complete equilibration of a vitrified liquid (amorphous solid). In particular, the stress noise may include a virtually frozen component leading to an incomplete relaxation. Therefore, the static shear modulus, $G(q, t \gg \tau_s)$, remains finite in this case, while the analogous *equilibrium* shear modulus must vanish since a *complete* equilibration after a small shear deformation of a glassy system must relax the shear stress due to the amorphous structure of the system [37]. (Note that we do not consider here a *permanently* crosslinked network whose equilibrium shear modulus is, of course, finite.) As a result, the equilibrium elasticity tensor can be expressed in terms of just two material functions: the equilibrium longitudinal, $L_e(q)$, and transverse, $M_e(q)$, elastic moduli (cf Equation (106)). These moduli, by their definition, provide a linear stress response (after a complete relaxation of the system) to a weak imposed longitudinal strain.

9. The most general relation between the stress-correlation (C) and elasticity (E) tensors is given in Equation (115). It is noteworthy that this equation was derived and is valid at $q \neq 0$. It cannot be generally applied for $q = 0$ since the stress-correlation

function C is ensemble-dependent in this case [32,37]. It is also remarkable that, based on Equation (115), we not only arrive at Equations (62)–(64) linking the shear, longitudinal and transverse components of C- and E-tensors, but also obtain two additional exact relations (116) involving other components of these tensors. The whole set of these relations then allows to obtain all components of the correlation tensor in terms of the elasticity tensor and vice versa. Strictly speaking, all the relations, Equations (62)–(64) and (116), are valid both for liquid systems (above the glass transition) and for amorphous solids (vitrified liquids), provided that they are completely equilibrated thermodynamically (this condition refers to the fact that the derivation of these relations assumed an equilibrium ensemble). Nevertheless, these relations are also valid for metastable glassy systems (trapped in a metabasin), provided that the lifetime of the metastable state is much longer than $1/s$ [32] and with the reservation that some q-dependent constants may have to be added in the rhs of Equations (116), cf ref. [32]. These constants are due to the presence of frozen stresses in glassy systems, reflecting their metastable nature (within a given metabasin); they must disappear upon averaging over the full equilibrium ensemble of metastable states.

10. There is a subtle problem associated with the new Equations (116): they involve two new memory functions, $N(q,t)$ and $P(q,t)$, which cannot be obtained based on the stress response to a deformation, and, therefore, apparently cannot be measured experimentally. One may wonder if these functions can be obtained based on the 'classical' relaxation moduli, $G(q,t)$, $L(q,t)$, $M(q,t)$. Our view is that while their exact prediction is generally impossible, the new functions N and P can be still predicted approximately at low q. As argued in Section 6 (see end of Section 6.2 and the beginning of Section 6.4), the elasticity tensor becomes nearly isotropic at low q, so that $N(q,t) \simeq L(q,t)$ and $P(q,t) \simeq M(q,t)$ at $qb \ll 1$ (cf Equations (122)). Replacing N with L in the first Equation (116) leads to an approximate equation, which was derived and rather thoroughly tested in refs. [32,37] using simulation data for a 2D system of polydisperse LJ particles. A very good agreement (with an accuracy of 1–2%) between C_N and its approximate prediction was observed at $qb \lesssim 0.5$ (where b is the interaction range) [32]. Here, we devised two more precise approximations for $N(q,t)$ valid for 2D systems (see Equations (127) and (130)). All the approximations have been tested at $t=0$ for a wide range of q for the same system. The comparison (between the exact and approximate $N(q) = N(q,0)$) is shown in Figure 1. It demonstrates that the basic (0th) approximations still work for $qb < 0.5$, while the new approximations are accurate in much wider q-regions: the first one is valid at $qb < 1.5$, the second at $qb < 3$.

11. To summarize, let us highlight the main new results presented in the paper:

(i) We provide a rigorous derivation of Equations (63) and (64) using FDT-based arguments (cf Section 4). These equations have been previously stated in refs. [32,37], but their detailed derivation was not worked out (note that Equation (63) was also stated in ref. [59]). Importantly, in Section 4, we provide a derivation of the unique fully tensorial equation (Equation (59)) from which the general FDT relations, Equation (62) (which is well-known [57,63,64]) and Equations (63) and (64), simply follow in a trivial way.

(ii) We derived approximate Equations (123) and (124) (valid for $qb \ll 1$) using FDT and the concept of stress noise (cf Section 6.3). Note that the derivation of Equation (123) was only hinted at previously (in ref. [37]), while the same equation was simply claimed in ref. [32].

(iii) Building upon the concept of stress noise, a key result of our work is Equation (115), relating the tensor of stress correlations $C_{\alpha\beta\gamma\delta}(\vec{q},t)$ with the tensor of elastic moduli $E_{\alpha\beta\gamma\delta}(\vec{q},t)$. Noteworthily, the form of Equation (115) agrees with Equation (45) of ref. [49], which establishes a relation between the memory kernel $M_{\alpha\beta\gamma\delta}(\vec{q},t)$ from the Zwanzig–Mori projection operator formalism and $C_{\alpha\beta\gamma\delta}(\vec{q},t)$ for monodisperse Brownian particles. This suggests the intriguing possibility of a deeper connection between $M_{\alpha\beta\gamma\delta}(\vec{q},t)$ and $E_{\alpha\beta\gamma\delta}(\vec{q},t)$, which is an interesting topic for future studies.

(iv) For 2D systems, we, for the first time, derived more precise equations (as compared to Equation (123)) defining the stress correlation function $C_N(q,t)$ in terms of the generalized relaxation moduli (cf the first Equation (116) and Equations (127) and (130)).

Author Contributions: Conceptualization, A.S.; investigation, All authors; writing—original draft preparation, All authors. All authors have read and agreed to the published version of the manuscript.

Funding: This research received no external funding.

Institutional Review Board Statement: Not applicable.

Data Availability Statement: The original contributions presented in the study are included in the article, further inquiries can be directed to the corresponding author.

Acknowledgments: Fruitful discussions with J.P.Wittmer are gratefully acknowledged. A grant of computer time at the HPC computing cluster of the University of Strasbourg is acknowledged as well. We also thank L.Klochko who produced simulation data used in Figure 1 during her PhD studies with us.

Conflicts of Interest: The authors declare no conflicts of interest.

Appendix A. Proof of Equation (109)

To simplify notations, let us define any component of the stress tensor at $t = 0$ (taking either Re or Im part of it) as X, $X = (\Re|\Im)\sigma_{\alpha'\beta'}(q,0)$, and similarly, Y corresponds to a later moment t: $Y = (\Re|\Im)\sigma_{\alpha\beta}(q,t)$, while Z is defined as a component of the current J at $t = 0$: $Z = (\Re|\Im)J_\alpha(q,0)$. More precisely, let us consider Z as a vector with $2d$ components, $\underline{Z} = \{Z_i, i = 1.\overline{2}d\}$, where d is the space dimension. Next, we observe that (i) X, Y and Z_i are collective variables (fluctuation amplitudes), so their joint distribution must be nearly Gaussian (with high accuracy, relative error $\sim 1/N$), (ii) for $q \neq 0$

$$\langle X \rangle = \langle Y \rangle = \langle Z_i \rangle = 0 \tag{A1}$$

and (iii) X and Z_i are uncorrelated at equilibrium,

$$\langle XZ_i \rangle = 0 \tag{A2}$$

To justify the last statement, it is enough to recall the time reversibility of the dynamics and to note that X is invariant, while Z_i changes sign once the time is reversed, $t \to -t$.

The correlation functions in Equations (107) and (108) can be considered as linear combinations of terms like $\langle XY \rangle_r$ (for Equation (107)) and $\langle XY \rangle$ (for Equation (108)). Therefore, to prove the validity of Equation (109), it is enough to show that

$$\langle XY \rangle_r = \langle XY \rangle \tag{A3}$$

for all components X and Y. Here, $\langle XY \rangle$ means the unconstrained average of XY (with totally equilibrium state at $t = 0$), and

$$\langle XY \rangle_r \equiv \langle XY \rangle_{\underline{Z}=0} \tag{A4}$$

implies the averaging under the condition that the initial state (at $t = 0$) was prepared with the restricted dynamics, which boils down to the condition $\underline{Z} = 0$.

Let us treat the rhs of Equation (A3). It is useful to represent the unconstrained joint distribution density of $(X, Y, \underline{Z})$ as

$$\rho(X, Y, \underline{Z}) = \rho(X)\rho(\underline{Z}|X)\rho(Y|X, \underline{Z}) \tag{A5}$$

where $\rho(X)$ is the distribution density of X considered separately, $\rho(\underline{Z}|X)$ is the conditional distribution of $\underline{Z}$ for a given value of the variable X, and $\rho(Y|X,\underline{Z})$ is a similar conditional distribution of Y for the given X and $\underline{Z}$ (note that the ρ-functions are distinguished according to their variables). Thus, we obtain

$$\langle XY \rangle = \int XY\rho(X,Y,\underline{Z})\mathrm{d}X\mathrm{d}Y\mathrm{d}^{2d}\underline{Z} = \int X\langle Y \rangle_X \rho(X)\mathrm{d}X \tag{A6}$$

where
$$\langle Y \rangle_X = \int \langle Y \rangle_{X,\underline{Z}} \rho(\underline{Z}|X) \mathrm{d}^{2d}\underline{Z}, \quad \langle Y \rangle_{X,\underline{Z}} = \int Y \rho(Y|X,\underline{Z}) \mathrm{d}Y \tag{A7}$$

are the mean values of Y for a given X, and for the given X and $\underline{Z}$, respectively. Introducing $\rho(X,Y|\underline{Z})$ as the conditional probability distribution of X, Y for a prescribed $\underline{Z}$, and similarly $\rho(X|\underline{Z})$, one finds

$$\langle XY \rangle_r = \int XY \rho(X,Y|\underline{Z}=0)\mathrm{d}X\mathrm{d}Y = \int X\rho(X|\underline{Z}=0) \langle Y \rangle_{X,\underline{Z}=0} \mathrm{d}X \tag{A8}$$

Here, all the probability distribution (ρ-) functions are Gaussian. The following obvious relation is useful:
$$\langle Y \rangle_{X,\underline{Z}} = \alpha X + \beta_i Z_i \tag{A9}$$

as follows from Equation (A1) and the Gaussian nature of the distributions (α and β_i are some constants; the sum over i is assumed). In a similar way, we find (cf Equations (A7) and (A9)):

$$\langle Y \rangle_X = \alpha X + \beta_i \langle Z_i \rangle_X \tag{A10}$$

and
$$\langle Z_i \rangle_X = \gamma_i X \tag{A11}$$

Finally, noting that
$$\langle X Z_i \rangle \equiv \langle X \langle Z_i \rangle_X \rangle = \gamma_i \langle X^2 \rangle \tag{A12}$$

and using Equation (A2), we find $\gamma_i = 0$, so that $\langle Y \rangle_X = \alpha X$ and

$$\langle XY \rangle = \langle X \langle Y \rangle_X \rangle = \alpha \langle X^2 \rangle \tag{A13}$$

Turning to the lhs of Equations (A3) and (A4), corresponding to the case with constraint at $t=0$, in order to obtain $\langle XY \rangle_r$, we must impose the condition $\underline{Z} = 0$ in Equation (A9) leading immediately to $\langle Y \rangle_{X,\underline{Z}=0} = \alpha X$ and, therefore, to (cf Equation (A8))

$$\langle XY \rangle_r = \left\langle X \langle Y \rangle_{X,\underline{Z}=0} \right\rangle_r = \alpha \int X^2 \rho(X|\underline{Z}=0) \mathrm{d}X \tag{A14}$$

Now, we again take into account that X and $\underline{Z}$ are not correlated (Equation (A2)); hence, $\rho(X|\underline{Z})$ is independent of $\underline{Z}$, $\rho(X|\underline{Z}) = \rho(X)$. Therefore, we obtain (cf (A14))

$$\langle XY \rangle_r = \alpha \langle X^2 \rangle \tag{A15}$$

Equations (A13) and (A15) directly lead to Equation (A3) and, therefore, to Equation (109).

Appendix B. The Relevant Properties of Isotropic Tensor Fields

Tensor fields (like $C_{\alpha\beta\gamma\delta}(\underline{q},t)$) characterizing isotropic systems are isotropic in the sense that they are invariant with respect to a simultaneous rotation of the coordinate frame and the vector argument $\underline{q}$ [44]. Such isotropic tensor functions can be written in the general case as [35,37,41,44] (here $\hat{q} \equiv \underline{q}/q, q \equiv |\underline{q}| \neq 0$):

$$C_{\alpha\beta\gamma\delta}(\underline{q},t) = a(q,t)\delta_{\alpha\beta}\delta_{\gamma\delta} + b(q,t)(\delta_{\alpha\gamma}\delta_{\beta\delta} + \delta_{\alpha\delta}\delta_{\beta\gamma}) + c(q,t)(\hat{q}_\alpha\hat{q}_\beta\delta_{\gamma\delta} + \hat{q}_\gamma\hat{q}_\delta\delta_{\alpha\beta})$$

$$+ d(q,t)(\hat{q}_\alpha\hat{q}_\gamma\delta_{\beta\delta} + \hat{q}_\alpha\hat{q}_\delta\delta_{\beta\gamma} + \hat{q}_\beta\hat{q}_\gamma\delta_{\alpha\delta} + \hat{q}_\beta\hat{q}_\delta\delta_{\alpha\gamma}) + e(q,t)\hat{q}_\alpha\hat{q}_\beta\hat{q}_\gamma\hat{q}_\delta \tag{A16}$$

provided that (for any $\underline{q},t$), the C-tensor obeys both minor and major symmetries:

$$C_{\alpha\beta\gamma\delta} = C_{\beta\alpha\gamma\delta}, \quad C_{\alpha\beta\gamma\delta} = C_{\alpha\beta\delta\gamma}, \quad C_{\alpha\beta\gamma\delta} = C_{\gamma\delta\alpha\beta} \tag{A17}$$

The first two (minor) symmetries obviously apply to the stress-correlation tensor (cf Equation (48)): they come from the well-known symmetry of the stress tensor, $\sigma_{\alpha\beta} = \sigma_{\beta\alpha}$. The last (major) symmetry comes from the time-reversibility of the dynamics, the isotropy of the tensor field $C_{\alpha\beta\gamma\delta}(q,t)$ and its minor symmetries (cf ref. [37]). The same symmetries also apply to the tensor of generalized elastic constants (according to Equation (105)).

For q parallel to the x-axis, there is no need to rotate the coordinate frame: it is already 'natural' in this case (ie it coincides with the NRC frame). Then, using Equation (A16) for such q, we obtain:

$$a(q,t) = C_P(q,t), \; b(q,t) = (C_N - C_P)/2, \; c(q,t) = C_M - C_P,$$

$$d(q,t) = C_G - (C_N - C_P)/2, \; e(q,t) = C_L + C_N - 2C_M - 4C_G \tag{A18}$$

where

$$C_N(q,t) = C_{2222}(q,t), \; C_P(q,t) = C_{2233}(q,t) \tag{A19}$$

and the arguments (q,t) are omitted in the rhs.

For 2D systems ($d = 2$), the a, b, c and d terms of the general Equation (A16) become entangled due to the mathematical identity

$$\hat{q}_\alpha \hat{q}_\gamma \delta_{\beta\delta} + \hat{q}_\alpha \hat{q}_\delta \delta_{\beta\gamma} + \hat{q}_\beta \hat{q}_\gamma \delta_{\alpha\delta} + \hat{q}_\beta \hat{q}_\delta \delta_{\alpha\gamma} =$$

$$2(\hat{q}_\alpha \hat{q}_\beta \delta_{\gamma\delta} + \hat{q}_\gamma \hat{q}_\delta \delta_{\alpha\beta}) - 2\delta_{\alpha\beta}\delta_{\gamma\delta} + (\delta_{\alpha\gamma}\delta_{\beta\delta} + \delta_{\alpha\delta}\delta_{\beta\gamma}), \; d=2 \tag{A20}$$

As a result, Equation (A16) can be simplified as

$$E_{\alpha\beta\gamma\delta}(q,t) = a_E(q,t)\delta_{\alpha\beta}\delta_{\gamma\delta} + b_E(q,t)(\delta_{\alpha\gamma}\delta_{\beta\delta} + \delta_{\alpha\delta}\delta_{\beta\gamma})$$

$$+c_E(q,t)(\hat{q}_\alpha \hat{q}_\beta \delta_{\gamma\delta} + \hat{q}_\gamma \hat{q}_\delta \delta_{\alpha\beta}) + e_E(q,t)\hat{q}_\alpha \hat{q}_\beta \hat{q}_\gamma \hat{q}_\delta, \; d=2 \tag{A21}$$

where we replaced C with E, a with a_E, etc.

References

1. Fuchs, M. Nonlinear Rheological Properties of Dense Colloidal Dispersions Close to a Glass Transition Under Steady Shear. *Adv. Polym. Sci.* **2010**, *236*, 55.
2. Nicolas, A.; Ferrero, E.E.; Martens, K.; Barrat, J.-L. Deformation and flow of amorphous solids: Insights from elastoplastic models. *Rev. Mod. Phys.* **2018**, *90*, 045006. [CrossRef]
3. Ferry, J.D. *Viscoelastic Properties of Polymers*, 3rd ed.; Wiley: New York, NY, USA, 1980.
4. Bird, R.B.; Armstrong, R.C.; Hassager, O. *Dynamics of Polymeric Liquids: Fluid Mechanics*; Wiley: New York, NY, USA 1987.
5. Berthier, L.; Biroli, G. Theoretical perspective on the glass transition and amorphous materials. *Rev. Mod. Phys.* **2011**, *83*, 587. [CrossRef]
6. Bobbili, S.V.; Milner, S.T. Simulation study of entanglement in semiflexible polymer melts and solutions. *Macromolecules* **2020**, *53*, 3861. [CrossRef]
7. Flenner, E.; Szamel, G. Viscoelastic shear stress relaxation in two-dimensional glass-forming liquids. *Proc. Natl. Acad. Sci. USA* **2019**, *116*, 2015. [CrossRef]
8. Fritschi, S.; Fuchs, M. Elastic moduli of a Brownian colloidal glass former. *J. Phys. Condens. Matter.* **2018**, *30*, 024003. [CrossRef]
9. Janssen, L.M.C. Mode-Coupling Theory of the Glass Transition: A Primer. *Front. Phys.* **2018**, *6*, 97. [CrossRef]
10. Kapteijns, G.; Richard, D.; Bouchbinder, E.; Schroder, T.B.; Dyre, J.C.; Lerner, E. Does mesoscopic elasticity control viscous slowing down in glassforming liquids? *J. Chem. Phys.* **2021**, *155*, 074502. [CrossRef] [PubMed]
11. Kremer, K.; Grest, G.S. Dynamics of entangled linear polymer melts: A molecular dynamics simulation. *J. Chem. Phys.* **1990**, *92*, 5057. [CrossRef]
12. Lerbinger, M.; Barbot, A.; Vandembroucq, D.; Patinet, S. Relevance of shear transformations in the relaxation of supercooled liquids. *Phys. Rev. Lett.* **2022**, *129*, 195501. [CrossRef]
13. Likhtman, A.E. *Polymer Science: A Comprehensive Reference, Volume 1, Chapter on 'Viscoelasticity and Molecular Rheology'*; Elsevier: Amsterdam, The Netherlands, 2012; pp. 133–179.

14. Likhtman, A.E.; McLeish, T.C.B. Quantitative theory for linear dynamics of linear entangled polymers. *Macromolecules* **2002**, *35*, 6332. [CrossRef]
15. Martins, M.L.; Zhao, X.; Demchuk, Z.; Luo, J.; Carden, G.P.; Toleutay, G.; Sokolov, A.P. Viscoelasticity of polymers with dynamic covalent bonds: Concepts and misconceptions. *Macromolecules* **2023**, *56*, 8688. [CrossRef]
16. Müller, M. Memory in the relaxation of a polymer density modulation. *J. Chem. Phys.* **2022**, *156*, 124902. [CrossRef] [PubMed]
17. Novikov, V.N.; Sokolov, A.P. Temperature dependence of structural relaxation in glass-forming liquids and polymers. *Entropy* **2022**, *24*, 1101. [CrossRef] [PubMed]
18. Voigtmann, T. Nonlinear glassy rheology. *Curr. Opin. Colloid Interf. Sci.* **2014**, *19*, 549. [CrossRef]
19. Takeru, H.; Tetsuya, K. Effect of liquid elasticity on nonlinear pressure waves in a viscoelastic bubbly liquid. *Phys. Fluids* **2023**, *35*, 043309.
20. Tadmor, E.B.; Miller, R.E. *Modeling Materials*; Cambridge University Press: Cambridge, UK, 2011.
21. Hansen, J.P.; McDonalds, I.R. *Theory of Simple Liquids*, 3rd ed.; Academic Press: San Diego, CA, USA, 2006.
22. Debenedetti, P.G.; Stillinger, F.H. Supercooled liquids and the glass transition. *Nature* **2001**, *410*, 259–267. [CrossRef] [PubMed]
23. Gelin, S.; Tanaka, H.; Lemaître, A. Anomalous phonon scattering and elastic correlations in amorphous solids. *Nat. Mater.* **2016**, *15*, 1177. [CrossRef]
24. Hassani, M.; Zirdehi, E.M.; Kok, K.; Schall, P.; Fuchs, M.; Varnik, F. Long-range strain correlations in 3d quiescent glass forming liquids. *Europhys. Lett.* **2018**, *124*, 18003. [CrossRef]
25. Illing, B.; Fritschi, S.; Hajnal, D.; Klix, C.; Keim, P.; Fuchs, M. Strain pattern in supercooled liquids. *Phys. Rev. Lett.* **2016**, *117*, 208002. [CrossRef] [PubMed]
26. Chacko, R.N.; Landes, F.P.; Biroli, G.; Dauchot, O.; Liu, A.; Reichman, D.R. Elastoplasticity mediates dynamical heterogeneity below the mode coupling temperature. *Phys. Rev. Lett.* **2021**, *127*, 048002. [CrossRef]
27. Jung, G.; Biroli, G.; Berthier, L. Predicting dynamic heterogeneity in glass-forming liquids by physics-inspired machine learning. *Phys. Rev. Lett.* **2023**, *130*, 238202. [CrossRef]
28. Ozawa, M.; Biroli, G. Elasticity, facilitation, and dynamic heterogeneity in glass-forming liquids. *Phys. Rev. Lett.* **2023**, *130*, 138201. [CrossRef] [PubMed]
29. Furukawa, A.; Tanaka, H. Direct evidence of heterogeneous mechanical relaxation in supercooled liquids. *Phys. Rev. E* **2011**, *84*, 061503. [CrossRef] [PubMed]
30. Puscasu, R.M.; Todd, B.D.; Daivis, P.J.; Hansen, J.S. Nonlocal viscosity of polymer melts approaching their glassy state. *J. Chem. Phys.* **2010**, *133*, 144907. [CrossRef] [PubMed]
31. Furukawa, A.; Tanaka, H. Nonlocal Nature of the Viscous Transport in Supercooled Liquids: Complex Fluid Approach to Supercooled Liquids. *Phys. Rev. Lett.* **2009**, *103*, 135703. [CrossRef]
32. Klochko, L.; Baschnagel, J.; Wittmer, J.P.; Meyer, H.; Benzerara, O.; Semenov, A.N. Theory of length-scale dependent relaxation moduli and stress fluctuations in glass-forming and viscoelastic fluids. *J. Chem. Phys.* **2022**, *156*, 164505. [CrossRef]
33. Puscasu, R.M.; Todd, B.D.; Daivis, P.J.; Hansen, J.S. Viscosity kernel of molecular fluids: Butane and polymer melts. *Phys. Rev. E* **2010**, *82*, 011801. [CrossRef] [PubMed]
34. Semenov, A.N.; Farago, J.; Meyer, H. Length-scale dependent relaxation shear modulus and viscoelastic hydrodynamic interactions in polymer liquids. *J. Chem. Phys.* **2012**, *136*, 244905. [CrossRef]
35. Maier, M.; Zippelius, A.; Fuchs, M. Stress auto-correlation tensor in glass-forming isothermal fluids: From viscous to elastic response. *J. Chem. Phys.* **2018**, *149*, 084502. [CrossRef]
36. Maier, M.; Zippelius, A.; Fuchs, M. Emergence of long-ranged stress correlations at the liquid to glass transition. *Phys. Rev. Lett.* **2017**, *119*, 265701. [CrossRef]
37. Klochko, L.; Baschnagel, J.; Wittmer, J.P.; Semenov, A.N. Long-range stress correlations in viscoelastic and glass-forming fluids. *Soft Matter* **2018**, *14*, 6835. [CrossRef] [PubMed]
38. Chowdhury, S.; Abraham, S.; Hudson, T.; Harrowell, P. Long range stress correlations in the inherent structures of liquids at rest. *J. Chem. Phys.* **2016**, *144*, 124508. [CrossRef] [PubMed]
39. Wu, B.; Iwashita, T.; Egami, T. Anisotropic stress correlations in two-dimensional liquids. *Phys. Rev. E* **2015**, *91*, 032301. [CrossRef] [PubMed]
40. Lemaître, A. Structural relaxation is a scale-free process. *Phys. Rev. Lett.* **2014**, *113*, 245702. [CrossRef] [PubMed]
41. Lemaître, A. Tensorial analysis of Eshelby stresses in 3D supercooled liquids. *J. Chem. Phys.* **2015**, *143*, 164515. [CrossRef]
42. Lemaître, A. Inherent stress correlations in a quiescent two-dimensional liquid: Static analysis including finite-size effects. *Phys. Rev. E* **2017**, *96*, 052101. [CrossRef] [PubMed]
43. Lemaître, A. Stress correlations in glasses. *J. Chem. Phys.* **2018**, *149*, 104107. [CrossRef] [PubMed]
44. Wittmer, J.P.; Semenov, A.N.; Baschnagel, J. Correlations of tensor field components in isotropic systems with an application to stress correlations in elastic bodies. *Phys. Rev. E* **2023**, *108*, 015002. [CrossRef]
45. Steffen, D.; Schneider, L.; Müller, M.; Rottler, J. Molecular simulations and hydrodynamic theory of nonlocal shear-stress correlations in supercooled fluids. *J. Chem. Phys.* **2022**, *157*, 064501. [CrossRef]
46. Ebert, F.; Dillmann, P.; Maret, G.; Keim, P. The experimental realization of a two-dimensional colloidal model system. *Rev. Sci. Instrum.* **2009**, *80*, 083902. [CrossRef] [PubMed]

47. Klix, C.; Ebert, F.; Weysser, F.; Fuchs, M.; Maret, G.; Keim, P. Glass elasticity from particle trajectories. *Phys. Rev. Lett.* **2012**, *109*, 178301. [CrossRef]
48. Klix, C.L.; Maret, G.; Keim, P. Discontinuous Shear Modulus Determines the Glass Transition Temperature. *Phys. Rev. X* **2015**, *5*, 041033. [CrossRef]
49. Landau, L.D.; Lifshitz, E.M. *Theory of Elasticity*; Pergamon Press: New York, NY, USA, 1959.
50. Wittmer, J.P.; Xu, H.; Baschnagel, J. Shear-stress relaxation and ensemble transformation of shear-stress autocorrelation functions. *Phys. Rev. E* **2015**, *91*, 022107. [CrossRef]
51. Wittmer, J.P.; Xu, H.; Benzerara, O.; Baschnagel, J. Fluctuation-dissipation relation between shear stress relaxation modulus and shear stress autocorrelation function revisited. *Mol. Phys.* **2015**, *113*, 2881. [CrossRef]
52. Landau, L.D.; Lifshitz, E.M. *Statistical Physics*; Pergamon: Oxford, UK, 1998.
53. Doi, M.; Edwards, S.F. *The Theory of Polymer Dynamics*; Oxford University: New York, NY, USA, 1986.
54. Frenkel, D.; Smit, B. *Understanding Molecular Simulation-from Algorithms to Applications*, 2nd ed.; Academic Press: San Diego, CA, USA, 2002.
55. Lutsko, J.F. Generalized expressions for the calculation of elastic constants by computer simulation. *J. Appl. Phys.* **1989**, *65*, 2991. [CrossRef]
56. Goldstein, H.; Poole, C.P.; Safko, J.L. *Classical Mechanics*, 3rd ed.; Addison-Wesley: Boston, MA, USA, 2001.
57. Balucani, U.; Zoppi, M. *Dynamics of the Liquid State*; Oxford University Press: Oxford, UK, 1995.
58. Klochko, L.; Baschnagel, J.; Wittmer, J.P.; Semenov, A.N. Relaxation moduli of glass-forming systems: Temperature effects and fluctuations. *Soft Matter* **2021**, *17*, 7867. [CrossRef] [PubMed]
59. Ruscher, C.; Semenov, A.N.; Baschnagel, J.; Farago, J.J. Anomalous sound attenuation in Voronoi liquid. *Chem. Phys.* **2017**, *146*, 144502. [CrossRef] [PubMed]
60. Klochko, L.; Baschnagel, J.; Wittmer, J.P.; Semenov, A.N. General relations to obtain the time-dependent heat capacity from isothermal simulations. *J. Chem. Phys.* **2021**, *154*, 164501. [CrossRef]
61. Rubinstein, M.; Colby, R. *Polymer Physics*; Oxford University: New York, NY, USA, 2003.
62. Shi, K.; Smith, E.R.; Santiso, E.E.; Gubbins, K.E. A perspective on the microscopic pressure (stress) tensor: History, current understanding, and future challenges. *J. Chem. Phys.* **2023**, *158*, 040901. [CrossRef]
63. Evans, D.J.; Morriss, G.P. *Statistical Mechanics of Nonequilibrium Liquids*; Academic Press: London, UK, 1990.
64. Götze, W. *Complex Dynamics of Glass-Forming Liquids: A Mode-Coupling Theory*; Oxford University Press: New York, NY, USA, 2009.
65. Vogel, F.; Fuchs, M. Stress correlation function and linear response of Brownian particles. *Eur. Phys. J. E* **2020**, *43*, 70. [CrossRef] [PubMed]
66. Klochko, L.; Baschnagel, J.; Wittmer, J.P.; Benzerara, O.; Ruscher, C.; Semenov, A.N. Composition fluctuations in polydisperse liquids: Glasslike effects well above the glass transition. *Phys. Rev. E* **2020**, *102*, 042611. [CrossRef] [PubMed]

Disclaimer/Publisher's Note: The statements, opinions and data contained in all publications are solely those of the individual author(s) and contributor(s) and not of MDPI and/or the editor(s). MDPI and/or the editor(s) disclaim responsibility for any injury to people or property resulting from any ideas, methods, instructions or products referred to in the content.

Article

Effectiveness of the Use of Polymers in High-Performance Concrete Containing Silica Fume

Alya Harichane [1,*], Nadhir Toubal Seghir [1], Paweł Niewiadomski [2], Łukasz Sadowski [2] and Michał Cisiński [3]

1. Institute of Science, University Center of Tipaza, Tipaza 42000, Algeria; toubalseghir.nadhir@cu-tipaza.dz
2. Department of Materials Engineering and Construction Processes, Wroclaw University of Science and Technology, Wybrzeże Wyspiańskiego 27, 50-370 Wroclaw, Poland; pawel.niewiadomski@pwr.edu.pl (P.N.); lukasz.sadowski@pwr.edu.pl (Ł.S.)
3. Department of Advanced Material Technologies, Wroclaw University of Science and Technology, Wybrzeże Wyspiańskiego 27, 50-370 Wroclaw, Poland; 240524@student.pwr.edu.pl
* Correspondence: harichane.alya@cu-tipaza.dz; Tel.: +213-696074103

Abstract: The incorporation of polycarboxylate ether superplasticizer (PCE)-type polymers and silica fume (SF) in high-performance concretes (HPC) leads to remarkable rheological and mechanical improvements. In the fresh state, PCEs are adsorbed on cement particles and dispersants, promoting the workability of the concrete. Silica fume enables very well-compacted concrete to be obtained, which is characterized by high mechanical parameters in its hardened state. Some PCEs are incompatible with silica fume, which can result in slump loss and poor rheological behavior. The main objective of this research is to study the influence of three types of PCEs, which all have different molecular architectures, on the rheological and mechanical behavior of high-performance concretes containing 10% SF as a partial replacement of cement. The results show that the carboxylic density of PCE has an influence on its compatibility with SF.

Keywords: polycarboxylate ether superplasticizer (PCE); rheology; molecular architecture; high performance concrete; carboxylic groups

Citation: Harichane, A.; Seghir, N.T.; Niewiadomski, P.; Sadowski, Ł.; Cisiński, M. Effectiveness of the Use of Polymers in High-Performance Concrete Containing Silica Fume. Polymers 2023, 15, 3730. https://doi.org/10.3390/polym15183730

Academic Editors: Alexandre M. Afonso, Luís L. Ferrás and Célio Pinto Fernandes

Received: 17 August 2023
Revised: 4 September 2023
Accepted: 5 September 2023
Published: 11 September 2023

Copyright: © 2023 by the authors. Licensee MDPI, Basel, Switzerland. This article is an open access article distributed under the terms and conditions of the Creative Commons Attribution (CC BY) license (https://creativecommons.org/licenses/by/4.0/).

1. Introduction

According to the World Cement Association (WCA), the cement industry is responsible for 5–6% of global carbon dioxide emissions [1]. This association requires a significant reduction in the carbon footprint of the cement industry. One approach to this is to partially move away from cement in favor of "mineral additions", which can often be waste products from other industries. However, substituting cement with mineral additions such as silica fume strongly modifies the rheological and mechanical properties of concrete [2,3]. Therefore, in some cases, the composition of concrete mixes has to be adjusted with regard to these additives in order to maintain the quality and parameters of the concrete. The use of polycarboxylate ether superplasticizers (PCEs) is absolutely necessary for concretes that contain mineral additives, e.g., silica fume, in order to make these additives "work". The compatibility of PCEs with silica fumes is currently the subject of many studies [4–7].

PCEs are currently the most effective at reducing the water content (up to 40%) and improving the workability and fluidity of concrete compared to other types of superplasticizers. PCEs are copolymers and have a comb structure [8,9]. The main chain of a PCE carries a negatively charged carboxylic group (COO–), which promotes the adsorption of the polymer onto the positively charged surfaces of cement particles or individual hydrate phases, primarily ettringite [$Ca_3Al(OH)_6$ $12H_2O]_2(SO_4)_3$ $2H_2O$ [10]. This occurs based on the electrostatic interaction. While adsorbed, the polymer promotes the dispersion of the cement particles due to the steric hindrance produced by its side chains containing polyethylene glycol (PEG) structures [11–21]. As a result, the water entrapped in the agglomerated binder particles can be released, increasing the flowability and reducing the viscosity of cement paste [22–25].

The adsorption and dispersing capacity of PCEs in cement-based materials depends on their chemical structure as well as their molecular weight, carboxylic groups, density, and length of the side chain [26–28]. Some PCEs are incompatible with silica fume in high-performance concretes, which in turn can result in a loss of slump and poor rheological behavior. This is due to the prior adsorption of the PCE on the surface of the silica fume [22,29–31], increasing the dispersion of the SF particles. Therefore, the adverse effect of PCEs on cement hydration could be mitigated by adding SF, which provides additional surfaces for PCE absorption and, as a result, reduces PCE adsorption on cement surfaces [6].

To the best of our knowledge, the effectiveness of PCEs in high-performance concrete (HPC) containing silica fume (SF) as a partial replacement of cement has not been fully investigated. There are several knowledge gaps regarding this matter, especially concerning the interaction of SF and PCE (with different molecular weights and carboxyl densities) in HPC. To fill this research gap, the authors of this paper conducted studies of the interaction between PCE and silica fume in cement slurries as well as in high-performance concretes (HPC) that have a low w/b ratio. Three PCEs with different molecular compositions were used to study their impact on the properties of concrete (HPC). Therefore, the fluidity and mechanical properties of HPC were evaluated. The impact of the PCE on the shear stresses, plastic viscosity, and zeta potential of cement slurries (containing 10% silica fume as a cement substitute) was also measured.

2. Materials and Methods

2.1. Materials

2.1.1. Superplasticizer

Three types of superplasticizers produced in Algeria were used: Polyflow SR 3600—designated as PCE1, Polyflow SR 5400—designated as PCE2, and Medaflow 30—designated as PCE3. The three polymers are new-generation, non-chlorinated, and based on modified polycarboxylic ether. The molecular weight (Mn, Mw) and the polydispersity indices (PDI) of the superplasticizers were measured using multi-detector steric exclusion chromatography (SEC) in France (laboratory for catalysis, polymerisation; processes and materials C2P2). The analytical conditions are presented in Table 1. The characteristics of the superplasticizers, according to their technical data sheets, are presented in Table 2.

Table 1. The analytical conditions of the SEC.

System	Column	Col Temp (C)	Solvent	Flow Rate (mL/min)	Inj Vol (μL)
SEC THF	SDV	35.00	THF	1.0000	100.0

Table 2. Properties of the PCEs.

Property	Type of Superplasticizer		
	PCE$_1$	PCE$_2$	PCE$_3$
Dry extract	22 ± 1%	30 ± 1%	30 ± 1%
pH	5.5	4.46	3.35
Density (g/cm^3)	1.06 ± 0.02	1.07 ± 0.02	1.07 ± 1
Mn (Da)	4036	13,621	4771
Mw (Da)	4386	15,959	5157
PDI	1.087	1.172	1.081
Carboxyl content (mmol/g)	1.80	1.87	1.95

PDI = Polydispersity index. Mn = Number average molecular weight of PCE. Mw = Weight average molecular weight of PCE.

2.1.2. Portland Cement and Silica Fume

The Portland cement (PC) that was used in the experimental study was CEM I 52.5 R. It is commercially produced by the Lafarge Company, whose manufacturing site is located in the city of Msila (in the eastern region of Algeria). The microsilica (SF) sample was obtained from Granitex, Algeria. Its marketing name is Medaplast HP, and its specific surface is 23,000 m²/kg. The chemical composition of the PC and SF is presented in Table 3. The particle size distributions of the PC and SF were prepared using a Malvern Mastersizer 2000 analyzer (liquid:Hydro, 2000MU). The results are presented in Figure 1.

Table 3. Chemical composition of the PC and SF.

	PC	SF
Oxide content (%)		
Silicon dioxide (SiO_2)	22.18	95.5
Aluminum oxide (Al_2O_3)	3.72	1.0
Iron oxide (Fe_2O_3)	0.17	1.0
Calcium oxide (CaO)	66.55	0.4
Magnesium oxide (MgO)	1.75	0.5
Sulfur trioxide (SO_3)	2.53	0.1
Potassium oxide (K_2O)	0.52	-
Sodium oxide (Na_2O)	0.07	0.6
Sodium oxide Na_2O Equivalent	0.41	0.4
Mineralogical composition of the cement determined using the Bogue equation		
Tricalcium silicate (C3S)	77.07	-
Dicalcium silicate (C2S)	5.52	-
Tricalcium aluminate (C3A)	9.57	-
Tetracalcium aluminoferrite (C4AF)	0.52	-
Physical properties		
Specific gravity (g/cm³)	3.07	2.3
Bulk density (g/cm³)	0.97	0.25
Blaine fineness (cm²/g)	3711.89	230.000

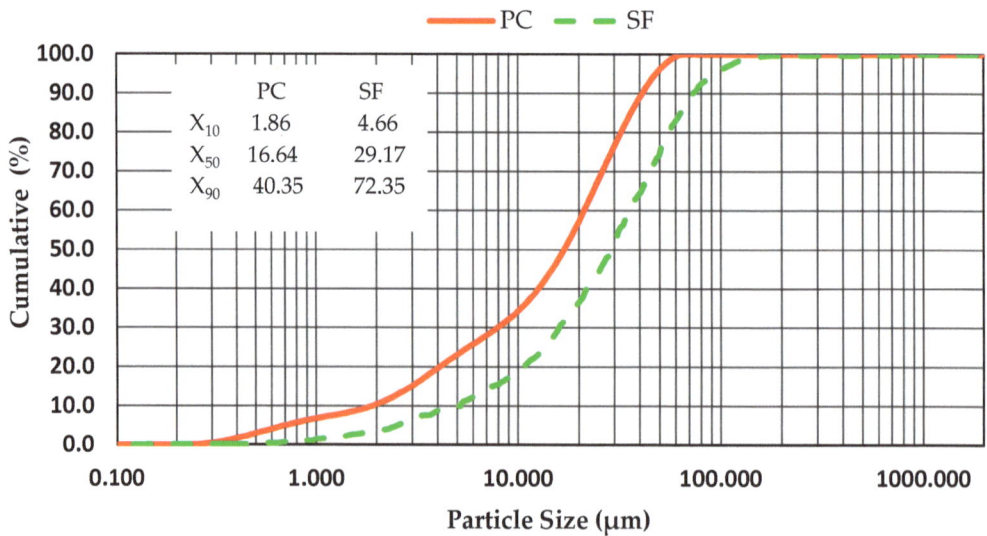

Figure 1. Particle size distribution curves of the PC and SF.

2.1.3. Aggregates

The properties of the aggregates are summarized in Table 4, and their particle size distribution curves are given in Figure 2.

Table 4. The properties of the aggregates.

Aggregates		Fineness Modulus	Sand Equivalent (%)	Apparent Density (kg/m³)
Fine Aggregate	Sand 0/1	1.14	83.5	1590
	Sand 0/4	2.64	74.5	1540
Coarse Aggregate	Gravel 8/15	----	----	1480
	Gravel 15/25	----	----	1470

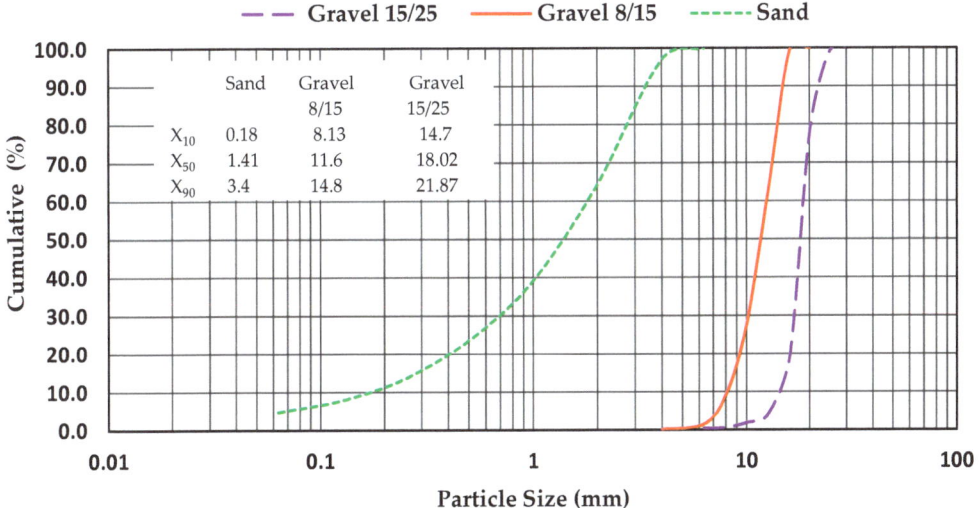

Figure 2. Particle size distribution curves of the used aggregates.

2.2. Experimental Measurements

2.2.1. Evaluation of the Content of Carboxyl Groups Using the Titration Method

The acid-base titration method was used to measure the content of carboxyl groups in each polymer sample [32]. The measurements involved the determination of the amount of NaOH solution, which in turn was expressed as the product of its molar concentration and the volume that was needed to titrate a sample. The determined amount of NaOH (number of moles) is directly related to the content of carboxyl groups in polymers because COO– and Na⁺ ions, which come from carboxyl groups and NaOH, react with each other in a stoichiometric ratio. Therefore, the more NaOH needed to titrate the sample, the more carboxyl groups it contains. The content of carboxyl groups can then be calculated as follows:

$$C(COO-) = [C_{NaOH} \times V_{NaOH}] \times m/100$$

where

$C(COO-)$ = content of carboxyl groups (mmol/100 g);
C_{NaOH} = molar concentration of NaOH (mmol/L);
V_{NaOH} = volume of the NaOH solution used (L); and m = weight of the sample.

2.2.2. The Composition of the Concrete Mix and the Performances of the Polymers in the High-Performance Concrete (HPC)

The recipe of the concrete was determined on the basis of the Dreux–Gorisse method while taking into account the ratio of its components (water, cement, sand, and aggregates), as shown in Table 5. The prepared concrete had a mechanical compressive strength of 55 MPa and a slump of 175 mm. The manufactured concrete mixes had a constant water-to-binder ratio of 0.35. The amount of PCE added to the concrete (in the presence of SF) in relation to the binder was adopted based on viscosimetric measurements (using a VT 550 viscometer). The ratio was assumed to be PCE/b = 1.75%.

Table 5. The formulations of the concrete mixes.

PC (kg/m^3)	SF (kg/m^3)	Coarse Aggregate (kg/m^3)	Fine Aggregate (kg/m^3)	Water (kg/m^3)	PCE (kg/m^3)	w/b
382.5	42.5	1031.5	810.5	157.5	7.45 PCE1	0.35
382.5	42.5	1031.5	810.5	157.5	7.45 PCE2	0.35
382.5	42.5	1031.5	810.5	157.5	7.45 PCE3	0.35

2.2.3. Fluidity

The slump test was used to evaluate the fluidity of the concrete mix in the case of all the formed HPCs. The dosage of superplasticizer (PC/b) was 1.75%, and it was determined using the VT 550 viscometer. The initial fluidity of the HPC was measured 5 min after mixing all the components. The concrete mix was then placed in a container, which was covered with a plastic film in order to prevent any water evaporation. The second slump test measurement was made after 1 h of concrete maturation. In each case, the concrete mix was stirred for one minute before taking the measurements.

2.2.4. Rheology

In order to evaluate the rheology of the cement paste containing 10% SF (characterized by a Blaine-specific surface of 4650 cm^2/g) as a partial replacement of cement, the water-to-binder ratio (w/b) was assumed to be 0.35. In this case, for an amount of water equal to 35 g, the mix contains 90 g of cement and 10 g of SF, as well as an added superplasticizer (PCE) in an amount of 0–2.5 wt% of the total cement-based binder. The test was performed at a temperature of 20 ± 1 °C. The rheology of the cement paste was evaluated using a VT 550 viscometer, which is a rotational viscometer. Using this viscometer, the following parameters of cement pastes can be measured: shear stresses (τ) and viscosities (μ), which are obtained as a function of shear rates ($\dot{\gamma}$).

2.2.5. Zeta Potential

The zeta potential represents the measurement of the intensity of electrostatic repulsion or attraction between particles. Therefore, its measurement provides an understanding of the causes of dispersion or flocculation of cement grains in suspensions. The zeta potentials of the cement pastes (containing PCEs at saturation dosages) were determined using a Malvern ZETASIZER 2000 in materials, processes and environment research unit; Algeria. In a standard experiment, 1 cm^3 of the cementitious suspension is diluted in 30 cm^3 of distilled water, after which 5 mL of this suspension is injected into the analyzer.

2.2.6. Compressive Strength

In order to determine the compressive strength, six cylindrical concrete specimens (150 × 300 mm^2) were prepared according to the EN 206-1 protocol. These specimens were kept for 24 h in molds. After demolding, all the specimens were stored in tap water at 20 ± 1 °C for 3, 7, and 28 days.

3. Results and Discussion

3.1. Molecular Weight Analysis

The molecular properties of the PCEs are shown in Figure 3 and Table 2. The results show that the PDI values for these polymers are slightly above 1.00, which in turn indicates a small dispersion of molecular weights in the polymers. Furthermore, the PDI value of PCE3 is the closest to 1 compared to the other two polymers. This means that PCE3 consists of the most similar macromolecules (regarding their length), which have a similar molar mass. In terms of molecular weight, expressed both as Mn and Mw, PCE2 obtained the highest values, while PCE1 obtained the lowest ones.

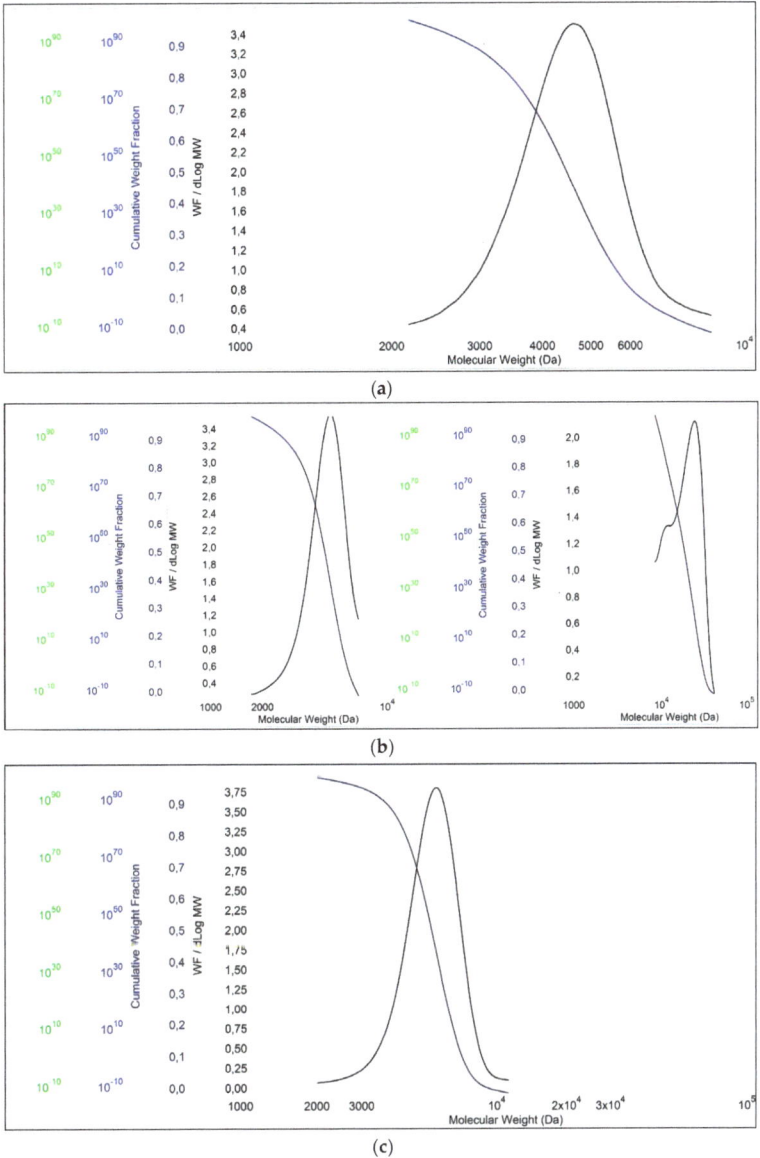

Figure 3. Chromatogram of polymers: (a) PCE1, (b) PCE2, and (c) PCE3.

3.2. Analysis of the Content of the Carboxyl Groups

The content of the carboxyl groups of the PCEs is shown in Table 2. This content directly influences the carboxyl density of the PCEs. It should be mentioned that both of these values increase simultaneously—the increase in one causes a rise in the other. Therefore, PCE3 has the highest carboxyl density, whereas PCE1 has the lowest one. This is due to the fact that the content of the carboxyl groups in PCE3 and PCE1, expressed for 1 g of the polymer, amounts to 1.95 and 1.80 mmol, respectively. The density of carboxyl groups in the main chain has a significant influence on the adsorption effect of PCE on cement particles. First, the increased density of carboxyl groups can enhance the adsorption capacity of PCEs [12,20]. However, different molecular structures will have different optimal carboxyl densities.

3.3. Fluidity

The results of the initial fluidity (slump) and the fluidity assessed after 1 h (slump retention) are shown in Figure 4. PCE3, with the highest charge density and a moderate molecular weight, exhibited a higher value of initial fluidity and a higher value of fluidity after 1 h compared to PCE1 and PCE2. The COO– group on the backbone of the PCE was adsorbed on the surface of cement particles by complexation with Ca^{2+} or by electrostatic attraction. An increase in the number of COO– groups increases the adsorption capacity of PCEs [32]. It means that the flowability of the cement paste increased with a rise in the content of the COO– groups. Therefore, very good compatibility was found between PCE3 and the SF, as they both retained the fluidity of the cement pastes after 1 h.

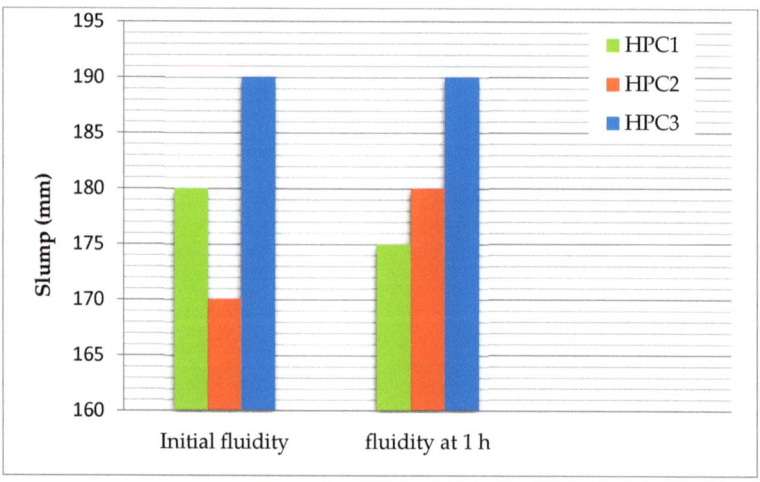

Figure 4. Slump values of the HPC.

PCE1, with the lowest carboxylic density and molecular weight, exhibited a poor slump and poor slump retention due to the prior adsorption of the PCE on the silica fume surfaces [22,29]. The SF provided additional surfaces for the adsorption of the PCE, which consequently reduced PCE adsorption on the cement surfaces. Moreover, it reduced the performance of the concrete, and it can be stated that there is an incompatibility between PCE1 and the SF.

PCE2 (with the highest molecular weight) had a bad initial slump and the highest slump retention. The amount of absorbed PCE on the surface of the cement paste increased with a decrease in molecular weight. Therefore, the PCE with a high molecular weight significantly delayed the hydration of the cement. This is due to the fact that the main chain of PCEs adsorbs on different cement particles simultaneously, making them agglomerated and, in turn, hindering the hydration process [33].

3.4. Rheology

The saturation point is the value of dosage beyond which the superplasticizer has no effect on the rheological properties of the cement paste. It was determined using the viscometer VT 550. Each of the polymers (PCE1, 2, and 3) had the same saturation point value, i.e., 1.75, at the tested water-to-binder ratio (w/b). Figures 5 and 6 show that shear stress and viscosity decreased with an increase in the dosage of the superplasticizer. The cement pastes became more fluid, with their flow approaching the characteristics of Newtonian flow. The shear stress and viscosity depend on the applied stress. The shear stress increases with a rise in the shear rate, whereas the viscosity decreases with an increase in the shear rate. After adding the PCEs, the yield stress was reduced [34,35]. The cement paste with PCE that has a moderate molecular weight and a high content of carboxylic groups exhibits the lowest shear stress and the lowest plastic viscosity. The carboxylic density and molecular weight of PCEs have a great effect on their dispersing performance. As the carboxylic density increases, the dispersing capability of a PCE improves [10]. PCE3 had the highest carboxylic density and a moderate molecular weight. Due to this, it exhibited the best adsorption behavior on cement grains, which in turn resulted in the lowest viscosity and shear stress of the samples with this polymer. PCE3, which had the highest anionic charge density and a moderate molecular weight, was more compatible with the SF when compared to PCE1 and PCE2.

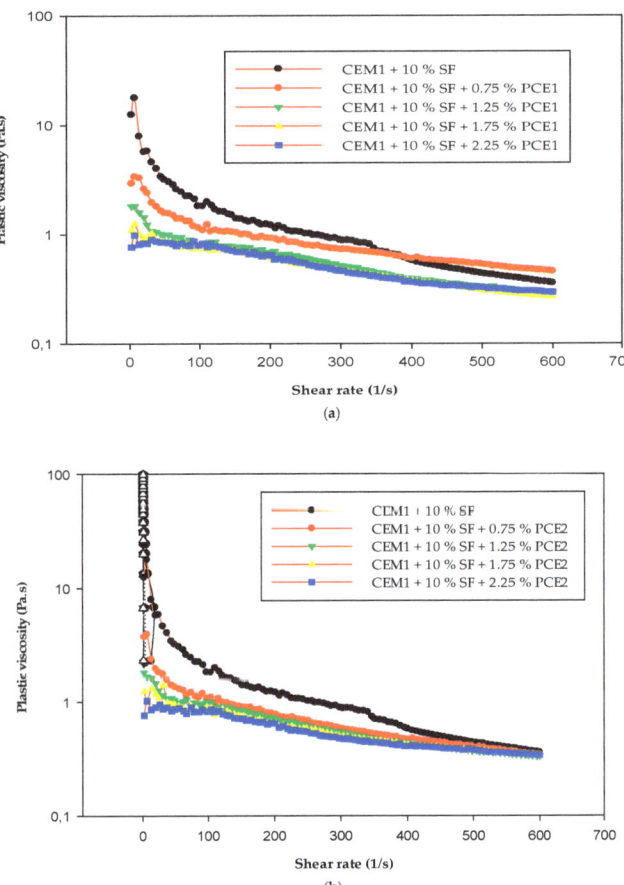

Figure 5. *Cont.*

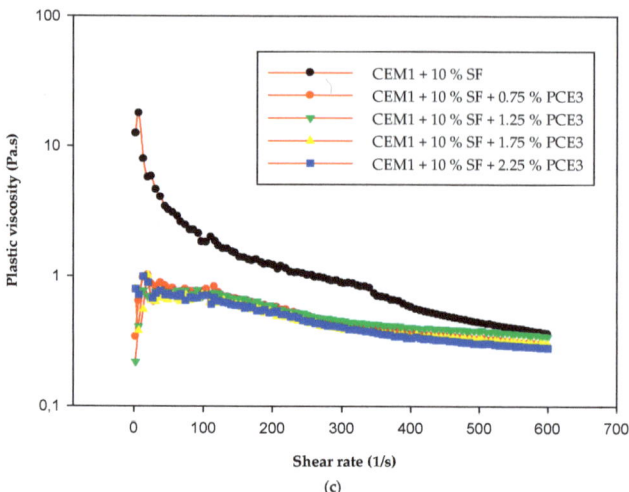

(c)

Figure 5. Plastic viscosity as a function of the shear rate of the cement paste: (**a**) PCE1, (**b**) PCE2, (**c**) PCE3; w/b = 0.35.

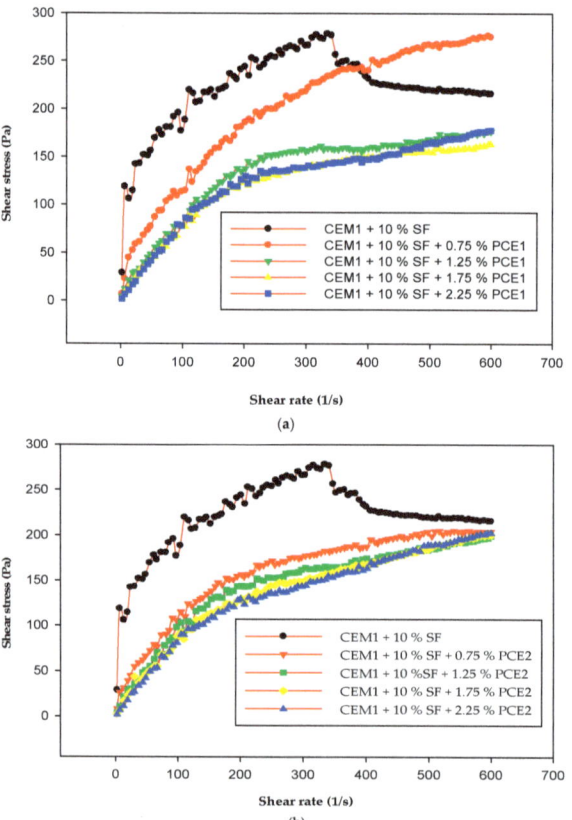

Figure 6. Cont.

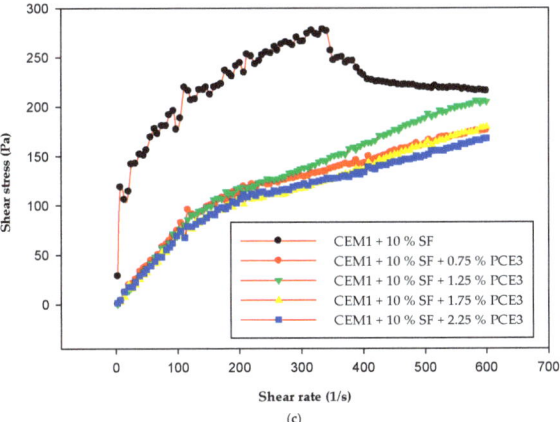

Figure 6. Shear stress as a function of the shear rate of the cement paste: (**a**) PCE1, (**b**) PCE2, (**c**) PCE3; w/b = 0.35.

3.5. Zeta Potential

The zeta potentials of the cement pastes prepared with the addition of the PCEs (with different carboxylic densities and molecular weights) are shown in Figure 7. The order of the zeta potentials is as follows: reference sample < PCE1 < PCE2 < PCE3. This indicates that there is a good correlation between the zeta potential and carboxylic density. The results show that the zeta potential of the cement paste initially exhibited a positive value (3.035 mV) but then changed to a negative value after adding the PCE. This is because there is adsorption of the anionic groups (COO–) from the PCE on the surface of the hydration products of cementitious materials (via electrostatic adsorption by complexation with Ca^{2+}). PCE3 had a high carboxyl content (1.95 mmol/g) and a moderate molecular weight, whereas its polydispersity index (5157 Da, 1.081) presented a high absolute value of zeta potential in the cement paste (9.7735 mV). This is due to its having the best adsorption capacity with regard to cement particles.

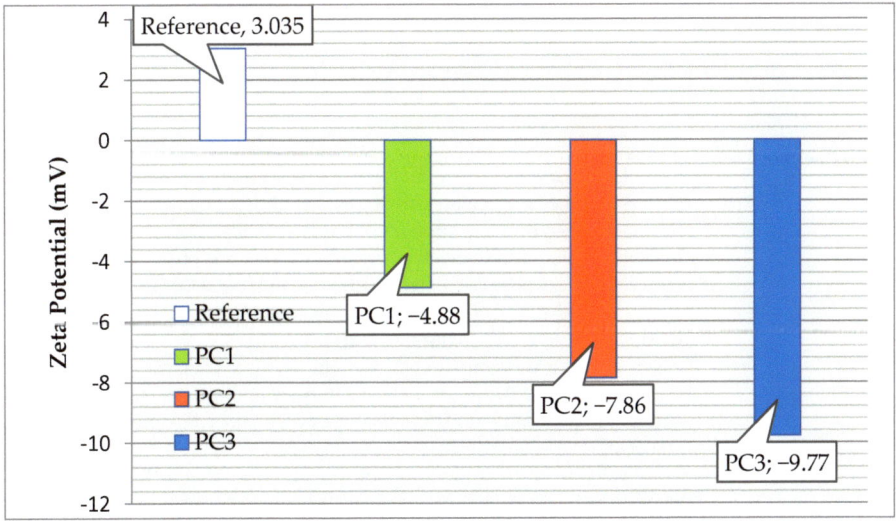

Figure 7. Zeta potential of the cement paste as a function of the PCE type.

3.6. Compressive Strength

The compressive strengths at 3, 7, and 28 days of HPC without superplasticizers are, respectively, 40.2, 45.1, and 50.9 MPa. They are lower than the compressive strength of the HPC with PCEs (Figure 8). Therefore, the three superplasticizers significantly improve the mechanical behavior of HPC. The compressive strength values of the hardened HPC were found to be the highest for the samples containing PCE3. The compressive strength values obtained for these samples, measured after 3, 7, and 28 curing days, were 50, 58.8, and 66 MPa, respectively. Generally, PCEs with high zeta potential exhibit good compressive strength values, which in turn are correlated with better adsorption of superplasticizer molecules [36].

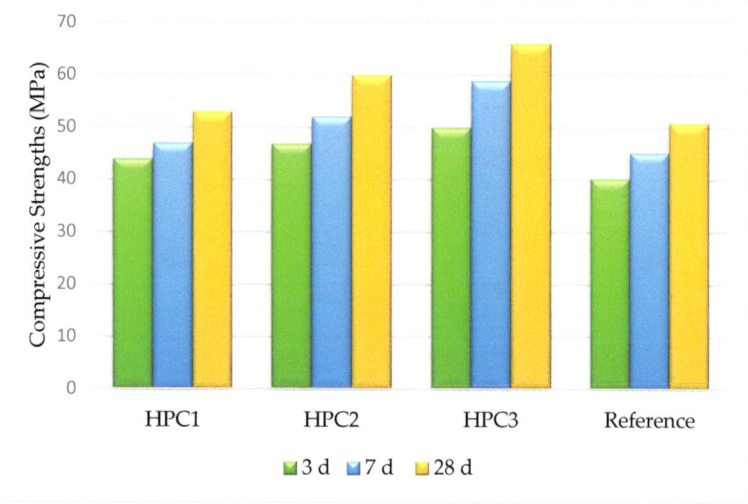

Figure 8. Compressive strengths of the HPC (containing superplasticizers PCE1, 2, and 3) after different times of hydration.

The aim of this work is to study the compatibility of a polymer with silica fume in HPC. The results show that the chemical structure (molecular weight; carboxylic group) of the polymer influences this compatibility. The best chemical structure is a polymer with a moderate molecular weight and high carboxyl density.

4. Conclusions

- The PCE with a moderate molecular weight and the highest content of carboxylic groups has the best dispersion (high value of zeta potential), the lowest viscosity, and the highest compressive strength of hardened HPC. Therefore, it can be considered to be compatible with SF. Such compatibility was not observed in the case of the PCE with the low carboxyl density, which was due to the fact that this PCE was adsorbed on the SF and reduced the ability of the cement particles to disperse. Therefore, the workability of the HPC was also reduced.
- There is a clear relationship between the zeta potential and rheological properties of cement paste and the anionic charge density and molecular weight of PCEs. A good correlation was also found between zeta potential, carboxylic density, and compressive strength.
- The results from the retention slump tests show that a PCE with a high molecular weight can be applied to precast concrete, whose consistency has to be retained over a longer period of time (until this concrete is delivered to a construction site).

- The molecular weight of polycarboxylate superplasticizer has a great impact on the properties of cementitious systems. Therefore, PCE3 can be seen to be a perfect polymer in which all the macromolecules have the same length and the same molar mass (PDI close to 1). This, in turn, makes it the most efficient admixture, which was proved by the results of the research conducted.

Author Contributions: Conceptualization, A.H.; Methodology, A.H.; Formal analysis, P.N.; Investigation, N.T.S. and M.C.; Resources, P.N.; Data curation, N.T.S. and Ł.S.; Writing—original draft, A.H.; Writing—review & editing, M.C. All authors have read and agreed to the published version of the manuscript.

Funding: This research received no external funding.

Institutional Review Board Statement: Not applicable.

Data Availability Statement: Not applicable.

Conflicts of Interest: The authors declare no conflict of interest.

References

1. World Cement Association Urges Climate Action. Available online: https://unfccc.int/news/world-cement-association-urges-climate-action (accessed on 17 February 2023).
2. Mazloom, M.; Ramezanianpour, A.A.; Brooks, J.J. Effect of silica fume on mechanical properties of high-strength concrete. *Cem. Concr. Compos.* **2004**, *26*, 347–357. [CrossRef]
3. Saad, M.; Abo-El-Enein, S.; Hanna, G.; Kotkata, M. Effect of temperature on physical and mechanical properties of concrete containing silica fume. *Cem. Concr. Res.* **1996**, *26*, 669–675. [CrossRef]
4. Plank, J.; Schroefl, C.; Gruber, M.; Lesti, M.; Sieber, R. Effectiveness of Polycarboxylate Superplasticizers in Ultra-High Strength Concrete: The Importance of PCE Compatibility with Silica Fume. *J. Adv. Concr. Technol.* **2009**, *7*, 5–12. [CrossRef]
5. Hommer, H. Interaction of polycarboxylate ether with silica fume. *J. Eur. Ceram. Soc.* **2009**, *29*, 1847–1853. [CrossRef]
6. Meng, W.; Kumar, A.; Khayat, K. Effect of silica fume and slump-retaining polycarboxylate-based dispersant on the development of properties of portland cement paste. *Cem. Concr. Compos.* **2019**, *99*, 181–190. [CrossRef]
7. Meng, W.; Lunkad, P.; Kumar, A.; Khayat, K. Influence of Silica Fume and Polycarboxylate Ether Dispersant on Hydration Mechanisms of Cement. *J. Phys. Chem. C* **2016**, *120*, 26814–26823. [CrossRef]
8. Harichane, A.; Benmounah, A. Influence of Polycarboxylic Ether-based Superplasticizers (PCE) on the Rheological Properties of Cement Pastes. *J. Mater. Eng. Struct.* **2021**, *8*, 325–339.
9. Erzengin, S.G.; Kaya, K.; Özkorucuklu, S.P.; Özdemir, V.; Yıldırım, G. The properties of cement systems superplasticized with methacrylic ester-based polycarboxylates. *Constr. Build. Mater.* **2018**, *166*, 96–109. [CrossRef]
10. Lei, L.; Hirata, T.; Plank, J. 40 years of PCE superplasticizers—History, current state-of-the-art and an outlook. *Cem. Concr. Res.* **2022**, *157*, 106826. [CrossRef]
11. Marchon, D.; Boscaro, F.; Flatt, R.J. First steps to the molecular structure optimization of polycarboxylate ether superplasticizers: Mastering fluidity and retardation. *Cem. Concr. Res.* **2019**, *115*, 116–123. [CrossRef]
12. Chen, S.; Sun, S.; Chen, X.; Zhong, K.; Shao, Q.; Xu, H.; Wei, J. Effects of core-shell polycarboxylate superplasticizer on the fluidity and hydration behavior of cement paste. *Colloids Surf. A* **2020**, *590*, 124464. [CrossRef]
13. Ma, B.; Qi, H.; Tan, H.; Su, Y.; Li, X.; Liu, X.; Li, C.; Zhang, T. Effect of aliphatic-based superplasticizer on rheological performance of cement paste plasticized by polycarboxylate superplasticizer. *Constr. Build. Mater.* **2020**, *233*, 117181. [CrossRef]
14. Harichane, A.; Benmounah, A.; Plank, J. Effect of Molecular Weight and Carboxylic Density of Polycarboxylates Ether Superplasticizer on Its performance in Cement Pastes. *J. Mater. Eng. Struct.* **2023**, *10*, 283–292.
15. Wang, X.; Yang, Y.; Shu, X.; Ran, Q.; Liu, J. Effects of polycarboxylate architecture on flow behaviour of cement paste. *Adv. Cem. Res.* **2021**, *33*, 49–58. [CrossRef]
16. Sha, S.; Wang, M.; Shi, C.; Xiao, Y. Influence of the structures of Polycarboxylate superplasticizer on its performance in cement-based materials-a review. *Constr. Build. Mater.* **2020**, *233*, 117257. [CrossRef]
17. Kai, K.; Heng, Y.; Yingbin, W. Effect of chemical structure on dispersity of polycarboxylate superplasticiser in cement paste. *Adv. Cem. Res.* **2019**, *32*, 456–464. [CrossRef]
18. Ezzat, M.; Xu, X.; El Cheikh, K.; Lesage, K.; Hoogenboom, R.; De Schutter, G. Structure property relationships for polycarboxylate ether superplasticizers by means of RAFT polymerization. *J. Colloid Interface Sci.* **2019**, *553*, 788–797. [CrossRef] [PubMed]
19. Cook, R.; Ma, H.; Kumar, A. Mechanism of tricalcium silicate hydration in the presence of polycarboxylate polymers. *SN Appl. Sci.* **2019**, *1*, 145. [CrossRef]
20. Feng, H.; Feng, Z.; Wang, W.; Deng, Z.; Zheng, B. Impact of polycarboxylate superplasticizers (PCEs) with novel molecular structures on fluidity, rheological behavior and adsorption properties of cement mortar. *Constr. Build. Mater.* **2021**, *292*, 123285. [CrossRef]

21. Li, R.; Lei, L.; Plank, J. Impact of metakaolin content and fineness on the behavior of calcined clay blended cements admixed with HPEG PCE superplasticizer. *Cem. Concr. Compos.* **2022**, *133*, 104654. [CrossRef]
22. Zhang, Q.; Shu, X.; Yang, Y.; Wang, X.; Liu, J.; Ran, Q. Preferential adsorption of superplasticizer on cement/silica fume and its effect on rheological properties of UHPC. *Constr. Build. Mater.* **2022**, *359*, 129519. [CrossRef]
23. Li, P.P.; Yu, Q.L.; Brouwers, H.J.H. Effect of PCE-type superplasticizer on early-age behaviour of ultra-high performance concrete (UHPC). *Constr. Build. Mater.* **2017**, *153*, 740–750. [CrossRef]
24. Park, S.H.; Kim, D.J.; Ryu, G.S.; Koh, K.T. Tensile behavior of Ultra High Performance Hybrid Fiber Reinforced Concrete. *Cem. Concr. Compos.* **2012**, *34*, 172–184. [CrossRef]
25. Wang, M.; Yao, H. Effects of polycarboxylate ether grafted silica fume on flowability, rheological behavior and mechanical properties of cement-silica fume paste with low water-binder ratio. *Constr. Build. Mater.* **2021**, *272*, 121946. [CrossRef]
26. Zhang, Y.; Kong, X. Correlations of the dispersing capability of NSF and PCE types of superplasticizer and their impacts on cement hydration with the adsorption in fresh cement pastes. *Cem. Concr. Res.* **2015**, *69*, 1–9. [CrossRef]
27. Zhang, Y.-R.; Kong, X.-M.; Lu, Z.-B.; Lu, Z.-C.; Hou, S.-S. Effects of the charge characteristics of polycarboxylate superplasticizers on the adsorption and the retardation in cement pastes. *Cem. Concr. Res.* **2015**, *67*, 184–196. [CrossRef]
28. Yang, H.; Plank, J.; Sun, Z. Investigation on the optimal chemical structure of methacrylate ester based polycarboxylate superplasticizers to be used as cement grinding aid under laboratory conditions: Effect of anionicity, side chain length and dosage on grinding efficiency, mortar workability and strength development. *Constr. Build. Mater.* **2019**, *224*, 1018–1025.
29. Schröfl, C.; Gruber, M.; Plank, J. Preferential adsorption of polycarboxylate superplasticizers on cement and silica fume in ultra-high performance concrete (UHPC). *Cem. Concr. Res.* **2012**, *42*, 1401–1408. [CrossRef]
30. Ji, Y.I.; Cahyadi, J.H. Effects of densified silica fume on microstructure and compressive strength of binary cement pastes. *Cem. Concr. Res.* **2003**, *33*, 1543–1548.
31. Mitchell, D.R.G.; Hinczak, I.; Day, R.A. Interaction of silica fume with calcium hydroxide solutions and hydrated cement pastes. *Cem. Concr. Res.* **1998**, *28*, 1571–1584. [CrossRef]
32. Ran, Q.; Liu, J.; Yang, Y.; Shu, X.; Zhang, J.; Mao, Y. Effect of molecular weight of polycarboxylate superplasticizer on its dis-persion, adsorption, and hydration of a cementitious system. *J. Mater. Civ. Eng.* **2016**, *28*, 04015184. [CrossRef]
33. Papo, A.; Piani, L. Effect of various superplasticizers on the rheological properties of Portland cement pastes. *Cem. Concr. Res.* **2004**, *34*, 2097–2101. [CrossRef]
34. Hallal, A.; Kadri, E.; Ezziane, K.; Kadri, A.H.; Khelafi, H. Combined effect of mineral admixtures with superplasticizers on the fluidity of the blended cement paste. *Constr. Build. Mater.* **2010**, *24*, 1418–1423. [CrossRef]
35. Ferrari, L.; Kaufmann, J.; Winnefeld, F.; Plank, J. Multi-method approach to study influence of superplasticizers on cement suspensions. *Cem. Concr. Res.* **2011**, *41*, 1058–1066. [CrossRef]
36. Plank, J.; Sachsenhauser, B. Impact of molecular structure on zeta potential and adsorbed conformation of α-allyl-ω-methoxypolyethylene glycol-maleic anhydride superplasticizers. *J. Adv. Concr. Technol.* **2006**, *4*, 233–239. [CrossRef]

Disclaimer/Publisher's Note: The statements, opinions and data contained in all publications are solely those of the individual author(s) and contributor(s) and not of MDPI and/or the editor(s). MDPI and/or the editor(s) disclaim responsibility for any injury to people or property resulting from any ideas, methods, instructions or products referred to in the content.

Review

The Elasticity of Polymer Melts and Solutions in Shear and Extension Flows

Andrey V. Subbotin [1,2], Alexander Ya. Malkin [1,*] and Valery G. Kulichikhin [1]

1. A.V. Topchiev Institute of Petrochemical Synthesis, Russian Academy of Sciences, Leninskii prosp. 29, Moscow 119991, Russia
2. A.N. Frumkin Institute of Physical Chemistry and Electrochemistry, Russian Academy of Sciences, Leninskii prosp. 31, Moscow 119071, Russia
* Correspondence: alex_malkin@mig.phys.msu.ru

Abstract: This review is devoted to understanding the role of elasticity in the main flow modes of polymeric viscoelastic liquids—shearing and extension. The flow through short capillaries is the central topic for discussing the input of elasticity to the effects, which are especially interesting for shear. An analysis of the experimental data made it possible to show that the energy losses in such flows are determined by the Deborah and Weissenberg numbers. These criteria are responsible for abnormally high entrance effects, as well as for mechanical losses in short capillaries. In addition, the Weissenberg number determines the threshold of the flow instability due to the liquid-to-solid transition. In extension, this criterion shows whether deformation takes place as flow or as elastic strain. However, the stability of a free jet in extension depends not only on the viscoelastic properties of a polymeric substance but also on the driving forces: gravity, surface tension, etc. An analysis of the influence of different force combinations on the shape of the stretched jet is presented. The concept of the role of elasticity in the deformation of polymeric liquids is crucial for any kind of polymer processing.

Keywords: polymers; viscoelasticity; shear; extension; jet

Citation: Subbotin, A.V.; Malkin, A.Y.; Kulichikhin, V.G. The Elasticity of Polymer Melts and Solutions in Shear and Extension Flows. *Polymers* **2023**, *15*, 1051. https://doi.org/10.3390/polym15041051

Academic Editors: Célio Pinto Fernandes, Luís L. Ferrás and Alexandre M. Afonso

Received: 3 February 2023
Revised: 17 February 2023
Accepted: 18 February 2023
Published: 20 February 2023

Copyright: © 2023 by the authors. Licensee MDPI, Basel, Switzerland. This article is an open access article distributed under the terms and conditions of the Creative Commons Attribution (CC BY) license (https://creativecommons.org/licenses/by/4.0/).

1. Introduction

Elasticity is obviously an inherent property of polymers. This property determines the huge usage of polymers in rubber industry. In this area, two approaches are naturally combined: the mechanics of large reversible deformations and the physics of interactions and deformations at the molecular level.

However, what is the level of our understanding of the role of elasticity in the flow of polymer solutions and melts? It is intuitively clear that in shear and tension there is a superposition of reversible and irreversible deformations, which is formulated in many constitutive equations proposed for polymer melts and solutions. Nevertheless, the role of elasticity is important not only for the flow, but for the emergence of new unexpected effects associated with the elastic instability of elastic liquids. Therefore, in this review, we wanted to collect and describe those phenomena that are directly caused by the elasticity of polymeric liquids. At the same time, as in the case of rubbers, we wanted to collect both macroscopic experimental facts and phenomena related to the orientation of individual macromolecules under one roof, which is especially important for the expansion of polymeric liquids.

There are several basic concepts in rheology that form its foundation. These are non-Newtonian flow, viscoelasticity, thixotropy, and viscoplasticity. All of these concepts have been the subject of extensive research, summarized in a large number of monographs and reviews. The elasticity of rheologically complex liquids that is inherent in solutions and melts of polymers to the greatest extent was usually considered a special case of viscoelasticity. However, consideration of the behavior of polymeric liquids demonstrates various effects associated specifically with the elasticity of these media. Next, we will try to review the current state of research in this area, not limited to the effects associated with the elasticity of polymer solutions and melts: the physics of macromolecular deformations responsible for the observed phenomena would also be considered. The analysis of the elasticity of polymeric liquids should be based on the existing fundamental concepts in polymer physics in order to have a common approach to understanding any new experimental fact.

These are:

- The ratio of the scales of internal time and observation time called the Deborah number, De, (Reiner, 1928 [1]), and the ratio of the scales of relaxation rate to deformation rate called the Weissenberg number, Wi [2];
- The consequence of this fundamental approach—the concept of the time (frequency)—temperature superposition approves that the same type of relaxation state (or rheological behavior) can be reached by varying either the rate of deformation (frequency) or temperature (Ferry [3]), Tobolsky [4]);
- The concept of the transition from linear to non-linear mechanical behavior in increasing the deformation rate [5,6].

The following pictures (Figure 1) illustrate these concepts for the domains of linear and non-linear viscoelasticity of polymers.

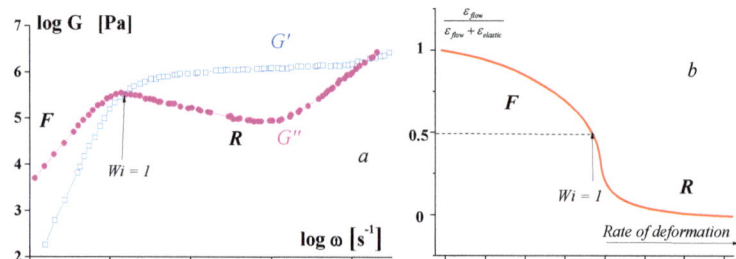

Figure 1. Typical behavior of polymer melts. *a*: relaxation states of polymers: frequency dependencies of the storage G' and loss moduli G'', F—flow (terminal) and R—rubbery states. *b*: ratio between flow and elastic deformations [7].

As for the transition to non-linearity in extension, Figure 2 shows a typical graph. The solid black line in this Figure corresponds to the linear limit of viscoelastic behavior and the colored lines show deviation of the linearity, higher rates correspond to lower deviation times [5,8].

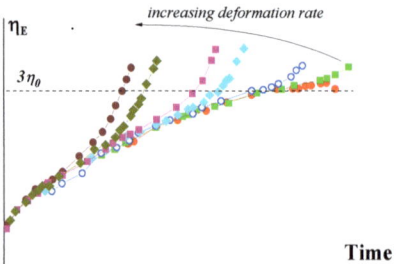

Figure 2. The linear-to-non-linear behavior transition in increasing the deformation rate under extension.

Under no circumstance can the ratio be considered as some "apparent elongation viscosity" since no point in this graph corresponds to a steady flow (which is an obligatory condition in the definition of "viscosity") and moreover, it may generally correspond to a flow-to-elastic deformation transition. This mistakable approach sometimes appears in discussions of experimental data.

Some recent reviews on the rheology of extension were published [9–11]. However, this line of research was outlined very quickly and the new experimental data and theoretical arguments have already accumulated and require analysis. This was conducted in a review based mainly on the publications of the last 5–7 years.

Peculiarities of extension of polymeric liquids have rather significant value when we consider either melts or dilute (or semi-dilute) solutions. This is due to the different technological applications of these two groups of liquids. In the first case, we meet with the processing of polymers mainly by molding or extrusion. In the second case, we deal mainly with fiber spinning. Blow molding of films occupies the intermediate position. Therefore, it seems reasonable to separate this review into two parts devoted to melts and solutions.

2. The Role of Elasticity in Polymer Processing

Any deformation in the polymer processing inevitably leads to molecular orientation that can be treated in terms of elasticity (stored energy) wherein normal stresses (under the extension flow) create a much higher effect than shear stresses (under the shear flow) that is clearly demonstrated in the simplest model of the deformation of a liquid drop inside a surrounding liquid under different modes of the flow [12–14]. Elastic (recoverable) force for a liquid droplet is surface tension while for polymers and in particular for polymer blends, the source of elasticity is molecular motion directed to the recovery of the equilibrium conformation and having the statistic (entropic) nature. Therefore, the inherent link between elasticity and elongation flow of viscoelastic polymeric liquids (solutions or melts) exists.

In this section, we will consider the effects due to the elasticity of polymeric liquids, which are observed when flowing through capillaries (channels). In this regard, the original study [15] complements the presented review.

The role of elasticity associated with the technological practice of spinning fibers and blow molding is of independent interest. This is a separate topic that is beyond the scope of this review. Some individual aspects of this problem are considered in [16,17]. Molecular understanding of this issue will be considered in the next part of this paper.

The role of extension is the most pronounced and important in shear flows through channels with variable cross-sections due to an obligatory change in the elongation velocity (i.e., the emergence of the rate of deformation). An evident example of such a case is extrusion. Figure 3 shows how velocity changes in the flow from a barrel with polymer melt, through a forming die (perhaps simply through a capillary), and in post-extrusion operations. There are two zones: the entrance into the die and the exit from it where elongation flow takes place.

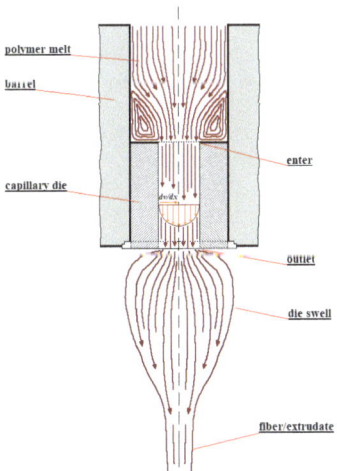

Figure 3. The velocity distribution in the flow through a cylinder die (capillary).

As was said above, large deformations in the elongation flow of viscoelastic polymeric liquids relate to their elasticity, which manifests itself in the transition flows taking place at the entrance and exit zones. Figure 4 illustrates two characteristic effects related to the flow of polymeric liquids in these zones: the emergence of secondary flow (Figure 4a) and die swelling (Figure 4b) [18]. The quantitative manifestations of these effects depend on the deformation conditions and the nature of the polymer (see, for example, [19,20])

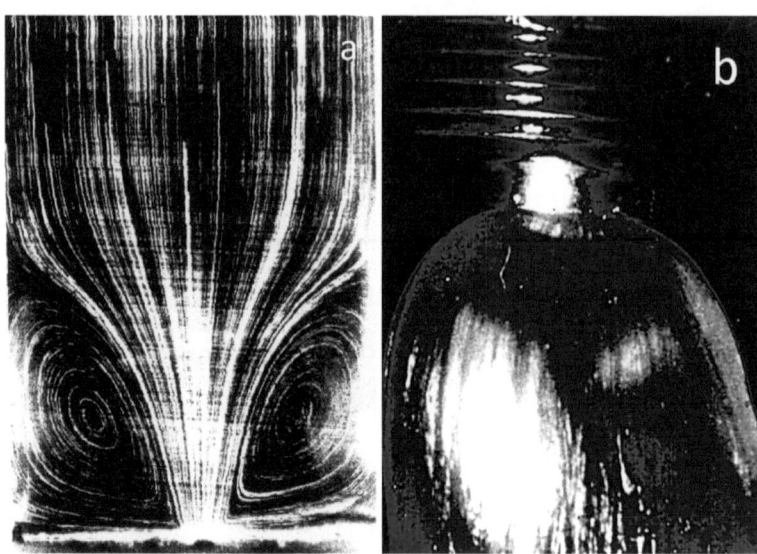

Figure 4. Typical effects associated with the elasticity of polymer melts: secondary flows at the entrance to a die (**a**) [18] and die swell after exit from a die (**b**) (authors' photo).

The role of these two zones becomes dominant for short dies, since it is in these cases that the main part of losses (energy dissipation) happens due to transient (viscoelastic) regimes of the flow associated with extension. This suggestion was confirmed by the possibility of considering the dimensional pressure losses as a universal function of the Deborah number, De [21], where De is defined as the ratio of characteristic time of the segmental movement of macromolecules to the residence time in the channel. The generalized result of this approach is shown in Figure 5. Here, the shear stress is reduced by the plateau modulus G in the rubbery, R, zone and De is determined via the relaxation time in the flow, F (terminal), zone (as in Figure 1a). The existence of a common dependence on the shear stress on the characteristic relaxation time (built in the reduced coordinates) indicates the decisive role of elasticity in the flow through short dies.

Secondary flows are a phenomenon first examined by L. Prandtl (1926) for Newtonian liquids, which are characterized in terms of the cross-plane component of the mean kinetic energy. Nowadays this effect is primarily considered by numerical methods [22,23]. This problem has received the new content for viscoelastic fluids, in which the Reynolds number is not the determining factor, and the elasticity of the liquid is expressed by the Weissenberg number [24–26]. Secondary flows in elastic liquids appear at very low Reynolds numbers and are associated with elastic turbulence. Their quantitative description depends on the choice of the rheological model due to the different approaches for the characterization of elastic non-linearity, and solutions of dynamic problems (such as in a simpler case of Newtonian liquids) are reached using numerical analysis methods. In the limits of this review, it is essential that secondary flows always increase the hydrodynamic resistance [27].

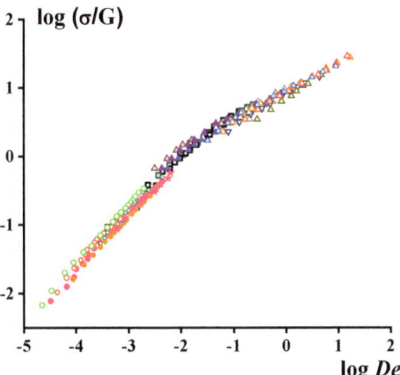

Figure 5. The reduced dependence of the shear stress on the Deborah number for different polymer melts and solutions—PAN solutions in DMSO, LDPE, polybutadiene, SAN, solutions of PIB in toluene—presented by different symbols (according to [21]).

In rheological measurements and in technological practice, the total entrance phenomena (including secondary flows and extension) are usually characterized by an end correction, n, as a measure of some conventional increase in the length of the die. Then, the true shear stress, σ, at the wall of the capillary is expressed as

$$\sigma = \frac{\Delta P R}{2(L + nR)} = \frac{\Delta P}{2(L/R + n)} \tag{1}$$

where ΔP is the difference in pressure at the entrance and exit of a capillary and R and L are the radius and the length of a capillary, respectively.

Sometimes, the gradient of the elongation velocity is used for calculating a conditional "elongation viscosity" $\eta^+ = \sigma_E/\dot{\varepsilon}$, although (as explained above) this value cannot be treated as "viscosity". Based on experimental evidence, it was shown that this value does not have any reasonable meaning for the capillary flows of viscoelastic polymer melts [28].

The end correction (summing all entrance additional energy losses) depends on the viscoelastic properties, and its relative contribution to the total pressure loss is determined by the ratio between L/R and n, i.e., the capillary length (as seen from Equation (1)). In this sense, in addition to the experimental data presented in Figure 5, the n(Wi) dependence is shown in Figure 6.

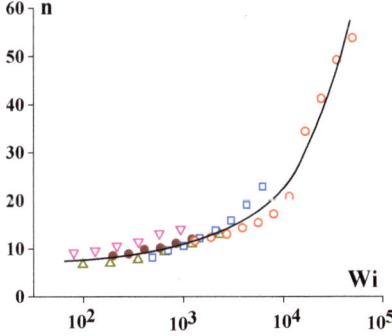

Figure 6. Dependence of end correction on the Weissenberg number. Different symbols correspond to various compositions of low- and high- molecular-weight polyethylenes (according to [28]).

The physics behind the increase in end correction alongside an increase in the *Wi* number is most likely related to the increase in elastic deformations depending on shear stress (or shear rate), but not with hydrodynamic reasons as in Cogsell's model. This occurs simultaneously with the development of the non-Newtonian effect. It is then reasonable that elasticity correlates with the degree of non-Newtonian effect. Indeed, it was found that the ratio of the apparent viscosity, $\eta(\dot{\gamma})$, to the initial (maximal) Newtonian viscosity, η_0, can be considered as a function of the stored elastic energy, *W*, i.e., the elasticity of the liquid [29] and expressed as

$$\frac{\eta(\dot{\gamma})}{\eta_0} = e^{\beta W/RT} \qquad (2)$$

where *R* is the universal gas constant, *T* is absolute temperature, and β is an individual constant of an elastic liquid. The stored elastic energy is calculated as

$$W(\dot{\gamma}) = \int_0^{\gamma_{el}} \sigma(\dot{\gamma}) d\gamma_{el} \qquad (3)$$

Here, $\sigma(\dot{\gamma})$ is the dependence of the shear stress, σ, on the shear rate, $\dot{\gamma}$, in a stationary flow (i.e., the flow curve) and $\gamma_{el} = \gamma_{el}(\dot{\gamma})$ is the elastic deformation at this shear rate. Independent measurements of the values in the right and left sides of Equation (2) for many different polymeric liquids confirmed the correction of this relationship.

Elastic deformations stored at the inlet due to the convergent flow relax after leaving the capillary and lead to the die swelling (Figure 4b). This phenomenon depends on the capillary length since the additional stresses at least partially relax when passing through the capillary, and thus the jet diameter depends on the value of *L/R*. This is illustrated by the photo (Figure 7) wherein two jets obtained at the same given volume output are shown but with long (left) and short (right) dies [28].

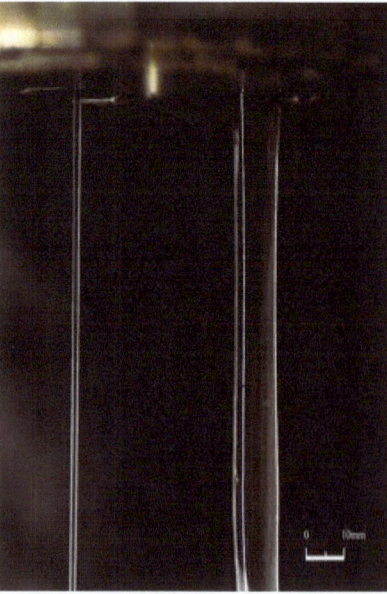

Figure 7. Comparison of two jects obtained at the same volume output but with different lengths of capillaries, long (**left**) and short (**right**), authors' photo.

The physics behind the die swell is rather evident. This is because of the release of stored elastic energy due to the recoverable conformation of the entangled polymeric chains. As said above, the quantitative measure of this effect depends on the length of the die (capillary). The die swell should be taken into account in the design of the processing equipment [30], and this is especially important when designing spinnerets in fiber spinning since the dies in these devices are always very short.

The mechanics of die swell was considered in many publications by numerical methods based primarily on the analysis of rather complex rheological equations [31,32]. This means that the theoretical base for such calculations is ready. However, its practical application requires knowledge of a large amount of information about the rheological properties of the processed polymer. Then, it is reasonable to apply the theoretical models in large-scale industrial production. Therefore, the problem of measuring die swell continues and many authors try to carry out the direct measurement for certain applications in extrusion polymer profiles [33,34]. A new aspect of this problem is associated with the extrusion of filaments for 3D printing (additive technology) [35,36].

The effect of extension (longitudinal deformations) at the capillary entrance becomes rather evident when we observe the flow of two liquids forming emulsions. This takes place in the flow of blends of immiscible polymer melts. The results of a model experiment are shown in Figure 8, where the deformation of a liquid droplet during the transition from a wide to narrow channel is illustrated [37].

Figure 8. Successive stages of the deformation of a drop during the transition from a wide volume to a narrow capillary. Initial (**a**), intermediate (**b**), and final (**c**) stages of the process (reproduced from [37] with permission.

The effect of self-oscillation during the high-speed extrusion of polymer melts (Figure 9) is a well-known and well-documented phenomenon [38]. Its physics, associated with the elastic rupture at the point of singularity on the exit section of a capillary, was qualitatively described by the old Cogswell model [39]. Self-oscillations represent the initial stage of instability in the flow of viscoelastic fluids. A detailed consideration of this issue is presented in the review [40].

This effect can be understood as an analog of the spurt effect. At high shear stresses, melt becomes elastic, and the edge of the capillary exit plays a role of a scrapper sliding along the rubbery surface (Figure 10). This is a reverse picture of the movement of a rubbery-like melt at the edge of the capillary. Although the appearance of periodic surface self-oscillating defects on the surface of extrudates (also known as the shark skin) associated with elastic ruptures at a singular point was described in many publications, its quantitative theory is still absent.

The next case of the influence of elasticity on the shear flow is associated with the interaction of a polymer fluid with the wall. This is a well-known effect of spurt associated with wall slip. This is also the case with using a rather mild measuring system in rotary rheometers [41]. A rigorous solution of this problem would be also interesting.

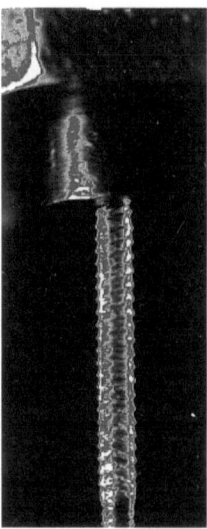

Figure 9. Self-oscillations at the exit of a capillary.

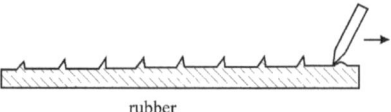

Figure 10. Self-oscillations due to the movement of a scribe along the rubbery surface.

We can also observe the consequences of elastic deformations of polymer melts preserving frozen residual stresses and shape memory in articles obtained by molding or extrusion. It is quite evident that residual stresses and memory effects are associated with orientation depending on the stress in shear flow (see, e.g., Figure 25 in [42]) and related to its elasticity. One can find a description of frozen stresses, their direct observations, and the influence of this phenomenon on the performance of final products in numerous publications (e.g., [43–45]) and there is no doubt in its practical importance [46]. However, the general theoretical model of this phenomenon is absent although some attempts for stimulating calculations for frozen stresses are known [47]. One of the earlier attempts to solve this problem, by constructing a rigorous system of equations, clearly demonstrated the correct way to do this as well as numerous difficulties encountered [48]. Indeed, the practical application of thermoviscoelastic problems requires a large amount of experimental information about the temperature dependences of rather complicated rheological properties of the material, as well as overcoming computational difficulties in solving a system of non-linear or integral equations. Nevertheless, we can be sure that this problem will attract the attention of professionals due to its practical importance in the processing and application of engineering plastics.

Although the elasticity of polymer liquids is their inherent property and is important at all stages of the traditional processing, it is extension that is the mode of deformation, where elasticity plays a decisive role due to direct correlation with the orientation of macromolecules and the influence of this factor on the technical properties of spun fibers. In fact, the extension of fibers in different stages of the technological process occurs due to the uncoiling of macromolecules associated with elastic deformations. Then, the increase in ultimate strain (at break) λ^*, as well as the strength of a matter, correlates with the draw rate. This should be an elastic drawing and a further increase in the draw rate can lead to the deformation-induced glass transition with a decrease in λ^*. This is shown in Figure 11, and the shape of the curve is obviously similar to the right part of the envelope curve in Figure 2.

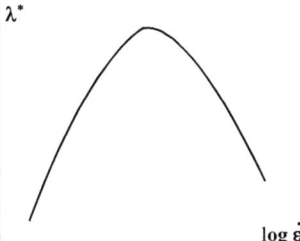

Figure 11. Correlation between the draw rate and the ultimate strain.

The universal modeling of extension presented in Figure 2 is in accordance with the generalized model [6] and is valid for homogeneous stretching. The extension of polymeric liquids in the elastic domain of deformations can occur with neck formation similarly to necking in solid polymers and further stretching entails the elastic yielding with the transition of homogeneous filament to the neck. This yielding effect happens at critical strain and finally results in the elastic breakup at non-uniform extension [49]. The concept of necking under extension was analyzed by stability analysis of the stretching process, which made it possible to obtain criteria of this effect corresponding to the existing phenomenological models of a non-linear viscoelastic liquid [50]. In some publications, the appearance of a plateau in the stress vs. deformation curve near the breakup point was described. This plateau takes place for linear polymers [51] as well as for ring macromolecules with an unusual sharp increase in the apparent viscosity [52]. The nature of this effect is not evident. The authors of original publications treat it as flow, although possibly this yielding happens due to the necking phenomenon.

Modeling the polymeric liquid bridges leading to failure during extension was discussed in [53]. However, it is necessary to keep in mind that the mechanism of fracture of polymeric liquids under tension depends on the draw rate, since the latter is determined by the relaxation state of the polymer and in particular the liquid flow-to-elastic state transition.

At rather low draw ratios, the classical Rayleigh–Plateau breakup of liquid jets due to surface tension disturbances is observed [54]. In the rubber-like state (in the medium range of draw rate), the breakup occurs quite in the same mode as for usual rubbers. However, at high deformation rates, the breakup is initiated by the appearance of the simultaneous propagation of multiple cracks while their position is random, similarly to how it happens in the rupture of different solids [55,56].

A rather different understanding of the mechanism of rupture in extension was proposed in [57,58]. The authors assume that there are only two different states associated with the regimes of deformations: the liquid and elastic solid. They proposed the cohesive failure model based on the entropic fracture hypothesis. According to this model, the rupture of the bond in the main macromolecular chain was assumed as the basic mechanism of the brittle breakup of a filament. This approach was criticized in [59], where contrary to the hypothesis that chains are fully uncoiled and scission in melt rupture is due to an "entropic fracture" mechanism, it was declared that sufficient enthalpic changes associated with conformational distortions at the bond level take place.

3. Elasticity in the Dynamics of Extension of Polymer Solutions
3.1. General Equations

In the previous part, we considered the features of the elastic behavior of polymeric liquids during technological processing, including both shear and extension. In recent years, significant progress has also been made in studying polymer solution behavior under extension. In this part, we mainly focus on elucidating the role of elasticity using the theoretical methods. We will consider two cases: a thread (bridge) connecting two droplets and self-thinning under the action of capillary forces (Figure 12), and a stationary jet stretched under the action of an external force after the solution leaves the orifice with a fixed flow rate. The external force can be mechanical and applied to the free end of the jet, as in the case of fiber drawing (Figure 13), as well as gravitational or electrical (electrospinning). In the latter case, the force is applied to the entire jet.

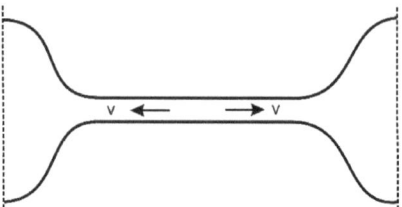

Figure 12. Schematic picture of a thread (bridge) connecting two droplets.

Figure 13. Stretching of a polymer solution jet by an external force F_{ext} during continuous fiber spinning.

First, let us formulate the basic equation for the balance of forces in the volume of liquid in the general case, including the inertial, viscoelastic, gravitational, and electrostatic forces. Denoting the density of the solution as ρ and assuming that the electric field inside the fluid is $\mathbf{E}_i(\mathbf{x}, t)$ and the velocity is $\mathbf{v}(\mathbf{x}, t)$, the momentum equation is written as [54,60–62]

$$\rho \frac{\partial \mathbf{v}}{\partial t} + \rho \mathbf{v} \cdot \nabla \mathbf{v} - \nabla \cdot (\Sigma - p\mathbf{I}) - \rho \mathbf{g} - q\mathbf{E}_i = 0 \quad (4)$$

where the velocity $\mathbf{v}(\mathbf{x}, t)$ obeys the incompressibility condition $\nabla \cdot \mathbf{v} = 0$. Here $\mathbf{x} = (x_1, x_2, x_3)$ is the coordinate, t is the time, p is the pressure, ∇ is the gradient operator, Σ is the stress tensor, $\rho \mathbf{g}$ is the gravity force density, $q\mathbf{E}_i$ is the electric force density, and $(q = e(n_+ - n_-)$ is the free charge density, where n_+ and n_- are concentrations of positively and negatively charged monovalent ions having the charge e respectively) and $\mathbf{I}$ is the unit tensor. The differential Equation (4) should be supplemented by the boundary condition on the free surface. This condition implies the balance of the viscoelastic, capillary, and electric forces.

$$p_s \mathbf{n} - \Sigma \cdot \mathbf{n} - \gamma C \mathbf{n} + \mathbf{F} = 0 \tag{5}$$

Here, p_s is the pressure at the surface, $\mathbf{n}$ is a normal vector to the surface, $C = \text{div}\,\mathbf{n}$ is the total surface curvature, and γ is the surface tension. The electric force $\mathbf{F}$ acting per unit area is given by:

$$\mathbf{F} = \varepsilon_0 (E_{o,n}\mathbf{E}_o - \varepsilon E_{i,n}\mathbf{E}_i) - \frac{\varepsilon_0}{2}\left(E_o^2 - \varepsilon E_i^2\right)\mathbf{n} \tag{6}$$

where $\mathbf{E}_o$ is the electric fields outside the liquid. The electric fields $\mathbf{E}_i$ and $\mathbf{E}_o$ are found from the electrostatic equations [60–63].

To study the dynamics of the rectilinear jet (thread), a cylindrical system of coordinates will be used. Assuming that the jet surface is described by an axisymmetric function $a = a(z,t)$, the normal (**n**) and tangential ($\boldsymbol{\tau}$) vectors to the surface are given by:

$$\mathbf{n} = -\frac{a'_z}{\sqrt{1+a'^2_z}}\mathbf{e}_z + \frac{1}{\sqrt{1+a'^2_z}}\mathbf{e}_r,\ \boldsymbol{\tau} = \frac{1}{\sqrt{1+a'^2_z}}\mathbf{e}_z + \frac{a'_z}{\sqrt{1+a'^2_z}}\mathbf{e}_r \tag{7}$$

Here, $\mathbf{e}_z$ and $\mathbf{e}_r$ are the unit vectors directed along and perpendicular to the jet axis, respectively.

Analysis of the three-dimensional momentum, Equation (4), with the boundary conditions (5) is a very difficult mathematical problem. In the case of axisymmetric rectilinear jet, the problem can be simplified using a slender body approximation since the profile of the jet slowly changes along the extension axis z, $|a'_z| \ll 1$ ($a'_z = \frac{\partial a}{\partial z}$). To derive the corresponding one-dimensional momentum equation, let us consider the jet section $[z, z+dz]$ [64]. After multiplying the Equation (1) by the vector $\mathbf{e}_z$ and integration over this section, we obtain the equation for the velocity component v_z:

$$\int_0^{a(z,t)} r dr \left(\rho \frac{\partial v_z}{\partial t} - \rho g - qE_z\right) + \frac{\partial}{\partial z}\left[\int_0^{a(z,t)} r dr \left(\rho v_z^2 + p - \Sigma_{zz}\right)\right] + a\sqrt{1+a'^2_z}(\rho \mathbf{v}\mathbf{v} + p_s\mathbf{I} - \Sigma) \cdot \mathbf{n}\mathbf{e}_z = 0 \tag{8}$$

where the gravity acts along the z-axis. The incompressibility condition $\nabla \cdot \mathbf{v} = 0$ after integration over the section $[z, z+dz]$ and subsequent use of the kinematic equation $\frac{\partial a}{\partial t} + v_z a'_z - v_r = 0$ reduces to the mass conservation equation:

$$\frac{\partial a^2}{\partial t} + \frac{\partial}{\partial z}\left(a^2 v_z\right) = 0 \tag{9}$$

Elimination of the pressure from Equation (8) (note, within the framework of the slender body approximation $p \simeq p_s$) by taking into account the boundary condition (5) we arrive at the well-known form of the momentum equation:

$$\frac{\partial}{\partial t}\left(\rho a^2 v_z\right) + a^2 \frac{\partial}{\partial z}(\gamma C - F_n) + \frac{\partial}{\partial z}\left[a^2 \left(\rho v_z^2 + \Sigma_{nn} - \Sigma_{zz}\right)\right] \simeq 2aF_\tau + qa^2 E_z + \rho g a^2 \tag{10}$$

where $F_\tau = \mathbf{F}\cdot\boldsymbol{\tau}$, $F_n = \mathbf{F}\cdot\mathbf{n}$ and $E_z = \mathbf{E}\cdot\mathbf{e}_z$. Furthermore, depending on the system under consideration, it is necessary to determine the stress tensor, and in the case of electrospinning, to add electrostatic and charge balance equations.

For the Newtonian liquid the stress tensor is $\Sigma = \eta\left(\nabla\mathbf{v} + (\nabla\mathbf{v})^T\right)$ where η is the viscosity and

$$(\nabla\mathbf{v})_{ij} = \frac{\partial v_j}{\partial x_i},\ (\nabla\mathbf{v})^T_{ij} = \frac{\partial v_i}{\partial x_j},\ i,j = 1,2,3 \tag{11}$$

are the velocity gradients. When considering polymer solutions exhibiting viscoelastic behavior, additional equations are required to determine the stress tensor [64–66]. Two approaches are possible here: phenomenological and molecular. In the phenomenological approach, the stress tensor is determined by the constitutive equation. The Maxwell, Oldroyd B, and FENE-P equations are often used to describe extension of polymer solutions without entanglements. The polymer chains in these rheological models are described by elastic dumbbells with constant friction. The most general is the FENE-P model. It captures the viscoelastic effects, as well as those in strong elongational flows when the finite extensibility of the polymer chains is important [64,65]. The polymer chain is modeled by a non-Hookean dumbbell with the extension force $\mathbf{f} = \frac{3k_B T}{R_0^2} \frac{\mathbf{R}}{1 - \mathbf{R}^2/L^2}$ which is related to the elastic energy $F_{el} = -\frac{3k_B T L^2}{2R_0^2} \ln\left(1 - \frac{\mathbf{R}^2}{L^2}\right)$. Here, k_B is the Boltzmann constant, T is the temperature, $\mathbf{R}$ is the distance between the beads, L is the maximum spring length, and $R_0^2 \propto L$ is the mean-square equilibrium distance between the beads.

The FENE-P model equations are formulated in terms of a conformation tensor $\mathbf{A} = \langle \mathbf{RR} \rangle$ where the angular brackets denote averaging over the distribution of the vector $\mathbf{R}$. The stress tensor $\mathbf{\Sigma}$ is a sum of the solvent stress $\mathbf{\Sigma}_s = \eta_s \left(\nabla \mathbf{v} + (\nabla \mathbf{v})^T \right)$ where η_s is the solvent viscosity and the polymer stress $\mathbf{\Sigma}_p$:

$$\mathbf{\Sigma} = \mathbf{\Sigma}_s + \mathbf{\Sigma}_p, \quad \mathbf{\Sigma}_p = G \frac{\mathbf{A}/R_0^2 - \mathbf{I}}{1 - \mathrm{tr}\mathbf{A}/L^2} \quad (12)$$

with $\mathbf{A}$ obeying

$$\tau \left[\frac{\partial \mathbf{A}}{\partial t} + (\mathbf{v} \cdot \nabla)\mathbf{A} - (\nabla \mathbf{v})^T \cdot \mathbf{A} - \mathbf{A} \cdot \nabla \mathbf{v} \right] + \frac{\mathbf{A} - R_0^2 \mathbf{I}}{1 - \mathrm{tr}\mathbf{A}/L^2} = 0 \quad (13)$$

Here, the elastic modulus is $G = 3nk_B T$ where is the concentration of polymer chains (springs) and τ is the relaxation time. The linear viscosity of the polymer component is expressed by means of the scaling relation $\eta_p \simeq G\tau$. The Oldroyd B model assumes infinite extensible polymer chains ($L \to \infty$) and the Maxwell model also does not take into account the solvent ($\mathbf{\Sigma}_s = 0$).

3.2. Capillary Thinning of a Polymer Solution Thread

One of the important and most studied systems is the liquid bridge connecting two droplets, Figure 12. The bridge can form, for example, after the separation of two planes containing liquid in the gap. Then, it becomes thinner due to the action of capillary forces. The breakup dynamics of a Newtonian liquid bridge (the normal stress difference in this case is $\Sigma_{zz} - \Sigma_{rr} = 3\eta \frac{\partial v_z}{\partial z}$) is related to its Ohnesorge number $\mathrm{Oh} = \eta/\sqrt{\rho\gamma a}$. If the bridge is thick enough ($\mathrm{Oh} \ll 1$), the inertial and capillary forces dominate, and the inertia–capillary regime or IC regime is realized. In this case, the minimum thread radius (the radius of the neck) obeys the scaling law $a_{\min}(t) = A(\gamma/\rho)^{1/3}(t_b - t)^{2/3}$ [66,67], where t_b is the putative breakup time. Different values for the prefactor A were proposed and used: $A \approx 0.4$ [68], $A \approx 0.64$ [69], and $A \approx 0.717$ [70]. The characteristic breakup time of the thread is $\tau_I \simeq 2.9\sqrt{\rho a^3/\gamma}$ [71,72] and the local Reynolds number is large in this regime: $\mathrm{Re} \sim 1/\mathrm{Oh} \gg 1$. At high Ohnesorge numbers, $\mathrm{Oh} \gg 1$, another visco-capillary regime, or VC regime, arises. It occurs in highly viscous liquids or in relatively thin threads. Inertial effects are negligible in this regime: $\mathrm{Re} \ll 1$, and the breakup time is $\tau_V = 6\eta a/\gamma$ [67,73]. The neck radius decreases linearly in time, $a(t) = 0.07(\gamma/\eta)(t_b - t)$ [74,75]. The Ohnesorge number reflects the ratio of two timescales, τ_V and τ_I: $\mathrm{Oh} \sim \tau_V/\tau_I$. Both regimes fail close to the breakup point, and a new visco-inertial regime emerges wherein both the inertia and the viscosity are equally important while the local Reynolds number is close to one [73].

The break-up of a polymer solution proceeds in a much more complicated way due to viscoelasticity. Early experimental [76–78] and theoretical [79,80] studies have revealed an important role of elasticity associated with the transition of polymer chains to an elongated state. The addition of high-molecular weight polymers in a low-viscosity solvent leads to the formation of long-lived bridges between the droplets even at very low polymer concentrations [81,82]. The dynamics of the bridges are described by two additional modes associated with the elasticity and finite extensibility of polymer chains [83–85]. The elasto-capillary (EC) regime is associated with the unfolding of polymer coils and the predominance of viscoelastic and capillary forces. The terminal quasi-Newtonian visco-capillary (TVC) regime is characterized by the almost complete orientation of macromolecules along the stretching axis [68,83,84]. Unfolding of polymer coils can already start in the IC regime. Both in the IC and VC regimes, the rate of stretching of the thread increases according to the law $\dot{\varepsilon}(t) = -\frac{2\dot{a}}{a} \propto (t_b - t)^{-1}$. This leads to an increase in the Weissenberg number $Wi = \dot{\varepsilon}\tau$ where τ is the characteristic relaxation time of the quiescent polymer solution. The transition to the EC regime occurs at $Wi \sim 1$. The EC regime was widely studied theoretically using the force balance equations, and the viscoelasticity of the polymer solutions was taken into account mainly on the basis of the classical constitutive equations of the Oldroyd-B and FENE-P models [79,80,85–90]. According to these theories, the radius of the thread a in the EC regime decreases as $a(t) \propto e^{-t/3\tau}$. The exponential law was observed in many experiments with dilute, semi-dilute, and concentrated polymer solutions using CaBER, DoS, and ROJER rheometry including visualization of the thinning dynamics [67,69,91–96].

The dynamics of the bridge in the EC regime can be described by Equation (7) after elimination the gravity and electrostatic forces. Assuming that the curvature $C \simeq 1/a$, Formula (6) obtains

$$\rho \frac{\partial v_z}{\partial t} + \rho v_z \frac{\partial v_z}{\partial z} = \frac{1}{\pi a^2} \frac{\partial}{\partial z} \left(\pi \gamma a + \pi a^2 (\Sigma_{zz} - \Sigma_{rr}) \right) \qquad (14)$$

This equation should be supplemented with appropriate boundary conditions in the transition region from the thread to the droplet. These conditions are determined through the thread tensile force, which is the sum of the surface and body forces: $T = 2\pi\gamma a + \pi a^2 (\Sigma_{zz} - p)$ [88]. This force generally differs from the net capillary force $2\pi\gamma a$ and can be written as [M2] $T = 2\pi\gamma a X$ where X depends on the ratio Σ_{zz}/p [96,97]. The pressure is found from the boundary condition $p = \gamma/a + \Sigma_{rr}$, hence

$$\gamma/a + \Sigma_{zz} - \Sigma_{rr} = T/(\pi a^2) \qquad (15)$$

In the EC regime, the radius of the thread is nearly constant along the axis, i.e., $a \simeq a(t)$ and the axial stress obey inequalities $GN \gg \Sigma_{zz}^p \gg G$ and $\Sigma_{zz}^p \gg \Sigma_{rr}^p$, at that contribution from the solvent, can be omitted: $\Sigma_{zz} - \Sigma_{rr} \simeq \Sigma_{zz}^p \simeq G(R_z^2/R_0^2)$. Therefore, Equations (9) and (10) are simplified in the EC regime:

$$\tau \frac{d}{dt}\Sigma_{zz}^p - 2\dot{\varepsilon}\tau\Sigma_{zz}^p + \Sigma_{zz}^p = 0 \qquad (16)$$

In Equation (16) $\frac{d\Sigma_{zz}^p}{dt} = \frac{\partial \Sigma_{zz}^p}{\partial t} + v_z \frac{d\Sigma_{zz}^p}{dz}$ and $\dot{\varepsilon} = -\frac{2}{a}\frac{\partial a}{\partial t}$. Based on the use of various theoretical methods, it is shown that the thread tension force in the EC mode is equal to $T = 3\pi\gamma a$, and the force balance equation in EC regime is written as $\Sigma_{zz} - \Sigma_{rr} \simeq \Sigma_{zz}^p \simeq 2\gamma/a$ [88,97,98]. The evolution of the thread radius is found from Equation (16) It changes over time as $a(t) = a_0 \left(\frac{\Sigma_0 a_0}{2\gamma} \right)^{1/3} e^{-t/3\tau}$ where $a_0 = a(0)$ is the initial radius and Σ_0 is the initial stress, $\Sigma_0 \geq G$ [88]. The experimental measurements of the stresses acting in the capillary bridge connecting the droplets were performed by Bazilevskii et al. [99–101].

The Weissenberg number in the EC regime is constant, $Wi_{EC} = \dot{\varepsilon}\tau = 2/3$, whereas in the IC regime it increases in time as $Wi_{IC} = \frac{4\tau}{3(t_b-t)}$, and in VC regime as $Wi_{VC} = \frac{\tau}{(t_b-t)}$. The transition from the VC to EC regime is associated with the beginning of coil unfolding, whereas the transition from IC to EC is determined by the change in the balance of forces. The unfolding of chains in the latter case begins already in the IC mode. The value of the Weissenberg number at the transition point is estimated from the condition that the viscoelastic force becomes the order of the capillary force, i.e., $\Sigma_{zz} \simeq 2\gamma/a$ where the stress component Σ_{zz} is found from Equation (13) with $\dot{\varepsilon} = \frac{4}{3(t_b-t)}$: $\Sigma_{zz} \simeq G\left(\frac{8\tau}{3(t_b-t)}\right)^{8/3}$ where $t_b - t = A^{-3/2}(\rho a^3/\gamma)^{1/2}$. The Weissenberg number at the transition point follows from the force balance equation: $Wi^* \sim \left(\frac{\gamma^2 \rho \tau}{\eta_p^3}\right)^{1/6}$. After the IC to EC transition point, the Weissenberg number must decrease to the value $Wi_{EC} = 2/3$, i.e., Wi first increases in the inertial regime as $Wi \propto (t_b - t)^{-1}$, and after passing through the maximum it decreases. The non-monotonic behavior of $\dot{\varepsilon}$ with time was observed in ref [69].

When the macromolecules become almost fully elongated ($A_{zz} \simeq L^2$), the EC regime transformed to the TVC regime with $\Sigma_p \simeq \Sigma_{zz}^p \simeq 2\eta_p(L^2/R_0^2)\dot{\varepsilon}$ and the radius of the thread decreases linearly in time, $a(t) \sim (\gamma/\eta_{eff})(t_b - t)$ [68]. The effective viscosity η_{eff} in the TVC regime is $\eta_{eff} \sim \eta_p N$.

Experiments show that the apparent relaxation time τ coming from fitting $a(t)$ in the EC regime significantly increases with a concentration in the dilute solution regime ($c \ll c^*$ or $\phi \ll \phi^*$ where ϕ is the volume fraction of polymer) [102]. These results are at odds with the Rouse–Zimm theory for dilute solutions, in which the relaxation time depends on the molecular weight, and the concentration dependence due to hydrodynamic interactions between the chains is weak [103]. This contradiction triggered questions on how to define a dilute solution and how interchain interactions affect the rheology of solutions in extensional flow [104–106].

The effect of hydrodynamic interactions on the thread dynamics can be taken into account using the molecular approach. One such approach in the case of a semi-flexible chain solution was formulated in the ref. [107]. The relaxation of the semi-flexible chain of contour length L, diameter d, and the Kuhn segment length l ($d \ll l \ll L$) in dilute solutions in the presence of a flow can be described by the equation on the orientational (stretching) parameter $s = R_z/L$ where R_z is the end-to-end distance of the chain [107], taking into account the hydrodynamic interactions:

$$\tau_R(1-s^2)^2\left(\frac{ds}{dt} - \dot{\varepsilon}s\right) = -1 - \frac{1}{3}(s^4 - 2s^2) \tag{17}$$

where $\tau_R = \frac{\pi}{18}\frac{\eta_s l L^2}{k_H T}$ is the Rouse relaxation time. At equilibrium, the orientational parameter is $s_0 \simeq R_0/L = \sqrt{l/L} \ll 1$. According to Equation (17), polymer coils begin to unfold if the condition $\tau_Z \dot{\varepsilon} > 1$ is satisfied where $\tau_Z = \tau_R s_0 \sim \frac{\pi}{18}\frac{\eta_s}{T}R_0^3$ is the Zimm relaxation time. Notably, the elasticity of a semi-flexible chain is approximately described by a non-Hookean dumbbell with elastic energy $F_{el} \simeq \frac{3k_B T \mathbf{R}^2}{2R_0^2}\frac{(1-\mathbf{R}^2/3L^2)}{1-\mathbf{R}^2/L^2}$.

In the EC regime, the polymer part of the axial stress Σ_{zz}^p exceeds the radial component Σ_{rr}^p, $\Sigma_{rr}^p \ll \Sigma_{zz}^p$, and $1 - s \ll 1$, therefore, the normal stress difference is $\Sigma_p \equiv \Sigma_{zz}^p - \Sigma_{rr}^p \simeq \frac{3ck_B T}{N}\frac{R_z^2}{R_0^2}$. Notably, this expression is similar to that in the Oldroyd B model with $A_{zz} = R_z^2$ since $R_z \gg R_0$. The radius of the thread $a(t)$ and the axial end-to-end distance $R_z(t)$ in the EC regime are found in Equation (17) after taking into account the force balance equation $\Sigma_p \sim 2\gamma/a$:

$$R_z \sim \frac{L}{3}(t/\tau_R), \quad a \sim a_1(\tau_R/t)^2, \quad \dot{\varepsilon} \simeq 4/t \tag{18}$$

Here, $t \leq \tau_R$ and $a_1 = \frac{3\pi}{4}\frac{\gamma l d^2}{\phi T}$ [107]. The power law $a \propto t^{-2}$ arises due to a linear dependence of the friction force on the longitudinal size of the chain that is a consequence of hydrodynamic interactions. It should prevail for dilute polymer solutions with concentration $c \ll c^*$. The rate of extension, $\dot{\varepsilon}(t)$, in the thinning process shows a non-monotonic time-dependence: it first increases as $\dot{\varepsilon} = (4/3)/(t_b - t)$ in the inertial regime but then decreases as $\dot{\varepsilon} \simeq 4/t$ in the viscoelastic regime.

To explain the discrepancy between the above theory and experiments with dilute solutions in the EC mode, which show an exponential thinning of the thread, the formation of transition bonds between monomers of different chains upon their contact was proposed [107]. Such bonds can exist, for example, in aqueous solutions of PAM [83,84] or PEO [108–110]. If the lifetime τ_b of a bond is long, $\tau_b \gg \eta_s d^3/T$, the polymer chain dynamics become Rouse-like with a high effective friction per chain which is proportional to the bond lifetime and the number of bonds $n_b \sim \phi N$, i.e., it is proportional to the number of monomers. Therefore, the chain relaxation time is $\tau_R^* \sim \tau_b \phi^2 N^2 \gg \tau_R$ and the chain dynamics should be the Rouse type [107,111]. The increase in the relaxation time τ_R^* with the polymer concentration is in qualitative agreement with the experiment. However, experimentally, a weaker dependence is observed [68,69,85,91,102,104].

3.3. Blistering Instability

One of the interesting phenomena observed in polymer threads is the appearance of pearling or blistering structures at the end of the exponential thinning regime when the polymer chains are highly stretched [78,112–120], Figure 14.

Figure 14. Blistering structure formed by continuous drawing out of a 25% solution of PAN (M_w = 85,000) in DMSO. Authors' photo.

These hierarchical droplets sequences strung on a polymer string have been identified in PEO, PAN, and PEM solutions. This type of instability differs from classical Rayleigh–Plateau pinching [67,71]. The formation of satellite droplets upon thinning of the thread formed by Oldroyd B liquid was studied in ref. [86,89]. Nevertheless, the proposed recursive relationship between filament diameters for successive generations in [86] does not fit correctly with the experimental data [112,113]. Numerical simulation of Bhat et al. [95] revealed the decisive role of inertia in the formation of satellite droplets. In the above theoretical works, the liquid was considered as a homogeneous medium, which does not allow a sufficient description of blister instabilities in polymer filaments. It is important that polymer solutions are characterized by concentration inhomogeneities, which under certain conditions, can grow and lead to separation into a solvent and a polymer-rich phase. Several mechanisms have been proposed to account for this effect.

One of the mechanisms is based on the flow-induced phase separation of the polymer solution into a polymer-rich phase and a solvent-rich phase [121–123]. This is due to the dependence of the interaction energy of macromolecules on their conformation, in particular, on their orientation parameters. The interaction energy of semiflexible chains in solution at the third virial approximation is given by $f_{int} = \frac{1}{2} B_2 c^2 + \frac{1}{3} B_3 c^3$, where c is the concentration of the polymer segments and B_2 and B_3 are the second and the third virial coefficients, respectively. Within standard approximation, the second virial coefficient includes the contribution from the steric repulsion and van der Waals attraction (i.e., $B_2 \simeq \frac{\pi}{2} l_1^2 d k$, where $k = I(s) - \frac{\Theta}{T}$ and Θ is Θ-temperature). The function $I(s)$ takes the steric repulsion between the segments into account and strongly depends on their orientation parameters. The third virial coefficient is $B_3 \simeq \frac{3\pi^2}{32} l^3 d^3 I(s)$ [121–123]. The steric repulsion between the extended chains in the EC regime decreases with an increase in their orientation, therefore the balance between repulsive and attractive interactions is shifted toward attractions as the Weissenberg number $Wi = \dot{\varepsilon} \tau$ increases. The polymer/solvent phase separation occurs when the second virial coefficient $B_2 < 0$ or $k = I(s) - \frac{\Theta}{T} < 0$. The volume fraction of the polymer in the polymer-rich phase ϕ_c is obtained from the equality of osmotic pressure to zero: $\phi_c \simeq |k|/I(s) \simeq T|k|/\Theta$, $\phi_c \gg \phi$ when $|k| \ll 1$. The kinetics of phase separation was analyzed in ref. [115,116]. On the first stage of spinodal decomposition, the oriented domains with characteristic size $\xi \sim \left(\frac{ld}{\phi |k|} \right)^{1/2}$ in the cross-section of the thread are formed which then collapse laterally with the formation of a network of highly-oriented and stiff fibrils having diameter $d_f \sim (ld)^{1/2}/|k|$, $d_f \ll \xi$, and the longitudinal size $\xi_z \sim \frac{d}{\phi|k|}$. On the final stage, the network of fibrils then tends to compress by squeezing out the solvent to the surface. The characteristic time of the first two stages $t_c \sim \tau_R \left(\frac{d}{\phi L|k|}\right)^2$ is much shorter than τ_R, therefore the phase separation could fully develop during the stretching regime. The formed annular solvent layer is unstable with respect to undulations [124], which should lead to the appearance of droplets. Methods of molecular dynamics modeling also confirm the formation of fibrillar structures by elongated PEO oligomers in an aqueous solution due to a decrease in the number of hydrogen bonds between PEO and water [125,126].

As mentioned above, the pearling structures are often observed after the elastocapillary regime for PEO solutions [78,112–116]. Deblais, et al. [116] showed that temperature significantly affects the dynamics of thread thinning and the onset of pearling instability, which confirms the idea of phase separation. The period of the pearling structure is close to the period of the droplet structure that occurs when a filament of an inviscid liquid breaks: $\lambda \simeq 2\pi\sqrt{2}a_0$ [115]. A close value was also obtained in the analysis of the instability of a thin annular solvent layer on a wire [124].

Another mechanism leading to the instability of polymer solution thread is related to chain migration into thinner regions with a higher concentration due to the (SCC) stress–concentration coupling effect [127]. However, the SCC theory [128–130] does not predict flow-induced phase separations in unentangled polymer solutions. In the case of an extensional flow, the SCC effect is always much weaker than the flow-induced thermodynamic interaction effect [131].

Recently, a capillary mechanism for the formation of annular droplets in the TVC regime was proposed [107,131–133]. It occurs when the radius of the thread is smaller than the macromolecular contour length L. Such a mechanism was considered both in threads of solutions of rodlike macromolecules [131,132], where the droplet formation was found to be an activated process, and in threads of dilute solutions of semi-flexible polymers, where the droplet formation proceeds without any energy barrier [107,133]. In the last system, the solvent droplets are formed spontaneously as a hierarchical process when new solvent beads are constantly emerging on the polymer strings connecting the existing droplets during capillary-induced thinning of the polymer core and the string radius decreases linearly with time. The resulting highly polydisperse system of droplets is characterized by a self-similar (fractal) size distribution. This picture agrees with experimental observations concerning pearling instabilities and blistering patterns. The capillary mechanism of the pearling instability may be important for PAM solutions whose thinning does not depend on temperature, in contrast to PEO solutions [116]. The contour length of the PAM chains used in the experiment is $L \sim 80$ μm (the monomer length is $l_1 \approx 0.4$ nm and $M_w \sim 15 \times 10^6$ g/mol), so the critical radius should be on the order of or less than 8 μm, which is consistent with experimental data [116].

3.4. Stretching a Polymer Solution Jet by an External Load

Fiber formation usually occurs by pulling a stream of polymer solution flowing out of the nozzle, Figure 13. Let us assume that the force F_{ext} stretching the jet is localized at the take-up device. If the polymer solution flowing out of the nozzle of radius a_0 with the flow rate Q, then Equation (6) in the stationary regime of flow can be written as

$$\frac{d}{dz}\left[-\gamma a + a^2\left(\rho v_z^2 - \Sigma_{zz} + \Sigma_{rr}\right)\right] = 0, \quad \Sigma_{zz} - \Sigma_{rr} = 3\eta_s \frac{dv_z}{dz} + \Sigma_p \quad (19)$$

where the flow velocity inside the jet is $v_z = \frac{Q}{\pi a^2}$. The boundary condition at the end of the jet ($z = H$, $a(H) = a_H$) implies the balance between the applied force and the jet tensile force:

$$\pi a_H^2 (\gamma/a_H + \Sigma_{zz} - \Sigma_{rr}) = F_{ext} \quad (20)$$

Integration of Equation (19) using (20) yields

$$\frac{6Q\eta_s}{\pi a}\frac{da}{dz} - a^2 \Sigma_p = \gamma a + \frac{\rho Q^2}{\pi^2}\left(\frac{1}{a_H^2} - \frac{1}{a^2}\right) - \frac{F_{ext}}{\pi} \qquad (21)$$

Omitting the inertia and assuming $F_{ext} \gg 2\pi a_0 \gamma$ in the case of Newtonian liquid ($\Sigma_p = 3\eta_p \frac{dv_z}{dz}$) we obtain an exponentially decaying jet profile: $a(z) = a_0 exp\left(-\frac{zF_{ext}}{6\pi Q(\eta_s + \eta_p)}\right)$. If the stretching of the jet occurs in the elastic regime when the polymer axial stress is dominant, $\Sigma_p > G$, then Equation (21) is reduced to $\Sigma_p = F_{ext}/(\pi a^2)$. Since the axial stress is $\Sigma_{zz}^p \simeq \Sigma_p$, the jet profile is found from Equations (16) where $\frac{d\Sigma_{zz}^p}{dt} = v_z \frac{d\Sigma_{zz}^p}{dz}$ and $\dot{\varepsilon} = \frac{dv_z}{dz} = -\frac{2Q}{\pi a^3}\frac{da}{dz}$: $a(z) = a_0\left(1 + \frac{\pi a_0^2 z}{Q\tau}\right)^{-1/2}$ [120]. Thus, the thinning of the jet occurs according to a power law. In this case, the Weissenberg number is constant along the jet, $Wi = 1$ ($\dot{\varepsilon} = \tau^{-1}$), and the orientational parameter of the chains increases as $s = R_z/L \sim s_0\left(\frac{F_{ext}z}{Q\eta_p}\right)^{1/2}$ ($s \lesssim 0.5$). This means that the steric repulsion between the chains decreases and spinodal decomposition of the solution with the release of the solvent is possible. Such an effect was observed with PAN solutions [120].

3.5. Effect of Gravity

The shape of a falling jet and the critical length before its disintegration into drops were the subjects of long-term interest. The length of a falling get can be very long [134] and greatly exceed the limit predicted by the Rayleigh–Plateau values [135]. The general explanation connects this effect to the transition from the capillary dominating regime of flow to the viscous regime. Clarke presented the complete formulation of the dynamic equation for the shape of a falling jet formed by a Newtonian liquid [136,137]. The validity of the general solution obtained by Clarke was examined rather carefully in [138] for the micro-flow device where a jet is formed between a feeding capillary and a suction cell. The results of studying the gravitational flow in a wide Reynolds number range confirmed the validity of the universal solution to the dynamic problem. The formation of a stable jet happens after the transition from periodic dripping to jetting along with increasing the velocity of a fluid [139]. Theoretical analysis of the behavior of the free-falling viscoelastic liquid jet allowed for establishing the instability boundary connected with the influence of surface effects [140,141]. The shape of free-falling stable jets created by viscoelastic concentrated polyacrylonitrile solutions were studied in [142], where the superposition of viscoelastic, capillary, and inertial forces for fluids with different rheological properties were analyzed. At a low polymer concentration, the jet profile is determined by the balance of capillary, inertial, and gravitational forces, while at higher velocities and highly viscous solutions, the balance of viscous, inertial, and gravitational forces becomes dominant. At very high concentrations, the role of elasticity increases, but the Weissenberg number remains below the critical value corresponding to the unfolding of polymer chains [142].

3.6. Electrospinning

The driving force of the jet flow in this case is of an electrostatic nature. Experiments show that if a voltage exceeding a critical value is applied to the meniscus of a liquid, it assumes a conical shape, also known as a Taylor cone, the top of which emits a thin jet [143–145], Figure 15.

Taylor was the first to show that perfectly conducting liquid forms a conical shape with the apex semi-angle $\theta_T = 49.3°$ due to a balance between the electrostatic and capillary forces [143]. The conical surfaces were also predicted for the ideal dielectric liquids whose dielectric constant ε exceeds some critical value $\varepsilon > \varepsilon_c \approx 17.6$ [146]. The cone half-angle θ in this case depends on ε and varies in the range $0 < \theta < 49.3°$. The surface of the dielectric cone carries only the polarization charge and the emanation of the jet from the cone apex is impossible similary to the case of the perfectly conducting liquid forming the Taylor cone. Recently, self-similar conical structures different from the hydrostatic Taylor cone and capable of emitting charges were described [62,147–151]. There are two families of micro-cones carrying surface charges [62]. The first family constitutes needle-like micro-cones having a small apex angle, $0 < \theta_c(\varepsilon) < 27°$, $\varepsilon > 1$, and $\theta_c(\varepsilon) \to 0$ at $\varepsilon \to \infty$. The micro-cones from the second family appear at $\varepsilon > 12.6$ and have the apex angles $36° < \theta_c(\varepsilon) < 49.3°$ where the upper boundary corresponds to the Taylor value $2\theta_c = 98.6°$ (at $\varepsilon \to \infty$). Based on the Onsager principle, it was shown that needle-like micro-cones are more stable [62,148].

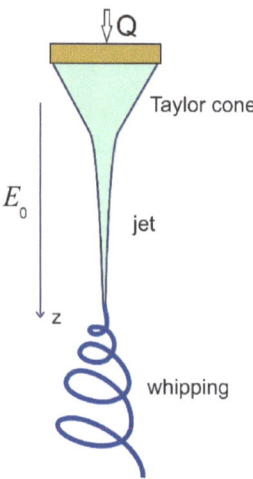

Figure 15. A typical electrospinning jet pattern includes a Taylor cone, a straight jet, and a whipping jet. Authors' drawing.

The behavior and the shape of the electrospinning jets were widely studied experimentally [152–156]. For the theoretical analysis of the shape of rectilinear jets, the slender body approximation is often used. The force balance Equation (7) for a stationary jet carrying only surface charge is written as

$$\frac{d}{dz}\left[a^2\left(\rho v_z^2 + \Sigma_{rr} - \Sigma_{zz}\right) - \gamma a\right] \simeq 2aF_\tau + a^2\frac{dF_n}{dz} \qquad (22)$$

The electrostatic field inside the jet is determined from equation [146]

$$E_z \simeq E_0 - \ln\frac{1}{|a'_z|}\left(\frac{1}{\varepsilon_0}\frac{d(a\sigma_f)}{dz} - \frac{(\varepsilon-1)}{2}\frac{d^2(E_z a^2)}{dz^2}\right) \qquad (23)$$

Here, E_0 is the external field generated by the electrode, ε_0 is the dielectric permittivity of the vacuum, and σ_f is the density of the free charges on the jet surface. The polarization charge is $\sigma_p \simeq -\frac{(\varepsilon-1)\varepsilon_0}{2a}\frac{d(E_z a^2)}{dz}$, so the normal component of the electric field is $E_n = \frac{\sigma_p}{(\varepsilon-1)\varepsilon_0}$. The flow inside the jet is characterized by the average velocity $Q/(\pi a^2)$, where Q is the volume of liquid which is issued from the nozzle per unit of time. The electric current inside the jet is a sum of the bulk current I_b, convective current I_Q, and surface current I_s:

$I = I_b + I_Q + I_s$ where $I_b \simeq \pi a^2 K E_z$ (K is the bulk conductivity of the liquid), $I_Q \simeq \frac{2Q\sigma_f}{a^2}$, and $I_s \simeq 2\pi a \sigma_f \mu E_z$ (μ is the mobility of the surface ions) [62]. At high flow rates, the surface current I_s can be neglected. Different electrospinning regimes of the Newtonian liquid were studied numerically [157–160]. The asymptotic shape of the jet is determined by the balance of inertial and electrical forces: $a(z) = \left(\frac{\rho Q^3}{2\pi^2 I E_0 z}\right)^{1/4}$ [161]. The effect of polymer elasticity on the jet profile with using Oldroyd-B and FINE-P models was considered by Carroll and Joo [162]. However, only low Weissenberg numbers were considered. Experiments show that the extension rate in the cone/jet transition region exceeds the inverse chain relaxation time ($Wi > 1$) which results in unfolding of the polymer chains, such that they are stretched inside the jet [155,163,164].

At relatively low flow rates, $Q\tau_E \ll D^3$, where D is the characteristic size of the meniscus and $\tau_E = \frac{\varepsilon \varepsilon_0}{K}$ is the charge relaxation time, the meniscus takes a conical shape and its stability is mainly determined by the balance of electrostatic and capillary forces, $F_n = \frac{\varepsilon \varepsilon_0 E_\tau^2}{2} + \frac{(\sigma_f + \sigma_p)^2}{2\varepsilon_0} \sim \frac{\gamma}{a}$, and the surface charge density is estimated as $\sigma_f \sim (\gamma \varepsilon_0/a)^{1/2}$ [165]. The electric field inside the cone/jet transition zone of radius a_0 is mainly generated by the surface charges of the cone, $E_z \sim \left(\frac{\gamma}{\varepsilon_0 a_0}\right)^{1/2}$, whereas on a distance $z \gg D$, $E_z \simeq E_0$. If $E_0 \ll \left(\frac{\gamma}{\varepsilon_0 a_0}\right)^{1/2}$, the electric field in the entire region $z > 0$ can be approximately represented as the sum $E_z \simeq E_0 + \kappa \left(\frac{\gamma}{\varepsilon_0(a_0+z)}\right)^{1/2}$, where κ is a numerical factor which depends on the geometry of the cone. The bulk current dominates inside the meniscus, whereas in the jet it is determined by the convective current I_Q. In the transition zone $I_b \simeq I_Q \simeq I/2$. From here we find the radius of the transition zone $a_0 \sim (Q\varepsilon_0/K)^{1/3}$ and the electric current $I \sim (\gamma K Q)^{1/2}$.

Next, let us focus on the regime when the Weissenberg number $Wi \gtrsim 1$ in the cone/jet transition zone. In this case, the polymer chains are stretched inside the jet and the stress difference can be written as $\Sigma_{zz} - \Sigma_{rr} \simeq \Sigma_0 \frac{a_0^4}{a^4}$ [166] where $\Sigma_0 \gtrsim G$ is the stress in the cone/jet transition zone. Substitution of this formula in Equation (22) and integration yields ($z \gg a_0$)

$$\left(\frac{\rho Q^2}{2\pi^2} - \frac{\Sigma_0 a_0^4}{2}\right)\left(\frac{1}{a^4} - \frac{1}{a_0^4}\right) + \gamma\left(\frac{1}{a} - \frac{1}{a_0}\right) \simeq \frac{I}{Q}\left(E_0 z + 2\kappa\sqrt{\frac{\gamma z}{\varepsilon_0}}\right) \quad (24)$$

This equation facilitates recovery of the main asymptotes of the jet profile which were found experimentally [154,156]: $a(z) \propto z^{-1/2}$ when $a_0 \ll z \ll D$ and $a(z) \propto (z + z_0)^{-1}$ where $z_0 \sim D\sqrt{\frac{\gamma}{\varepsilon_0 E_0^2 D}}$ when $z \gg D$; $a(z) \propto (z)^{-1/4}$ at $z \to \infty$ and $\frac{\rho Q^2}{2\pi^2} > \frac{\Sigma_0 a_0^4}{2}$. It is interesting to note that for low flow rates when $\frac{\rho Q^2}{2\pi^2} < \frac{\Sigma_0 a_0^4}{2}$, the straight jet has a finite length $\sim \frac{Q\gamma}{IE_0 a^*}$, where $a^* \sim \left(\pi^2 \Sigma_0 a_0^4 - \rho Q^2\right)^{1/3} \gamma^{-1/3}$. A similar result was obtained in numerical calculations [166,167]. In this case, the orientation parameter increases along the jet axis as $s \simeq s_0 a_0^2/a^2$, and reaches its maximum value at the end of the straight section of the jet. For $z > H$, the behavior of the chain should be unstable. The length of the rectilinear section of the jet increases with an increase in the flow rate and decreases with an increase in the field strength, which agrees with the experiment [168,169]. For large flow rates, $\frac{\rho Q^2}{2\pi^2} > \frac{\Sigma_0 a_0^4}{2}$, the jet is rectilinear. In this mode, the order parameter changes non-monotonically along the jet: first, it increases up to a certain maximum value (at which the Weissenberg number reduces to $Wi \sim 1$) and then decreases [166,167]. The decrease is associated with the relaxation of the polymer chains.

The formation of fibers from the jet occurs as a result of the chain orientation, aggregation, and evaporation of the solvent. Numerical calculations have shown that during electrospinning, polymer chains can be strongly elongated along the flow in the rectilinear section of the jet, so that their orientational order parameter reaches the value $s \gtrsim 0.5$ [166,167]. The high orientation of polymer chains can lead to a phase separation of the polymer solution [166] with the emergence of string-like structures. These structures were identified experimentally [170,171].

With a further decreasing flow rate, the convective current decreases and therefore, another regime with $Q < \mu_+ E b^2$ is realized. The cone/jet transition is determined by the equality of the bulk and surface currents, $I_b \simeq I_s$. The radius of the transition zone here is $b = b_3 \sim (\mu_+/K)^{2/3}(\gamma\varepsilon_0)^{1/3}$. The surface current is dominated at $b \ll b_3$ as well as the meniscus issued needle-like micro-cones in this case [148]. These micro-cones were identified in near-field electrospinning and can be used to create a nanoscale fiber [172].

4. Conclusions

Elasticity is the immanent property of polymers due to the flexibility and anisotropy of macromolecular chains. We discussed the role of elasticity in different flow modes of polymeric liquids (solutions and melts) considering both sides of the story: macroscopic effects and the input of elasticity into the formulation and solution of basic dynamic equations of a continuum medium. In all cases, the Weissenberg number (Wi) is a crucial factor in determining the liquid-to-solid-like type of polymer behavior (at Wi ~ 1).

We have considered the general picture of the flow of elastic liquids associated with their elasticity through short capillaries. The effects of the inlet vortex and the die swelling are well known. The relaxation phenomenon determines apparent high values of end correction in the capillary flow. Then, the role of the length of a capillary is related to the duration of relaxation. The hydrodynamic resistance of short capillaries is the universal function of the Deborah number. The die swelling also depends on the capillary length due to the partial relaxation and swelling decrease for long capillaries in comparison with short ones. In addition to the initiation of secondary flows, the Weissenberg number is also responsible for the transition to the elastic instability of the stream inside the capillary and the periodic oscillation of the jet at the capillary exit. Indeed, both effects are observed at Wi $\gtrsim$ 1.

Elasticity also affects the shape of the jet leaving the capillary. However, examination of the behavior of jets should be based on the analysis of fundamental dynamic equations which requires taking into account the other factors such as the gravity force, surface tension, and finally the electric forces in the case of electrospinning. The theoretical analysis correlated with the experimental fact showed that the polymer elasticity becomes dominating at high extension rates (Wi $\gtrsim$ 1) when the polymer coils unfold. In this case, the nature of the extension force plays an important role. The Weissenberg number correlates with the transition from the viscous flow of the jet to its solid-like behavior.

Author Contributions: Conceptualization, and Section 4—V.G.K.; Sections 1 and 2—A.Y.M. Section 3—A.V.S. All authors have read and agreed to the published version of the manuscript.

Funding: Grant No. 17-79-30108 of the Russian Science Foundation.

Institutional Review Board Statement: Not applicable.

Data Availability Statement: Not applicable.

Conflicts of Interest: The authors declare no conflict of interest.

References

1. Reiner, M. The Deborah Number. *Phys. Today* **1964**, *17*, 62. [CrossRef]
2. Weissenberg, K. Proceedings of the First International Congress on Rheology, Scheveningen, The Netherlands, 21–24 September 1948; Publ. North-Holland Publishing Co.: Amsterdam, The Netherlands, 1949; p. 29.
3. Ferry, J.D. *Viscoelastic Properties of Polymers*, 3rd ed.; John Wiley & Sons: Chichester, UK; Brisbane, QLD, Australia; Toronto, ON, Canada; Singapore, 1980; ISBN 0-471-04894-1.
4. Tobolsky, A.V. *Properties and Structure of Polymers*; John Wiley & Sons, Inc.: New York, NY, USA; London, UK, 1962; ASIN: B000NOCV46.
5. Vinogradov, G.V.; Malkin, A.Y. *Rheology of Polymers*; Springer: Berlin/Heidelberg, Germany, 1980.
6. Malkin, A.Y.; Petrie, C.J.S. Some conditions for rupture of polymer liquids in extension. *J. Rheol.* **1997**, *41*, 1–25. [CrossRef]
7. Baumgaertel, M.; Schausberger, A.; Winter, H.H. The relaxation of polymers with linear flexible chains of uniform length. *Rheol. Acta* **1990**, *29*, 400–408. [CrossRef]
8. Münstedt, H. Extensional Rheology and Processing of Polymeric Materials. *Int. Polym. Process.* **2018**, *33*, 594–618. [CrossRef]

9. Malkin, A.; Arinstein, A.; Kulichikhin, V. Polymer extension flows and instabilities. *Prog. Polym. Sci.* **2014**, *39*, 959–978. [CrossRef]
10. Malkin, A.Y.; Subbotin, A.V.; Kulichikhin, V.G. Stability of polymer jets in extension: Physicochemical and rheological mechanisms. *Russ. Chem. Rev.* **2020**, *89*, 811–823. [CrossRef]
11. Huang, Q. When Polymer Chains Are Highly Aligned: A Perspective on Extensional Rheology. *Macromolecules* **2022**, *55*, 715–727. [CrossRef]
12. Kékesi, T.; Amberg, G.; Wittberg, L.P. Drop deformation and breakup in flows with shear. *Chem. Eng. Sci.* **2016**, *140*, 319–329. [CrossRef]
13. Ha, J.-W.; Leal, L.G. An experimental study of drop deformation and breakup in extensional flow at high capillary number. *Phys. Fluids* **2001**, *13*, 1568–1576. [CrossRef]
14. Prieto, J.L. Viscoelastic Effects on Drop Deformation Using a Machine Learning-Enhanced, Finite Element Method. *Polymers* **2020**, *12*, 1652. [CrossRef]
15. Saengow, C.; Giacomin, A.J. Fluid Elasticity in Plastic Pipe Extrusion: Loads on Die Barrel. *Int. Polym. Process.* **2017**, *32*, 648–658. [CrossRef]
16. Papanastasiou, T.S.; Alaie, S.M.; Chen, Z. High-Speed, Non-Isothermal Fiber Spinning. *Int. Polym. Process.* **1994**, *9*, 148–158. [CrossRef]
17. Zatloukal, M.; Vlček, J. Modeling of the film blowing process by using variational principles. *J. Non-Newton. Fluid Mech.* **2004**, *123*, 201–213. [CrossRef]
18. Boger, D.V.; Walters, K. *Rheological Phenomena in Focus*; Elsevier: Amsterdam, The Netherlands, 1993; pp. 53–68. ISBN 9780444600684.
19. Mitsoulis, E.; Hatzikiriakos, S.G. Annular Extrudate Swell of a Fluoropolymer Melt. *Int. Polym. Process.* **2012**, *27*, 535–546. [CrossRef]
20. Mitsoulis, E. Effect of Viscoelasticity in Fountain Flow of Polyethylene Melts. *Int. Polym. Process.* **2009**, *24*, 439–451. [CrossRef]
21. Malkin, A.; Ilyin, S.; Vasilyev, G.; Arinina, M.; Kulichikhin, V. Pressure losses in flow of viscoelastic polymeric fluids through short channels. *J. Rheol.* **2014**, *58*, 433–448. [CrossRef]
22. Vinuesa, R.; Schlatter, P.; Nagib, H.M. Secondary flow in turbulent ducts with increasing aspect ratio. *Phys. Rev. Fluids* **2018**, *3*, 054606. [CrossRef]
23. Nikitin, N.V.; Popelenskaya, N.V.; Stroh, A. Prandtl's Secondary Flows of the Second Kind. Problems of Description, Prediction, and Simulation. *Fluid Dyn.* **2021**, *56*, 513–538. [CrossRef]
24. Letelier, M.F.; Siginer, D.A. Secondary flows of viscoelastic liquids in straight tubes. *Int. J. Solids Struct.* **2003**, *40*, 5081–5095. [CrossRef]
25. Alves, M.; Oliveira, P.; Pinho, F. Numerical Methods for Viscoelastic Fluid Flows. *Annu. Rev. Fluid Mech.* **2021**, *53*, 509–541. [CrossRef]
26. Sun, X.; Wang, S.; Zhao, M. Viscoelastic flow in a curved duct with rectangular cross section over a wide range of Dean number. *Phys. Fluids* **2021**, *33*, 033101. [CrossRef]
27. Browne, C.A.; Datta, S.S. Elastic turbulence generates anomalous flow resistance in porous media. *Sci. Adv.* **2021**, *7*, eabj2619. [CrossRef] [PubMed]
28. Malkin, A.Y.; Kulichikhin, V.G.; Gumennyi, I.V. Comparing flow characteristics of viscoelastic liquids in long and short capillaries (entrance effects). *Phys. Fluids* **2021**, *33*, 013105. [CrossRef]
29. Triliskii, K.K.; Ischuk, Y.L.; Malkin, A. On the nature of the flow of various plastic dispersions and polymers. *Kolloid Zh.* **1988**, *50*, 535–541. (In Russian)
30. Siegbert, R.; Behr, M.; Elgeti, S. Die swell as an objective in the design of polymer extrusion dies. *AIP Conf. Proc.* **2016**, *1769*, 140003. [CrossRef]
31. Ganvir, V.; Gautham, B.; Pol, H.; Bhamla, M.S.; Sclesi, L.; Thaokar, R.; Lele, A.; Mackley, M. Extrudate swell of linear and branched polyethylenes: ALE simulations and comparison with experiments. *J. Non-Newton. Fluid Mech.* **2011**, *166*, 12–24. [CrossRef]
32. Comminal, R.; Pimenta, F.; Hattel, J.H.; Alves, M.A.; Spangenberg, J. Numerical simulation of the planar extrudate swell of pseudoplastic and viscoelastic fluids with the streamfunction and the VOF methods. *J. Non-Newton. Fluid Mech.* **2018**, *252*, 1–18. [CrossRef]
33. Alzarzouri, F.; Deri, F. Evaluation of Die Swell Behavior During Capillary Extrusion of Poly(lactic acid)/High density polyethylene Blend Melts). *Technium* **2020**, *2*, 34–42. [CrossRef]
34. Tammaro, D.; Walker, C.; Lombardi, L.; Trommsdorff, U. Effect of extrudate swell on extrusion foam of polyethylene terephthalate. *J. Cell. Plast.* **2020**, *57*, 911–925. [CrossRef]
35. Yousefi, A.-M.; Smucker, B.; Naber, A.; Wyrick, C.; Shaw, C.; Bennett, K.; Szekely, S.; Focke, C.; Wood, K.A. Controlling the extrudate swell in melt extrusion additive manufacturing of 3D scaffolds: A designed experiment. *J. Biomater. Sci. Polym. Ed.* **2017**, *29*, 195–216. [CrossRef]
36. Wang, Z.; Smith, D.E. Rheology Effects on Predicted Fiber Orientation and Elastic Properties in Large Scale Polymer Composite Additive Manufacturing. *J. Compos. Sci.* **2018**, *2*, 10. [CrossRef]
37. Patlazhan, S.A.; Kravchanko, I.; Poldushov, M.; Miroshnikov, Y.; Kulichikhin, V. Deformation behavior of drops in the flow through a channel with sharp confinement. *Kolloid Zh.* **2022**, *84*, 186–191. (In Russian) [CrossRef]

38. Burghelea, T.I.; Griess, H.J.; Münstedt, H. Comparative investigations of surface instabilities ("sharkskin") of a linear and a long-chain branched polyethylene. *J. Non-Newton. Fluid Mech.* **2010**, *165*, 1093–1104. [CrossRef]
39. Cogswell, F.N. Stretching flow instabilities at the exits of extrusion dies. *J. Non-Newton. Fluid Mech.* **1977**, *2*, 37–47. [CrossRef]
40. Malkin, A.Y. Flow instability in polymer solutions and melts. *Polym. Sci. Ser. C* **2006**, *48*, 21–37. [CrossRef]
41. Skvortsov, I.Y.; Malkin, A.Y.; Kulichikhin, V.G. Self-Oscillations Accompanying Shear Flow of Colloidal and Polymeric Systems. Reality and Instrumental Effects. *Colloid J.* **2019**, *81*, 176–186. [CrossRef]
42. Kulichikhin, V.G.; Malkin, A.Y. The Role of Structure in Polymer Rheology: Review. *Polymers* **2022**, *14*, 1262. [CrossRef]
43. Swain, D.; Philip, J.; Pillai, S.; Ramesh, K. A Revisit to the Frozen Stress Phenomena in Photoelasticity. *Exp. Mech.* **2016**, *56*, 903–917. [CrossRef]
44. Isaza, C.A.V.; Posada, J.C.; Sierra, M.J.D.; Castro-Caicedo, A.J.; Botero-Cadavid, J.F. Analysis of Residual Stress of Injected Plastic Parts: A Multivariable Approach. *Res. J. Appl. Sci. Eng. Technol.* **2021**, *18*, 43–58. [CrossRef]
45. Tedde, G.M.; Santo, L.; Bellisarbo, D.; Quandini, F. Frozen Stresses in Shape Memory Polymer Composites. *Matriale Plast.* **2022**, *55*, 2018494. [CrossRef]
46. Tan, N.; Lin, L.; Deng, T.; Dong, Y. Evaluating the Residual Stress and Its Effect on the Quasi-Static Stress in Polyethylene Pipes. *Polymers* **2022**, *14*, 1458. [CrossRef]
47. Weng, C.; Ding, T.; Zhou, M.; Liu, J.; Wang, H. Formation Mechanism of Residual Stresses in Micro-Injection Molding of PMMA: A Molecular Dynamics Simulation. *Polymers* **2020**, *12*, 1368. [CrossRef] [PubMed]
48. Zhenga, R.; Kennedy, P.; Phan-Thien, N.; Fan, X.-J. Thermoviscoelastic simulation of thermally and pressure-induced stresses in injection molding for the prediction of shrinkage and warpage for fibre-reinforced thermoplastics. *J. Non-Newton. Fluid Mech.* **1999**, *84*, 159–190. [CrossRef]
49. Wang, Y.; Boukany, P.; Wang, S.-Q.; Wang, X. Elastic Breakup in Uniaxial Extension of Entangled Polymer Melts. *Phys. Rev. Lett.* **2007**, *99*, 237801. [CrossRef] [PubMed]
50. Hoyle, D.; Fielding, S. Necking after extensional filament stretching of complex fluids and soft solids. *J. Non-Newton. Fluid Mech.* **2017**, *247*, 132–145. [CrossRef]
51. Huang, Q.; Rasmussen, H.K. Extensional flow dynamics of polystyrene melt. *J. Rheol.* **2019**, *63*, 829–835. [CrossRef]
52. Huang, Q.; Ahn, J.; Parisi, D.; Chang, T.; Hassager, O.; Panyukov, S.; Rubinstein, M.; Vlassopoulos, D. Unexpected Stretching of Entangled Ring Macromolecules. *Phys. Rev. Lett.* **2019**, *122*, 208001. [CrossRef]
53. Hassager, O.; Wang, Y.; Huang, Q. Extensional rheometry of model liquids: Simulations of filament stretching. *Phys. Fluids* **2021**, *33*, 123108. [CrossRef]
54. Eggers, J. Nonlinear dynamics and breakup of free-surface flows. *Rev. Mod. Phys.* **1997**, *69*, 865–930. [CrossRef]
55. Huang, Q.; Alvarez, N.J.; Shabbir, A.; Hassager, O. Multiple Cracks Propagate Simultaneously in Polymer Liquids in Tension. *Phys. Rev. Lett.* **2016**, *117*, 087801. [CrossRef]
56. Huang, Q.; Hassager, O. Polymer liquids fracture like solids. *Soft Matter* **2017**, *13*, 3470–3474. [CrossRef] [PubMed]
57. Wagner, M.H.; Narimissa, E.; Huang, Q. On the origin of brittle fracture of entangled polymer solutions and melts. *J. Rheol.* **2018**, *62*, 221–233. [CrossRef]
58. Wagner, M.H.; Narimissa, E.; Huang, Q. Response to "Letter to the Editor: 'Melt rupture unleashed by few chain scission events in fully stretched strands'" [J. Rheol. 63, 105 (2018)]. *J. Rheol.* **2019**, *63*, 419–421. [CrossRef]
59. Wang, S.-Q. Letter to the Editor: Melt rupture unleashed by few chain scission events in fully stretched strands. *J. Rheol.* **2019**, *63*, 105–107. [CrossRef]
60. Melcher, J.R.; Taylor, G.I. Electrohydrodynamics: A Review of the Role of Interfacial Shear Stresses. *Annu. Rev. Fluid Mech.* **1969**, *1*, 111–146. [CrossRef]
61. Saville, D.A. ELECTROHYDRODYNAMICS: The Taylor-Melcher Leaky Dielectric Model. *Annu. Rev. Fluid Mech.* **1997**, *29*, 27–64. [CrossRef]
62. Subbotin, A.V.; Semenov, A.N. Electrohydrodynamics of stationary cone-jet streaming. *Proc. R. Soc. A Math. Phys. Eng. Sci.* **2015**, *471*, 20150290. [CrossRef]
63. Subbotin, A.; Stepanyan, R.; Chiche, A.; Slot, J.J.M.; Brinke, G.T. Dynamics of an electrically charged polymer jet. *Phys. Fluids* **2013**, *25*, 103101. [CrossRef]
64. Bird, R.B.; Armstrong, R.C.; Hassager, O. *Dynamics of Polymeric Fluids*; Wiley: New York, NY, USA, 1987.
65. Larson, R.G. *Constitutive Equations for Polymer Melts and Solutions*; Butterworths: London, UK, 1988; ISBN 0409901199.
66. Day, R.F.; Hinch, E.J.; Lister, J.R. Self-Similar Capillary Pinchoff of an Inviscid Fluid. *Phys. Rev. Lett.* **1998**, *80*, 704–707. [CrossRef]
67. Eggers, J.; Villermaux, E. Physics of liquid jets. *Rep. Prog. Phys.* **2008**, *71*, 036601. [CrossRef]
68. Dinic, J.; Sharma, V. Macromolecular relaxation, strain, and extensibility determine elastocapillary thinning and ex-tensional viscosity of polymer solutions. *Proc. Natl. Acad. Sci. USA* **2019**, *116*, 8766–8774. [CrossRef] [PubMed]
69. Sur, S.; Rothstein, J. Drop breakup dynamics of dilute polymer solutions: Effect of molecular weight, concentration, and viscosity. *J. Rheol.* **2018**, *62*, 1245–1259. [CrossRef]
70. Wee, H.; Anthony, C.R.; Basaran, O.A. Breakup of a low-viscosity liquid thread. *Phys. Rev. Fluids* **2022**, *7*, L112001. [CrossRef]
71. Rayleigh, L. On the Instability of Jets. *Proc. Lond. Math. Soc.* **1878**, *1*, 4–13. [CrossRef]
72. Chen, Y.-J.; Steen, P.H. Dynamics of inviscid capillary breakup: Collapse and pinchoff of a film bridge. *J. Fluid Mech.* **1997**, *341*, 245–267. [CrossRef]

73. Li, Y.; Sprittles, J.E. Capillary breakup of a liquid bridge: Identifying regimes and transitions. *J. Fluid Mech.* **2016**, *797*, 29–59. [CrossRef]
74. Papageorgiou, D.T. On the breakup of viscous liquid threads. *Phys. Fluids* **1995**, *7*, 1529–1544. [CrossRef]
75. Papageorgiou, D.T. Analytical description of the breakup of liquid jets. *J. Fluid Mech.* **1995**, *301*, 109–132. [CrossRef]
76. Bazilevskii, A.V.; Voronkov, S.I.; Entov, V.M.; Rozhkov, A.N. Orientation effects in the breakup of jets and threads of dilute polymer solutions. *Sov. Phys. Dokl.* **1981**, *26*, 333–336.
77. Bazilevskii, A.V.; Entov, V.M.; Lerner, M.M.; Rozhkov, A.N. Failure of polymer solution filaments. *Polym. Sci. Ser. A* **1997**, *39*, 316–324.
78. Christanti, Y.; Walker, L.M. Surface tension driven jet break up of strain-hardening polymer solutions. *J. Non-Newton. Fluid Mech.* **2001**, *100*, 9–26. [CrossRef]
79. Yarin, A.L. *Free Liquid Jets and Films: Hydrodynamics and Rheology*; John Wiley & Sons: New York, NY, USA, 1993.
80. Entov, V.M.; Hinch, E.J. Effect of a spectrum of relaxation times on the capillary thinning of a filament of elastic liquid. *J. Non-Newton. Fluid Mech.* **1997**, *72*, 31–53. [CrossRef]
81. Amarouchene, Y.; Bonn, D.; Meunier, J.; Kellay, H. Inhibition of the Finite-Time Singularity during Droplet Fission of a Polymeric Fluid. *Phys. Rev. Lett.* **2001**, *86*, 3558–3561. [CrossRef] [PubMed]
82. Deblais, A.; Herrada, M.A.; Eggers, J.; Bonn, D. Self-similarity in the breakup of very dilute viscoelastic solutions. *J. Fluid Mech.* **2020**, *904*, R2. [CrossRef]
83. Stelter, M.; Brenn, G.; Yarin, A.L.; Singh, R.P.; Durst, F. Validation and application of a novel elongational device for polymer solutions. *J. Rheol.* **2000**, *44*, 595–616. [CrossRef]
84. Stelter, M.; Brenn, G.; Yarin, A.L.; Singh, R.P.; Durst, F. Investigation of the elongational behavior of polymer solutions by means of an elongational rheometer. *J. Rheol.* **2002**, *46*, 507–527. [CrossRef]
85. Bazilevskii, A.V.; Entov, V.M.; Rozhkov, A.N. Breakup of an Oldroyd liquid bridge as a method for testing the rhe-ological properties of polymer solutions. *Polym. Sci. Ser. A Ser. B* **2001**, *43*, 716–726.
86. Chang, H.-C.; Demekhin, E.A.; Kalaidin, E. Iterated stretching of viscoelastic jets. *Phys. Fluids* **1999**, *11*, 1717–1737. [CrossRef]
87. Li, J.; Fontelos, M.A. Drop dynamics on the beads-on-string structure for viscoelastic jets: A numerical study. *Phys. Fluids* **2003**, *15*, 922–937. [CrossRef]
88. Clasen, C.; Eggers, J.; Fontelos, M.A.; Li, J.; McKINLEY, G.H. The beads-on-string structure of viscoelastic threads. *J. Fluid Mech.* **2006**, *556*, 283. [CrossRef]
89. Bhat, P.P.; Appathurai, S.; Harris, M.T.; Pasquali, M.; McKinley, G.H.; Basaran, O.A. Formation of beads-on-a-string structures during break-up of viscoelastic filaments. *Nat. Phys.* **2010**, *6*, 625–631. [CrossRef]
90. Turkoz, E.; Lopez-Herrera, J.M.; Eggers, J.; Arnold, C.B.; Deike, L. Axisymmetric simulation of viscoelastic filament thinning with the Oldroyd-B model. *J. Fluid Mech.* **2018**, *851*, R2. [CrossRef]
91. Dinic, J.; Zhang, Y.; Jimenez, L.N.; Sharma, V. Extensional Relaxation Time of Dilute, Aqueous, Polymer Solutions. *ACS Macro Lett.* **2015**, *4*, 804–808. [CrossRef] [PubMed]
92. Dinic, J.; Jimenez, L.N.; Sharma, V. Pinch-off dynamics and dripping-onto-substrate (DoS) rheometry of complex fluids. *Lab A Chip* **2017**, *17*, 460–473. [CrossRef] [PubMed]
93. Jimenez, L.N.; Dinic, J.; Parsi, N.; Sharma, V. Extensional Relaxation Time, Pinch-Off Dynamics, and Printability of Semidilute Polyelectrolyte Solutions. *Macromolecules* **2018**, *51*, 5191–5208. [CrossRef]
94. Keshavarz, B.; Sharma, V.; Houze, E.C.; Koerner, M.R.; Moore, J.R.; Cotts, P.M.; Threlfall-Holmes, P.; McKinley, G.H. Studying the effects of elongational properties on atomization of weakly viscoelastic solutions using Rayleigh Ohnesorge Jetting Extensional Rheometry (ROJER). *J. Non-Newton. Fluid Mech.* **2015**, *222*, 171–189. [CrossRef]
95. Mathues, W.; Formenti, S.; McIlroy, C.; Harlen, O.G.; Clasen, C. CaBER vs ROJER—Different time scales for the thinning of a weakly elastic jet. *J. Rheol.* **2018**, *62*, 1135–1153. [CrossRef]
96. McKinley, G.H.; Tripathi, A. How to extract the Newtonian viscosity from capillary breakup measurements in a filament rheometer. *J. Rheol.* **2000**, *44*, 653–670. [CrossRef]
97. Semenov, A.; Nyrkova, I. Capillary Thinning of Viscoelastic Threads of Unentangled Polymer Solutions. *Polymers* **2022**, *14*, 4420. [CrossRef]
98. Zhou, J.; Doi, M. Dynamics of viscoelastic filaments based on Onsager principle. *Phys. Rev. Fluids* **2018**, *3*, 084004. [CrossRef]
99. Bazilevskii, A.V. Dynamics of horizontal viscoelastic fluid filaments. *Fluid Dyn.* **2013**, *48*, 97–108. [CrossRef]
100. Bazilevskii, A.V.; Rozhkov, A.N. Dynamics of Capillary Breakup of Elastic Jets. *Fluid Dyn.* **2014**, *49*, 827–843. [CrossRef]
101. Bazilevskii, A.V.; Rozhkov, A.N. Dynamics of the Capillary Breakup of a Bridge in an Elastic Fluid. *Fluid Dyn.* **2015**, *50*, 800–811. [CrossRef]
102. Tirtaatmadja, V.; McKinley, G.H.; Cooper-White, J.J. Drop formation and breakup of low viscosity elastic fluids: Effects of molecular weight and concentration. *Phys. Fluids* **2006**, *18*, 043101. [CrossRef]
103. Muthukumar, M.; Freed, K.F. Theory of Concentration Dependence of Polymer Relaxation Times in Dilute Solutions. *Macromolecules* **1978**, *11*, 843–852. [CrossRef]
104. Clasen, C.; Plog, J.P.; Kulicke, W.-M.; Owens, M.; Macosko, C.; Scriven, L.E.; Verani, M.; McKinley, G.H. How dilute are dilute solutions in extensional flows? *J. Rheol.* **2006**, *50*, 849–881. [CrossRef]

105. Prabhakar, R.; Gadkari, S.; Gopesh, T.; Shaw, M.J. Influence of stretching induced self-concentration and self-dilution on coil-stretch hysteresis and capillary thinning of unentangled polymer solutions. *J. Rheol.* **2016**, *60*, 345–366. [CrossRef]
106. Prabhakar, R.; Sasmal, C.; Nguyen, D.A.; Sridhar, T.; Prakash, J.R. Effect of stretching-induced changes in hydro-dynamic screening on coil-stretch hysteresis of unentangled polymer solutions. *Phys. Rev. Fluids* **2017**, *2*, 011301. [CrossRef]
107. Subbotin, A.V.; Semenov, A.N. Dynamics of Dilute Polymer Solutions at the Final Stages of Capillary Thinning. *Macromolecules* **2022**, *55*, 2096–2108. [CrossRef]
108. Ho, D.L.; Hammouda, B.; Kline, S.R. Clustering of poly(ethylene oxide) in water revisited. *J. Polym. Sci. Part B Polym. Phys.* **2003**, *41*, 135–138. [CrossRef]
109. Hammouda, B.; Ho, D.L.; Kline, S.R. SANS from Poly(ethylene oxide)/Water Systems. *Macromolecules* **2002**, *35*, 8578–8585. [CrossRef]
110. James, D.F.; Saringer, J.H. Extensional flow of dilute polymer solutions. *J. Fluid Mech.* **1980**, *97*, 666–671. [CrossRef]
111. Rubinstein, M.; Semenov, A.N. Thermoreversible gelation in solutions of associating polymers: 2. Linear dynamics. *Macromolecules* **1998**, *31*, 1386–1397. [CrossRef]
112. Oliveira, M.; McKinley, G. Iterated stretching and multiple beads-on-a-string phenomena in dilute solutions of highly extensible flexible polymers. *Phys. Fluids* **2005**, *17*, 071704. [CrossRef]
113. Oliveira, M.S.; Yeh, R.; McKinley, G.H. Iterated stretching, extensional rheology and formation of beads-on-a-string structures in polymer solutions. *J. Non-Newton. Fluid Mech.* **2006**, *137*, 137–148. [CrossRef]
114. Sattler, R.; Wagner, C.; Eggers, J. Blistering Pattern and Formation of Nanofibers in Capillary Thinning of Polymer Solutions. *Phys. Rev. Lett.* **2008**, *100*, 164502. [CrossRef]
115. Sattler, R.; Gier, S.; Eggers, J.; Wagner, C. The final stages of capillary break-up of polymer solutions. *Phys. Fluids* **2012**, *24*, 023101. [CrossRef]
116. Deblais, A.; Velikov, K.P.; Bonn, D. Pearling Instabilities of a Viscoelastic Thread. *Phys. Rev. Lett.* **2018**, *120*, 194501. [CrossRef]
117. Kibbelaar, H.V.M.; Deblais, A.; Burla, F.; Koenderink, G.H.; Velikov, K.P.; Bonn, D. Capillary thinning of elastic and viscoelastic threads: From elastocapillarity to phase separation. *Phys. Rev. Fluids* **2020**, *5*, 092001. [CrossRef]
118. Semakov, A.V.; Kulichikhin, V.G.; Tereshin, A.K.; Antonov, S.V.; Malkin, A.Y. On the nature of phase separation of polymer solutions at high extension rates. *J. Polym. Sci. Part B Polym. Phys.* **2015**, *53*, 559–565. [CrossRef]
119. Malkin, A.Y.; Semakov, A.V.; Skvortsov, I.Y.; Zatonskikh, P.; Kulichikhin, V.G.; Subbotin, A.V.; Semenov, A.N. Spinnability of Dilute Polymer Solutions. *Macromolecules* **2017**, *50*, 8231–8244. [CrossRef]
120. Kulichikhin, V.G.; Skvortsov, I.Y.; Subbotin, A.V.; Kotomin, S.V.; Malkin, A.Y. A Novel Technique for Fiber Formation: Mechanotropic Spinning—Principle and Realization. *Polymers* **2018**, *10*, 856. [CrossRef] [PubMed]
121. Subbotin, A.V.; Semenov, A.N. Phase separation in dilute polymer solutions at high-rate extension. *J. Polym. Sci. Part B Polym. Phys.* **2016**, *54*, 1066–1073. [CrossRef]
122. Semenov, A.N.; Subbotin, A.V. Phase Separation Kinetics in Unentangled Polymer Solutions Under High-Rate Ex-tension. *J. Polym. Sci. Part B Polym.* **2017**, *55*, 623–637. [CrossRef]
123. Subbotin, A.V.; Semenov, A.N. Phase Separation in Polymer Solutions under Extension. *Polym. Sci. Ser. C* **2018**, *60*, 106–117. [CrossRef]
124. Goren, S.L. The instability of an annular thread of fluid. *J. Fluid Mech.* **1962**, *12*, 309–319. [CrossRef]
125. Donets, S.; Sommer, J.-U. Molecular Dynamics Simulations of Strain-Induced Phase Transition of Poly(ethylene oxide) in Water. *J. Phys. Chem. B* **2018**, *122*, 392–397. [CrossRef]
126. Donets, S.; Guskova, O.; Sommer, J.-U. Flow-Induced Formation of Thin PEO Fibers in Water and Their Stability after the Strain Release. *J. Phys. Chem. B* **2020**, *124*, 9224–9229. [CrossRef]
127. Eggers, J. Instability of a polymeric thread. *Phys. Fluids* **2014**, *26*, 033106. [CrossRef]
128. Helfand, E.; Fredrickson, G.H. Large fluctuations in polymer solutions under shear. *Phys. Rev. Lett.* **1989**, *62*, 2468–2471. [CrossRef]
129. Doi, M.; Onuki, A. Dynamic coupling between stress and composition in polymer solutions and blends. *J. De Phys. II* **1992**, *2*, 1631–1656. [CrossRef]
130. Milner, S.T. Dynamical theory of concentration fluctuations in polymer solutions under shear. *Phys. Rev. E* **1993**, *48*, 3674–3691. [CrossRef]
131. Subbotin, A.V.; Semenov, A.N. Capillary-Induced Phase Separation in Ultrathin Jets of Rigid-Chain Polymer Solutions. *JETP Lett.* **2020**, *111*, 55–61. [CrossRef]
132. Subbotin, A.V.; Semenov, A.N. Multiple droplets formation in ultrathin bridges of rigid rod dispersions. *J. Rheol.* **2020**, *64*, 13–27. [CrossRef]
133. Subbotin, A.V.; Semenov, A.N. Dynamics of annular solvent droplets under capillary thinning of non-entangled polymer solution. *J. Rheol.* **2023**, *67*, 53–65. [CrossRef]
134. Senchenko, S.; Bohr, T. Shape and stability of a viscous thread. *Phys. Rev. E* **2005**, *71*, 056301. [CrossRef]
135. Javadi, A.; Eggers, J.; Bonn, D.; Habibi, M.; Ribe, N.M. Delayed Capillary Breakup of Falling Viscous Jets. *Phys. Rev. Lett.* **2013**, *110*, 144501. [CrossRef]
136. Clarke, N.S. A differential equation in fluid mechanics. *Mathematika* **1966**, *13*, 51–53. [CrossRef]
137. Clarke, N.S. Two-dimensional flow under gravity in a jet of viscous liquid. *J. Fluid Mech.* **1968**, *31*, 481–500. [CrossRef]

138. Montanero, J.M.; Herrada, M.A.; Ferrera, C.; Vega, E.J.; Ganan-Calvo, A. On the validity of a universal solution for viscous capillary jets. *Phys. Fluids* **2011**, *23*, 122103. [CrossRef]
139. Clanet, C.; Lasheras, J.C. Transition from dripping to jetting. *J. Fluid Mech.* **1999**, *383*, 307–326. [CrossRef]
140. Sauter, U.S.; Buggisch, H.W. Stability of initially slow viscous jets driven by gravity. *J. Fluid Mech.* **2005**, *533*, 237–257. [CrossRef]
141. Alhushaybari, A.; Uddin, J. Convective and absolute instability of viscoelastic liquid jets in the presence of gravity. *Phys. Fluids* **2019**, *31*, 044106. [CrossRef]
142. Subbotin, A.V.; Skvortsov, I.Y.; Kuzin, M.S.; Gerasimenko, P.S.; Kulichikhin, V.G.; Malkin, A.Y. The shape of a falling jet formed by concentrated polymer solutions. *Phys. Fluids* **2021**, *33*, 083108. [CrossRef]
143. Taylor, G.I. Disintegration of water drops in an electric field. *Proc. R. Soc. London Ser. A Math. Phys. Sci.* **1964**, *280*, 383–397. [CrossRef]
144. Fernández de la Mora, J. The fluid dynamics of Taylor cones. *Annu. Rev. Fluid Mech.* **2007**, *39*, 217–243. [CrossRef]
145. Lauricella, M.; Succi, S.; Zussman, E.; Pisignano, D.; Yarin, A.L. Models of polymer solutions in electrified jets and solution blowing. *Rev. Mod. Phys.* **2020**, *92*, 035004. [CrossRef]
146. Ramos, A.; Castellanos, A. Conical points in liquid-liquid interfaces subjected to electric fields. *Phys. Lett. A* **1994**, *184*, 268–272. [CrossRef]
147. Subbotin, A. Electrohydrodynamics of cones on the surface of a liquid. *JETP Lett.* **2015**, *100*, 657–661. [CrossRef]
148. Subbotin, A.; Semenov, A.N. Electrohydrodynamics of a Cone–Jet Flow at a High Relative Permittivity. *JETP Lett.* **2015**, *102*, 815–820. [CrossRef]
149. Subbotin, A.V.; Semenov, A.N. Volume-Charged Cones on a Liquid Interface in an Electric Field. *JETP Lett.* **2018**, *107*, 186–191. [CrossRef]
150. Subbotin, A.V.; Semenov, A.N. Micro-cones on a liquid interface in high electric field: Ionization effects. *Phys. Fluids* **2018**, *30*, 022108. [CrossRef]
151. Belyaev, M.; Zubarev, N.; Zubareva, O. Space-charge-limited current through conical formations on the surface of a liquid with ionic conductivity. *J. Electrost.* **2020**, *107*, 103478. [CrossRef]
152. Yarin, A.L.; Koombhongse, S.; Reneker, D.H. Taylor cone and jetting from liquid droplets in electrospinning of nanofibers. *J. Appl. Phys.* **2001**, *90*, 4836–4846. [CrossRef]
153. Yu, J.H.; Fridrikh, S.V.; Rutledge, G.C. The role of elasticity in the formation of electrospun fibers. *Polymer* **2006**, *47*, 4789–4797. [CrossRef]
154. Helgeson, M.E.; Grammatikos, K.N.; Deitzel, J.M.; Wagner, N.J. Theory and kinematic measurements of the mechanics of stable electrospun polymer jets. *Polymer* **2008**, *49*, 2924–2936. [CrossRef]
155. Han, T.; Yarin, A.L.; Reneker, D.H. Viscoelastic electrospun jets: Initial stresses and elongational rheometry. *Polymer* **2008**, *49*, 1651–1658. [CrossRef]
156. Greenfeld, I.; Fezzaa, K.; Rafailovich, M.H.; Zussman, E. Fast X-ray Phase-Contrast Imaging of Electrospinning Polymer Jets: Measurements of Radius, Velocity, and Concentration. *Macromolecules* **2012**, *45*, 3616–3626. [CrossRef]
157. Feng, J.J. The stretching of an electrified non-Newtonian jet: A model for electrospinning. *Phys. Fluids* **2002**, *14*, 3912–3926. [CrossRef]
158. Hohman, M.M.; Shin, M.; Rutledge, G.; Brenner, M.P. Electrospinning and electrically forced jets. II. Applications. *Phys. Fluids* **2001**, *13*, 2221–2236. [CrossRef]
159. Higuera, F.J. Stationary viscosity-dominated electrified capillary jets. *J. Fluid Mech.* **2006**, *558*, 143–152. [CrossRef]
160. Reznik, S.N.; Zussman, E. Capillary-dominated electrified jets of a viscous leaky dielectric liquid. *Phys. Rev. E* **2010**, *81*, 026313. [CrossRef] [PubMed]
161. Kirichenko, N.; Petryanov-Sokolov, I.V.; Suprun, N.N.; Shutov, A.A. Asymptotic radius of a slightly conducting liquid jet in an electric field. *Sov. Phys. Dokl.* **1986**, *31*, 611–613.
162. Carroll, C.P.; Joo, Y.L. Electrospinning of viscoelastic Boger fluids: Modeling and experiments. *Phys. Fluids* **2006**, *18*, 053102. [CrossRef]
163. Wang, C.; Hashimoto, T.; Wang, Y.; Lai, H.-Y.; Kuo, C.-H. Formation of Dissipative Structures in the Straight Segment of Electrospinning Jets. *Macromolecules* **2020**, *53*, 7876–7886. [CrossRef]
164. Chen, G.; Lai, H.; Lu, P.; Chang, Y.; Wang, C. Light Scattering of Electrospinning Jet with Internal Structures by Flow-Induced Phase Separation. *Macromol. Rapid Commun.* **2022**, *44*, 2200273. [CrossRef]
165. De La Mora, J.F.; Loscertales, I.G. The current emitted by highly conducting Taylor cones. *J. Fluid Mech.* **1994**, *260*, 155–184. [CrossRef]
166. Subbotin, A.; Kulichikhin, V. Orientation and Aggregation of Polymer Chains in the Straight Electrospinning Jet. *Materials* **2020**, *13*, 4295. [CrossRef]
167. Subbotin, A.V. Features of the Behavior of a Polymer Solution Jet in Electrospinning. *Polym. Sci. Ser. A* **2021**, *63*, 172–179. [CrossRef]
168. Shin, Y.M.; Hohman, M.M.; Brenner, M.P.; Rutledge, G.C. Experimental characterization of electrospinning: The electrically forced jet and instabilities. *Polymer* **2001**, *42*, 09955–09967. [CrossRef]
169. Xin, Y.; Reneker, D.H. Hierarchical polystyrene patterns produced by electrospinning. *Polymer* **2012**, *53*, 4254–4261. [CrossRef]

170. Wang, C.; Hashimoto, T. Self-Organization in Electrospun Polymer Solutions: From Dissipative Structures to Ordered Fiber Structures through Fluctuations. *Macromolecules* **2018**, *51*, 4502–4515. [CrossRef]
171. Wang, C.; Hashimoto, T. A Scenario of a Fiber Formation Mechanism in Electrospinning: Jet Evolves Assemblies of Phase-Separated Strings That Eventually Split into As-spun Fibers Observed on the Grounded Collector. *Macromolecules* **2020**, *53*, 9584–9600. [CrossRef]
172. Choi, S.; Shin, D.; Chang, J. Nanoscale Fiber Deposition via Surface Charge Migration at Air-to-Polymer Liquid Interface in Near-Field Electrospinning. *ACS Appl. Polym. Mater.* **2020**, *2*, 2761–2768. [CrossRef]

Disclaimer/Publisher's Note: The statements, opinions and data contained in all publications are solely those of the individual author(s) and contributor(s) and not of MDPI and/or the editor(s). MDPI and/or the editor(s) disclaim responsibility for any injury to people or property resulting from any ideas, methods, instructions or products referred to in the content.

Article

Analytical Solutions to the Unsteady Poiseuille Flow of a Second Grade Fluid with Slip Boundary Conditions

Evgenii S. Baranovskii

Department of Applied Mathematics, Informatics and Mechanics, Voronezh State University, 394018 Voronezh, Russia; esbaranovskii@gmail.com

Abstract: This paper deals with an initial-boundary value problem modeling the unidirectional pressure-driven flow of a second grade fluid in a plane channel with impermeable solid walls. On the channel walls, Navier-type slip boundary conditions are stated. Our aim is to investigate the well-posedness of this problem and obtain its analytical solution under weak regularity requirements on a function describing the velocity distribution at initial time. In order to overcome difficulties related to finding classical solutions, we propose the concept of a generalized solution that is defined as the limit of a uniformly convergent sequence of classical solutions with vanishing perturbations in the initial data. We prove the unique solvability of the problem under consideration in the class of generalized solutions. The main ingredients of our proof are a generalized Abel criterion for uniform convergence of function series and the use of an orthonormal basis consisting of eigenfunctions of the related Sturm–Liouville problem. As a result, explicit expressions for the flow velocity and the pressure in the channel are established. The constructed analytical solutions favor a better understanding of the qualitative features of time-dependent flows of polymer fluids and can be applied to the verification of relevant numerical, asymptotic, and approximate analytical methods.

Keywords: Poiseuille flow; Rivlin–Ericksen fluids; second grade fluids; slip boundary conditions; analytical solutions; existence and uniqueness theorem; Sturm–Liouville problem; Abel's criteria

Citation: Baranovskii, E.S. Analytical Solutions to the Unsteady Poiseuille Flow of a Second Grade Fluid with Slip Boundary Conditions. *Polymers* **2024**, *16*, 179. https://doi.org/ 10.3390/polym16020179

Academic Editors: Alexandre M. Afonso, Luís L. Ferrás and Célio Pinto Fernandes

Received: 2 December 2023
Revised: 28 December 2023
Accepted: 3 January 2024
Published: 7 January 2024

Copyright: © 2024 by the author. Licensee MDPI, Basel, Switzerland. This article is an open access article distributed under the terms and conditions of the Creative Commons Attribution (CC BY) license (https:// creativecommons.org/licenses/by/ 4.0/).

1. Introduction

It is well known that simulations of flows of polymer solutions and melts produce many challenging mathematical problems. The widespread use of polymeric materials in numerous industrial and engineering fields requires an elaborate mathematical theory of polymer media. In the past few decades, different complex models of polymer dynamics have been proposed (see [1–4] and the references therein), and it cannot be said that a particular model is dominant over others.

Many polymer fluids belong to the class of *fluids of complexity N*, which obey the following constitutive relation (see [5,6]):

$$\mathbb{T} = -p\mathbb{I} + \mathbb{F}(\mathbb{A}_1,\ldots,\mathbb{A}_N),$$

where

- $\mathbb{T}$ is the Cauchy stress tensor;
- $\mathbb{I}$ is the identity tensor;
- $\mathbb{F}$ is a frame indifferent response function;
- $\mathbb{A}_1,\ldots,\mathbb{A}_N$ are the first N Rivlin–Ericksen tensors:

$$\mathbb{A}_1 \stackrel{\text{def}}{=} \nabla v + (\nabla v)^\top,$$

$$\mathbb{A}_j \stackrel{\text{def}}{=} \frac{d}{dt}\mathbb{A}_{j-1} + \mathbb{A}_{j-1}\nabla v + (\nabla v)^\top \mathbb{A}_{j-1},\ j=2,\ldots,N;$$

- v is the flow velocity;
- ∇v denotes the velocity gradient;
- $(\nabla v)^\top$ denotes the transpose of the velocity gradient;
- the differential operator d/dt is the material time derivative defined by

$$\frac{d}{dt}\mathbb{S} \stackrel{\text{def}}{=} \frac{\partial}{\partial t}\mathbb{S} + (v \cdot \nabla)\mathbb{S};$$

- p is the pressure.

Recall that if $\mathbb{F}$ is a polynomial of degree N, then the fluid is called a *fluid of grade N*. The classical incompressible Newtonian fluid satisfies the constitutive relation

$$\mathbb{T} = -p\mathbb{I} + \mu \mathbb{A}_1, \quad \mu = \text{const} > 0,$$

and is a fluid of grade 1, while fluids with shear-dependent viscosity modeling by

$$\mathbb{T} = -p\mathbb{I} + \mu(\mathbb{A}_1)\mathbb{A}_1, \quad \mu(\mathbb{A}_1) > 0,$$

belong to the class of fluids of complexity 1.

In this paper, we deal with the second grade fluids:

$$\mathbb{T} = -p\mathbb{I} + \mu \mathbb{A}_1 + \alpha_1 \mathbb{A}_2 + \alpha_2 \mathbb{A}_1^2, \quad (1)$$

where

- μ is the viscosity coefficient, $\mu > 0$;
- α_1 and α_2 are the normal stress moduli.

Relation (1) is well suited to describe the dynamics of some nonlinear viscoelastic fluids, for example, diluted polymer suspensions.

Previous studies have shown that a viscoelastic fluid modeled by (1) is compatible with the thermodynamic laws and stability principles if the following restrictions hold (see [7,8]):

$$\alpha_1 \geq 0, \quad \alpha_1 + \alpha_2 = 0. \quad (2)$$

For more details on the physical background of fluids of differential type, we refer to the review paper [9].

In the sequel, it is assumed that both conditions from (2) are fulfilled. Using the notation

$$\alpha \stackrel{\text{def}}{=} \alpha_1 = -\alpha_2,$$

one can rewrite relation (1) as follows:

$$\mathbb{T} = -p\mathbb{I} + \mu \mathbb{A}_1 + \alpha \mathbb{A}_2 - \alpha \mathbb{A}_1^2. \quad (3)$$

Clearly, a Newtonian fluid can be considered as the limit case of second grade fluids as $\alpha \to 0^+$.

In this paper, we consider an initial-boundary value problem modeling the unsteady flow of a second grade fluid in a plane channel with impermeable solid walls. It is assumed that the flow is unidirectional and driven by a constant pressure gradient along the channel. Our main aim is to investigate the unique solvability of this problem and find its analytical solution under weak regularity requirements on a function describing the velocity distribution at initial time.

One of features of our work is that *Navier-type slip boundary conditions* [10] are used on the channel walls instead of the standard no-slip boundary condition $v = 0$. The importance of interfacial (slip) constitutive laws and their influence on various flow characteristics (especially for non-Newtonian fluid flows) are mentioned by many researchers (see, for example, [11–16] and the literature cited therein). In particular, as stated in [16], the study of wall slip is significant since it can be used to establish the true rheology of materials by correcting data related to slip effects and explaining mismatches in rheological data obtained from rheometers that have different geometries.

Another important feature of the present work is that analytical solutions are constructed not just for the motion equations with boundary conditions but for an *initial-boundary value problem* for these equations with initial data from a very wide class of functions.

The plan of the paper is as follows. In the next section, we give the preliminaries that are needed for mathematical handling of the problem. In Section 3, we provide a description of the mathematical model under consideration and rigorously formulate the initial-boundary value problem for the Poiseuille flow with slip boundary conditions. Section 4 is devoted to deriving an explicit expression for the pressure under the assumption that the velocity field is known. With this expression in hand, we can focus on finding the velocity field, which is performed in Section 5. We consider both classical and generalized formulations of an initial-boundary value boundary problem related to a third-order partial differential equation describing the flow velocity. It is shown that the classical solution is unique and satisfies an energy equality (Theorem 1). But, in general, the classical formulation is inconvenient because, in this framework, our problem is not solvable for a wide class of natural initial data. In order to overcome this difficulty, we introduce the concept of a generalized solution that is defined as the limit of a uniformly convergent sequence of classical solutions with vanishing perturbations in initial data. We prove the unique solvability of the problem under consideration in the class generalized solutions (Theorem 2). The main ingredients of our proof are a generalized Abel criterion for uniform convergence of function series (Proposition 3) and the use of an orthonormal basis consisting of eigenfunctions of the related Sturm–Liouville problem. As a result, we arrive at explicit formulas for the flow velocity and pressure in the channel, assuming that the initial velocity belongs to a suitable Sobolev space.

The present paper is a continuation of the works [17–19], in which various slip problems were considered for the case of steady-state unidirectional motion. It should be mentioned that analytical solutions for both steady and time-dependent flows of second grade fluids were investigated by many authors. Some exact solutions related to flows of this class of viscoelastic fluids through straight channels or pipes under a constant pressure gradient were obtained by Ting [20]. The main object of his work was to find the role of the material constant α_1. Ting in particular showed the unboundedness of solutions for the case $\alpha_1 < 0$. For partial differential equations (PDEs) describing time-dependent shearing flows of second grade fluids with $\alpha_1 < 0$, a qualitative theory (instability, uniqueness, and nonexistence theorems) was developed by Coleman et al. [21]. Analytical solutions for the flow velocity corresponding to the second problem of Stokes were obtained in work [22] by applying the Laplace transform technique. Using the Caputo–Fabrizio time fractional derivative, Fetecau et al. [23] performed an analytical investigation of the magnetohydrodynamic (MHD) flow of second grade fluids over a moving infinite flat plate. In work [24], it was shown that the governing equations for velocity and non-trivial shear stress related to some isothermal MHD unidirectional flows of second grade fluids in a porous medium have identical forms. Shankar and Shivakumara [25] investigated the temporal stability of the plane Poiseuille and Couette flows of a Navier–Stokes–Voigt-type viscoelastic fluid. Recently, Fetecau and Vieru [26] obtained the general solutions for MHD flows of incompressible second grade fluids between infinite horizontal parallel plates which are embedded in a porous medium. Analysis of the second grade hybrid nanofluid flow over a stretching flat plate is given in [27]. A semi-analytical approach for the investigation of rivulet flows of non-Newtonian fluids was developed in [28,29]. Finally, we mention mathematical studies devoted to the existence and uniqueness of solutions to PDEs describing second grade fluid flows [30–36] as well as to optimal flow control and controllability problems [37–40].

The number of articles about second grade fluids and other types of viscoelastic media is constantly increasing, and this provides deeper understanding of relevant physical effects and support for technological advances, in particular, in the polymer industry. On the other hand, the existing research and challenging problems in this field motivate the development of new approaches for PDE analysis.

2. Mathematical Preliminaries

In this section, we collect the auxiliary statements that are needed to obtain the main results.

2.1. Function Spaces

$C^0[0,h]$ denotes the space of all continuous real-valued functions defined on $[0,h]$ and endowed with the max-norm

$$\|v\|_{C^0[0,h]} \stackrel{\text{def}}{=} \max_{y \in [0,h]} |v(y)|.$$

As usual, $H^1[0,h]$ denotes the Sobolev space $W^{1,2}[0,h]$. The scalar product and the norm in this space are defined as follows:

$$(v,w)_{H^1[0,h]} \stackrel{\text{def}}{=} \int_0^h v(y)w(y)\,dy + \int_0^h v'(y)w'(y)\,dy,$$

$$\|v\|_{H^1[0,h]} \stackrel{\text{def}}{=} (v,v)_{H^1[0,h]}^{\frac{1}{2}},$$

where the symbol $'$ denotes the derivative with respect to y.

Recall that $H^1[0,h]$ is compactly embedded in $C^0[0,h]$. Therefore, if $\{w_i\}_{i=1}^\infty \subset H^1[0,h]$, $w_0 \in H^1[0,h]$, and

$$w_i \to w_0 \text{ in the space } H^1[0,h] \text{ as } i \to \infty,$$

then

$$w_i \rightrightarrows w_0 \text{ on the interval } [0,h] \text{ as } i \to \infty.$$

For $\kappa > 0$, we introduce the space $H^1_\kappa[0,h]$, consisting of functions that belong to $H^1[0,h]$ with the following scalar product:

$$(v,w)_{H^1_\kappa[0,h]} \stackrel{\text{def}}{=} \int_0^h v'(y)w'(y)\,dy + \kappa v(h)w(h) + \kappa v(0)w(0).$$

It is easy to show that this scalar product is well defined and the associated Euclidean norm is equivalent to the standard H^1-norm.

2.2. Special Basis Constructed from Eigenfunctions of a Sturm–Liouville Problem

Applying the approach from [41], one can derive the following statement.

Proposition 1 (Special orthonormal basis). *Let $\kappa > 0$. Suppose that $\{\lambda_i\}_{i=1}^\infty$ is the increasing sequence of eigenvalues of the following Sturm–Liouville problem:*

$$\begin{cases} -v''(y) = \lambda v(y), \ y \in (0,h), \\ -v'(0) + \kappa v(0) = 0, \\ v'(h) + \kappa v(h) = 0, \end{cases}$$

and $\{v_i\}_{i=1}^\infty$ is the associated sequence of eigenfunctions such that $\|v_i\|_{H^1_\kappa[0,h]} = 1$, for all $i \in \mathbb{N}$. Then $\{v_i\}_{i=1}^\infty$ is an orthonormal basis of the space $H^1_\kappa[0,h]$.

2.3. Abel's Criterion for Series of Functions of Various Variables

Let us recall one well-known statement arising in the theory of numerical series.

Proposition 2 (Abel's lemma). *Let $\alpha_1, \ldots, \alpha_m, \beta_1, \ldots, \beta_m$ be real numbers and*

$$B_s \stackrel{\text{def}}{=} \sum_{i=1}^{s} \beta_i, \quad s \in \{1, \ldots, m\}.$$

Suppose the sequence $\{\alpha_i\}_{i=1}^{m}$ is monotone and L is a number such that

$$\max_{s \in \{1, \ldots, m\}} |B_s| \leq L,$$

then

$$\left| \sum_{i=1}^{m} \alpha_i \beta_i \right| \leq L(|\alpha_1| + 2|\alpha_m|).$$

Below, we present a generalization of Abel's uniform convergence test for the case where functions depend on various variables.

Proposition 3 (Generalized Abel's criterion). *Let $\mathcal{X}$ and $\mathcal{Y}$ be subsets of the set of real numbers. Suppose the functions $a_n : \mathcal{X} \to \mathbb{R}$ and $b_n : \mathcal{Y} \to \mathbb{R}$, with $n \in \mathbb{N}$, satisfy the following three conditions:*

(C.1) *the sequence $\{a_n(x)\}_{n=1}^{\infty}$ is monotone for any $x \in \mathcal{X}$;*
(C.2) *there exists a number K such that $|a_n(x)| \leq K$ for any $x \in \mathcal{X}$ and $n \in \mathbb{N}$;*
(C.3) *the function series $\sum_{n=1}^{\infty} b_n(y)$ is uniformly convergent on the set $\mathcal{Y}$.*

Then the function series $\sum_{n=1}^{\infty} a_n(x) b_n(y)$ is uniformly convergent on the Cartesian product $\mathcal{X} \times \mathcal{Y}$.

Proof. Let

$$\widetilde{\alpha}_{n,i}(x) \stackrel{\text{def}}{=} a_{n+i}(x), \qquad \widetilde{\beta}_{n,i}(y) \stackrel{\text{def}}{=} b_{n+i}(y),$$

$$B_{n,i}(y) \stackrel{\text{def}}{=} \sum_{j=1}^{i} \widetilde{\beta}_{n,j}(y), \qquad S_{n,m}(x,y) \stackrel{\text{def}}{=} \sum_{j=n+1}^{n+m} a_j(x) b_j(y),$$

for any $n, m, i \in \mathbb{N}$, $x \in \mathcal{X}$, and $y \in \mathcal{Y}$. Note that

$$\begin{aligned} S_{n,m}(x,y) &= \sum_{i=1}^{m} a_{n+i}(x) b_{n+i}(y) \\ &= \sum_{i=1}^{m} \widetilde{\alpha}_{n,i}(x) \widetilde{\beta}_{n,i}(y). \end{aligned} \quad (4)$$

In view of condition (C.3), for an arbitrary positive number ε, there exists an integer N_ε such that

$$\left| \sum_{j=1}^{m} b_{n+j}(y) \right| < \frac{\varepsilon}{3K},$$

for any $n \in \{N_\varepsilon, N_\varepsilon + 1, N_\varepsilon + 2, \ldots\}$, $m \in \mathbb{N}$, and $y \in \mathcal{Y}$.
Therefore, we have

$$\begin{aligned} |B_{n,i}(y)| &= \left| \sum_{j=1}^{i} \widetilde{\beta}_{n,j}(y) \right| \\ &= \left| \sum_{j=1}^{i} b_{n+j}(y) \right| \leq \frac{\varepsilon}{3K}, \end{aligned}$$

for any $n \in \{N_\varepsilon, N_\varepsilon + 1, N_\varepsilon + 2, \ldots\}$, $i \in \mathbb{N}$, and $y \in \mathcal{Y}$.

Moreover, from condition (C.1) it follows that, for any $n \in \mathbb{N}$ and $x \in \mathcal{X}$, the sequence $\{\widetilde{\alpha}_{n,i}(x)\}_{i=1}^\infty$ is monotone.

Using Abel's lemma, condition (C.2), and (4), we derive

$$|\mathcal{S}_{n,m}(x,y)| = \left|\sum_{i=1}^m \widetilde{\alpha}_{n,i}(x)\widetilde{\beta}_{n,i}(y)\right|$$
$$\leq \frac{\varepsilon}{3K}(|\alpha_{n,1}(x)| + 2|\alpha_{n,m}(x)|)$$
$$= \frac{\varepsilon}{3K}(|a_{n+1}(x)| + 2|a_{n+m}(x)|) \leq \varepsilon,$$

for any $n \in \{N_\varepsilon, N_\varepsilon + 1, N_\varepsilon + 2, \ldots\}$, $m \in \mathbb{N}$, and $(x,y) \in \mathcal{X} \times \mathcal{Y}$. This yields that the function series $\sum_{n=1}^\infty a_n(x)b_n(y)$ is uniformly convergent on the set $\mathcal{X} \times \mathcal{Y}$. Thus, Proposition 3 is proved. □

3. Description of the Mathematical Model

3.1. Flow Configuration

We will consider the unidirectional motion of a second grade fluid between the horizontal parallel plates $y = 0$ and $y = h$, assuming that the flow is driven by a constant pressure gradient

$$\frac{\partial p}{\partial x} = -\xi, \quad \xi = \text{const}, \quad \xi > 0. \tag{5}$$

In other words, we deal with the *plane Poiseuille flow*.

Figure 1 shows the chosen coordinate system and the flow geometry.

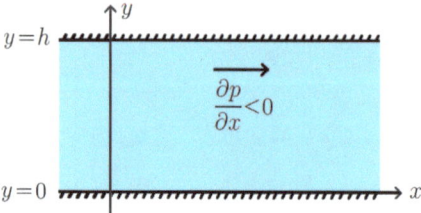

Figure 1. Sketch of the plane Poiseuille flow.

3.2. Governing Equations

As is well known, the unsteady flow of a fluid with constant density is governed by the following system of equations:

The momentum equation: $\rho\left(\dfrac{\partial v}{\partial t} + (v \cdot \nabla)v\right) = \text{div}\,\mathbb{T} + \rho g,$ (6)

The continuity equation: $\nabla \cdot v = 0,$ (7)

where there are the following variables:
- ρ is the fluid density, $\rho > 0$;
- $v = (v_1, v_2, v_3)^\top$ is the velocity vector;
- $\mathbb{T}$ is the Cauchy stress tensor;
- $g = (g_1, g_2, g_3)^\top$ is the external force per unit mass;
- the operators div and ∇ are the divergence and the gradient, respectively (with respect to the space variables x, y, z).

Assume that
$$g = (0, -g, 0)^\top, \tag{8}$$
where g is the value of acceleration due to gravity.

Within the framework of the Poiseuille flow, for the velocity components v_1, v_2, v_3, we have
$$v_1 = u, \quad v_2 = 0, \quad v_3 = 0,$$
where $u = u(y,t)$ is an unknown function. Then the following equalities hold:
$$\nabla \cdot v = 0, \quad (v \cdot \nabla)v = \mathbf{0}. \tag{9}$$

In view of the second relation from (9), Equation (6) reduces to
$$\rho \frac{\partial v}{\partial t} = \operatorname{div} \mathbb{T} + \rho g. \tag{10}$$

For simplicity, we assume the density ρ is equal to 1. Since the fluid obeys the constitutive relation (3), we can rewrite (10) in the form
$$\frac{\partial v}{\partial t} = \operatorname{div}(\mu \mathbb{A}_1 + \alpha \mathbb{A}_2 - \alpha \mathbb{A}_1^2) - \nabla p + g. \tag{11}$$

3.3. Statement of Initial-Boundary Value Problem for the Poiseuille Flow

Let us introduce the notation:
$$\Pi_h \stackrel{\text{def}}{=} \{(x,y,z) \in \mathbb{R}^3 : 0 < y < h\}, \quad \Gamma_a \stackrel{\text{def}}{=} \{(x,y,z) \in \mathbb{R}^3 : y = a\}.$$

We use (11) for handling the unsteady Poiseuille flow of a second grade fluid in the channel Π_h at a given time interval $[0, T]$. Of course, in order to obtain physically relevant solutions, Equation (11) must be supplemented with suitable boundary and initial conditions for the velocity field. Taking into account slip effects, we arrive at the following initial-boundary value problem for the Poiseuille flow (IBVP-PF).

For a given function $u_0 \colon [0,h] \to \mathbb{R}$ (initial data), find functions $u \colon [0,h] \times [0,T] \to \mathbb{R}$ and $p \colon \overline{\Pi}_h \times [0,T] \to \mathbb{R}$ such that $v = (u,0,0)^\top$ and p satisfy (5) and (11) in $\Pi_h \times (0,T)$ and the following three conditions hold:

(i) the impermeability boundary condition
$$v \cdot \mathbf{n} = 0 \quad \text{on } (\Gamma_0 \cup \Gamma_h) \times (0,T); \tag{12}$$

(ii) the slip boundary condition
$$\mu(\mathbb{A}_1 \mathbf{n})_{\tan} = -k v_{\tan} \quad \text{on } (\Gamma_0 \cup \Gamma_h) \times (0,T), \tag{13}$$

where $\mathbf{n}$ is the unit outward normal vector to the channel walls and k is the slip coefficient, $k > 0$;

(iii) the initial condition
$$v|_{t=0} = (u_0, 0, 0)^\top \quad \text{in } \Pi_h. \tag{14}$$

Sections 4 and 5 are devoted to finding an analytical solution to the IBVP-PF.

Note that the limit case of boundary condition (13) when $k = 0$ corresponds to the perfect slip condition [42–45], which means that the fluid actually does not interact with the walls of the channel. On the contrary, if $k \to +\infty$, then relations (12) and (13) tend to the classical no-slip boundary condition $v = 0$. Thus, Navier's slip boundary condition can be considered a homotopy transformation connecting the no-slip condition on the one hand with the perfect slip condition on the other hand. For modeling slip effects on surfaces that have portions with different physical properties, it is convenient to use a variable slip coefficient. In this regard, see, for example, the work [46], dealing with a position-dependent Navier-type slip boundary condition, which is formulated in terms of the operator **curl**.

4. Explicit Expression for the Pressure in Terms of the Velocity Gradient

First, we calculate the Rivlin–Ericksen tensors $\mathbb{A}_1$ and $\mathbb{A}_2$:

$$\mathbb{A}_1 = \begin{bmatrix} 0 & \frac{\partial u}{\partial y} & 0 \\ \frac{\partial u}{\partial y} & 0 & 0 \\ 0 & 0 & 0 \end{bmatrix}, \quad \mathbb{A}_2 = \begin{bmatrix} 0 & \frac{\partial^2 u}{\partial y \partial t} & 0 \\ \frac{\partial^2 u}{\partial y \partial t} & 2\left(\frac{\partial u}{\partial y}\right)^2 & 0 \\ 0 & 0 & 0 \end{bmatrix}.$$

Next, using these equalities, we rewrite (11) in the form

$$\frac{\partial v}{\partial t} = \operatorname{div} \begin{bmatrix} -\alpha\left(\frac{\partial u}{\partial y}\right)^2 & \mu\frac{\partial u}{\partial y} + \alpha\frac{\partial^2 u}{\partial y \partial t} & 0 \\ \mu\frac{\partial u}{\partial y} + \alpha\frac{\partial^2 u}{\partial y \partial t} & \alpha\left(\frac{\partial u}{\partial y}\right)^2 & 0 \\ 0 & 0 & 0 \end{bmatrix} - \nabla p + g.$$

Taking into account (8), it is easy to see that the last equation is equivalent to the following system:

$$\frac{\partial u}{\partial t} - \mu \frac{\partial^2 u}{\partial y^2} - \alpha \frac{\partial^3 u}{\partial y^2 \partial t} + \frac{\partial p}{\partial x} = 0, \quad (15)$$

$$-\alpha \frac{\partial}{\partial y}\left[\left(\frac{\partial u}{\partial y}\right)^2\right] + \frac{\partial p}{\partial y} = -g, \quad (16)$$

$$\frac{\partial p}{\partial z} = 0. \quad (17)$$

Note that (15)–(17) can be considered as a starting point for solving the IBVP-PF.

Now, we are ready to find the pressure p under the assumption that the velocity field is known. In view of (17), the pressure is independent of z, that is, $p = p(x, y, t)$. Moreover, taking into account (5), we conclude that the pressure should be sought in the form

$$p(x, y, t) = -\xi x + \phi(y, t) \quad (18)$$

with an unknown function $\phi = \phi(y, t)$.

From (16) and (18), it follows that

$$\phi(y, t) = \alpha \left(\frac{\partial u}{\partial y}\right)^2 - gy + C(t), \quad (19)$$

where $C = C(t)$ is an arbitrary function.

Substituting (19) into (18), we obtain the resulting formula for the pressure in the channel:

$$p(x, y, t) = -\xi x + \alpha \left(\frac{\partial u}{\partial y}\right)^2 - gy + C(t). \quad (20)$$

Remark 1. *In the next section, we obtain an explicit formula for the function u, and then one can return to (20) for the calculation of the pressure p.*

5. Initial-Boundary Value Problem Related to the Flow Velocity

In order to find the x-component of the velocity field, we must solve the initial-boundary value problem

$$\begin{cases} \dfrac{\partial u}{\partial t} - \mu \dfrac{\partial^2 u}{\partial y^2} - \alpha \dfrac{\partial^3 u}{\partial t \partial y^2} = \xi, & y \in (0, h), \ t \in (0, T), \\ \mu \dfrac{\partial u}{\partial y} = ku, & y = 0, \ t \in (0, T), \\ \mu \dfrac{\partial u}{\partial y} = -ku, & y = h, \ t \in (0, T), \\ u = u_0, & y \in (0, h), \ t = 0. \end{cases} \quad (21)$$

For the derivation of system (21), we used (12)–(15) and the equality $\partial p / \partial x = -\xi$ in $\Pi_h \times (0, T)$.

5.1. Classical Solutions

Definition 1 (Classical solution). *We shall say that a function $u \colon [0, h] \times [0, T] \to \mathbb{R}$ is a classical solution of initial-boundary value problem (21) if the following two conditions hold:*

(i) *the functions u, $\dfrac{\partial u}{\partial t}$, $\dfrac{\partial u}{\partial y}$, $\dfrac{\partial^2 u}{\partial y^2}$, $\dfrac{\partial^3 u}{\partial t \partial y^2}$ belong to the space $C^0([0, h] \times [0, T])$;*

(ii) *the function u satisfies system (21) in the usual sense.*

It is clear that a necessary condition for the solvability of problem (21) in the classical formulation is the fulfillment of the following equalities:

$$\mu u_0'(0) = ku_0(0), \quad \mu u_0'(h) = -ku_0(h),$$

which are sometimes called the *compatibility conditions*. However, these conditions can be omitted when the generalized formulation is used (see Section 5.2).

Theorem 1 (Uniqueness and energy equality for classical solutions). *Let u be a classical solution of problem (21). Then this solution is unique and satisfies the energy equality*

$$\int_0^h u^2(y, t)\, dy + \alpha \int_0^h \left(\dfrac{\partial u}{\partial y}(y, t) \right)^2 dy + \dfrac{\alpha k}{\mu} \left(u^2(h, t) + u^2(0, t) \right)$$

$$+ 2k \int_0^t \left(u^2(h, s) + u^2(0, s) \right) ds + 2\mu \int_0^t \int_0^h \left(\dfrac{\partial u}{\partial y}(y, s) \right)^2 dy\, ds \quad (22)$$

$$= 2\xi \int_0^t \int_0^h u(y, s)\, dy\, ds + \int_0^h u_0^2(y)\, dy + \alpha \int_0^h (u_0'(y))^2\, dy + \dfrac{\alpha k}{\mu} \left(u_0^2(h) + u_0^2(0) \right),$$

for any $t \in [0, T]$.

Proof. Let u_1 and u_2 be classical solutions of problem (21) and $w \stackrel{\text{def}}{=} u_1 - u_2$. It is easy to see that

$$\dfrac{\partial w}{\partial t} - \mu \dfrac{\partial^2 w}{\partial y^2} - \alpha \dfrac{\partial^3 w}{\partial t \partial y^2} = 0, \quad y \in (0, h), \ t \in (0, T), \quad (23)$$

$$\mu \dfrac{\partial w}{\partial y} = kw, \quad y = 0, \ t \in (0, T), \quad (24)$$

$$\mu \frac{\partial w}{\partial y} = -kw, \quad y = h, \ t \in (0, T), \tag{25}$$

$$w = 0, \quad y \in (0, h), \ t = 0. \tag{26}$$

Let us multiply both sides of (23) by w and then integrate the obtained equality with respect to y from 0 to h:

$$\int_0^h \frac{\partial w}{\partial t} w \, dy = \mu \int_0^h \frac{\partial^2 w}{\partial y^2} w \, dy + \alpha \int_0^h \frac{\partial^3 w}{\partial t \partial y^2} w \, dy. \tag{27}$$

Using the relation $(f^2)' = 2ff'$ and integration by parts, we derive from (27) that

$$\frac{1}{2} \int_0^h \frac{\partial}{\partial t} (w^2) \, dy = -\mu \int_0^h \left(\frac{\partial w}{\partial y}\right)^2 dy + \mu \frac{\partial w}{\partial y}(h, t) w(h, t) - \mu \frac{\partial w}{\partial y}(0, t) w(0, t)$$

$$- \frac{\alpha}{2} \int_0^h \frac{\partial}{\partial t}\left(\left(\frac{\partial w}{\partial y}\right)^2\right) dy + \alpha \frac{\partial^2 w}{\partial t \partial y}(h, t) w(h, t) - \alpha \frac{\partial^2 w}{\partial t \partial y}(0, t) w(0, t).$$

Taking into account boundary conditions (24) and (25), we obtain

$$\frac{1}{2} \int_0^h \frac{\partial}{\partial t} (w^2) \, dy = -\mu \int_0^h \left(\frac{\partial w}{\partial y}\right)^2 dy - kw^2(h, t) - kw^2(0, t)$$

$$- \frac{\alpha}{2} \int_0^h \frac{\partial}{\partial t}\left(\left(\frac{\partial w}{\partial y}\right)^2\right) dy - \frac{\alpha k}{2\mu} \frac{\partial w^2}{\partial t}(h, t) - \frac{\alpha k}{2\mu} \frac{\partial w^2}{\partial t}(0, t).$$

Integrating this equality with respect to t from 0 to τ and using (26), we arrive at the following inequality:

$$\int_0^h w^2(y, \tau) \, dy \leq 0, \ \tau \in [0, T],$$

which implies that w is the zero function. Thus, we have $u_1 = u_2$.

In order to derive the energy equality that holds for the classical solution, we multiply both parts of the first equality from system (21) by the function u. Further, applying the integration by parts formula and then integrating the obtained equality with respect to time from 0 to t, we arrive at (22). Thus, Theorem 1 is proved. □

5.2. Generalized Solutions

The proposed concept of classical solutions is inconvenient because, in this framework, problem (21) is not solvable for a wide class of natural initial data. For example, if

$$u_0 \in C^1[0, h] \setminus C^2[0, h], \tag{28}$$

then problem (21) has no classical solutions. Indeed, assuming that u is a classical solution, we have

$$u_0 = u(\cdot, 0) \in C^2[0, h],$$

which contradicts (28). Therefore, we introduce the concept of generalized solutions and then prove the corresponding existence and uniqueness theorem.

Definition 2 (Generalized solution). *We shall say that a function $u\colon [0,h] \times [0,T] \to \mathbb{R}$ is a generalized solution to problem (21) if there exist sequences $\{u^{(m)}\}_{m=1}^{\infty}$ and $\{u_0^{(m)}\}_{m=1}^{\infty}$ such that the following three conditions hold:*

(i) *for any $m \in \mathbb{N}$, the function $u^{(m)}$ is a classical solution to the problem*

$$\begin{cases} \dfrac{\partial v}{\partial t} - \mu \dfrac{\partial^2 v}{\partial y^2} - \alpha \dfrac{\partial^3 v}{\partial t \partial y^2} = \zeta, & y \in (0,h),\ t \in (0,T), \\[4pt] \mu \dfrac{\partial v}{\partial y} = kv, & y=0,\ t \in (0,T), \\[4pt] \mu \dfrac{\partial v}{\partial y} = -kv, & y=h,\ t \in (0,T), \\[4pt] v = u_0^{(m)}, & y \in (0,h),\ t=0; \end{cases}$$

(ii) *the sequence $\{u^{(m)}\}_{m=1}^{\infty}$ converges to the function u uniformly on the set $[0,h] \times [0,T]$ as $m \to \infty$;*

(iii) *the sequence $\{u_0^{(m)}\}_{m=1}^{\infty}$ converges to the function u_0 in the Sobolev space $H^1[0,h]$ as $m \to \infty$.*

Clearly, if a function u is a classical solution to problem (21), then this function is a generalized solution to the one. The converse statement is not true. For example, a generalized solution can have smoothness that is not enough to be a classical solution.

The concept of generalized solutions has a clear physical meaning: a generalized solution is the uniform limit of some sequence of classical solutions of the model under consideration with small perturbations in the initial velocity data when these perturbations uniformly tend to the zero function.

The following theorem gives our main results.

Theorem 2 (Existence, uniqueness, and explicit expressions for generalized solutions). *Suppose a function u_0 belongs to the Sobolev space $H^1[0,h]$. Then:*

(i) *problem (21) has a unique generalized solution;*

(ii) *the following formula can be used to calculate the generalized solution:*

$$u(y,t) = u_*(y) + \sum_{m=1}^{\infty} C_m q_m(y) \varphi_m(t), \tag{29}$$

where

$$u_*(y) \stackrel{\text{def}}{=} -\frac{\zeta}{2\mu} y(y-h) + \frac{\zeta h}{2k} \quad \text{(the steady-state component)}, \tag{30}$$

$$q_m(y) \stackrel{\text{def}}{=} \frac{h}{\eta_m}\left(\frac{h}{2} + \frac{\mu k h^2}{\eta_m^2 \mu^2 + k^2 h^2}\right)^{-\frac{1}{2}} \sin\left(\frac{\eta_m}{h} y + \psi_m\right), \tag{31}$$

$$\varphi_m(t) \stackrel{\text{def}}{=} \exp\left(-\frac{\eta_m^2 \mu}{h^2 + \alpha \eta_m^2} t\right), \quad \psi_m \stackrel{\text{def}}{=} \arctan\left(\frac{\mu \eta_m}{kh}\right), \tag{32}$$

$$C_m \stackrel{\text{def}}{=} \int_0^h (u_0'(y) - u_*'(y)) q_m'(y)\, dy \\ + \frac{k}{\mu}(u_0(0) - u_*(0)) q_m(0) + \frac{k}{\mu}(u_0(h) - u_*(h)) q_m(h), \tag{33}$$

and the numbers $\{\eta_1, \eta_2, \ldots\}$ are positive roots of the transcendental equation

$$\cot \eta = \frac{\mu \eta}{2kh} - \frac{kh}{2\mu \eta} \tag{34}$$

with respect to η.

Proof. Following [47], we will construct a generalized solution of (21) using the reduction of the original problem to a Sturm–Liouville problem with Robin-type boundary conditions. The proof of Theorem 2 is derived in five steps.

Step 1: Passing to a new unknown function. Consider the function u_*, which is defined in (30). Using the representation

$$u(y,t) = \tilde{u}(y,t) + u_*(y),$$

we reduce (21) to a new problem:

$$\begin{cases} \dfrac{\partial \tilde{u}}{\partial t} - \mu \dfrac{\partial^2 \tilde{u}}{\partial y^2} - \alpha \dfrac{\partial^3 \tilde{u}}{\partial t \partial y^2} = 0, & y \in (0,h),\ t \in (0,T), \\[4pt] \mu \dfrac{\partial \tilde{u}}{\partial y} = k\tilde{u}, & y = 0,\ t \in (0,T), \\[4pt] \mu \dfrac{\partial \tilde{u}}{\partial y} = -k\tilde{u}, & y = h,\ t \in (0,T), \\[4pt] \tilde{u} = u_0 - u_*, & y \in (0,h),\ t = 0 \end{cases} \quad (35)$$

with respect to a new unknown function $\tilde{u}$.

Step 2: Separation of variables.

Let us construct nontrivial solutions of problem (35) in the form

$$\tilde{u}(y,t) = \varphi(t)\, q(y).$$

From the first relation of (35), we derive

$$\varphi'(t)\big(q(y) - \alpha q''(y)\big) = \mu\, \varphi(t)\, q''(y). \quad (36)$$

We are interested in the case when $q - \alpha q'' \not\equiv 0$. Indeed, if $q - \alpha q'' \equiv 0$, then from (36) it follows that $q'' \equiv 0$, which is not compatible with the used boundary conditions when $q \neq 0$.

Let y_0 be a point such that $q(y_0) - \alpha q''(y_0) \neq 0$ and

$$\lambda \stackrel{\text{def}}{=} \frac{q''(y_0)}{q(y_0) - \alpha q''(y_0)}.$$

Substituting $y = y_0$ into (36), we obtain

$$\varphi'(t)\big(q(y_0) - \alpha q''(y_0)\big) = \mu\, \varphi(t)\, q''(y_0),$$

whence

$$\varphi'(t) = \lambda\, \mu\, \varphi(t). \quad (37)$$

From (36) and (37) it follows that

$$\lambda\big(q(y) - \alpha q''(y)\big) = q''(y),$$

which is equivalent to

$$-q''(y) + \frac{\lambda}{1 + \lambda\, \alpha}\, q(y) = 0. \quad (38)$$

Moreover, in order to satisfy the boundary conditions from system (35), the following relations must be satisfied:

$$-\mu q'(0) + kq(0) = 0, \quad (39)$$
$$\mu q'(h) + kq(h) = 0. \quad (40)$$

System (38)–(40) is a particular case of the well-known Sturm–Liouville problem. The eigenvalues related to problems (38)–(40) are the numbers $(\eta_m/h)^2$, with $m \in \mathbb{N}$, where η_m is a positive root of Equation (34), while the eigenfunctions are defined in (31). Note that these eigenfunctions are chosen such that their norms in the space $H^1_{k/\mu}[0,h]$ are equal to 1 for any $m \in \mathbb{N}$.

Step 3: *Finding the decay-in-time components of a solution.*
The values of the parameter λ are determined from the relation

$$-\frac{\lambda_m}{1+\lambda_m \alpha} = \left(\frac{\eta_m}{h}\right)^2, \quad m \in \mathbb{N},$$

whence

$$\lambda_m = -\frac{\eta_m^2}{h^2 + \alpha \eta_m^2}, \quad m \in \mathbb{N}.$$

Substituting these values into (37), we arrive at the formula

$$\varphi_{m,C}(t) = C \exp\left(-\frac{\eta_m^2 \mu}{h^2 + \alpha \eta_m^2} t\right), \quad C = \text{const}, \quad m \in \mathbb{N},$$

for the decay-in-time components of a solution to the problem under consideration.

Step 4: *Constructing a generalized solution in the series form.* Setting $\varphi_m \stackrel{\text{def}}{=} \varphi_{m,1}$, we introduce a function $\tilde{u}_m : [0,h] \times [0,T] \to \mathbb{R}$,

$$\tilde{u}_m(y,t) \stackrel{\text{def}}{=} q_m(y)\varphi_m(t), \quad m \in \mathbb{N}, \quad (41)$$

which satisfies the first three relations of system (35).

Using the function sequence $\{\tilde{u}_m\}_{m=1}^{\infty}$, one can construct a generalized solution of problem (35). For this purpose, we consider a function series

$$\sum_{m=1}^{\infty} \ell_m \tilde{u}_m(y,t) \quad (42)$$

with

$$\ell_m \stackrel{\text{def}}{=} (u_0 - u_*, q_m)_{H^1_{k/\mu}[0,h]}, \quad m \in \mathbb{N}. \quad (43)$$

Let us show that series (42) is uniformly convergent on the set $[0,h] \times [0,T]$. Note that the following statements hold.

- From Proposition 1 it follows that the sequence $\{q_m\}_{m=1}^{\infty}$ is an orthonormal basis of the space $H^1_{k/\mu}[0,h]$.
- The following series

$$\sum_{i=1}^{\infty} \ell_m q_m(y)$$

is convergent in the Sobolev space $H^1[0,h]$, and hence this series is uniformly convergent on $[0,h]$.
- For any $t \in [0,T]$, the sequence $\{\varphi_m(t)\}_{m=1}^{\infty}$ is monotone.
- For any $t \in [0,T]$ and $m \in \mathbb{N}$, we have $0 < \varphi_m(t) \leq 1$.

Applying the generalized Abel criterion (see Proposition 3), we establish the uniform convergence of series (42) on $[0,h] \times [0,T]$.

Further, we define the function $\tilde{u}\colon [0,h]\times[0,T]\to\mathbb{R}$ by the formula

$$\tilde{u}(y,t) \stackrel{\text{def}}{=} \sum_{m=1}^{\infty} \ell_m \tilde{u}_m(y,t)$$

and show that this function is a generalized solution of problem (35). Let

$$\tilde{u}^{(m)}(y,t) \stackrel{\text{def}}{=} \sum_{i=1}^{m} \ell_i \tilde{u}_i(y,t). \tag{44}$$

Clearly, we have the following:

(a) for any $m\in\mathbb{N}$, the function $\tilde{u}^{(m)}$ satisfes the first three relations of system (35);
(b) the sequence $\{\tilde{u}^{(m)}\}_{m=1}^{\infty}$ converges to the function $\tilde{u}$ uniformly on $[0,h]\times[0,T]$ as $m\to\infty$.

Moreover, by equalities (41), (43) and (44), one can obtain

$$\begin{aligned}\tilde{u}^{(m)}(y,0) &= \sum_{i=1}^{m} \ell_i \tilde{u}_i(y,0) = \sum_{i=1}^{m} \ell_i q_i(y) \\ &= \sum_{i=1}^{m}(u_0-u_*,q_i)_{H^1_{k/\mu}[0,h]}q_i(y).\end{aligned} \tag{45}$$

Since the sequence $\{q_i\}_{i=1}^{\infty}$ is an orthonormal basis of the space $H^1_{k/\mu}[0,h]$, from (45) it follows that the sequence $\{\tilde{u}^{(m)}(\cdot,0)\}_{m=1}^{\infty}$ converges to the function u_0-u_* in the space $H^1_{k/\mu}[0,h]$ as $m\to\infty$. Hence,

$$\tilde{u}^{(m)}(\cdot,0)\to u_0-u_* \text{ in the space } H^1[0,h] \text{ as } m\to\infty. \tag{46}$$

Combining (46) with statements (a) and (b), we deduce that the function $\tilde{u}$ is a generalized solution of problem (35). Therefore, it can easily be checked that the function u defined in (29) is a generalized solution of problem (21).

Step 5: Uniqueness. The proof of the uniqueness of the obtained generalized solution can be performed by standard techniques (see, for example, [48]), and hence we omit details.

Thus, the proof of Theorem 2 is complete. □

Remark 2. *Using Formulas (29)–(33), it is easy to calculate and visualize the x-component of the velocity for the unsteady Poiseuille flow of both second grade fluids and Newtonian fluids; see, for example, Figure 2.*

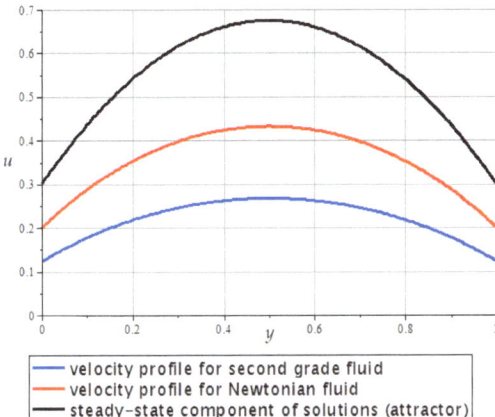

Figure 2. Velocity profiles for the plane Poiseuille flow with $h = 1, \mu = 1, \xi = 3, k = 5, u_0 \equiv 0, \alpha = 0.2$ (second grade fluid), $\alpha = 0$ (Newtonian fluid) at time $t = 0.2$.

6. Conclusions

In this paper, we investigated the initial-boundary value problem describing the unsteady Poiseuille flow of a second grade fluid in the channel $0 \leq y \leq h$ with impermeable solid walls under slip boundary conditions. Applying the concept of generalized solutions, we proved the unique solvability of this problem for any initial data from the Sobolev space $H^1[0, h]$. Our proof uses the method of separation of variables with a special orthonormal basis consisting of eigenfunctions of the related Sturm–Liouville problem as well as the generalized Abel criterion for uniform convergence of function series. As a result, the problem was solved in a closed form. Namely, the explicit formulas for the flow velocity and the pressure in the channel were established. The obtained analytical solutions contribute to a better understanding of the qualitative features of unsteady flows of polymer fluids and can be used for testing relevant numerical, asymptotic, and approximate analytical methods. Finally, note that the proposed approach is quite universal and can be adapted to solving many other problems related to the channel flows of non-Newtonian fluids with various types of boundary conditions.

Funding: This research received no external funding.

Institutional Review Board Statement: Not applicable.

Data Availability Statement: Data are contained within the article.

Conflicts of Interest: The author declares no conflict of interest.

References

1. Bird, R.B.; Curtiss, C.; Amstrong, R.; Hassager, O. *Dynamics of Polymeric Liquids*, 2nd ed.; Wiley: New York, NY, USA, 1987; Volume 2.
2. Doi, M.; Edwards, S.F. *The Theory of Polymer Dynamics*; Oxford University Press: Oxford, UK, 1988.
3. Pokrovskii, V.N. *The Mesoscopic Theory of Polymer Dynamics*; Springer: Dordrecht, The Netherlands, 2010. [CrossRef]
4. Le Bris, C.; Lelièvre, T. Micro-macro models for viscoelastic fluids: Modelling, mathematics and numerics. *Sci. China Math.* **2012**, *55*, 353–384. [CrossRef]
5. Rivlin, R.S.; Ericksen, J.L. Stress-deformation relations for isotropic materials. *J. Ration. Mech. Anal.* **1955**, *4*, 323–425. [CrossRef]
6. Cioranescu, D.; Girault, V.; Rajagopal, K.R. *Mechanics and Mathematics of Fluids of the Differential Type*; Springer: Cham, Switzerland, 2016. [CrossRef]
7. Dunn, J.E.; Fosdick, R.L. Thermodynamics, stability and boundedness of fluids of complexity 2 and fluids of second grade. *Arch. Ration. Mech. Anal.* **1974**, *56*, 191–252. [CrossRef]
8. Fosdick, R.L.; Rajagopal, K.R. Anomalous features in the model of "second order fluids". *Arch. Ration. Mech. Anal.* **1979**, *70*, 145–152. [CrossRef]

9. Dunn, J.E.; Rajagopal, K.R. Fluids of differential type: Critical review and thermodynamic analysis. *Int. J. Eng. Sci.* **1995**, *33*, 689–729. [CrossRef]
10. Navier, C.L.M.H. Mémoire sur le lois du mouvement des fluides. *Mém. l'Acad. Sci. l'Inst. France* **1823**, *6*, 389–416.
11. Denn, M.M. Extrusion instabilities and wall slip. *Annu. Rev. Fluid Mech.* **2001**, *33*, 265–287. [CrossRef]
12. Rajagopal, K.R. On some unresolved issues in non-linear fluid dynamics. *Russ. Math. Surv.* **2003**, *58*, 319–330. [CrossRef]
13. Xu, H.; Cui J. Mixed convection flow in a channel with slip in a porous medium saturated with a nanofluid containing both nanoparticles and microorganisms. *Int. J. Heat Mass Transf.* **2018**, *125*, 1043–1053. [CrossRef]
14. Wang, G.J.; Hadjiconstantinou, N.G. Universal molecular-kinetic scaling relation for slip of a simple fluid at a solid boundary. *Phys. Rev. Fluids* **2019**, *4*, 064201. [CrossRef]
15. Wilms, P.; Wieringa, J.; Blijdenstein, T.; van Malssen, K.; Hinrichs, J. Wall slip of highly concentrated non-Brownian suspensions in pressure driven flows: A geometrical dependency put into a non-Newtonian perspective. *J. Non-Newton. Fluid Mech.* **2020**, *282*, 104336. [CrossRef]
16. Ghahramani, N.; Georgiou, G.C.; Mitsoulis, E.; Hatzikiriakos, S.G.J.G. Oldroyd's early ideas leading to the modern understanding of wall slip. *J. Non-Newton. Fluid Mech.* **2021**, *293*, 104566. [CrossRef]
17. Hron, J.; Le Roux, C.; Malek, J.; Rajagopal, K.R. Flows of incompressible fluids subject to Navier's slip on the boundary. *Comput. Math. Appl.* **2008**, *56*, 2128–2143. [CrossRef]
18. Baranovskii, E.S.; Artemov, M.A. Steady flows of second-grade fluids in a channel. *Vestn. S.-Peterb. Univ. Prikl. Mat. Inf. Protsessy Upr.* **2017**, *13*, 342–353. [CrossRef]
19. Baranovskii, E.S. Exact solutions for non-isothermal flows of second grade fluid between parallel plates. *Nanomaterials* **2023**, *13*, 1409. [CrossRef]
20. Ting, T.W. Certain unsteady flows of second grade fluids. *Arch. Ration. Mech. Anal.* **1963**, *14*, 1–26. [CrossRef]
21. Coleman, B.D.; Duffin, R.J.; Mizel, V.J. Instability, uniqueness, and nonexistence theorems for the equation $u_t = u_{xx} - u_{xtx}$ on a strip. *Arch. Ration. Mech. Anal.* **1965**, *19*, 100–116. [CrossRef]
22. Nazar, M.; Corina, F.; Vieru, D.; Fetecau, C. New exact solutions corresponding to the second problem of Stokes for second grade fluids. *Nonlinear Anal. Real World Appl.* **2010**, *11*, 584–591. [CrossRef]
23. Fetecau, C.; Zafar, A.A.; Vieru, D.; Awrejcewicz, J. Hydromagnetic flow over a moving plate of second grade fluids with time fractional derivatives having non-singular kernel. *Chaos Solit. Fractals* **2020**, *130*, 109454. [CrossRef]
24. Fetecau, C.; Vieru, D. On an important remark concerning some MHD motions of second grade fluids through porous media and its applications. *Symmetry* **2022**, *14*, 1921. [CrossRef]
25. Shankar, B.M.; Shivakumara, I.S. Stability of plane Poiseuille and Couette flows of Navier–Stokes–Voigt fluid. *Acta Mech.* **2023**, *234*, 4589–4609. [CrossRef]
26. Fetecau, C.; Vieru, D. General solutions for some MHD motions of second-grade fluids between parallel plates embedded in a porous medium. *Symmetry* **2023**, *15*, 183. [CrossRef]
27. Arif, M.; Saeed, A.; Suttiarporn, P.; Khan, W.; Kumam, P.; Watthayu, W. Analysis of second grade hybrid nanofluid flow over a stretching flat plate in the presence of activation energy. *Sci. Rep.* **2022**, *12*, 21565. [CrossRef] [PubMed]
28. Ershkov, S.V.; Leshchenko, D. Note on semi-analytical nonstationary solution for the rivulet flows of non-Newtonian fluids. *Math. Methods Appl. Sci.* **2022**, *45*, 7394–7403. [CrossRef]
29. Ershkov, S.V.; Prosviryakov, E.Y.; Leshchenko, D. Marangoni-type of nonstationary rivulet-flows on inclined surface. *Int. J. Non-Linear Mech.* **2022**, *147*, 104250. [CrossRef]
30. Cioranescu, D.; Ouazar, E.H. Existence and uniqueness for fluids of second-grade. *Nonlinear Partial. Differ. Equ. Appl.* **1984**, *109*, 178–197.
31. Cioranescu, D.; Girault, V. Weak and classical solutions of a family of second grade fluids. *Int. J. Non-Linear Mech.* **1997**, *32*, 317–335. [CrossRef]
32. Le Roux, C. Existence and uniqueness of the flow of second-grade fluids with slip boundary conditions. *Arch. Ration. Mech. Anal.* **1999**, *148*, 309–356. [CrossRef]
33. Kloviene, N.; Pileckas, K. Nonstationary Poiseuille-type solutions for the second-grade fluid flow. *Lith. Math. J.* **2012**, *52*, 155–171. [CrossRef]
34. Baranovskii, E.S. Existence results for regularized equations of second-grade fluids with wall slip. *Electron. J. Qual. Theory Differ. Equ.* **2015**, *2015*, 91. [CrossRef]
35. Baranovskii, E.S. Weak solvability of equations modeling steady-state flows of second-grade fluids. *Differ. Equ.* **2020**, *56*, 1318–1323. [CrossRef]
36. Jaffal-Mourtada, B. Global existence of the 3D rotating second grade fluid system. *Asymptot. Anal.* **2021**, *124*, 259–290. [CrossRef]
37. Chemetov, N.V.; Cipriano, F. Optimal control for two-dimensional stochastic second grade fluids. *Stoch. Proc. Appl.* **2018**, *128*, 2710–2749. [CrossRef]
38. Cipriano, F.; Pereira, D. On the existence of optimal and ϵ-optimal feedback controls for stochastic second grade fluids. *J. Math. Anal. Appl.* **2019**, *475*, 1956–1977. [CrossRef]
39. Ngo, V.S.; Raugel, G. Approximate controllability of second-grade fluids. *J. Dyn. Control Syst.* **2021**, *27*, 531–556. [CrossRef]
40. Almeida, A.; Chemetov, N.V.; Cipriano, F. Uniqueness for optimal control problems of two-dimensional second grade fluids. *Electron. J. Differ. Equ.* **2022**, *2022*, 22. [CrossRef]

41. Baranovskii, E.S. Global solutions for a model of polymeric flows with wall slip. *Math. Meth. Appl. Sci.* **2017**, *40*, 5035–5043. [CrossRef]
42. Kaplický, P.; Tichý, J. Boundary regularity of flows under perfect slip boundary conditions. *Cent. Eur. J. Math.* **2013**, *11*, 1243–1263. [CrossRef]
43. Burmasheva, N.V.; Larina, E.A.; Prosviryakov, E.Y. A Couette-type flow with a perfect slip condition on a solid surface. *Vestn. Tomsk. Gos. Univ. Mat. Mekh.* **2021**, *74*, 79–94. [CrossRef]
44. Mácha, V.; Schwarzacher, S. Global continuity and BMO estimates for non-Newtonian fluids with perfect slip boundary conditions. *Rev. Mat. Iberoam.* **2021**, *37*, 1115–1173. [CrossRef]
45. Gkormpatsis, S.D.; Housiadas, K.D.; Beris, A.N. Steady sphere translation in weakly viscoelastic UCM/Oldroyd-B fluids with perfect slip on the sphere. *Eur. J. Mech. B Fluids* **2022**, *95*, 335–346. [CrossRef]
46. Baranovskii, E.S. The Navier–Stokes–Voigt equations with position-dependent slip boundary conditions. *Z. Angew. Math. Phys.* **2023**, *74*, 6. [CrossRef]
47. Artemov, M.A.; Baranovskii, E.S. Unsteady flows of low concentrated aqueous polymer solutions through a planar channel with wall slip. *Eur. J. Adv. Eng. Technol.* **2015**, *2*, 50–54.
48. Petrovsky, I.G. *Lectures on Partial Differential Equations*; Interscience Publishers: New York, NY, USA, 1954.

Disclaimer/Publisher's Note: The statements, opinions and data contained in all publications are solely those of the individual author(s) and contributor(s) and not of MDPI and/or the editor(s). MDPI and/or the editor(s) disclaim responsibility for any injury to people or property resulting from any ideas, methods, instructions or products referred to in the content.

Article

Improved Approach for ab Initio Calculations of Rate Coefficients for Secondary Reactions in Acrylate Free-Radical Polymerization

Fernando A. Lugo [1], Mariya Edeleva [2], Paul H. M. Van Steenberge [1] and Maarten K. Sabbe [1,*]

[1] Laboratory for Chemical Technology (LCT), Department of Materials, Textiles, and Chemical Engineering, Ghent University, Technologiepark-Zwijnaarde 125, 9052 Ghent, Belgium; fernando.lugo@ugent.be (F.A.L.); paul.vansteenberge@ugent.be (P.H.M.V.S.)

[2] Center for Polymer and Material Technology (CPMT), Department of Materials, Textiles, and Chemical Engineering, Ghent University, Technologiepark-Zwijnaarde 130, 9052 Ghent, Belgium; mariya.edeleva@ugent.be

* Correspondence: maarten.sabbe@ugent.be

Citation: Lugo, F.A.; Edeleva, M.; Van Steenberge, P.H.M.; Sabbe, M.K. Improved Approach for ab Initio Calculations of Rate Coefficients for Secondary Reactions in Acrylate Free-Radical Polymerization. *Polymers* 2024, 16, 872. https://doi.org/ 10.3390/polym16070872

Academic Editors: Célio Pinto Fernandes, Luís L. Ferrás and Alexandre M. Afonso

Received: 23 February 2024
Revised: 14 March 2024
Accepted: 17 March 2024
Published: 22 March 2024

Copyright: © 2024 by the authors. Licensee MDPI, Basel, Switzerland. This article is an open access article distributed under the terms and conditions of the Creative Commons Attribution (CC BY) license (https:// creativecommons.org/licenses/by/ 4.0/).

Abstract: Secondary reactions in radical polymerization pose a challenge when creating kinetic models for predicting polymer structures. Despite the high impact of these reactions in the polymer structure, their effects are difficult to isolate and measure to produce kinetic data. To this end, we used solvation-corrected M06-2X/6-311+G(d,p) ab initio calculations to predict a complete and consistent data set of intrinsic rate coefficients of the secondary reactions in acrylate radical polymerization, including backbiting, β-scission, radical migration, macromonomer propagation, mid-chain radical propagation, chain transfer to monomer and chain transfer to polymer. Two new approaches towards computationally predicting rate coefficients for secondary reactions are proposed: (*i*) explicit accounting for all possible enantiomers for reactions involving optically active centers; (*ii*) imposing reduced flexibility if the reaction center is in the middle of the polymer chain. The accuracy and reliability of the ab initio predictions were benchmarked against experimental data via kinetic Monte Carlo simulations under three sufficiently different experimental conditions: a high-frequency modulated polymerization process in the transient regime, a low-frequency modulated process in the sliding regime at both low and high temperatures and a degradation process in the absence of free monomers. The complete and consistent ab initio data set compiled in this work predicts a good agreement when benchmarked via kMC simulations against experimental data, which is a technique never used before for computational chemistry. The simulation results show that these two newly proposed approaches are promising for bridging the gap between experimental and computational chemistry methods in polymer reaction engineering.

Keywords: kinetics of radical polymerization; acrylates; secondary reactions; ab initio; kinetic Monte Carlo; pulsed-laser polymerization

1. Introduction

Polyacrylates find a wide range of applications, being important ingredients in coatings, paints, casts, adhesives, 3D printing resins, hydrogels, etc. [1–8], requiring control over the material properties (i.e., transparency, gloss, haziness, curing temperature and glass transition temperature). The wide range of applications has, in recent years, increased the interest in deeply understanding the kinetic parameters of acrylates [9,10]. These properties are determined by the molecular properties of the polymer chains, (e.g., the average molar mass and molar mass distribution, unsaturation (i.e., terminal-double-bond) level, and branching density) [11,12].

Due to the complex reaction mechanism of acrylate free-radical polymerization (FRP) (i.e., the presence of numerous secondary reactions (Figure 1)), the relation between the molecular structure of the acrylic polymers and reaction conditions is not straightforward. The active radical center of an alkyl acrylate polymer chain, a so-called end-chain radical (ECR), can undergo intramolecular chain transfer reactions (backbiting), which will transfer the radical center towards the middle of the chain, forming a mid-chain radical (MCR). This tertiary carbon-centered radical has a different chemical reactivity than its secondary radical counterpart: the ability to migrate the radical center through the backbone; chain breaking through a β-scission reaction; or further propagation at a slower rate. This last reaction produces branches in the polymer, which further complicates the structural analysis of the polymer product. The β-scission reaction reduces the average chain length and produces unsaturated polymer chains (macromonomers). The latter can propagate with another ECR or MCR, bonding longer chains and creating a branch. ECRs or MCRs can also undergo intermolecular chain transfer reactions, terminating a chain by abstracting a hydrogen atom from another chain and, hence, creating a new MCR in a random position in the polymer backbone. Additionally, if the reaction proceeds with a hydrogen atom of a monomer, the chain transfer reaction produces a new unimer radical, which will start a new chain. These rather rare reaction events during polymerization significantly affect the microstructure of the polymer chains [13] and consequently they determine the material properties. As a result, the design of the tailored material properties requires a profound knowledge of these rare secondary-reaction events and accurate values of their rate coefficients.

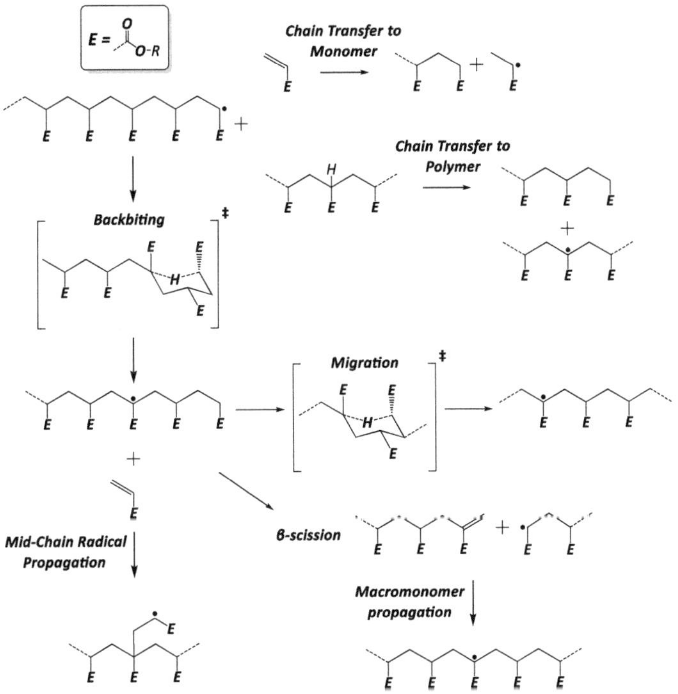

Figure 1. Most relevant secondary reactions in free-radical polymerization of acrylates.

Precise measurement of the rate coefficients in FRP reactions is challenging. For the k_p of propagation reactions, the IUPAC Working Party on "Modelling of kinetics and processes of polymerization" proposed the use of pulsed-laser polymerization (PLP) combined with molar mass distribution (MMD) analysis as a reliable methodology [14–18]. This method was applied to multiple monomer types by multiple research groups, producing a set of benchmark values for monomers as styrene, (meth)acrylic esters, including linear and cyclic alkyl substituents [6,17–25]. However, the applicability of PLP for secondary reactions is not straightforward.

Despite the advances in measuring techniques, the effect of secondary reactions on the MMD is still difficult to isolate from the rest of the reactions. Some research relies on the analysis of the polymers' molecular properties (branching degree, unsaturation level, M_n and MMD) to derive the values of the rate coefficients for the secondary reactions [26–33]. To analyze these properties, final-microstructure analysis methods, such as ^{13}C NMR, are useful for discerning the effects of the secondary reactions in the microstructure. However, each of these molecular properties is a consequence of a set of serial or parallel secondary reactions, making it difficult to link the structural observations directly to individual rate coefficients. For example, in acrylate radical polymerization, both backbiting and chain-transfer-to-polymer reactions create MCRs. These can propagate to form branches or undergo β-scission and split the chain in two. Therefore, the degree of branching cannot be directly attributed to a single rate coefficient.

To overcome this challenge of data extraction from experiments, computational methods are used to aid the analysis of the post-processed results. For example, kinetic Monte Carlo (kMC) methods can simulate polymer properties, such as the MMD, allowing for more detailed comparisons with the measured experimental data. In this way, the kMC method allows for identifying the effects of a particular secondary reaction on the polymer product by fitting simulations and experiments under specific reaction conditions. The reliability of determining the rate coefficients through comparison to experimental data via these simulations strongly depends on the sensitivity towards the rate coefficients. It has been shown that acrylate polymers properties are sensitive to several rate coefficients at the same time [13] and, consequently, the reliable determination of individual rate coefficients is a complicated matter.

First-principle methods allow for the prediction of the rate coefficients for elementary reaction steps, which is especially useful in cases where the experimental determination is not straightforward. FRP reactions have already been studied using ab initio methods by several authors [34–39] for propagation, as well as secondary reactions. Ab initio research on secondary reactions typically focuses on which type of model/basis set combination is most suitable to represent the reaction barriers, and on the size of the model molecule. However, most rate coefficients obtained via ab initio calculations span across a very wide range. For example, the backbiting reactivity varies over multiple orders of magnitude depending on the method and basis set and cannot be considered to be in acceptable agreement with the known experimental data. A direct comparison of ab initio and experimentally derived rate coefficients is not always straightforward, as first-principle rate coefficients always represent an individual elementary reaction, while experimentally derived rate coefficients can incorporate contributions from neglected reactions or be strongly cross correlated to other rate coefficients or other phenomena (e.g., diffusional limitations, which mask or disguise the intrinsic rate coefficient) [40].

The accuracy of ab initio predictions depends on the correct construction of the molecular model. A representative ab initio model starts by selecting an adequate molecular-structure representation. The molecular structure of this model should not be as large as actual polymers because of computational limitations, while an overly small model will not account for the steric and electronic effects imposed by long chains. Therefore, the molecular model is typically made as large as can be managed with a given computational method. Furthermore, FRP reactions occur in a condensed phase in which the solvation of the radicals highly impacts the secondary reactions. By default, ab initio methods consider their rate coefficient predictions in the gas phase, which incorrectly describes the reality. Hence, the addition of solvation effects by some additional model is required, such as PCM or COSMO-RS [41]. Because the size of the molecular model is significantly smaller than a real polymer molecule, the effects that arise from the macromolecular nature of polymers, such as the size of the molecule, tacticity and 3D folding of the polymer chain, need to be accounted for through different approaches. Firstly, the acrylate polymer backbone is atactic in the case of FRP of acrylates because the repeating units can be arranged in different manners depending on the chiralities of the substituents. Secondly, the segments in the middle of the polymer chain experience restricted motion, which affects the reactivity of MCRs.

In this work, we develop a complete set of rate coefficients of the secondary reactions in FRP of alkyl acrylates based on ab initio calculations, using the same computational level for all reactions to maintain a consistent model. We account for the solvation effects via the COSMO-RS theory. We propose two new approaches that reduce the gap between the predicted and experimental rate coefficients. The first approach accounts for all permutations of an atactic polymer chain by considering all optical isomers of the chain segment model for those reactions that involve considerable movements in the backbone of the chain. The second approach considers the flexibility reduction of a segment in a large polymer chain. We account for this by applying a geometry restriction to the model component's molecular structure to simulate the strain applied by the rest of the backbone. To demonstrate that both approaches are efficient, the resulting ab initio data set was tested by performing kMC simulations of (i) pulsed-laser polymerization–size exclusion chromatography (PLP-SEC) traces under three highly different experimental conditions, and (ii) a controlled degradation of ARGET ATRP polymers. We selected a broad range of temperatures, monomer concentrations and laser frequencies to achieve a high sensitivity towards a particular secondary-reaction rate coefficient in each simulation.

2. Computational Details

2.1. Level of Theory

Ab initio calculations were performed using the Gaussian 16_C.01 package [42]. All structures were optimized at the B3LYP/6-311+G(d,p) level. Thermal contributions were calculated using the harmonic-oscillator approach at the same computational level. All structures were confirmed to have zero imaginary frequencies in the case of reactants and products, and exactly one in the transition states. Electronic energies for all stationary points were determined using single-point M06-2X/6-311+G(d,p) calculations using the previously optimized B3LYP/6-311+G(d,p) geometries. Standard ideal gas statistical thermodynamics were used to calculate the enthalpies, entropies and Gibbs energies in the gas phase [43]. The optimized geometries used to calculate every reaction within this work are included in Supporting Information.

To include solvation effects, the Gibbs energies of solvation (ΔG_{solv}) were calculated using the COSMO-RS theory [44–46] as implemented in the COSMOtherm software package [47]. These solvation energies were calculated by COSMO-RS using BP86/TZVP single-point calculations based on the same B3LYP/6-311+G(d,p) geometry optimized before, using a methyl acrylate as the solvent with a density of 950 g L^{-1} at 298.15 K. The effect of the solvation energy on the rate coefficients is shown in Table S6 in Supporting Information. Afterwards, the ΔG_{solv} was included in the gas Gibbs energy with the following equation:

$$G^{\circ}_{cond} = G^{\circ}_{gas} + \Delta G_{solv} \quad (1)$$

Classical transition state theory was used to calculate the rate coefficients based on the Gibbs reaction barrier in the condensed phase via the following equation:

$$k(T) = \kappa(T) \frac{k_B T}{h} \left(C^0\right)^{-\Delta^{\ddagger} n} exp\left(-\frac{\Delta^{\ddagger} G^0_{cond}}{RT}\right) \quad (2)$$

where $\kappa(T)$ is the quantum-tunneling correction factor calculated through the Eckart method [48]; k_B is the Boltzmann constant; h is the Plank constant; C^0 is the standard concentration of 1 mol L^{-1}; $\Delta^{\ddagger} n$ is the difference in moles between the reactant and transition states; $\Delta^{\ddagger} G^0_{cond}$ is the temperature-dependent Gibbs energy barrier; and R is the universal gas constant. Lastly, Arrhenius parameters were regressed over a broad temperature range starting at 298.15 K till 413.15 K, as used in the simulations explained in the following section.

2.2. Molecular-Model Construction Considering Chiral Effects

To account for the influence of the polymer chain chirality, we predicted the reaction rate coefficient for each chiral permutation within the chosen polymer section and then averaged out the rate coefficients, including the weighting factors. For this purpose, the number of enantiomers within a molecular model must be clearly defined based on the number of chiral centers. Each radical molecular model with n units has $n-1$ chiral centers, considering that the unit possessing the radical lacks chirality. Then, the molecular model has several possible enantiomers equal to all chiral permutations, excluding mirror-image structures. For example, if an ECR molecular model possesses three units, the first two units possess chirality, and the last one does not because it has a radical. Hence, the structure has two possible chiral permutations: *RRM*/SSM** and *RSM*/SRM**. Afterwards, the rate coefficients for all chiral permutations are averaged out using weighting factors based on chiral propagation probabilities, which will be explained in detail in Section 3.1. These rules apply for all molecules, with a few exceptions, which will be explicitly explained.

2.3. Reduced-Flexibility Approach

Firstly, the molecular structures of the reactants, transition state and products were optimized. Secondly, a strain was applied on the reactant and transition-state chains by increasing the distance between the first and last carbon atoms in 0.2 Å steps. This process gradually increased the electronic energies of both the reactant and transition states, which impacted the Gibbs energy of each structure. Lastly, the Gibbs energy was recalculated at each step, and the reaction barrier trend was studied so that the effect induced by the reduced-flexibility effect could be distinguished.

2.4. Simulation Details

The kinetic Monte Carlo (*kMC*) modeling of the pulsed-laser polymerization–size exclusion chromatography (PLP-SEC) of *n*BuA was performed using the models reported by Marien et al. [49] and Vir et al. [50], starting from the well-established Gillespie algorithm [51] combined with tree-based data structures [52], advanced sampling algorithms [53], detailed reaction schemes and well-established experimentally determined rate coefficients. The modeling details are provided by Marien et al. [49] and Vir et al. [50] We simulated 3 sets of the PLP-SEC experimental conditions, which are sensitive to the values of backbiting and β-scission secondary reactions: (1) 306 K in bulk performed at a 500 Hz laser frequency, (2) 306 K with a solvent fraction (Φ_s) of 0.75 and a 50 Hz laser frequency and (3) 413 K in bulk performed at a 10 Hz laser frequency, being sensitive to ECR propagation, backbiting and β-scission, respectively.

Electron spray ionization mass spectrometry (ESI-MS) data for the synthesis of macromonomers (MMs) via the activation of bromine-capped poly(*n*-butyl acrylate) were simulated using the data set predicted in this work, including backbiting, radical migration, β-scission, macromonomer propagation and chain transfer to polymer. The rest of the model parameters were taken from Van Steenberge et al. [54] Because this experiment was conducted free of monomers in the mixture, the results were insensitive to reactions involving monomers, enhancing the effect of the aforementioned side reactions used in this simulation.

3. Results and Discussion

To develop the complete rate coefficient data set, each relevant reaction in the acrylate radical polymerization had to be accounted for. This included ECR propagation, backbiting or intramolecular chain transfer, β-scission, migration, macromonomer propagation, MCR propagation, chain transfer to monomer and, lastly, chain transfer to polymer or intermolecular chain transfer. For each reaction, a suitable molecular model was defined, and then ab initio predictions were performed and compared with the literature data. For every reaction, we used a specific molecular model for the ab initio calculations to achieve the maximal accuracy and minimal computational time. The relevance of the application of different molecular models arises from the structural factors, which affect a particular reaction. For example, the ECR propagation of acrylates is influenced by the molecular structure of the monomer and not by optical isomerism. Thus, we can use a simplistic model made up of two units instead of a more time-consuming larger model. At the same time, to simulate backbiting, we need to explicitly account for at least five monomer units to correctly represent the chiral effects of nearby units. Consequently, variation in the molecular model allows us to reduce the computational time without losing the accuracy of the predictions. To guide the reader, the molecular models are specified in the section corresponding to each secondary reaction. Finally, we applied kMC simulations to show that the set of rate coefficients derived via our improved methodology, which is a benchmark procedure never used before, provides an adequate representation of the experimental results.

3.1. Backbiting in Atactic Polymer Chains

The default approach in quantum chemistry is to calculate rate coefficients based on the minimum-energy structures of the reactant and transition states; however, this strategy is insufficient to predict backbiting rate coefficients. We realized, by testing different optical isomers for the methyl acrylate polymer chain, that the reaction barrier is strongly dependent on the tacticity of the chosen model. In the case of methyl acrylate, the pentamer model radical used has 16 possible different optical isomers due to the existence of four different chiral centers in the main chain, as shown in Figure 2.

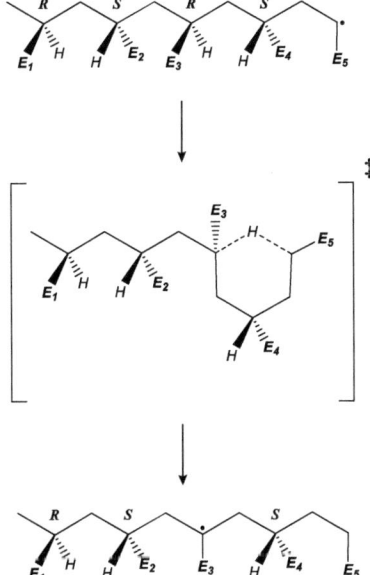

Figure 2. Backbiting reaction of the methyl acrylate pentamer: minimum-energy optical isomer (RSRS/SRSR).

A careful study of all the chiral permutations of these optical isomers reveals that the minimal-energy isomer corresponds to the structure with alternating chiralities (RSRS/SRSR), as this arrangement minimizes the lateral acrylate side-chain interactions. Nevertheless, the Gibbs energies in the condensed phase of the various optical isomers are, at most,

11 kJ mol^{-1} more than the minimum-energy isomer (see Supporting Information, Table S1, for all optical isomers), and therefore these other optical isomers are thermodynamically accessible as well. Although the energies indicate that these different optical isomers can exist, the actual chiral composition of the polymer chains is controlled by the propagation kinetics.

During the propagation of an ECR in acrylate polymerization, the achiral radical center turns into a chiral carbon atom. Depending on the formed chirality, the addition could follow two pathways: in the first, it produces a chiral center with the same chirality as the preceding monomer unit in the polymer chain (*meso* dyad), and, in the second, it produces a chiral atom with a different chirality than the preceding monomer unit (*racemo* dyad). For example, a dimer radical is formed by two units: one with a chiral center (R) and the other being an ECR. After the addition of the radical to a monomer, the resulting trimer radical can have a chirality of RR or RS, as shown in the following equation and the scheme in Figure 3.

$$R + M \xrightarrow{k_{pIC}} RR$$
$$R + M \xrightarrow{k_{pAC}} RS$$

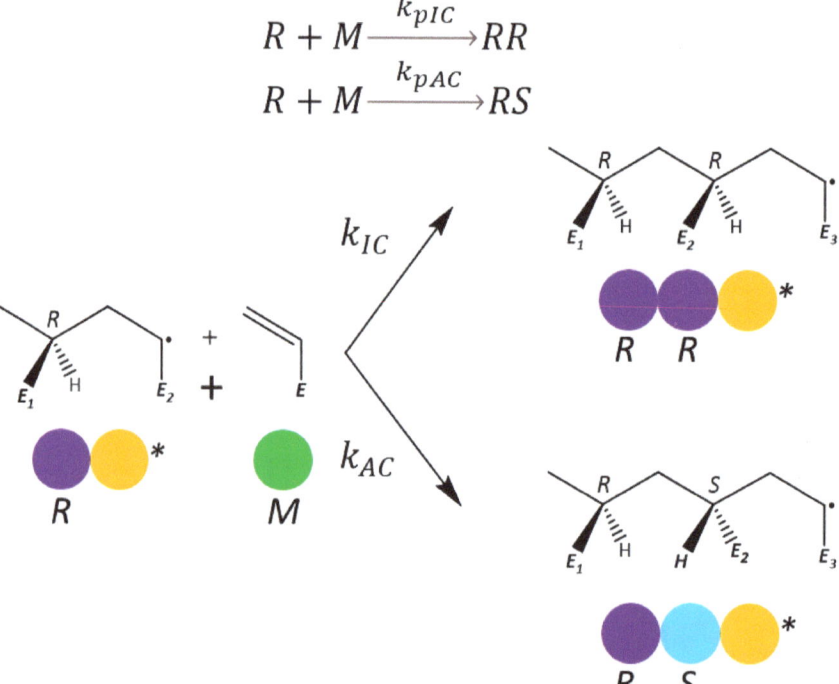

Figure 3. ECR propagation scheme showing identical-chirality (IC) propagation, producing a *meso* dyad, and alternating-chirality (AC) propagation, producing a *racemo* dyad.

In this equation, the rate coefficient for identical-chirality propagation is represented by k_{pIC}, and its counterpart, alternating-chirality propagation, is represented by k_{pAC}. The relation between these two rate coefficients and the overall propagation rate determines the probability of the next monomer addition to create a *meso* (P_{IC}) or *racemo* (P_{AC}) dyad, as defined in Equation (3):

$$P_{IC} = \frac{k_{pIC}}{k_{pIC} + k_{pAC}} \qquad P_{AC} = \frac{k_{pAC}}{k_{pIC} + k_{pAC}} \qquad (3)$$

These rate coefficients were predicted by using a dimer model, and the results are shown in Table S2 of Supporting Information. The addition probabilities are analogous to the tacticity of the polymer because the chirality determines the relative orientation of the acrylate groups around the polymer backbone. Experimental measurements found that methyl acrylate has an almost atactic structure with a slight preference towards the *meso* configuration [55] (Table 1), which translates to a slight preference for identical-chirality sequences. This work predicts a preference towards alternating chirality, implying more *racemo* combinations than experiments predict at a relation of 2:1. However, this preference does not indicate that the acrylate polymer is not atactic because the substituents are still disposed in a random manner. Moreover, the ab initio Gibbs energy reaction barriers between identical- and alternating-chirality propagations differ only by 1.6 kJ mol^{-1} at room temperature (see Supporting Information, Table S2), which, in terms of the ab initio accuracy, means that both reaction barriers are considered equivalent.

Table 1. Addition chirality probabilities for ECR propagation in methyl acrylate polymerization.

Addition Probabilities	Experimental Work [55]	This Work
Identical chirality (P_{IC})	0.52	0.34
Alternating chirality (P_{AC})	0.48	0.66

By knowing the probability to form an identical or alternating chiral center in the backbone upon addition, it is possible to calculate the probability that an n-length polymer segment exists within the backbone by multiplying the probabilities. As an example, a segment within the backbone composed by three units that are a *meso* and then a *racemo* dyad has a probability of existence of $P_{IC} \times P_{AC}$. This enables us to calculate the chiral composition of the polymer supposing a fixed-length polymer segment, which is useful for constructing complete ab initio molecular models.

For the backbiting molecular model, we calculated the chiral composition of each possible pentamer structure based on the addition chiral probabilities for each of the 16 different enantiomers. Table 2 includes the chiral composition based on our ab initio predictions and those reported in the literature based on experiments [55]. To have a consistent relation between the chirality of the atoms of the backbone in different molecular models, the atom chirality was determined assuming that the pentamer model structure is embedded in a long polymer chain (i.e., assuming the first unit in the pentamer model (unit 1 in Figure 2)). In the backbiting reaction of the methyl acrylate pentamer, the minimum-energy optical isomer (RSRS/SRSR) is attached to the rest of the chain. With this assumption, the priority order for the chirality determination considers the left-side backbone in Figure 1 as a higher priority in contrast to the right-side backbone, resulting in R or S carbon atoms, depending on the disposition of the acrylate substituent (see Supporting Information, Figure S1, for an explicit example of determining the chiralities in the pentamer model radical). As a result, this assignment of chiralities implies that an RRRR or SSSS sequence is isotactic and that an RSRS or SRSR sequence is syndiotactic.

Table 2. Statistical occurrences of chiral sequences in the applied methyl acrylate pentamer model, based on the addition ratios from Table 1. Experimental values are taken from Satoh et al. [55].

Enantiomer	Experiment (%)	This Work (%)
SSSS/RRRR	14.1	3.9
SSSR/RRRS	13.0	7.6
SSRS/RRSR	12.0	14.8
SRSS/RSRR	12.0	14.8
RSSS/SRRR	13.0	7.6
SSRR/RRSS	13.0	7.6
SRSR/RSRS	11.1	28.7
RSSR/SRRS	12.0	14.8

The statistical occurrences of the chiral sequences shown in Table 2 illustrate that no singular chiral sequence dominates the overall mixture for more than 30%. This observation remains consistent whether derived from the experimental tacticity or the computational probabilities calculated ab initio. The dominant configuration predicted in this study, namely, the RSRS/SRSR sequence, accounts for less than 30% of the sequences, while the most probable configuration identified through the experimental data, the RRRR/SSSS sequence, comprises less than 15% of the sequences. While these results may seem rather discrepant to the general reader, it must be stressed that this is the first time that these optical isomers have been explicitly calculated ab initio. These results underscore that the polymer exhibits a diverse enantiomeric composition rather than being accurately represented by a single-structure ab initio molecular model. It is evident that considering multiple configurations becomes pertinent in understanding the polymer's microstructure, while relying on a singular model might fail to capture the inherent complexity of the polymer's chiral composition, especially when the enantiomers display distinct reactivities.

Depending on the chirality sequence, we indeed found that the pentamer optical isomers display different backbiting reactivities, as shown in Table 3, in which the rate coefficient changes depending on the backbone chirality sequence. The arrangement of the side chains influences the energetics of the reactant radical and six-ring transition state for the backbiting. In general, isomers with alternating chiralities have higher stabilities, leading to higher reaction barriers for backbiting in comparison to isomers with identical tacticity sequences, which are typically less stable and therefore have lower backbiting barriers, as shown in Table S1. All these observations lead us to consider the reactivity of each statistically and energetically relevant isomer, rather than considering only the minimal-energy isomer, to calculate the rate coefficient. As such, the rate coefficient is calculated as the average of each optical isomer rate coefficient, weighted by the statistical occurrence of each optical isomer (see Equation (4) and Table 2). In Equation (4), $\bar{k}_{BB}$ represents the weighted-average backbiting rate coefficient, ω_i represents the weight for each optical isomer (i), k_i is the respective rate coefficient of the isomer (i) and n is the total number of optical isomers, which depends on the size and number of chiral atoms in the molecular model. In the backbiting case in which the model is a pentamer with four chiral atoms, which produce eight relevant optical isomers, n is equal to 8:

$$\bar{k}_{BB} = \sum_{i=1}^{n=8} \omega_i k_i \quad (4)$$

Table 3. Weighted-average approach applied to methyl acrylate backbiting using Equation (4).

Enantiomer	Simulated ω_i	k [s^{-1}] 298.15K	k [s^{-1}] 413.15K	A [s^{-1}]	E_a [kJ/mol]
SSSS/RRRR	0.039	2.40×10^2	3.96×10^4	2.29×10^{10}	45.8
SSSR/RRRS	0.076	5.64×10^2	7.87×10^4	2.96×10^{10}	44.3
SSRS/RRSR	0.148	2.35×10^2	2.16×10^4	2.75×10^9	40.5
SRSS/RSRR	0.148	3.16×10^1	1.04×10^3	1.23×10^7	33.8
RSSS/SRRR	0.076	1.08×10^3	2.94×10^5	5.63×10^{11}	49.4
SSRR/RRSS	0.076	3.38×10^2	5.90×10^4	3.92×10^{10}	46.2
SRSR/RSRS	0.287	4.92×10^1	1.55×10^4	4.88×10^{10}	51.6
RSSR/SRRS	0.148	2.89×10^2	4.82×10^4	2.85×10^{10}	45.8
$\bar{k}_{BB}$		2.57×10^2	4.95×10^4	4.09×10^{10} *	46.8 *

* If the weighted-average backbiting ($\bar{k}_{BB}$) is calculated by experimental weighting factors instead of ab initio results, the changes are negligible: the pre-exponential factor changes to 6.34×10^{10} and the activation energy changes to 46.9 kJ mol^{-1}.

The obtained weighted-average rate coefficient ($\bar{k}_{BB}$) is calculated from the rate coefficients of the individual optical isomers along with the individual rate coefficient (k_{BB}) for each optical isomer, as reported in Table 3. The data in Table 3 show that the difference between the rate coefficients for optical isomers could be up to two orders of magnitude. Therefore, if only one of the chiral permutations is chosen as a model radical to calculate the backbiting rate coefficient, it would be an incorrect description of the average rate coefficient for an actual mixture of ECRs. The weighted-average approach provides a better overall representation of the different rates at which backbiting occurs.

The main source for the differences in the rate coefficients (Table 3) is related to the steric hindrance between the side chains of the pentamer model. The energy of these large structures is mostly dependent on the arrangement of methyl acrylate units relative to each other, and it is a direct consequence of the tacticity of the main chain. In the reactants (pentamer ECRs), the main-chain energy is minimized in a straight zig-zag shape. However, if the chirality is identical in two sequential units (*meso dyad*), then the steric repulsion between the side chains makes it energetically more favorable to bend the main chain of the polymer so that the distance between the acrylate units is increased. This will result in a curled structure, which is more likely to undergo backbiting in comparison with its counterpart with the alternating-chirality structure (*racemo dyad*) because this chain will be straighter (see Supporting Information, Figure S2, for a detailed example). This curled structure is mentioned by Yu et al. [56] in their ab initio work as an intermediate state before the backbiting reaction. Furthermore, it is important to note that the tacticity also affects the formation of the transition state in diverse ways for each optical isomer. Otherwise, the reactivities would be equal for each optical isomer despite the relative reactant stability (see Supporting Information, Table S1). The relative differences between the transition states of optical isomers arise from two sources: the arrangement of the acrylate units within the ring (third, fourth and fifth carbon atoms in Figure 2), which vary between an axial or equatorial placement, and the shape of the remainder of the chain that is not part of the ring, related to the first and second monomer units shown in Figure 2. For a visual model representation of these structures, see Supporting Information, Figure S3.

Figure 4 compares our ab initio Arrhenius parameter prediction (black line) to the available literature data: the experimental median is the blue dashed line [28,32,50,57–61], the median is the grey dashed line, 25–75% of the experimental percentile zone is in light blue and the corresponding ab initio results are in light grey [34,35,56,62]. Figure 4 shows that our predictions are much closer to the experimental values than most reported ab initio predictions. Especially in the low-temperature range, our ab initio predictions agree almost perfectly with the experimentally measured values. However, there is no close agreement between our predicted values and the literature-based measured values in the high-temperature region. The higher activation energies predicted ab initio results in higher rate coefficients than the experiments indicate. Nevertheless, because all experimental works regarding alkyl acrylates are performed in the low-temperature range (273–333 K), these might be influenced by a strong tunneling effect (included in this work), which is negligible at high temperatures. Hamzehlou et al. [32] is the only experimental work focusing on high-temperature measurements, and their results produce a higher activation energy than the low-temperature measurements. Hence, data comparison is complicated because the extrapolation of low-temperature to high-temperature data should be performed accounting for the tunneling effect, which is not typically the case in the experimental literature.

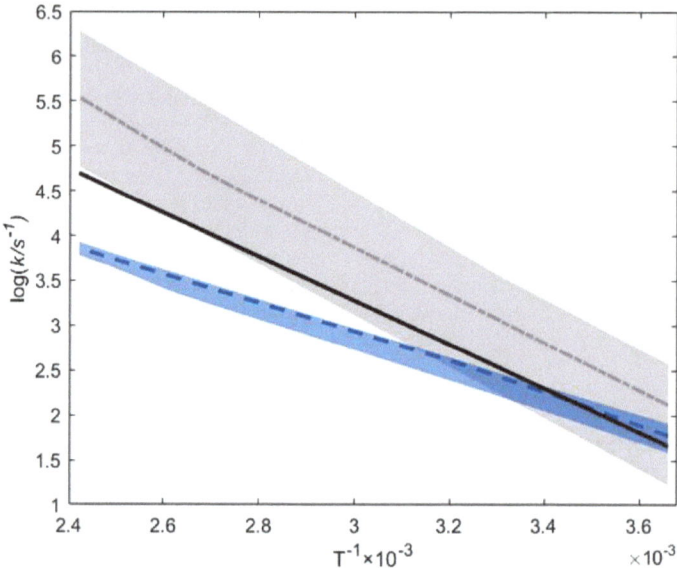

Figure 4. Comparison of the rate coefficients in this work with the literature values for the backbiting reaction. The colored sections represent the area between the 25% and 75% percentiles within a group of data: ■ this work's prediction; ▬ ▬ experimental median; ▬▬ ab initio median; 25–75% ab initio percentile area [34,35,56,62]; 25–75% experimental percentile area [28,32,50,57–61].

3.2. β-Scission

After backbiting or migration, the MCR formed can undergo a β-scission reaction, producing an ECR and a macromonomer. Depending on which bond around the MCR breaks, the reaction is called left or right β-scission. We chose a pentamer model to calculate the rate coefficient of the β-scission and took into account the influence of the monomer units adjacent to the MCR correctly (Figure 5). Our molecular model represents a left β-scission. It has been previously shown that the barriers for left and right β-scission are comparable [34]. The β-scission reaction is one of the most highly activated reactions in most polymerization models. A β-scission reaction similar to the used model, calculated via ab initio group additive values, shows a gas-phase activation energy around 100.3 kJ mol^{-1} [63].

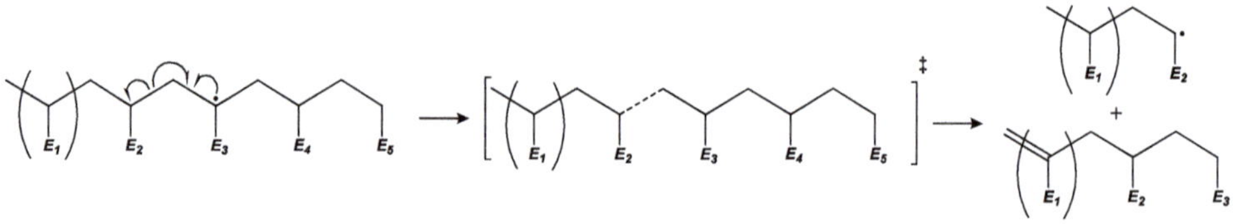

Figure 5. β-scission molecular model representing a left β-scission reaction.

Based on the pentamer model, we predicted the following Arrhenius parameters for the left β-scission of the methyl acrylate MCR:

$$k_\beta = \left(3.35 \times 10^{13} \text{ s}^{-1}\right) \exp\left(-\frac{115.97 \text{ kJ}}{RT}\right) \quad (5)$$

Our calculations resulted in a rate coefficient lower than those of any experimental results published [27,30–32,50,61]. However, it is in accordance with ab initio data published using similar methods [34,35,62], although no work before has predicted this rate coefficient using a basis set so large. A more detailed comparison of experimental and ab initio results from the literature data is included in Supporting Information, Table S4.

Despite the low rate coefficient predicted by ab initio results, experiments have thoroughly documented the existence of macromonomers [30,64–66], which are the product of β-scission; hence, the ab initio model might be misrepresenting the experimentally observed β-scission reaction. Knowing that other quantum chemical approaches predict similar rate coefficients, the discrepancy is therefore likely related to the inadequacy of the chosen pentamer model under actual polymer chain conditions; there must be an extra effect lowering the barrier, which is not included in a typical model structure. The above-mentioned approach of including all enantiomers will impact the β-scission rate coefficients less than backbiting because the spatial arrangement of the alkyl groups is similar in the reactant and transition states. Therefore, we explored other aspects (i.e., the reduced degree of freedom in an actual chain). In a real MCR, the chain segment where the β-scission reaction occurs is embedded in a longer polymer chain at both sides. This backbone reduces the degrees of freedom available for those monomer units involved in the reaction. Hence, the rest of the polymer will impose the strain, compression and limitation of the lateral movements over the segment. Thus, we investigated whether reducing the internal flexibility by constraining the geometry (i.e., mimicking the embedding in the polymer chain) can reduce the activation barrier and bring it closer to the experimental observed value.

If strain is applied to a straight polymer chain, the angles between the main-chain carbon atoms can be expected to become wider, and the covalent bonds between these carbon atoms can be expected to become slightly longer. From a thermodynamic point of view, if the covalent bond length increases, the structure's energy does as well, easing the bond-breaking reaction. And if this effect impacts the transition state less than the reactant structure, then it results in a lower reaction barrier. On the contrary, by the compression of the polymer chain, the structure bends to reduce steric interactions and produces a twisted, more compact arrangement. This geometry will increase the reaction barrier because the structure needs to stretch first to make the β-scission possible, causing an effect that will not dominate the rate coefficient. Additionally, not only longitudinal but also transversal movements can occur in the chain, and the real effect of the transversal movement on the structure will be like strain or compression, depending on the movement of the main chain. Therefore, the transversal-movement impact is considered by the strain and compression effects. Strain and compression are effects that ease and complicate the β-scission reaction. If the rate coefficient prediction considers both, the rate-enhancing effect will dominate the final rate coefficient. Hence, the strain effect will be included in the model to better represent an actual MCR embedded in a polymer chain.

To include strain in the model and mimic the interactions with the polymer backbone, we limit the flexibility of the polymer chain by imposing a maximum-elongation restriction on both the reactant- and transition-state structures. For this, we fix the distance between the first and last carbon atoms in the main chain of the pentamer model and increase the distance gradually. This geometry restriction increases the energy in both the reactant and transition states, as Figure 6 shows. The electronic energies of both structures become larger as the distance grows, yet the increase is much larger for the reactant structure than for the transition state. The rigid nature of the MCR structure causes the higher increase in energy, while, in the transition state, the increased flexibility of the breaking bond makes the energy less susceptible to elongation. Therefore, the effect of the flexibility reduction decreases the reaction barrier for the β-scission, as shown in Figure 6. Thus, we confirmed that the imposed strain lowers the barrier for the β-scission reaction and contributes to the better agreement of the calculated and experimental values of the rate coefficient.

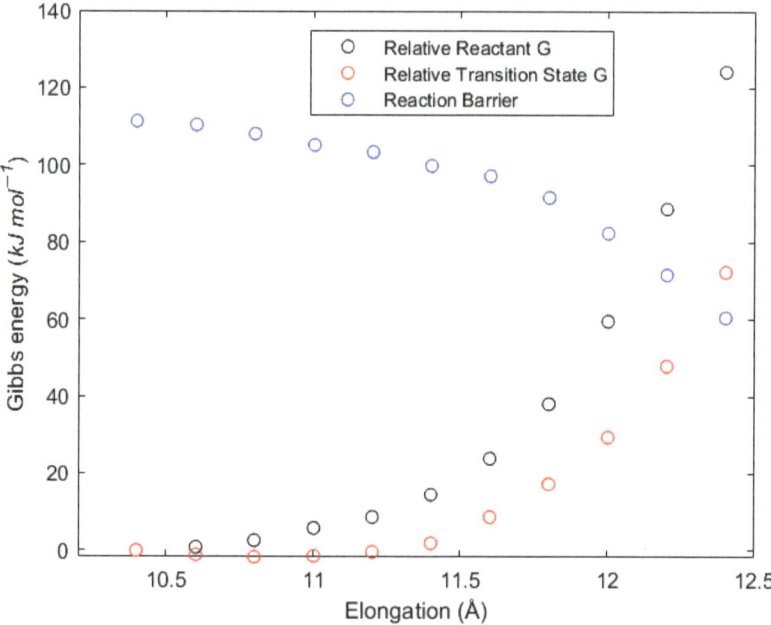

Figure 6. Reduced-flexibility effect on each structure involved in the reaction. The black circle (O) represent the Gibbs energy of a reactant structure with a certain elongation, relative to the non-elongated reactant molecular model. The red circle (O) represents the Gibbs energy of the transition-state structure with a certain elongation, relative to the non-elongated transition-state molecular model. The blue circle (O) represents the Gibbs reaction barrier at a certain elongation.

The actual strain present in a polymer chain will be a strongly time-dependent phenomenon, as the polymer chain mobility, chain length, temperature, dilution, stirring rates, etc., will affect it. If the reactant is stretched to the extreme, at some point, one main-chain bond will break, and the structure will resemble the products. Because no strict criterion exists for the optimal extension distance, the elongation that replicates an activation energy like the experimental reported values is chosen. Table 4 displays the rate coefficients and activation energies depending on the distance restriction. At 12.0 Å, an elongation of 20.8% of the reactant and 15.4% of the transition state is reached, and the activation energy decreases to 86.9 kJ mol^{-1}. The rate coefficient at said elongation is 86.3 s^{-1} at 383.15 K, very similar to the high-accuracy experimental values of Vir et al. [50] (71.5 ± 9.7 s^{-1}), reported in Table S4, indicating that the constrained geometry is a realistic representation of an actual polymer mid-chain radical. The rate coefficient based on an elongation of 12.0 Å was used in the kMC simulations (see Section 3.8).

Introducing strain to predict the rate coefficient of the β-scission is valid for radicals that are in the middle of the backbone, such as the products of radical migration or CTP. Those MCRs produced directly from ECR backbiting (bbMCR) have more freedom of movement on one side of the backbone, compared to the model we are discussing in this section. Hence, this bbMCR undergoes β-scission at a slower rate, similar to Equation (5). Despite this difference, the overall rate coefficient for the β-scission will be dominated by the middle-of-the-backbone MCR because its rate coefficient is several orders of magnitude higher than its bbMCR counterpart, even if the fraction is less than 0.1% of the total MCRs. Because the restricted-mobility approach gives good results for the acrylates, it can improve the accuracy of ab initio rate coefficient prediction for other monomers as well.

Table 4. Activation energies and rate coefficients for the β-scission reaction by restricting the distance between the first and last carbon atoms in the model structures.

Elongation between First and Last Carbon Atoms	E_a [kJ mol^{-1} s^{-1}]	k @ 383.15 K [s^{-1}]
Non-strained	116.0	5.14×10^{-3}
10.6	115.1	1.38×10^{-2}
10.8	112.6	2.89×10^{-2}
11.0	109.4	7.26×10^{-2}
11.2	108.0	1.26×10^{-1}
11.4	104.5	3.84×10^{-1}
11.6	101.2	8.83×10^{-1}
11.8	95.8	5.01×10^{0}
12.0	86.3	9.85×10^{1}
12.2	75.6	2.77×10^{3}
12.4	64.5	8.92×10^{4}

3.3. Migration

During the migration reaction, the MCR changes its position on the main chain through a hydrogen transfer mechanism, forming a six-membered ring structure similar to that formed via backbiting. A pentamer model was chosen to maintain consistency with the other secondary reactions. The reactant structure possesses the radical in the fourth unit, as Figure 7 shows, instead of in the fifth unit, as performed to study the backbiting (Figure 2). Afterwards, this radical transfers towards the second unit during the migration reaction.

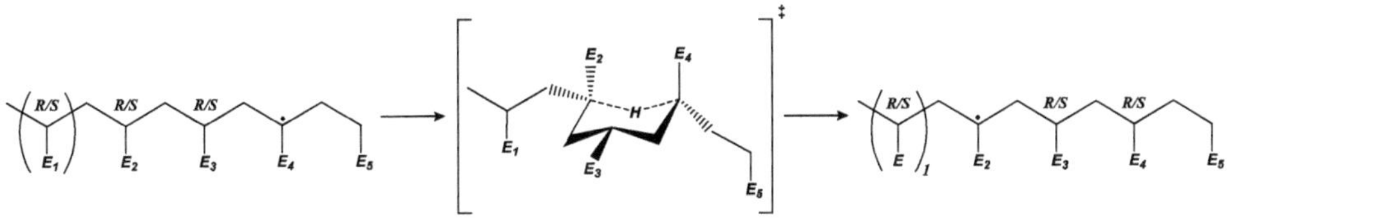

Figure 7. General methyl acrylate pentamer structure with the mid-chain radical on position 4, as used for the migration reaction model.

Similar to backbiting, for the migration reaction, the presence of multiple optical isomers can have a significant influence on the values of the predicted rate coefficients. Therefore, we applied the same weighted-average approach to evaluate the value of the k_{MIG}. For this case, only four optical isomers can be defined because the MCR eliminates one chiral center (Figure 7), making this model simpler than the eight permutations used for backbiting. Table 5 shows the rate coefficients, the weighting factors derived from the propagation probabilities and the final weighted averages for these optical isomers.

Table 5 shows that, as in the backbiting prediction, the rate coefficients for the migration range over different orders of magnitude depending on the chirality. According to these results, the rate coefficient decreases opposite to the number of alternating chiralities in the optical isomer. This means that the largest migration rate coefficient is observed for the isotactic sequence (RRR/SSS), which is the most unstable optical isomer and the most stable transition state. Conversely, the slowest rate coefficient is predicted for the syndiotactic sequence (RSR/SRS), which is the most stable enantiomer. Both structures with one alternating and one identical chirality (SRR/RSS and RRS/SSR) are within these two extremes, although the SRR/RSS enantiomer rate coefficient is similar to that of the syndiotactic sequence (RSR/SRS) (Table S3 in Supporting Information shows a thorough comparison of the energies). Eventually, we calculated the value of the rate coefficient as the weighted average of the rate coefficients for the individual isomers, in which the weighting factors are given by the abundances of the enantiomers presented in Table 1.

(Table 5; $k_{MIG} = 2.92 \times 10^5$ s^{-1} at 413 K). Interestingly, the isotactic sequence (RRR/SSS) dominates the weighted-average rate coefficients and Arrhenius parameters despite being the less probable isomer, showing the importance of considering all possible enantiomers within a reaction.

Table 5. Migration rate coefficients in high-temperature range (top) and comparison with literature values (bottom).

Enantiomer	Simulated ω_i	k [s^{-1}] 298.15K	k [s^{-1}] 413.15K	A [s^{-1}]	Ea [kJ mol^{-1}]
SSS/RRR	0.116	5.24×10^4	2.46×10^6	6.40×10^9	29.5
SSR/RRS	0.224	2.52×10^1	2.72×10^4	2.13×10^{10}	51.9
SRS/RSR	0.436	6.43×10^{-1}	1.07×10^3	2.13×10^9	55.4
RSS/SRR	0.224	1.58×10^0	2.31×10^2	9.05×10^6	39.1
k_{mig} (weighted average)		6.08×10^3	2.92×10^5	7.85×10^8	29.6
Van Steenberge et al. [54]			1.6×10^2		
Ballard et al. [67]			3×10^3		
Cuccato et al. [62]		6.24×10^0	6.62×10^{-4}	2.86×10^{10}	63.3

In contrast to backbiting, the rate coefficient predicted for migration using the weighted-average approach does not agree with the literature data. At the temperature of 413 K, Van Steenberge et al. [54] reported 1.6×10^2 s^{-1}, obtained from fitting the experimental data via kinetic Monte Carlo simulations, while Ballard et al. [67] reports 3×10^3 s^{-1}, obtained through experimental measurements. Both these reported rate coefficients disagree with the value predicted in this work, as they are from two to three orders of magnitude lower. The predicted rate coefficient for migration is even faster than the calculated rate coefficient for backbiting at said temperature. This is not what one would expect thermodynamically: the radical transfer from a tertiary-to-tertiary carbon atom, such as in migration, is expected to be slower than a radical transfer from the secondary to tertiary carbon atoms, as in backbiting. This inconsistency is addressed in the next paragraph.

In the calculations above, we allowed the free motion of the model polymer chain. However, similar to β-scission, the migration reaction proceeds in an MCR embedded in a polymer backbone on both sides. The rest of the chain will limit the flexibility of the model segment, making it an interesting case to test the procedure previously used for the β-scission. The main chain must contract to form the six-membered ring structure representing the transition state. In this process, the rest of the backbone will restrict the movement in the real polymer, while this effect lacks representation in the computational model. Figure 8 shows the effect of the reduced flexibility in the $G^{\ddagger}$ for the migration reaction. In this case, the Gibbs energy barrier increases as the flexibility is reduced, in contrast to the β-scission behavior, where the barrier decreases as the flexibility is reduced. Therefore, the reduced-flexibility approach correctly predicts the impact of the strain in the transition state, demonstrating that the influence of the backbone can explain the difference between the experimental and predicted rate coefficients. Similar to the β-scission reaction, the Gibbs energy graph lacks an objective criterion for selecting an elongation that can reproduce the correct barrier. For the migration reaction, an approach based on thermodynamic reasoning is proposed: the rate coefficient of the migration reaction should be in the same order of magnitude as that of the backbiting, as they are similar reactions, but the strain imposed by the backbone should make the rate coefficient reasonably smaller than that obtained via backbiting.

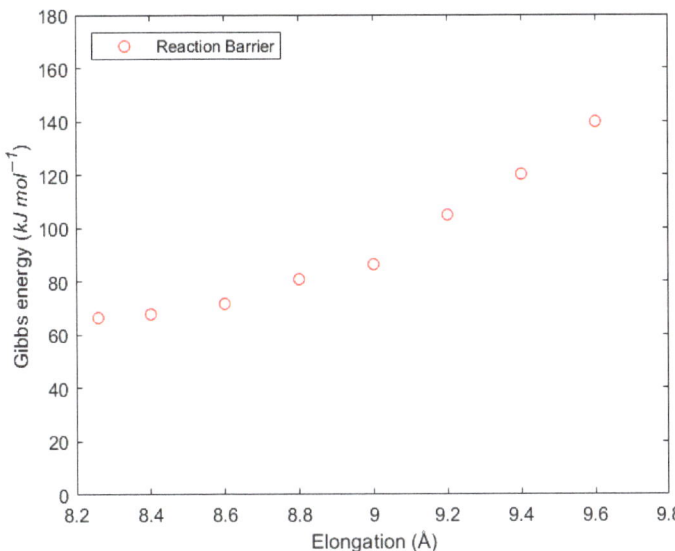

Figure 8. $\Delta^{\ddagger}G$ (○) vs. elongation of the radical migration reaction barrier applying the reduced-flexibility approach.

Finally, the elongation value chosen corresponds to 8.80 Å. This provides a $G^{\ddagger}$ value of 82.2 kJ mol^{-1} and a rate coefficient of 1.60×10^3 s^{-1} at 413.15 K, as seen in Table 6. An elongation of 6.5% provides a reasonable value for the rate coefficient instead of the 11.56% required for β-scission. This difference relates to the fact that the six-membered-ring transition state (Figure 6) for migration is rigid and not affected by the elongation. Therefore, the effects of elongation are distributed over less C-C bonds in comparison to β-scission, explaining the smaller elongation required.

Table 6. Activation energies and rate coefficients for the migration reaction by imposing a restriction on the distance between the first and last carbon atoms in the model structures. The values selected for the simulations are shown in bold.

Elongation between First and Last Carbon Atoms	$G^{\ddagger}$ 413.15 K [kJ mol^{-1} s^{-1}]	k 413.15 K [s^{-1}]
Non-strained	63.4	3.84×10^5
8.4	64.8	2.55×10^5
8.6	70.0	5.64×10^4
8.8	**82.2**	$\mathbf{1.60 \times 10^3}$
9.0	93.7	5.76×10^1
9.2	113.5	1.80×10^{-1}
9.4	139.1	1.03×10^{-4}
9.6	171.3	8.87×10^{-9}

3.4. Macromonomer Propagation

Macromonomers are the products of a β-scission reaction and are therefore chains that contain unsaturated ends. Their reactivities are similar but not identical to acrylate monomers, as they can undergo a propagation reaction that produces a polymer chain with an MCR and a side branch.

Figure 9 displays the molecular model, which consists of a unimer radical added to an unsaturated methyl acrylate tetramer. Because the backbone spatial disposition remains through the reaction, the influence of the chiralities of the nearby units on the reaction can be expected to be negligible. Consequently, no further considerations were made.

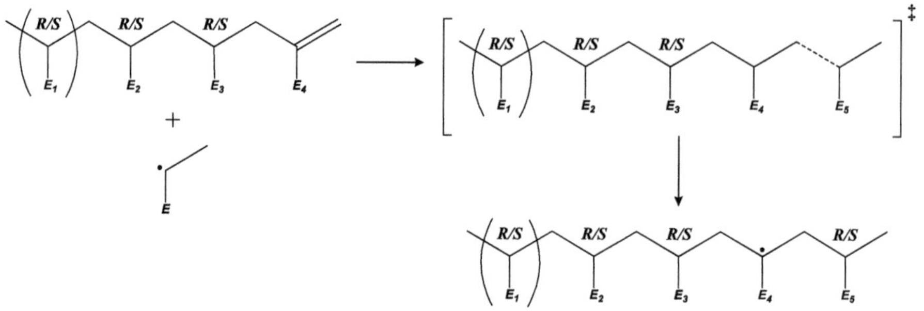

Figure 9. Macromonomer propagation model: unimer radical propagating to a tetramer macromonomer.

Table 7 shows the predicted rate coefficient at 413.15 K and the pre-exponential factor and activation energy for macromonomer propagation. This table includes, for comparison, the rate coefficient used by Van Steenberge et al. [54] in their kinetic Monte Carlo simulation based on the experimental electron spray ionization mass spectra measured by Junkers and Barner-Kowollik [25] for macromonomer synthesis. Additionally, we included the rate coefficient determined by Wang et al. [30] The comparison in Table 7 clearly shows that we predicted a rate coefficient 20% higher than those of other authors, which can be considered a good agreement. Moreover, it is within the same order of magnitude as the predicted ECR propagation rate coefficient: 2.18×10^5 vs. 3.02×10^5 L mol^{-1} s^{-1} for the ECR and macromonomer propagation, respectively. From a thermodynamic point of view, the macromonomer propagation is slightly faster because, in this reaction, a more stable MCR, (i.e., a tertiary carbon radical) is formed instead of the secondary radical formed during regular propagation. The associated more-negative-reaction enthalpy typically results in a decrease in the activation energy, following the Brønsted−Evans−Polanyi relation, as was also observed in this work.

Table 7. Macromonomer propagation Arrhenius parameters and comparison with ECR propagation.

Reaction	Source	k @ 413.15K [L mol^{-1} s^{-1}]	A [L mol^{-1} s^{-1}]	E_a [kJ mol^{-1}]
MM Propagation	This work	3.02×10^5	3.97×10^7	16.8
MM Propagation	Van Steenberge et al. [54]	2.5×10^5		
MM Propagation	Wang et al. [30]	6.63×10^4		
ECR Propagation	This work	2.18×10^5	1.77×10^8	23.0

3.5. MCR Propagation

The selected molecular model for the MCR propagation is a trimer structure with a radical in the central unit that can be added to a monomer, as shown in Figure 10. To account for the influence of optical isomerism, we considered two possibilities for the adjacent monomer units: one with the same chirality (R-MCR-R) and another with different chiralities (R-MCR-S). The results are shown in Table 8 along with the literature data, which include the rate coefficients from multiple acrylates.

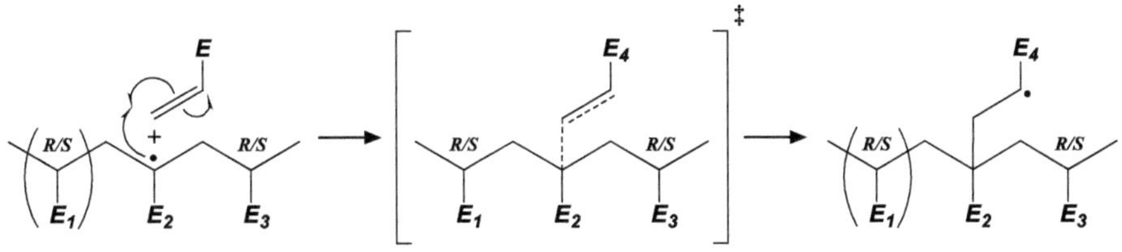

Figure 10. Mid-chain radical propagation model. A trimer structure with an MCR in the middle unit propagating to an MA monomer.

The experimental results agree quite well with those of different authors in the literature [27,50,58,60,68–70], independently of the method used to measure the rate coefficient. In this work, the predicted rate coefficient is in decent agreement with the experimental values because it is less than a factor of two larger than the mid-chain radical propagation for the methyl and n-butyl acrylate experimental values. Thus, our method accurately predicts the rate coefficient of MCR propagation.

Table 8. MCR propagation predicted rates and published results.

Acrylate	k @ 298.15K [L mol^{-1} s^{-1}]	k @ 413.15K [L mol^{-1} s^{-1}]	A [L mol^{-1} s^{-1}]	E_a [kJ mol^{-1}]	Source
methyl R-MCR-R	3.11×10^1	6.78×10^2	6.5×10^6	30.0	This work
R-MCR-S	3.66×10^1	1.06×10^3	2.0×10^6	27.5	
WA	3.36×10^1	8.49×10^2	3.7×10^6	28.8	
methyl	1.79×10^1	3.64×10^2	$8.9 \pm 0.5 \times 10^5$	26.8 ± 1.5	[60]
n-butyl	1.31×10^1	3.37×10^2	$1.52 \pm 0.14 \times 10^6$	28.9 ± 3.2	[58]
n-butyl	1.05×10^1	2.51×10^2	9.2×10^5	28.3	[71]
n-butyl	1.05×10^1	3.10×10^2	1.98×10^6 *	30.1 ± 9.7	[50]
t-butyl	2.87×10^0	3.21×10^1	1.68×10^4 **	21.5 ± 3.6	[69]
dodecyl	4.57×10^0	1.12×10^2	$4.5 \pm 0.8 \times 10^5$	28.5 ± 1.4	[60]

* Recalculated from the original report stating ln(A) = 14.5 ± 3.7. ** Recalculated from the original report stating ln(A) = 9.73 ± 0.70.

3.6. Chain Transfer to Monomer

Chain transfer to monomer (CTM) is a reaction in which a hydrogen is abstracted by an ECR from an acrylate monomer, leading to the termination of the growing radical polymer chain and the formation of a new unimer radical, as the monomer molecule contains several hydrogen atoms that can be abstracted (Figure 11). For carbon atoms possessing abstractable hydrogens in the CTM model of a butyl acrylate monomer molecule, the chain transfer to monomer can yield different types of products. In most of the experimental studies, it is not specified which H atom is being abstracted, as this is difficult to determine experimentally. Therefore, the rate coefficient is an overall rate coefficient considering all the abstraction reaction possibilities. To simulate this with ab initio methods, all reaction possibilities between these two species should be taken into consideration, and the k_{CTM} should be calculated as a sum of all the contributions.

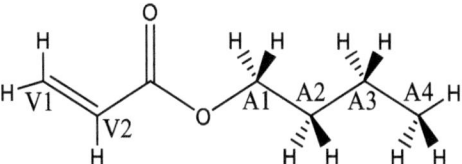

Figure 11. Carbons atoms possessing abstractable hydrogens in the CTM model of a butyl acrylate monomer molecule.

The *n*-butyl acrylate monomer possesses twelve hydrogen atoms in total that are available for abstraction (Figure 11): within the vinylic group, carbon atom V1 possesses two H atoms and V2 has one H atom; within the alkyl substituent, the secondary carbon atoms A1–A3 possess two H atoms each, and the primary carbon atom A4 has three hydrogen atoms. It should be noted that the molecular model for all previous reactions is represented with methyl acrylate for computational efficiency. However, for the chain-transfer-to-monomer reaction, we need to use a butyl substituent in our model to consider all the H atoms of the side chain in the chain-transfer-to-monomer reaction. The structure and side chain of the model radical representing the abstracting radical are less relevant, so a dimer methyl acrylate end-chain radical suffices. Table 9 shows the rate coefficients and Arrhenius parameters predicted by this work along with those in literature for H abstracting. As expected, the abstraction of hydrogens within the vinylic group (V1 and V2) has a high barrier because of the unstable vinylic radical products; hence, disregarding these reactions even at high temperatures induces little error. The results for abstractions from the *n*-butyl acrylate in atoms A1–A3 are similar, which is expected because of the resemblance between the reactions. The chain transfer to monomer of the primary carbon A4 has a higher activation energy than those of the secondary carbon abstractions (A1–A3), but the abstraction remains faster than those from V1 and V2. This reaction in atom A4 proceeds faster than those in the vinylic group. Consequently, the computational results show that the abstraction of hydrogen preferably occurs from carbon atoms A1 to A3.

Table 9. CTM kinetic parameters for the five different possibilities of hydrogen abstraction and the average sums.

CTM	k @ 333.15K [L mol^{-1} s^{-1}]	A [L mol^{-1} s^{-1}]	E_a [kJ mol^{-1}]
V1	2.43×10^{-8}	2.44×10^6	89.1
V2	8.11×10^{-8}	5.72×10^7	94.7
A1	3.82×10^{-2}	1.97×10^6	49.0
A2	6.17×10^{-2}	3.69×10^6	49.5
A3	2.24×10^{-2}	1.87×10^6	50.3
A4	3.59×10^{-4}	1.68×10^6	61.5
k_{CTM}	2.45×10^{-1}	1.82×10^7	49.6
Maeder and Gilbert [72]	2.24×10^0	$2.9 \pm 0.9 \times 10^5$	32.6 ± 0.8
Laki et al. [61]	1.48×10^1	4.88×10^6	35.2 ± 0.61

In Table 9, we also compare the literature values for the chain-transfer-to-monomer-reaction rate coefficients. The main difference between ab initio and experimentally derived data is that the former predict elementary reactions and the latter measure the apparent rate coefficients. To carefully consider each possible abstraction, each rate coefficient must be multiplied by the number of hydrogen atoms via Equation (5), except for the last three hydrogen atoms in the alkyl chain, as these will already be included in the threefold symmetry for the methyl rotation. The results are shown in Table 9.

$$k_{CTM} = k_{V1} + k_{V2} + 2 \times (k_{A1} + k_{A2} + k_{A3}) + k_{A4} \tag{6}$$

Other authors have performed ab initio predictions in the gas phase in chain transfer to monomer [36,38]. Their results display lower rate coefficients than our predictions in this work, probably due to the differences in the chosen computational levels, but the trend between the abstracted hydrogen atoms is maintained independently of the method/basis set or model chosen. Most of the experimental reported data show smaller rate coefficients in comparison to our predictions, while the most recent research shows similar rate coefficients. A table with a thorough comparison is shown in Supporting Information, Table S4.

3.7. Chain Transfer to Polymer

Chain transfer to polymer comprises an ECR abstracting a hydrogen atom from a random point in the middle of the polymer chain, producing an MCR and a dead-polymer chain. Figure 12 shows this reaction for the molecular model that we used for the calculations: a unimer ECR abstracting a hydrogen atom from a trimer methyl acrylate chain in the middle unit. Additionally, we predicted the kinetic parameters of the abstraction of hydrogens atoms from the alkyl chain (butyl acrylate), based on the rate coefficients predicted for the chain transfer to monomer. Hence, we calculated the chain-transfer-to-polymer rate coefficient (k_{CTP}) as the sum of the chain transfer to backbone (k_{CTB}), depicted in Figure 12, and the chain transfer to alkyl branch (k_{CTA}).

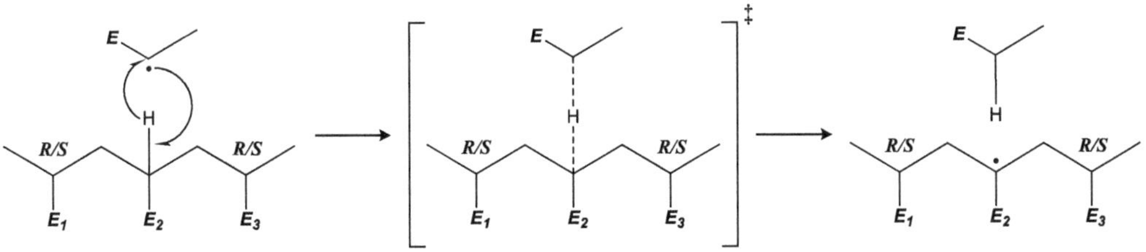

Figure 12. Chain-transfer-to-polymer model. An ECR monomer abstracting a hydrogen from the methyl acrylate trimer in the middle unit.

During chain transfer to polymer, a large polymer chain approaches the middle of another large chain. As the abstracted hydrogen bonds to a tertiary carbon in the middle of the chain, the steric interaction of nearby units impacts the space available for the reaction. Therefore, for this reaction, we considered all possible chiralities and weighted their average for the trimer molecular model. Four different chirality permutations exist for the trimer: RRR/SSS, RSR/SRS, RRS/SSR and SRR/RSS. Permutations RRS/SSR and SRR/RSS are symmetrically equivalent enantiomers, as Figure 12 shows (chain-transfer-to-polymer model). An ECR monomer abstracting a hydrogen from the methyl acrylate trimer in the middle unit shows because of the extra methyl group addition on unit 3. As a result, the rate coefficients of both permutations are equal in the weighted average. Additionally, Table 10 shows the results along with the weighted average of the chain transfer to backbone ($\bar{k}_{CTB}$), which was calculated by using Equation (3), and the weighting factors (ω_i) derived similarly as in the previous sections.

Table 10. Chain-transfer-to-polymer rate coefficients predicted via chirality considerations and inclusion of weighted averages for structures depending on ECR propagation rates.

Reaction	Simulated ω_i	k [s^{-1}] 298.15K	k [s^{-1}] 413.15K	A [s^{-1}]	E_a [kJ mol^{-1}]
RRR/SSS	0.1156	1.72×10^{-1}	1.92×10^{1}	3.91×10^{6}	42.0
RRS/SSR	0.4488	6.97×10^{-4}	3.25×10^{-1}	2.70×10^{6}	54.7
RSR/SRS	0.4356	4.67×10^{-4}	1.77×10^{-1}	8.48×10^{5}	52.9
Chain transfer to backbone $\bar{k}_{CTB}$		2.03×10^{-2}	2.43×10^{0}	5.97×10^{5}	42.6
Chain transfer to alkyl branch k_{CTA}		3.32×10^{-2}	8.47×10^{0}	1.50×10^{7}	49.5
$\bar{k}_{CTP}$		5.20×10^{-2}	1.07×10^{1}	1.05×10^{7}	47.4

Chain transfer to polymer is typically measured in terms of the side branches via ^{13}C NMR or SEC trace fitting. Any reaction that contributes to the number of side branches might add its effect to the apparent rate coefficients. Therefore, it is interesting to study all hydrogen atom abstractions from an embedded monomer unit and not only the abstraction from the tertiary carbon atom embedded in the main chain. For *n*-butyl acrylate, abstractions of hydrogen atoms from the ester substituent are possible. The rate coefficient for these abstractions can be extrapolated from chain transfer to monomer because their magnitude resembles the chain-transfer-to-polymer rate coefficient. Table 10 presents the rate coefficients of chain transfer to the alkyl side chain of the monomer (k_{CTA}).

This new secondary radical in the alkyl chain could propagate with a similar rate coefficient as ECR propagation and, after further propagation steps, resemble a side branch in the backbone. In their work, Boschmann and Vana [29] show an increase in the chain-transfer-to-polymer rate coefficient based on the number of carbons, from 0.3 to 7.1 L mol^{-1} s^{-1} between butyl and dodecyl acrylate. A possible reaction for this radical is intramolecular transfer to the main chain through a similar ring structure as in backbiting or migration, though it will be a seven-, eight- or nine-atom structure depending on the radical location. Preliminary predictions on the eight-ring structure yield a rate coefficient like chain transfer to monomer, so this radical is more likely to propagate, and its rate is excluded from this data set. In conclusion, the chain-transfer-to-polymer rate coefficient (k_{CTP}) is the sum of the transfer to backbone ($\bar{k}_{CTB}$) and transfer to alkyl side chain (k_{CTA}).

Many groups have studied chain transfer to polymer in the last two decades but have found no general agreement in terms of the rate coefficient values. Table S5 displays a clear summary of previous works [26,29,36,54,57,73,74]. The experimentally measured rate coefficients at a single temperature could differ by three orders of magnitude, and the differences in the activation energies are as high as 20 kJ mol^{-1}.

A comparison of this work's predicted rate coefficient for chain transfer to polymer yields a general agreement, in terms of order of magnitude, with the low-side rate coefficients of Plessis et al. [73] and Arzamendi [57]. However, the activation energies differ by over 10 kJ mol^{-1}, and the pre-exponential factors differ by two orders of magnitude. Nonetheless, an exact match exists between this work's rate coefficient and that of Boshmann and Vana's work [29] via Z-RAFT polymerization and C NMR analysis. When comparing it to the recent rate coefficient values of Ballard et al. [74] and Van Steenberge et al. [54], which are on the high end, this work's value is lower by an order of magnitude at 413.15 K. To the best of our knowledge, this manuscript is the first to report an ab initio chain-transfer-to-polymer rate coefficient using a proper model structure product of head-to-tail propagation.

Summarizing the different methodologies reported in the literature, various authors report rate coefficients that differ by multiple orders of magnitude. No clear conclusion exists on which rate coefficient is more likely than the other reported values. In conclusion, the chain-transfer-to-polymer reaction is difficult to measure and model because diverse reactions can create side branches, and the consideration of all of them is required to predict the correct overall rate coefficient. We will benchmark whether this approach provides a reasonable prediction in the simulations shown in the following section, where the chain-transfer-to-polymer rate coefficient is applied in the simulation of ESI-MS spectra.

3.8. kMC Simulations

To evaluate the effect of our calculated rate coefficients on the polymerization kinetics, the kMC model for acrylate polymerization from Vir et al. [50] was taken. This model works as a stochastic tool to solve the kinetic system that simulates the PLP-SEC experiment as an alternative to solving the set of kinetic differential equations. Then, the results from the simulations using the rate coefficients reported above were benchmarked against the experimental and simulated data of PLP-SEC reported by Vir et al. [49,50] and Marien et al. [75], as well as ESI-MS for the synthesis of macromonomers obtained via the activation of bromine-capped p-(*n*-butyl acrylate). In order to decouple the effect of each reaction as much as possible, we used a gradual approach for the selection of the experimental conditions to be simulated, with the objective of maximizing the sensitivity towards a single particular secondary-reaction rate coefficient. The values of the propagation, backbiting and β-scission were benchmarked against PLP-SEC data, whereas the values of the migration and chain-transfer-to-polymer rate coefficients, the effects of which are difficult to derive from the SEC trace, were benchmarked against ESI-MS.

Table 11 reports the data sets used for the simulation of the PLP-SEC experiments. We selected three sufficiently different experimental conditions: (1) 306 K, in bulk, at a laser frequency of 500 Hz; (2) 306 K, with a solvent fraction of 0.75 and at a laser frequency of 50 Hz; (3) 413 K, in bulk, at a laser frequency of 10 Hz. The first data set uses the reference parameters reported by Vir et al. [50] The second data set uses this work's predicted rate coefficients without including special considerations: the backbiting Arrhenius parameters are chosen with a single enantiomer (RSSS/SRRR) in a narrow low-temperature range and the β-scission excludes the reduced-flexibility approach. In the third data set, the backbiting rate coefficient is replaced by the weighted average of all the enantiomers in a narrow temperature range. Finally, in the fourth data set, the unconstrained β-scission rate coefficient replaces the reduced-flexibility one. Hence, the expectation is that Data Set 4 should approximate Data Set 1 the best, while Data Sets 2 and 3 should only provide moderate or qualitative agreement with the experiment.

Table 11. Reference data set and data sets used for three kMC simulations under different conditions. The pre-exponential factor units are L mol^{-1} s^{-1} or s^{-1} depending on the molecularity of the reaction. The activation energies units are kJ mol^{-1} K^{-1}.

	k_p		k_{BB}		$k_{\beta SC}$		$k_{p,MCR}$	
	A	E_a	A	E_a	A	E_a	A	E_a
Data Set 1: reference set (Vir et al. [50])	2.2×10^7	17.9	5.4×10^7	30.6	7.9×10^{12}	81.1	1.9×10^6	30.1
Data Set 2: no special approaches	1.8×10^8	23.0	3.1×10^{10}	49.4	1.2×10^{14}	112.9	3.7×10^6	28.8
Data Set 3: all enantiomers considered for backbiting–weighted-average (WA) approach	1.8×10^8	23.0	4.1×10^{10}	46.8	1.2×10^{14}	112.9	3.7×10^6	28.8
Data Set 4: reduced flexibility applied to β-scission	1.8×10^8	23.0	4.1×10^{10}	46.8	2.8×10^{13}	86.3	3.7×10^6	28.8

The first experiment (Figure 13) is sensitive to the value of the ECR propagation because the laser frequency is high, and it lowers the impact of all the secondary (slow) reactions. The fitted parameters published by Vir et al. [50] reproduce a perfect fit of the experimental SEC trace displayed in Figure 13a, including a perfect match for the first and second inflection points (the points in Figure 13a,b). Data Set 2, without special considerations, incorrectly predicts the SEC trace, as depicted in Figure 13a. The simulated trace (red line) loses the bimodality of the experimental trace because the backbiting reaction frequency (k_{bb}) is higher than the laser frequency. This allows backbiting to proceed even under these conditions, limiting the further propagation of radicals surviving the next pulse that produces the bimodality. This effect limits the chain size and impedes the second-peak formation in the SEC trace. Data Set 3 includes the rate coefficients predicted by the weighted average for backbiting, which massively improves the SEC trace (blue) simulation because the backbiting rate coefficient is now smaller than the pulsed-laser frequency. This data set predicts the first inflection point accurately, and the second inflection point is slightly lower than the experimental value. Data Set 4 includes reduced flexibility for β-scission reactions, and although it has no apparent effect on the SEC trace (purple), it slightly improves the prediction of the second-inflection-point prediction. In conclusion, including these methods shows improvement in the simulated SEC trace because they predict the experimental bimodality. Although they do not produce a perfect match, the inflection points shown in Figure 13b have minimal errors in their predictions.

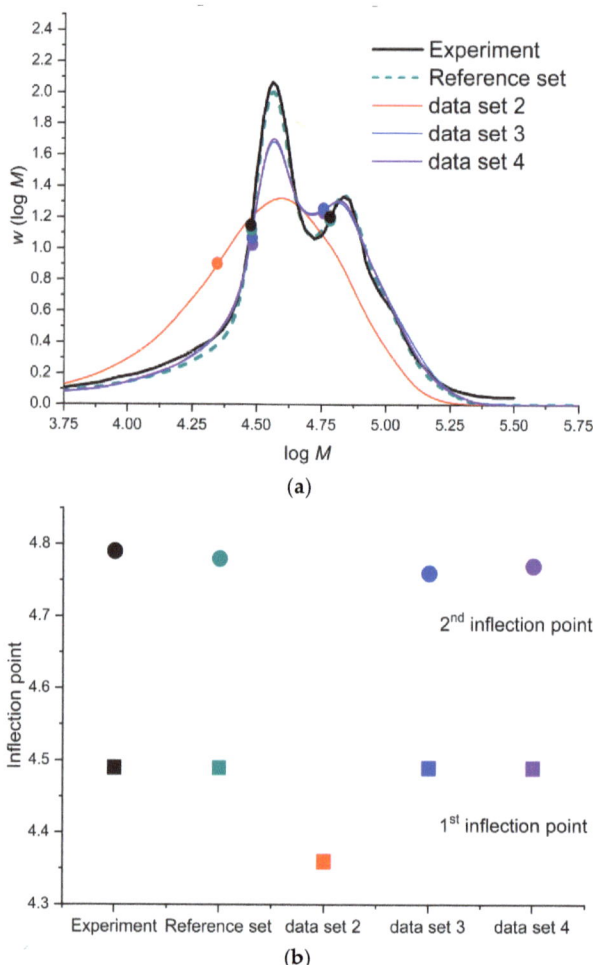

Figure 13. Experiment (1): PLP-SEC trace experiment of n-butyl acrylate at 306 K, in bulk, at laser frequency of 500 Hz. Comparison between experiment [49], kMC-simulated SEC trace with parameters derived from [50] and this work's predicted rate coefficients via different methods.

Experiment (2) provides the ideal conditions for enhanced the backbiting sensitivity to the SEC trace because it features a low pulse frequency and low monomer concentration. Figure 14 presents the simulation results. Experiment (2) involved a PLP-SEC trace experiment of n-butyl acrylate at 306 K, a solvent fraction of 0.75 and a laser frequency of 50 Hz. A comparison was made between the experiment [75] and kMC-simulated SEC trace with the parameters derived from Vir et al. [50] Data Set 2, without special considerations, underestimates the inflection point. Although the SEC trace (red) has the desired shape, it is shifted to the left, pointing towards a lower molecular mass induced by a higher backbiting rate, which slows the apparent propagation rate. Data Set 3 shows that including the consideration of all enantiomers for backbiting greatly improves the inflection point prediction, which is almost in perfect agreement with the experiment, although the SEC trace (blue) is slightly right-shifted. Data Set 4, including the β-scission reduced flexibility, seems to worsen the inflection point prediction, although it is within an acceptable range for ab initio prediction.

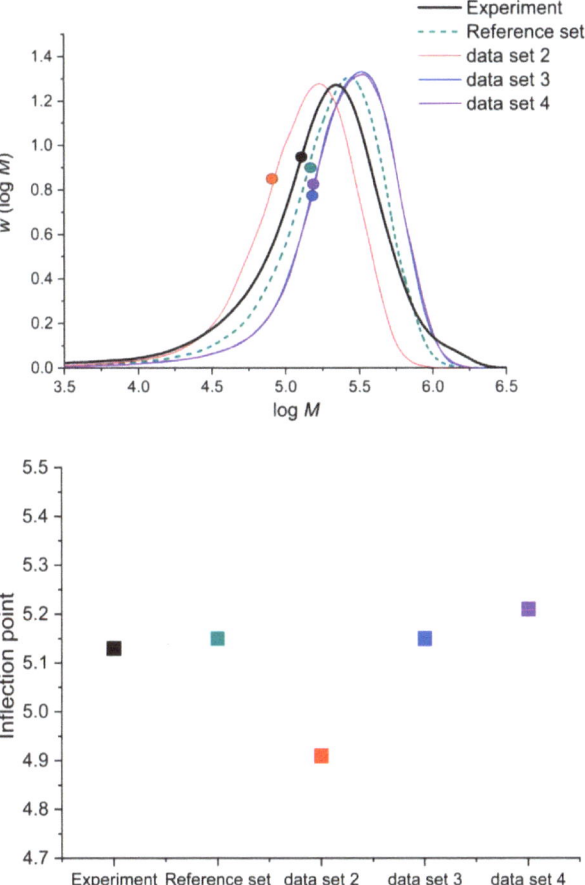

Figure 14. Experiment (2): PLP-SEC trace experiment of n-butyl acrylate at 306 K, a solvent fraction of 0.75 and a laser frequency of 50 Hz. Comparison between experiment [75] and kMC-simulated SEC trace with parameters derived from Vir et al. [50] (reference set) and this work's predicted rate coefficients via different methods (data 2, 3 and 4).

Figure 15 shows the simulation results compared to the experiment (3). The experimental design enhances the secondary reaction's effect by increasing the temperature and lowering the laser frequency. Data Set 2 simulates an SEC trace (red), which is far right-shifted to the experimental one and even possesses a bimodality that is nonexistent in the experimental results. This effect is probably the result of a slower β-scission, which reduces the chain-breaking effect, allowing the chains to grow further, which right-shifts the SEC trace. Data Set 3, which includes the consideration of all enantiomers for backbiting, also predicts a poor agreement between the ab initio and experiment results, shifting even more to the right of the SEC trace (blue) and increasing the inflection point even more. Data Set 4 includes the reduced flexibility for the β-scission, resulting in a decent agreement with the experiment. The approach applied in this data set recovers the unimodality of the SEC trace and lowers the inflection point at least to the same order of magnitude as the experimental result (black line). The SEC trace of Data Set 4 (purple) left-shifts the points towards lower molecular masses. This effect might be induced by the model predicting a higher β-scission rate coefficient, which is because the said reaction causes a chain-breaking effect that left-shifts the SEC trace. Further adjustment of the β-scission rate coefficient by multiplying it with a factor of 0.5, which is within the typical confidence interval for ab initio predicted rate coefficients, produces Data Set 4' shown in Figure 15 (purple dotted line for the SEC trace and purple void square for the inflection point). This improves the prediction greatly, as now both the SEC trace and the inflection point are in great agreement with the experiment, implying that the ab initio prediction is relatively good but slightly overestimates the β-scission rate coefficient. This overestimation of the rate coefficient could be related to the previously discussed difference between middle-of-the-backbone MCRs and recently backbit MCRs, where the former reacts much faster than the latter. Multiplying the β-scission rate coefficient with a factor of 0.5 is, in this interpretation, equivalent to assuming that 50% of the MCRs are "recently backbit MCRs" that decompose much slower. The large impact of this small change in the β-scission rate coefficient on the prediction of the SEC trace and inflection point prediction highlights the strong sensitivity of these reaction conditions to the β-scission rate coefficient.

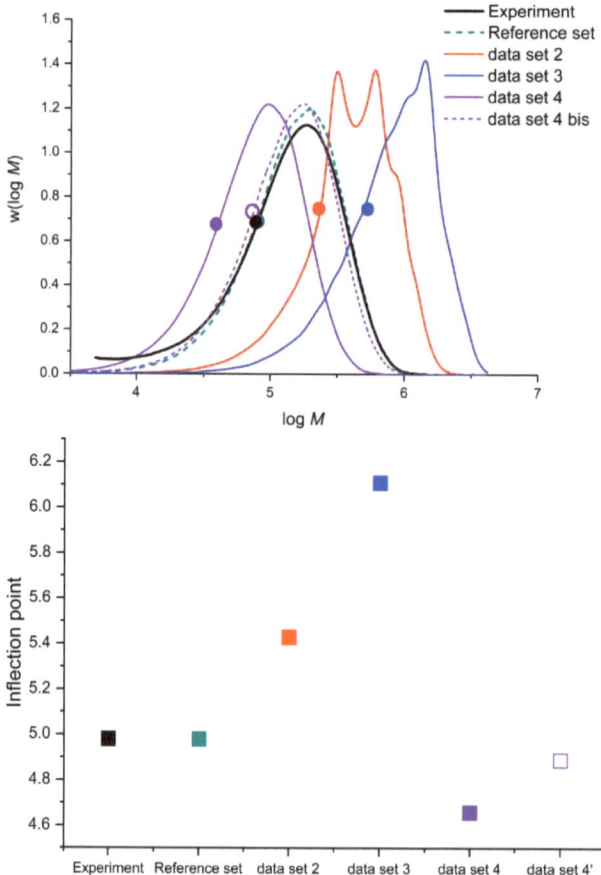

Figure 15. Experiment (3): PLP-SEC trace experiment of n-butyl acrylate at 413 K, in bulk, at laser frequency 10 Hz. Comparison between experiment [50] and kMC-simulated SEC trace with parameters derived from Vir et al. [50] (reference set) and this work's predicted rate coefficients via different methods (data 2, 3 and 4).

The last set of simulated experiments (i.e., the ESI-MS data for the synthesis of macromonomers (MMs) via the activation of bromine-capped poly(n-butyl acrylate)) is sensitive to the values of the macromonomer propagation, chain-transfer-to-polymer and migration rate coefficients. We compared the data sets for these rate coefficients used by Van Steenberge et al. [54] and report them here in Table 12. Figure 16 shows the ESI-MS simulation. In general, the simulations with our data set agree with the experimental behavior, except for the "dead"-polymer-fraction prediction.

Table 12. Rate coefficient comparison between Van Steenberge et al. [54] and this work for the electron spray ionization–mass spectrometry experiment. The simulation made with this work's data set includes rate coefficients used by Van Steenberge et al. [54] for activation, deactivation and termination.

Reaction	Rate Used in Van Steenberge et al. [54] k [L mol^{-1} s^{-1}] or [s^{-1}] @ 413.15K	This Work's Data Set k [L mol^{-1} s^{-1}] or [s^{-1}] @ 413.15K
Activation	4.0×10^3	4.0×10^3
Deactivation	1.0×10^6	1.0×10^6
Reduction	3.0×10^{-1}	3.0×10^{-1}
Backbiting	6.5×10^3	9.4×10^4
Migration	1.6×10^2	1.6×10^3
Chain transfer to polymer	6.0×10^2	1.1×10^1
β-scission	1.2×10^0	3.4×10^2
Macromonomer addition	2.5×10^5	3.0×10^5
Termination	1.0×10^8	1.0×10^8

As seen in Figure 16b, the "dead"-polymer red curve does not describe the experimental data points. The calculated rate coefficient with the largest uncertainty is the chain-transfer-to-polymer reaction because of the large spread within the experimental data and between the experimental and predicted data. The red curve is not sensitive to the chain-transfer-to-polymer rate coefficient, as an increase of 1000 times in the predicted rate coefficient barely modifies the red curve (see Supporting Information, Figure S4). Moreover, the light-blue curve shows a steeper decay in the ECR species with the parameters presented in this work. The difference relates to the faster backbiting and migration reactions, tilting the initiation equilibria towards more MCRs compared to ECRs at late reaction times. These MCRs in the present data set would rather undergo β-scission than chain transfer to polymer. This increases the amount of macromonomers, as depicted in Figure 16b, bottom right, and might be a key factor in the number of "dead" polymers in the mixture.

It is important to mention that the reliability of the ab initio prediction is enhanced by testing the results against the experimental data under very different conditions with sensitivities towards different types of reactions. This task could not be possible without the integration of the ab initio predictions into the kMC simulations, showing that the combination of both techniques could provide further insight into the polymerization kinetics. The results are far from a perfect fit but provide decent agreement.

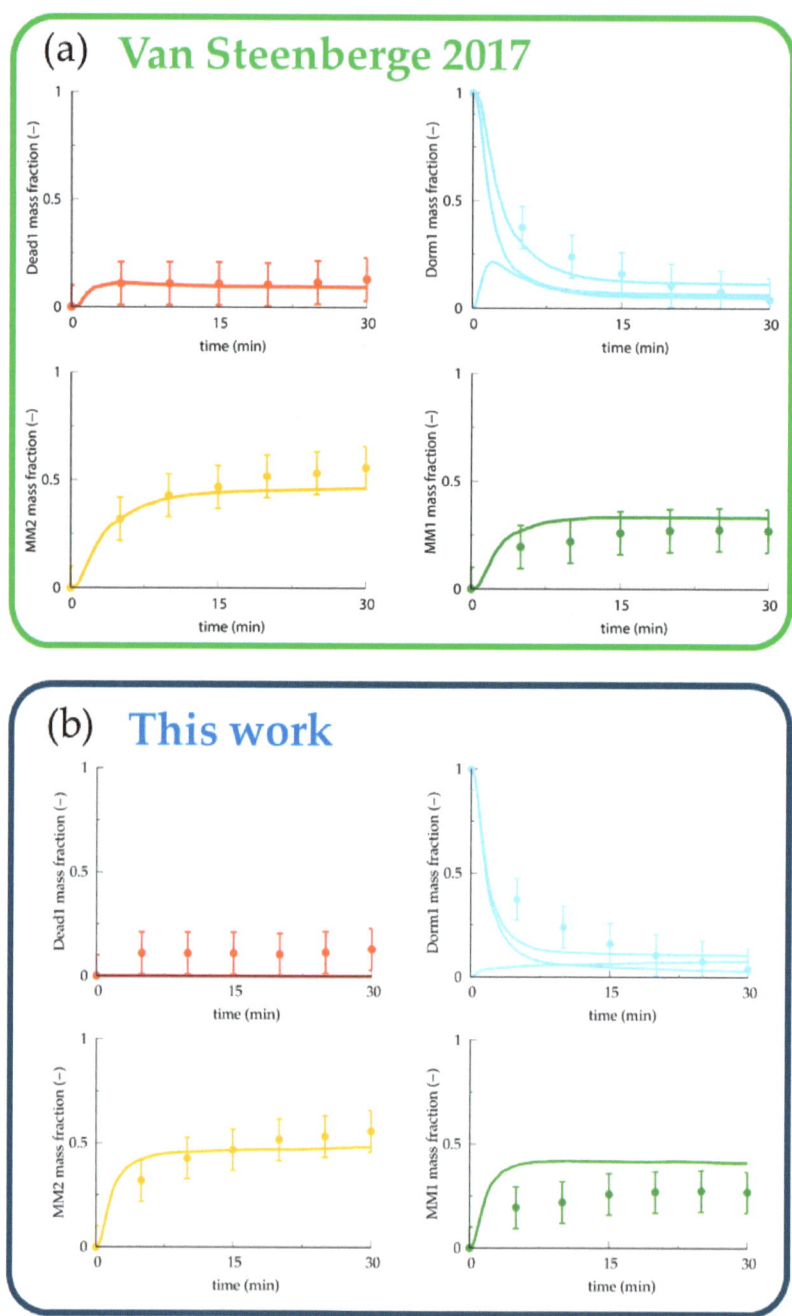

Figure 16. Simulated (full lines) and experimental (points) mass fractions corresponding to the main ESI-MS signals of the synthesis of macromonomers (MMs) via the activation of bromine-capped poly(n-butyl acrylate) in Van Steenberge et al. [54], top left, red; dead-polymer product of CTP, top right, blue; dormant Br-capped p-(n-BA) including the fraction of activated ECRs and MCRs, bottom left, yellow; MM product of right β-scission, bottom right, green; MM product of left β-scission.

4. Conclusions

In this article, we report a complete and consistent calculated set of rate coefficients predicted through ab initio methods for acrylate radical polymerization. To enhance the accuracy of the prediction, we apply two new approaches: (1) calculation of the weighted average for the reactions influenced by the optical isomerism of the molecular model, and (2) geometrical constrains to mimic the restricted motion of the segments embedded in the middle of the polymer chain. Instead of comparing the rate coefficients one on one to the literature values, the obtained rate coefficients were benchmarked against experimental data via kMC simulations. The actual experiments and applied experimental conditions varied widely in order to selectively address the isolation of the secondary reactions as much as possible, a methodology that proves suitable for comparing ab initio rate coefficients.

We showed that this work's data set simulates the MMDs of acrylate polymers with an acceptable error, although the predicted rate coefficients differ from those measured via experimental work or with model assistance. Additionally, our two new approaches provide a substantial improvement in the prediction of the experimental SEC traces and their inflection points.

In any secondary reaction involving significant changes in the chain's shape during the transition state, considering all enantiomers is crucial. The ab initio predictions of the rate coefficients for the backbiting and migration results show that the reactivities are dependent on the molecular-model chirality. Optical isomers with alternating chiralities have lower rate coefficients for secondary reactions thanks to the increased reactant stability, while, in general, sequential identical chiral permutations possess higher values. This strategy should prove useful for any type of atactic polymer, while, for polymers presenting tacticity, it should suffice to use a model describing the tacticity.

The reduced-flexibility approach shows an improvement in the rate coefficient prediction of the β-scission reaction. Without this approach, the predictions of the SEC trace and inflection points are far off the experimental results. The reduced-flexibility-approach applicability is easily extendable to any other type of macromolecular chemistry.

This work serves as an improvement in the techniques used to predict rate coefficients for the polymerization process that can aid in polymerization modeling. Further development is available for these methods because considering all the optical isomers requires testing against the experimental data of various atactic polymers. The reduction in flexibility requires a proper criterion for the geometry restriction, which could be studied by applying the method to different polymers and comparing the results to high-quality experimental measurements.

Supplementary Materials: The following supporting information can be downloaded at: https://www.mdpi.com/article/10.3390/polym16070872/s1, Figure S1: Determination of the chirality in a pentamer end-chain radical; Figure S2: Methyl acrylate pentamer structure: RRRR/SSSS optical isomer straight vs curled; Figure S3: Transition state structure for the backbiting reaction; Figure S4: ESI-MS of the synthesis of macromonomers via activation of bromine-capped poly(n-butyl acrylate) including chain transfer to polymer multiplied by a factor of 1000; Table S1: Relative energies of all end-chain radicals optical isomers pentamer structures and their respective transition state for the backbiting reaction; Table S2: Predicted rate coefficients for the propagation reaction of methyl acrylate in bulk; Table S3: Relative energies of all end-chain radicals optical isomers pentamer structures and their respective transition state for the migration reaction; Table S4: β-scission of mid-chain radicals experimental and predicted published rate coefficients; Table S5: CTM rate coefficients published in literature; Table S6: Chain transfer to polymer rate coefficient and Arrhenius parameters from previously published results by different authors. Table S7: Effect of the solvation energy on different secondary reactions; List of optimized geometries.

Author Contributions: Conceptualization, F.A.L., M.E., P.H.M.V.S. and M.K.S.; methodology, F.A.L. and M.K.S.; software, F.A.L., M.E., P.H.M.V.S. and M.K.S.; validation, M.E., P.H.M.V.S. and M.K.S.; formal analysis, F.A.L. and M.E.; investigation, F.A.L. and M.E.; resources, P.H.M.V.S. and M.K.S.; data curation, F.A.L. and M.E.; writing-original draft preparation, F.A.L.; writing-review and editing, M.E., P.H.M.V.S. and M.K.S.; visualization, F.A.L. and M.E.; supervision, M.E., P.H.M.V.S. and M.K.S.; project administration, F.A.L., M.E., P.H.M.V.S. and M.K.S.; funding acquisition, M.K.S. All authors have read and agreed to the published version of the manuscript.

Funding: Maarten Sabbe and Fernando Lugo acknowledge a starting grant from the Research Fund of Ghent University (BOF;01N03819).

Institutional Review Board Statement: Not applicable.

Data Availability Statement: The data presented in this study is publicly available in the supporting information. All the information required to reproduce our results is the optimized geometries of reactants structures and transition states, which are included in the supporting information. All shown experimental data is extracted from previous work published elsewhere.

Acknowledgments: The computational resources (Stevin Supercomputer Infrastructure) and services used in this work were provided by the VSC (Flemish Supercomputer Center), funded by Ghent University, FWO and the Flemish Government—EWI department.

Conflicts of Interest: The authors declare no conflict of interest.

References

1. Serrano-Aroca, Á.; Deb, S. *Acrylate Polymers for Advanced Applications*; IntechOpen: Rijeka, Croatia, 2020.
2. Agboluaje, M.; Refai, I.; Manston, H.H.; Hutchinson, R.A.; Dušička, E.; Urbanová, A.; Lacík, I. A comparison of the solution radical propagation kinetics of partially water-miscible non-functional acrylates to acrylic acid. *Polym. Chem.* **2020**, *11*, 7104–7114. [CrossRef]
3. Penzel, E.; Ballard, N.; Asua, J.M. Polyacrylates. In *Ullmann's Encyclopedia of Industrial Chemistry*; Wiley Online Library: Wiley-VCH, Weinheim, 2018; pp. 1–20.
4. Gómez, P.G. Development of Electrochemical (bio) Sensors and Microanalytical Systems: Application to the Wine Industry. Ph.D. Thesis, Universitat Autònoma de Barcelona, Barcelona, Spain, 2017.
5. Vandenbergh, T.J.J.; Olabisi, O.; Adewale, K. *Handbook of Thermoplastics*; CRC Press Taylor & Francis Group: Boca Raton, FL, USA, 2016.
6. Barner-Kowollik, C.; Beuermann, S.; Buback, M.; Castignolles, P.; Charleux, B.; Coote, M.L.; Hutchinson, R.A.; Junkers, T.; Lacík, I.; Russell, G.T.; et al. Critically evaluated rate coefficients in radical polymerization—7. Secondary-radical propagation rate coefficients for methyl acrylate in the bulk. *Polym. Chem.* **2014**, *5*, 204–212. [CrossRef]
7. Bakhshi, H.; Kuang, G.; Wieland, F.; Meyer, W. Photo-Curing Kinetics of 3D-Printing Photo-Inks Based on Urethane-Acrylates. *Polymers* **2022**, *14*, 2974. [CrossRef]
8. Konuray, O.; Morancho, J.M.; Fernández-Francos, X.; García-Alvarez, M.; Ramis, X. Curing kinetics of dually-processed acrylate-epoxy 3D printing resins. *Thermochim. Acta* **2021**, *701*, 178963. [CrossRef]
9. Ballard, N.; Asua, J.M. Radical polymerization of acrylic monomers: An overview. *Prog. Polym. Sci.* **2018**, *79*, 40–60. [CrossRef]
10. Pirman, T.; Ocepek, M.; Likozar, B. Radical Polymerization of Acrylates, Methacrylates, and Styrene: Biobased Approaches, Mechanism, Kinetics, Secondary Reactions, and Modeling. *Ind. Eng. Chem. Res.* **2021**, *60*, 9347–9367. [CrossRef]
11. Miasnikova, A.; Laschewsky, A. Influencing the phase transition temperature of poly(methoxy diethylene glycol acrylate) by molar mass, end groups, and polymer architecture. *J. Polym. Sci. Part A Polym. Chem.* **2012**, *50*, 3313–3323. [CrossRef]
12. Márquez, I.; Alarcia, F.; Velasco, J.I. Synthesis and Properties of Water-Based Acrylic Adhesives with a Variable Ratio of 2-Ethylhexyl Acrylate and n-Butyl Acrylate for Application in Glass Bottle Labels. *Polymers* **2020**, *12*, 428. [CrossRef]
13. Edeleva, M.; Marien, Y.W.; Van Steenberge, P.H.M.; D'Hooge, D.R. Impact of side reactions on molar mass distribution, unsaturation level and branching density in solution free radical polymerization of n-butyl acrylate under well-defined lab-scale reactor conditions. *Polym. Chem.* **2021**, *12*, 2095–2114. [CrossRef]
14. Olaj, O.F.; Bitai, I. The laser flash-initiated polymerization as a tool of evaluating (individual) kinetic constants of free radical polymerization, 3. Information from degrees of polymerization. *Die Angew. Makromol. Chem.* **1987**, *155*, 177–190. [CrossRef]
15. Davis, T.P.; O'Driscoll, K.F.; Piton, M.C.; Winnik, M.A. Determination of propagation rate constants using a pulsed laser technique. *Macromolecules* **1989**, *22*, 2785–2788. [CrossRef]
16. Buback, M.; Gilbert, R.G.; Russell, G.T.; Hill, D.J.T.; Moad, G.; O'Driscoll, K.F.; Shen, J.; Winnik, M.A. Consistent values of rate parameters in free radical polymerization systems. II. Outstanding dilemmas and recommendations. *J. Polym. Sci. Part A Polym. Chem.* **1992**, *30*, 851–863. [CrossRef]
17. Deady, M.; Mau, A.W.H.; Moad, G.; Spurling, T.H. Evaluation of the kinetic parameters for styrene polymerization and their chain length dependence by kinetic simulation and pulsed laser photolysis. *Die Makromol. Chem.* **1993**, *194*, 1691–1705. [CrossRef]

18. O'Driscoll, K.F.; Kuindersma, M.E. Monte Carlo simulation of pulsed laser polymerization. *Macromol. Theory Simul.* **1994**, *3*, 469–478. [CrossRef]
19. Buback, M.; Gilbert, R.G.; Hutchinson, R.A.; Klumperman, B.; Kuchta, F.-D.; Manders, B.G.; O'Driscoll, K.F.; Russell, G.T.; Schweer, J. Critically evaluated rate coefficients for free-radical polymerization, 1. Propagation rate coefficient for styrene. *Macromol. Chem. Phys.* **1995**, *196*, 3267–3280. [CrossRef]
20. Beuermann, S.; Buback, M.; Davis, T.P.; Gilbert, R.G.; Hutchinson, R.A.; Olaj, O.F.; Russell, G.T.; Schweer, J.; van Herk, A.M. Critically evaluated rate coefficients for free-radical polymerization, 2. Propagation rate coefficients for methyl methacrylate. *Macromol. Chem. Phys.* **1997**, *198*, 1545–1560. [CrossRef]
21. Beuermann, S.; Buback, M.; Davis, T.P.; Gilbert, R.G.; Hutchinson, R.A.; Kajiwara, A.; Klumperman, B.; Russell, G.T. Critically evaluated rate coefficients for free-radical polymerization, 3. Propagation rate coefficients for alkyl methacrylates. *Macromol. Chem. Phys.* **2000**, *201*, 1355–1364. [CrossRef]
22. Beuermann, S.; Buback, M.; Davis, T.P.; García, N.; Gilbert, R.G.; Hutchinson, R.A.; Kajiwara, A.; Kamachi, M.; Lacík, I.; Russell, G.T. Critically Evaluated Rate Coefficients for Free-Radical Polymerization, 4. *Macromol. Chem. Phys.* **2003**, *204*, 1338–1350. [CrossRef]
23. Asua, J.M.; Beuermann, S.; Buback, M.; Castignolles, P.; Charleux, B.; Gilbert, R.G.; Hutchinson, R.A.; Leiza, J.R.; Nikitin, A.N.; Vairon, J.-P.; et al. Critically Evaluated Rate Coefficients for Free-Radical Polymerization, 5. *Macromol. Chem. Phys.* **2004**, *205*, 2151–2160. [CrossRef]
24. Beuermann, S.; Buback, M.; Hesse, P.; Kuchta, F.-D.; Lacík, I.; Herk, A.M.v. Critically evaluated rate coefficients for free-radical polymerization Part 6: Propagation rate coefficient of methacrylic acid in aqueous solution (IUPAC Technical Report). *Pure Appl. Chem.* **2007**, *79*, 1463–1469. [CrossRef]
25. Junkers, T.; Barner-Kowollik, C. Optimum Reaction Conditions for the Synthesis of Macromonomers Via the High-Temperature Polymerization of Acrylates. *Macromol. Theory Simul.* **2009**, *18*, 421–433. [CrossRef]
26. Ahmad, N.M.; Heatley, F.; Lovell, P.A. Chain Transfer to Polymer in Free-Radical Solution Polymerization of n-Butyl Acrylate Studied by NMR Spectroscopy. *Macromolecules* **1998**, *31*, 2822–2827. [CrossRef]
27. Peck, A.N.F.; Hutchinson, R.A. Secondary Reactions in the High-Temperature Free Radical Polymerization of Butyl Acrylate. *Macromolecules* **2004**, *37*, 5944–5951. [CrossRef]
28. Plessis, C.; Arzamendi, G.; Alberdi, J.M.; van Herk, A.M.; Leiza, J.R.; Asua, J.M. Evidence of Branching in Poly(butyl acrylate) Produced in Pulsed-Laser Polymerization Experiments. *Macromol. Rapid Commun.* **2003**, *24*, 173–177. [CrossRef]
29. Boschmann, D.; Vana, P. Z-RAFT Star Polymerizations of Acrylates: Star Coupling via Intermolecular Chain Transfer to Polymer. *Macromolecules* **2007**, *40*, 2683–2693. [CrossRef]
30. Wang, W.; Nikitin, A.N.; Hutchinson, R.A. Consideration of Macromonomer Reactions in n-Butyl Acrylate Free Radical Polymerization. *Macromol. Rapid Commun.* **2009**, *30*, 2022–2027. [CrossRef]
31. Nikitin, A.N.; Hutchinson, R.A.; Wang, W.; Kalfas, G.A.; Richards, J.R.; Bruni, C. Effect of Intramolecular Transfer to Polymer on Stationary Free-Radical Polymerization of Alkyl Acrylates, 5—Consideration of Solution Polymerization up to High Temperatures. *Macromol. React. Eng.* **2010**, *4*, 691–706. [CrossRef]
32. Hamzehlou, S.; Ballard, N.; Reyes, Y.; Aguirre, A.; Asua, J.M.; Leiza, J.R. Analyzing the discrepancies in the activation energies of the backbiting and β-scission reactions in the radical polymerization of n-butyl acrylate. *Polym. Chem.* **2016**, *7*, 2069–2077. [CrossRef]
33. Lena, J.-B.; Deschamps, M.; Sciortino, N.F.; Masters, S.L.; Squire, M.A.; Russell, G.T. Effects of Chain Transfer Agent and Temperature on Branching and β-Scission in Radical Polymerization of 2-Ethylhexyl Acrylate. *Macromol. Chem. Phys.* **2018**, *219*, 1700579. [CrossRef]
34. Liu, S.; Srinivasan, S.; Grady, M.C.; Soroush, M.; Rappe, A.M. Backbiting and β-scission reactions in free-radical polymerization of methyl acrylate. *Int. J. Quantum Chem.* **2014**, *114*, 345–360. [CrossRef]
35. Cuccato, D.; Mavroudakis, E.; Moscatelli, D. Quantum Chemistry Investigation of Secondary Reaction Kinetics in Acrylate-Based Copolymers. *J. Phys. Chem. A* **2013**, *117*, 4358–4366. [CrossRef]
36. Moghadam, N.; Liu, S.; Srinivasan, S.; Grady, M.C.; Soroush, M.; Rappe, A.M. Computational Study of Chain Transfer to Monomer Reactions in High-Temperature Polymerization of Alkyl Acrylates. *J. Phys. Chem. A* **2013**, *117*, 2605–2618. [CrossRef]
37. Moghadam, N.; Liu, S.; Srinivasan, S.; Grady, M.C.; Rappe, A.M.; Soroush, M. Theoretical Study of Intermolecular Chain Transfer to Polymer Reactions of Alkyl Acrylates. *Ind. Eng. Chem. Res.* **2015**, *54*, 4148–4165. [CrossRef]
38. Mavroudakis, E.; Cuccato, D.; Moscatelli, D. Quantum Mechanical Investigation on Bimolecular Hydrogen Abstractions in Butyl Acrylate Based Free Radical Polymerization Processes. *J. Phys. Chem. A* **2014**, *118*, 1799–1806. [CrossRef] [PubMed]
39. Van Cauter, K.; Van Den Bossche, B.J.; Van Speybroeck, V.; Waroquier, M. Ab Initio Study of Free-Radical Polymerization: Defect Structures in Poly(vinyl chloride). *Macromolecules* **2007**, *40*, 1321–1331. [CrossRef]
40. D'hooge, D.R.; Reyniers, M.-F.; Marin, G.B. The Crucial Role of Diffusional Limitations in Controlled Radical Polymerization. *Macromol. React. Eng.* **2013**, *7*, 362–379. [CrossRef]
41. Edeleva, M.; Van Steenberge, P.H.M.; Sabbe, M.K.; D'hooge, D.R. Connecting Gas-Phase Computational Chemistry to Condensed Phase Kinetic Modeling: The State-of-the-Art. *Polymers* **2021**, *13*, 3027. [CrossRef] [PubMed]
42. Frisch, M.J.; Trucks, G.W.; Schlegel, H.B.; Scuseria, G.E.; Robb, M.A.; Cheeseman, J.R.; Scalmani, G.; Barone, V.; Petersson, G.A.; Nakatsuji, H.; et al. *Gaussian 16, Revision C.01*; Gaussian, Inc.: Wallingford, CT, USA, 2016.

43. McQuarrie, D.A. *Physical Chemistry: A Molecular Approach*; University Science Books: Sausalito, CA, USA, 1997.
44. Klamt, A. Conductor-like Screening Model for Real Solvents: A New Approach to the Quantitative Calculation of Solvation Phenomena. *J. Phys. Chem.* **1995**, *99*, 2224–2235. [CrossRef]
45. Klamt, A.; Jonas, V.; Bürger, T.; Lohrenz, J.C.W. Refinement and Parametrization of COSMO-RS. *J. Phys. Chem. A* **1998**, *102*, 5074–5085. [CrossRef]
46. Eckert, F.; Klamt, A. Fast solvent screening via quantum chemistry: COSMO-RS approach. *AIChE J.* **2002**, *48*, 369–385. [CrossRef]
47. COSMOlogic GmbH & Co. KG. *COSMOthermX, version 19.0.4*; COSMOlogic: Leverkusen, Germany, 2019.
48. Eckart, C. The Penetration of a Potential Barrier by Electrons. *Phys. Rev.* **1930**, *35*, 1303–1309. [CrossRef]
49. Marien, Y.W.; Van Steenberge, P.H.M.; Barner-Kowollik, C.; Reyniers, M.-F.; Marin, G.B.; D'hooge, D.R. Kinetic Monte Carlo Modeling Extracts Information on Chain Initiation and Termination from Complete PLP-SEC Traces. *Macromolecules* **2017**, *50*, 1371–1385. [CrossRef]
50. Vir, A.B.; Marien, Y.W.; Van Steenberge, P.H.M.; Barner-Kowollik, C.; Reyniers, M.-F.; Marin, G.B.; D'Hooge, D.R. From n-butyl acrylate Arrhenius parameters for backbiting and tertiary propagation to β-scission via stepwise pulsed laser polymerization. *Polym. Chem.* **2019**, *10*, 4116–4125. [CrossRef]
51. Gillespie, D.T. Exact stochastic simulation of coupled chemical reactions. *J. Phys. Chem.* **1977**, *81*, 2340–2361. [CrossRef]
52. Van Steenberge, P.H.M.; D'hooge, D.R.; Reyniers, M.F.; Marin, G.B. Improved kinetic Monte Carlo simulation of chemical composition-chain length distributions in polymerization processes. *Chem. Eng. Sci.* **2014**, *110*, 185–199. [CrossRef]
53. De Smit, K.; Marien, Y.W.; Edeleva, M.; Van Steenberge, P.H.M.; D'hooge, D.R. Roadmap for Monomer Conversion and Chain Length-Dependent Termination Reactivity Algorithms in Kinetic Monte Carlo Modeling of Bulk Radical Polymerization. *Ind. Eng. Chem. Res.* **2020**, *59*, 22422–22439. [CrossRef]
54. Van Steenberge, P.H.M.; Vandenbergh, J.; Reyniers, M.-F.; Junkers, T.; D'hooge, D.R.; Marin, G.B. Kinetic Monte Carlo Generation of Complete Electron Spray Ionization Mass Spectra for Acrylate Macromonomer Synthesis. *Macromolecules* **2017**, *50*, 2625–2636. [CrossRef]
55. Satoh, K.; Kamigaito, M. Stereospecific Living Radical Polymerization: Dual Control of Chain Length and Tacticity for Precision Polymer Synthesis. *Chem. Rev.* **2009**, *109*, 5120–5156. [CrossRef] [PubMed]
56. Yu, X.; Broadbelt, L.J. Kinetic Study of 1,5-Hydrogen Transfer Reactions of Methyl Acrylate and Butyl Acrylate Using Quantum Chemistry. *Macromol. Theory Simul.* **2012**, *21*, 461–469. [CrossRef]
57. Arzamendi, G.; Plessis, C.; Leiza, J.R.; Asua, J.M. Effect of the Intramolecular Chain Transfer to Polymer on PLP/SEC Experiments of Alkyl Acrylates. *Macromol. Theory Simul.* **2003**, *12*, 315–324. [CrossRef]
58. Nikitin, A.N.; Hutchinson, R.A.; Buback, M.; Hesse, P. Determination of Intramolecular Chain Transfer and Midchain Radical Propagation Rate Coefficients for Butyl Acrylate by Pulsed Laser Polymerization. *Macromolecules* **2007**, *40*, 8631–8641. [CrossRef]
59. Barth, J.; Buback, M. SP–PLP–EPR Investigations into the Chain-Length-Dependent Termination of Methyl Methacrylate Bulk Polymerization. *Macromol. Rapid Commun.* **2009**, *30*, 1805–1811. [CrossRef] [PubMed]
60. Kattner, H.; Buback, M. Termination, Propagation, and Transfer Kinetics of Midchain Radicals in Methyl Acrylate and Dodecyl Acrylate Homopolymerization. *Macromolecules* **2018**, *51*, 25–33. [CrossRef]
61. Laki, S.; AShamsabadi, A.; Riazi, H.; Grady, M.C.; Rappe, A.M.; Soroush, M. Experimental and Mechanistic Modeling Study of Self-Initiated High-Temperature Polymerization of Ethyl Acrylate. *Ind. Eng. Chem. Res.* **2020**, *59*, 2621–2630. [CrossRef]
62. Cuccato, D.; Mavroudakis, E.; Dossi, M.; Moscatelli, D. A Density Functional Theory Study of Secondary Reactions in n-Butyl Acrylate Free Radical Polymerization. *Macromol. Theory Simul.* **2013**, *22*, 127–135. [CrossRef]
63. Paraskevas, P.D.; Sabbe, M.K.; Reyniers, M.-F.; Papayannakos, N.G.; Marin, G.B. Group Additive Kinetics for Hydrogen Transfer between Oxygenates. *J. Phys. Chem. A* **2015**, *119*, 6961–6980. [CrossRef]
64. Zorn, A.-M.; Junkers, T.; Barner-Kowollik, C. Synthesis of a Macromonomer Library from High-Temperature Acrylate Polymerization. *Macromol. Rapid Commun.* **2009**, *30*, 2028–2035. [CrossRef]
65. Heidarzadeh, N.; Hutchinson, R.A. Maximizing macromonomer content produced by starved-feed high temperature acrylate/methacrylate semi-batch polymerization. *Polym. Chem.* **2020**, *11*, 2137–2146. [CrossRef]
66. Hirano, T.; Yamada, B. Macromonomer formation by sterically hindered radical polymerization of methyl acrylate trimer at high temperature. *Polymer* **2003**, *44*, 347–354. [CrossRef]
67. Ballard, N.; Veloso, A.; Asua, J.M. Mid-Chain Radical Migration in the Radical Polymerization of n-Butyl Acrylate. *Polymers* **2018**, *10*, 765. [CrossRef]
68. Barth, J.; Buback, M. SP-PLP-EPR—A Novel Method for Detailed Studies into the Termination Kinetics of Radical Polymerization. *Macromol. React. Eng.* **2010**, *4*, 288–301. [CrossRef]
69. Wenn, B.; Junkers, T. Kilohertz Pulsed-Laser-Polymerization: Simultaneous Determination of Backbiting, Secondary, and Tertiary Radical Propagation Rate Coefficients for tert-Butyl Acrylate. *Macromol. Rapid Commun.* **2016**, *37*, 781–787. [CrossRef]
70. Quintens, G.; Junkers, T. Pulsed laser polymerization–size exclusion chromatography investigations into backbiting in ethylhexyl acrylate polymerization. *Polym. Chem.* **2022**, *13*, 2019–2025. [CrossRef]
71. Barth, J.; Buback, M.; Hesse, P.; Sergeeva, T. Termination and Transfer Kinetics of Butyl Acrylate Radical Polymerization Studied via SP-PLP-EPR. *Macromolecules* **2010**, *43*, 4023–4031. [CrossRef]
72. Maeder, S.; Gilbert, R.G. Measurement of transfer constant for butyl acrylate free-radical polymerization. *Macromolecules* **1998**, *31*, 4410–4418. [CrossRef]

73. Plessis, C.; Arzamendi, G.; Leiza, J.R.; Schoonbrood, H.A.S.; Charmot, D.; Asua, J.M. Modeling of Seeded Semibatch Emulsion Polymerization of n-BA. *Ind. Eng. Chem. Res.* **2001**, *40*, 3883–3894. [CrossRef]
74. Ballard, N.; Hamzehlou, S.; Asua, J.M. Intermolecular Transfer to Polymer in the Radical Polymerization of n-Butyl Acrylate. *Macromolecules* **2016**, *49*, 5418–5426. [CrossRef]
75. Vir, A.B.; Marien, Y.W.; Van Steenberge, P.H.M.; Barner-Kowollik, C.; Reyniers, M.-F.; Marin, G.B.; D'Hooge, D.R. Access to the β-scission rate coefficient in acrylate radical polymerization by careful scanning of pulse laser frequencies at elevated temperature. *React. Chem. Eng.* **2018**, *3*, 807–815. [CrossRef]

Disclaimer/Publisher's Note: The statements, opinions and data contained in all publications are solely those of the individual author(s) and contributor(s) and not of MDPI and/or the editor(s). MDPI and/or the editor(s) disclaim responsibility for any injury to people or property resulting from any ideas, methods, instructions or products referred to in the content.

Article

Simulation and Analysis of the Loading, Relaxation, and Recovery Behavior of Polyethylene and Its Pipes

Furui Shi * and P.-Y. Ben Jar

Department of Mechanical Engineering, University of Alberta, 10-203 Donadeo Innovation Centre for Engineering, 9211-116 Street NW, Edmonton, AB T6G 1H9, Canada; ben.jar@ualberta.ca
* Correspondence: furui@ualberta.ca

Citation: Shi, F.; Jar, P.-Y.B. Simulation and Analysis of the Loading, Relaxation, and Recovery Behavior of Polyethylene and Its Pipes. *Polymers* **2024**, *16*, 3153. https://doi.org/10.3390/polym16223153

Academic Editors: Célio Pinto Fernandes, Luís Lima Ferrás, Alexandre M. Afonso and Arash Nikoubashman

Received: 21 August 2024
Revised: 15 October 2024
Accepted: 31 October 2024
Published: 12 November 2024

Copyright: © 2024 by the authors. Licensee MDPI, Basel, Switzerland. This article is an open access article distributed under the terms and conditions of the Creative Commons Attribution (CC BY) license (https://creativecommons.org/licenses/by/4.0/).

Abstract: Spring–dashpot models have long been used to simulate the mechanical behavior of polymers, but their usefulness is limited because multiple model parameter values can reproduce the experimental data. In view of this limitation, this study explores the possibility of improving uniqueness of parameter values so that the parameters can be used to establish the relationship between deformation and microstructural changes. An approach was developed based on stress during the loading, relaxation, and recovery of polyethylene. In total, 1000 sets of parameter values were determined for fitting the data from the relaxation stages with a discrepancy within 0.08 MPa. Despite a small discrepancy, the 1000 sets showed a wide range of variation, but one model parameter, $\sigma_{v,L}(0)$, followed two distinct paths rather than random distribution. The five selected sets of parameter values with discrepancies below 0.04 MPa were found to be highly consistent, except for the characteristic relaxation time. Therefore, this study concludes that the uniqueness of model parameter values can be improved to characterize the mechanical behavior of polyethylene. This approach then determined the quasi-static stress of four polyethylene pipes, which showed that these pipes had very close quasi-static stress. This indicates that the uniqueness of the parameter values can be improved for the spring–dashpot model, enabling further study using spring–dashpot models to characterize polyethylene's microstructural changes during deformation.

Keywords: relaxation; modeling; mechanical properties; polyethylene

1. Introduction

Polymers are widely used in our daily life [1,2], among which more than two-thirds are semi-crystalline polymers (SCPs) [3]. SCPs, such as polyethylene (PE), are a class of thermoplastics with complicated microstructures [4–8], which have attracted significant attention from many research groups [9–17]. In view of the fact that SCPs are increasingly used in various industrial sectors for fluid transportation [18], packaging [19], electronics [20], civil engineering [21], aerospace [22], medical devices [23], automotive components [24], etc., due to their chemical inertness and attractive mechanical properties [25–29], it is important to provide a proper characterization of their stress response to deformation. However, SCPs exhibit complex time-dependent behaviors, including relaxation and creep [4,30–35], which could significantly impact their performance in all applications. Therefore, a full characterization of SCPs for their mechanical behavior, which includes the time-dependent stress response to deformation, is essential to ensure reliable performance in their entire designed lifetime [36].

Stress relaxation under a constant deformation level has long been used to assess the performance of plastic pipes [37,38]. Moser and Folkman [38] demonstrated the usefulness of using stress relaxation tests to predict the long-term performance of plastic pipes and their interaction with soil systems [39]. In view of the fact that plastic pipes are designed to have a lifespan exceeding 50 years [40–45], with about 95% of plastic pipes made of PE [44,46–49], stress relaxation tests and the corresponding data analysis based on modeling have been widely used to study the long-term mechanical performance of PE and its pipes [44].

In a relaxation test at a constant deformation level, the stress decrease is very significant at the beginning, but eventually reaches an asymptotic limit [26,50]. The stress–time curve during the relaxation process is known to be influenced by the loading rate prior to relaxation [46], and a transition of the mechanism involved in the deformation process could be detected by characterizing the relaxation behavior before and after the transition [51,52]. Although the relaxation and recovery processes are known to give different stress responses to deformation, as the former is introduced after loading and the latter after unloading, both are carried out at a constant deformation level, with a bigger stress change in the former than in the latter [53]. At the same deformation level, the two processes are expected to reach the same stress level that is known as quasi-static stress. We have recently developed a test for characterizing SCPs' viscous behavior, named the multiple-relaxation–recovery test (RR test), in which a recovery process is generated right after a relaxation process at a similar deformation level, and the two processes are repeated multiple times with the increase in specimen displacement [54,55]. Compared to the multiple-relaxation test described in the literature [51], the RR test allows for the determination of the unloading stiffness of the materials and reveals the unusual stress response of recovery behavior.

Various models have been used to analyze the mechanical test results of SCPs [56–68], among which models consisting of springs and dashpots have been used to mimic the stress response to deformation. Basic spring–dashpot models are known as Maxwell [69] and Voigt models, which represent the basic relaxation and creep behaviors, but are insufficient for simulating SCPs' highly nonlinear behavior [64]. However, when Eyring's equation was used to govern the stress response of the dashpot element [70–77], some success was obtained. Recently, the three-branch model proposed by Hong et al. [78,79] and Izraylit et al. [80], with only one branch containing an Eyring dashpot, was successfully used to mimic relaxation behavior. However, this model was not applicable to the recovery behavior after unloading [81]. Our recent work [81] also showed that some three-branch spring–dashpot models are not able to provide a full description of the stress change during relaxation and recovery phases of the RR test, especially for the unusual stress drop detected during the recovery. A three-branch model with two Maxwell branches and one spring branch, on the other hand, has been able to simulate both relaxation and recovery behavior fairly accurately. Most of the works using a three-branch model [82–87] only provided a single set of parameter values to mimic the experimental data, even though it is commonly believed that multiple sets of the parameter values exist for a model to mimic the experimental data [88–90]. As a result, the use of a spring–dashpot model to reproduce the experimental data is often considered merely a curve-fitting exercise. The parameter values were not used to characterize the viscous part of the mechanical properties of SCPs. Therefore, it is necessary to develop a novel approach to improve the uniqueness of model parameters for the accurate simulation and characterization of SCPs' mechanical behavior. Such an approach is the subject of this paper.

In this work, an analysis method was developed based on global and local optimization to simulate the relaxation, recovery, and loading behaviors of PE and its pipes using a three-branch spring dashpot model based on Eyring's law. The model contains two time-dependent, viscous branches and one time-independent, quasi-static branch. Data from RR tests on cylindrical specimens and notched pipe ring (NPR) specimens were used in the simulation to generate 1000 sets of parameter values to mimic the stress drop at the relaxation stages. The range of variation for these parameter values was examined and discussed. The best five fits were selected to improve the uniqueness of the model parameter values. Then, the analysis method was applied to four PE types of pipes, and their quasi-static stress as a function of specimen displacement was determined and discussed.

2. Experiments

2.1. Materials

Cylindrical specimens of one type of HDPE [81] and NPR specimens of four different pipes were used in this study. The cylindrical specimen, named HDPE-b following the previous publication, has characteristics as detailed in the previous work [81]. The dimension of the cylindrical specimen is shown in Figure 1a, and Figure 1b shows the cylindrical specimen prepared for the tests. Figure 1c shows the dimensions of the NPR specimen cut from the PEX-a pipe, and a sample of the specimen is shown in Figure 1d. The four types of NPR specimens were obtained from four PE pipes, with their characteristics summarized in Table 1, which lists the materials of the four pipes, pipe name, density, yield strength, and hydrostatic design basis (HDB), defined in ASTM D2837 [91], representing the long-term hydrostatic strength of a pipe. All pipes have a ratio of pipe outer diameter to wall thickness (SDR) of 11.

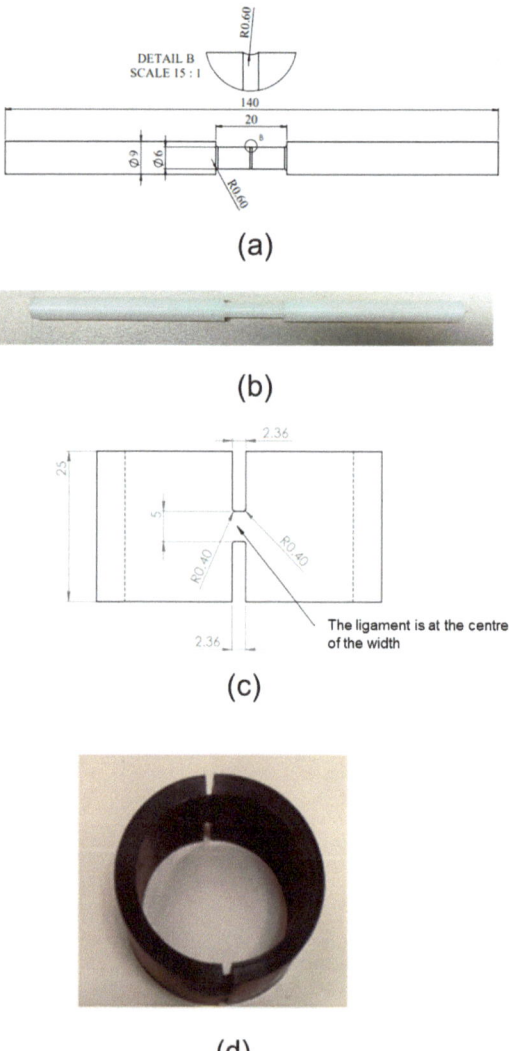

Figure 1. Specimens used in the RR tests: (**a**) dimensions of cylindrical specimen, (**b**) cylindrical specimen, (**c**) dimensions of NPR specimen (PE-Xa pipe as an example), and (**d**) NPR specimen. All units are in millimeters.

Table 1. Characteristics of pipes used in this study.

Material	Pipe Name	Density (kg/m³)	Yield Strength (MPa)	HDB @23 °C (MPa)
HDPE	PE4710-black	949	24.8	11.03
HDPE	PE4710-yellow	949	>24.1	11.03
PEX	PE-Xa	938	19	8.62
MDPE	PE2708	940	19.3	8.62

The set-up of the RR test in the universal test machine was depicted in refs. [92–96].

2.2. Mechanical Characterization

RR tests were carried out using a Qualitest Quasar 100 universal test machine (Qualitest, Lauderdale, FL, USA), with data collected by a personal computer [51]. The details of the RR tests were described in the previous work [81,96]. The RR test consists of six stages in one cycle: 1st loading, relaxation, 2nd loading, stabilization, unloading, and recovery. The maximum deformation introduced in the RR tests was set to exceed the yield point, at which approximately 30 cycles were generated [51]. The sample curves of the RR tests on cylindrical specimens are available in previous publications [81,96]. The crosshead speed was set to 1 mm/min, with 10,000 s allocated for each relaxation, stabilization, or recovery stage. To ensure repeatability and reliability, two specimens were tested for each material, except for the PE4710-black pipe, for which only one RR test was conducted due to the laboratory shutdown in the COVID-19 pandemic period.

3. Data Analysis

3.1. Three-Branch Model

In this study, the three-branch, spring–dashpot model employed for the simulation of the relaxation, recovery, and loading behaviors of the results from RR tests is depicted in Figure 2. This model is known as the Maxwell–Weichert model, which has been commonly used to mimic the stress response to deformation of a variety of materials [56,83,85,97–100]. As shown in Figure 2, the model incorporates three springs governed by Hooke's law [52,101–105] and two dashpots governed by Eyring's law [106–113]. The left, middle, and right branches represent long-term viscous stress, short-term viscous stress, and quasi-static stress, respectively, denoted by the subscripts L, S, and qs [114].

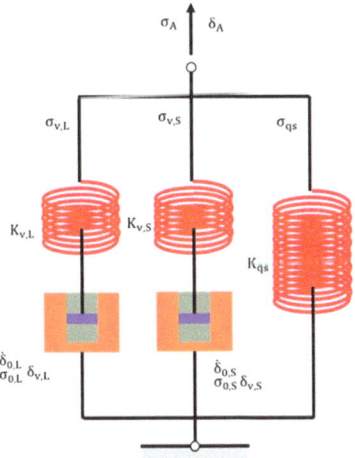

Figure 2. Three-branch spring–dashpot model used in this study.

From our previous publication on the three-branch model [96], the equations governing stress response as a function of time during the relaxation, recovery, and loading stages were derived. The stress change [107] during each relaxation or recovery stage can be expressed as follows:

$$\begin{aligned}\Delta\sigma_A &= \sigma_A(0) - \sigma_A(t) \\ &= \sigma_{v,L}(0) + \sigma_{v,S}(0) \\ &- 2\sigma_{0,L}\tanh^{-1}\{\tanh[\sigma_{v,L}(0)/(2\sigma_{0,L})]\exp(-t/\tau_{v,L})\} \\ &- 2\sigma_{0,S}\tanh^{-1}\{\tanh[\sigma_{v,S}(0)/(2\sigma_{0,S})]\exp(-t/\tau_{v,S})\}\end{aligned} \quad (1)$$

$$\tau_{v,i} = \sigma_{0,i} / \left(K_{v,i} \dot{\delta}_{0,i} \right) \quad (2)$$

where σ_A represents the applied engineering stress, t the time from the beginning of the stage, $\sigma_{v,i}(0)$ the viscous stress at the beginning of the stage, $\sigma_{0,i}$ the reference stress, $\tau_{v,i}$ the characteristic relaxation time, $K_{v,i}$ the spring stiffness, and $\dot{\delta}_{0,i}$ the reference stroke rate, for $i = L$ or S.

For each loading stage, the stress responses for the long-term and short-term branches were determined as follows:

$$\dot{\sigma}_{v,L} = K_{v,L}\dot{\delta}_A - (\sigma_{0,L}/\tau_{v,L})\sinh(\sigma_{v,L}/\sigma_{0,L}) \quad (3)$$

$$\dot{\sigma}_{v,S} = K_{v,S}\dot{\delta}_A - (\sigma_{0,S}/\tau_{v,S})\sinh(\sigma_{v,S}/\sigma_{0,S}) \quad (4)$$

where $\dot{\delta}_A$ is the crosshead speed of the test machine and $\dot{\sigma}_{v,i}$ is the first derivative of $\sigma_{v,i}$ with respect to time t, for $i = L$ or S.

To estimate values for the fitting parameters in Equations (1), (3), and (4), the inverse analysis method [115–127] was employed by simulating the experimental data of the RR tests.

3.2. Method for Data Analysis

This section describes a new analysis method for the simulation of the relaxation, recovery, and loading behavior of PE and its pipes. The analysis method uses a new optimization approach that combines global and local optimization techniques.

In our previous work [96], a genetic algorithm (GA) in MATLAB R2021b was used to determine model parameter values via the inverse approach. However, that method was constrained by several assumptions that limited its applicability to a specific type of loading range. For example, the method depends on the presence of a plateau region [51] of the stress–displacement curve to determine one of the model parameter values. For test data that do not have such a clear plateau region, the method could not be used.

In the current study, a method was developed without the requirement of a plateau region. Rather, the new method focuses solely on the minimization of the maximum difference between the experimental data and values generated by the model in Figure 2, based on the principle known as minimax in approximation theory [128,129]. Setiyoko et al. [130] reported minimax as an approach that contrasts the widely used least squares for determining values for parameters [82,85,131–138]. Many researchers have typically determined a single set of values for their model parameters [139–142], but whether the values for the model parameters are unique remains a challenging question. In our previous work [96], ten sets of values for the model parameters in Figure 2 were determined to examine variations in the values [96]; however, the time for determining the ten sets of values was long due to constraints imposed in the algorithms, such as the assumption of the plateau region. By removing these assumptions, it became possible to obtain 1000 sets of the parameter values within a reasonable timeframe.

All programs developed in this study were coded in MATLAB R2024a, and the values for the model parameters in Equations (1), (3), and (4) were determined using experimental data at the relaxation, the recovery, and the first loading stages of the RR tests. At each of the relaxation or recovery stages, the values for the parameters in Figure 2 were assumed to remain fixed as the material microstructure during the relaxation and recovery was deemed to remain unchanged [143]. At each of the first loading stages, values for $K_{v,L}$ and $K_{v,S}$ were assumed to remain fixed as the deformation range introduced at each of the first loading stages was deemed to be small enough to allow the values for $K_{v,L}$ and $K_{v,S}$ to remain constant. However, the values for $\sigma_{0,L}$, $\tau_{v,L}$, $\sigma_{0,S}$, and $\tau_{v,S}$ were allowed to vary at each of the first loading stages.

Figure 3 depicts the entire procedure used to determine values for the fitting parameters in Figure 2, including the initial 1000 sets of parameter values based on the experimental data at the relaxation stages, and then the 5 best sets of parameter values at each of the recovery and the 1st loading stages. The objective function of the analysis was to minimize the maximum difference between the experimental data and the values generated by the model.

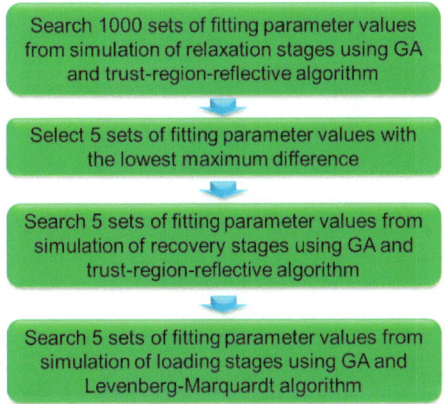

Figure 3. Procedure for the determination of fitting parameters in the relaxation, recovery, and loading stages of RR tests.

The programs for identifying the optimal fitting parameter values are based on the hybrid combination of the global optimization approach, GA, and the local optimization approach, lsqnonlin [144–147] in MATLAB R2024a, also known as the combined two-phase strategy [148]. GA was first used to identify the fitting parameter values, and the generated fitting parameter values were set as the initial guesses of lsqnonlin. The objective function of GA is the maximum difference in stress response between the experiments and the model, which needs to be minimized. The population size was set to be 200, and the maximum number of generations was 600. According to Renders and Flasse [149], global optimization inherently involves a fundamental conflict between accuracy, reliability, and computing time. As a result, Mahinthakumar and Sayeed [150] suggested that the strength of GA could be decreased when the population was converged to a narrow location in the search space and the difference between solutions was small. It was also reported that GA often requires extensive iterations and tends to converge slowly [151–153]. On the other hand, local optimization is more efficient in narrow search areas and thus is increasingly hybridized with GA to accelerate computation [154–158]. As a result, a hybrid global–local approach was developed, by using GA and lsqnonlin [144–147] in MATLAB R2024a for global and local optimization [149], respectively, to identify the fitting parameters for the relaxation stages of the RR tests.

In the first step, as illustrated in Figure 3, a numerical method was developed using the inverse approach to search for 1000 sets of values for the fitting parameters in the three-branch model in Figure 2 in order to mimic the experimental data at the relaxation.

The initial value ranges were set to be the same as those in the previous work [96], i.e., [0.1, 20] (in MPa) for $\sigma_{v,L}(0)$, [0.01, 2] (in MPa) for $\sigma_{0,L}$, [1000, 90,000] (in second) for $\tau_{v,L}$, [0.1, 20] (in MPa) for $\sigma_{v,S}(0)$, [0.01, 2] (in MPa) for $\sigma_{0,S}$, and [1, 900] (in second) for $\tau_{v,S}$. GA was used to identify the six fitting parameters in Equation (1), and the generated fitting parameter values were set as the initial guesses of lsqnonlin which were based on the trust-region-reflective algorithm [159]. In view of the fact that the speed of the computer program could be increased using parallel computing [160], parallel computing was implemented using 'parfor' in MATLAB R2024a, following the work in ref. [161], to speed up the simulation so that 1000 sets of model parameter values could be determined at the first step in a reasonable timeframe. In addition, the experimental data for the very first relaxation stage were ignored in the analysis, because it did not have any prior recovery stage, and thus did not possess the same deformation history as the relaxation stages in other cycles. In other words, the analysis conducted in this study always started from the relaxation stage in the second cycle of the RR test.

In the second step of Figure 3, five sets of fitting parameter values with the smallest maximum difference between the experimental data at the relaxation stages and the simulation results were selected. In the third step, each of the five sets of values from the second step was used to determine one set of fitting parameter values for the recovery stages at similar deformation levels. The initial values of the fitting parameters at the recovery stages, for example in the mth cycle of the RR tests, were set to be [0.01, α] (in MPa) for $\sigma_{v,L}(0)$, [0.001, β] (in MPa) for $\sigma_{0,L}$, [1000, 90,000] (in second) for $\tau_{v,L}$, [-20, -0.001] (in MPa) for $\sigma_{v,S}(0)$, [0.001, 2] (in MPa) for $\sigma_{0,S}$, and [1, 10,000] (in second) for $\tau_{v,S}$, where α and β are the values for $\sigma_{v,L}(0)$ in the relaxation stage of the mth cycle and the $\sigma_{0,L}$ values in the next relaxation stage, i.e., in the $(m+1)$th cycle. In view of the fact that the range of stress variation at the recovery stages was much less than that at the corresponding relaxation stages, it was deemed unnecessary to determine 1000 sets of parameter values for the simulation of the recovery stages.

The final step in Figure 3 is to determine five sets of fitting parameter values for Equations (3) and (4) to simulate the stress variation at the 1st loading stage in each cycle, based on the parameter values determined for the relaxation and recovery stages in steps 2 and 3, respectively. For this purpose, the method was similar to that used in our previous work [96], based on GA in MATLAB R2021b, but with the improvement of combining GA with lsqnonlin. However, in this case lsqnonlin automatically employed the Levenberg–Marquardt algorithm, as the original method [96] was designed to fit only one data point at a time but the trust-region-reflective algorithm requires the number of data points (equations) to be at least equal to the number of parameters (variables).

It should be noted that in the literature, many researchers [51,52,92,95] have used constant characteristic relaxation time for their simulation. However, as suggested in ref [162], the effect of the characteristic relaxation time on the determination of σ_{qs} should be evaluated and the characteristic relaxation time should be allowed to vary with deformation. The novelty of the proposed method, as described above, originates from its ability to allow change in the characteristic relaxation time during the deformation. The proposed method also enables the evaluation of the influence of the characteristic relaxation time on the determination of other model parameter values. In addition, the combination of global and local optimization also significantly reduced the searching time for the 1000 parameter values, allowing the selection of the best five sets of parameter values and thus evaluating the uniqueness of the parameter values for the characterization of the viscous behavior of SCPs.

3.3. Resolution of the Experimental Measurements

Many researchers have studied material properties using mechanical tests [108,141,163–169], but few have considered the resolution of the test data [170]. For example, Mulliken and Boyce [171] successfully predicted the stress response of polymers in tension and compression tests using a constitutive model, but the resolution of the experimental measurements was not reported to justify the quality of the prediction. According to Jar [165,172], the uncertainty of the experimental measurements affects the accuracy of the test results. Therefore, a model that provides a good fitting to the experimental data with a poor resolution does not provide a clear indication on the validity of the model. In view of this potential issue, the resolution of the stress measurements obtained from this study was determined to assess the accuracy of the test results.

For the cylindrical specimens, σ_A was calculated using the following expression:

$$\sigma_A = \frac{4F}{(\pi D^2)} \quad (5)$$

where F is the measured tensile force using the universal testing machine, and D the initial diameter of the gage section measured using a digital caliper. Therefore, the resolution of σ_A for the cylindrical specimens, $d\sigma_A$, can be expressed as follows [170,173]:

$$d\sigma_A = \left|\frac{4dF}{(\pi D^2)}\right| + \left|\frac{8FdD}{(\pi D^3)}\right| \quad (6)$$

where dF and dD are the resolutions of the force and diameter measurements, respectively.

Similarly, the resolutions for the NPR specimens can be calculated using the following equation.

$$d\sigma_A = \left|\frac{dF}{t_1w_1+t_2w_2}\right| + \left|\frac{Fw_1dt_1}{(t_1w_1+t_2w_2)^2}\right| + \left|\frac{Ft_1dw_1}{(t_1w_1+t_2w_2)^2}\right| + \left|\frac{Fw_2dt_2}{(t_1w_1+t_2w_2)^2}\right| + \left|\frac{Ft_2dw_2}{(t_1w_1+t_2w_2)^2}\right| \quad (7)$$

where t_j is the initial thickness of the gauge section j of the NPR specimens, and w_j is the corresponding initial width of the gauge section (j is 1 or 2, representing the two ligaments of the NPR specimens).

In this study, the resolution of the universal test machine for the force measurement was 0.5 N and the resolution of the digital caliper for the dimensional measurement was 0.01 mm. As an example, for a cylindrical specimen with D of 5.90 mm and the maximum force of 402.5 N, the resolution of its stress measurement, $d\sigma_A$, is

$$d\sigma_A = \left|\frac{4dF}{(\pi D^2)}\right| + \left|\frac{8FdD}{(\pi D^3)}\right| = 0.0682 \text{ MPa} \quad (8)$$

Similarly, the resolution for the stress measurement of NPR specimens from different PE pipes can be determined based on the dimensions and maximum force generated in the RR tests.

4. Results and Discussion

4.1. Accuracy of the Simulation

This section presents 1000 sets of parameter values for the simulation of the relaxation stages in the RR tests, including the maximum difference between the simulation and the experimental data and a comparison of the simulation results with the resolution of the experimental data.

In the previous study, we found that the three-branch model can accurately describe results at the relaxation, recovery, and loading stages of RR tests [96]. The previous analysis relied on several assumptions, such as constant $\tau_{v,L}$ and $\tau_{v,S}$ values [52,174], and considered the continuity of the parameter values with the increase in deformation. In this study, the method presented in Section 3.2 was used to generate 1000 sets of parameter values for the simulation of experimental data at the relaxation stages of the RR tests on cylindrical specimens and NPR specimens. Table 2 summarizes the resolution of the measured stress data and their maximum difference with the modeling results, the latter based on the 1000 sets of fitting parameter values.

Table 2. Resolution of the measured stress data and maximum difference in the stress response between the experiments and model during relaxation stages based on the 1000 sets of model parameter values.

Sample Specimens	Resolution of Experimental Measurement (MPa)	Max Difference of Stress Between Experimental Measurements and Model Simulation from the Study (MPa)
HDPE-b, cylindrical	0.0682	0.0618
PE-Xa, NPR pipe	0.0767	0.0759
PE4710-yellow, NPR pipe	0.0746	0.0666
PE4710-black, NPR pipe	0.0743	0.0591
PE2708, NPR pipe	0.0590	0.0524

Table 2 shows the values for the experimental resolution based on Equations (6) and (7) and the maximum difference in stress between the experimental data and the simulation data using the 1000 sets of parameter values for cylindrical and NPR specimens at the relaxation stages. This indicates that the values of experimental resolution are slightly larger than the values of maximum difference in stress between the experimental data and simulation data. From Table 2, it should be noted that the values of the maximum difference are less than 0.07 MPa, which is smaller than 0.08 MPa reported in our previous work [96]. In the literature, the maximum difference between the experiments and model was reported to be in the range from 0.17 to about 1 MPa [30,85,175–178]. In addition, the difference between the resolution of the test data and the value for the maximum difference is less than 0.01 MPa, with the maximum difference in HDPE-b being even smaller than the resolution of the test data. This indicates that the analysis method created in this study can provide good agreement between the model and experiments. This high accuracy was also achieved for the NPR specimens in Table 2, with the maximum differences being less than 0.08 MPa.

The results in Table 2 show the capability of the three-branch model based on the proposed analysis method presented in Section 3.2, which is consistent with the work in the literature [99]. Jar [179] further validated the close simulation of the three-branch model in a new test, named the MR test, which entails relaxation behavior at different deformation levels. However, none of the results in these works found the maximum difference between the experiments and model to be less than 0.08 MPa. Table 2 also suggests that since the inverse approach relies on the quality of the experimental measurements, further improvement of the simulation accuracy requires the improvement of the resolution for the experimental data.

Figure 4 illustrates 1000 sets of fitting parameter values for the simulation of the relaxation stages of different deformation levels of one RR test on an HDPE-b cylindrical specimen. As shown in Figure 4a, $\sigma_{v,L}(0)$ clearly follows two distinct paths with the increase in stroke, namely, an upper path and a lower path. Works in the literature always present a single path of the fitting parameters [92,108,174], even for our previous work, which showed ten sets of the fitting parameter values [96]. Note that Pyrz and Zaïri [180] identified 20 sets of parameter values but no pattern was identified for these values.

Figure 4 also suggests that a two-path pattern exists for the variation in $\sigma_{v,S}(0)$ and $\sigma_{0,S}$ with stroke in Figures 4b and 4d, respectively, though $\sigma_{0,L}$ in Figure 4c mainly shows a single path. With the consideration of the limited resolution for the experimental measurement, this two-path pattern for $\sigma_{v,L}(0)$, $\sigma_{v,S}(0)$, and $\sigma_{0,S}$ values indicates that the fitting parameters could show some identifiable variation with the increase in deformation, rather than the random distribution that has been believed in the past. Therefore, there is a possibility that these model parameters could be linked to microstructural changes in SCPs.

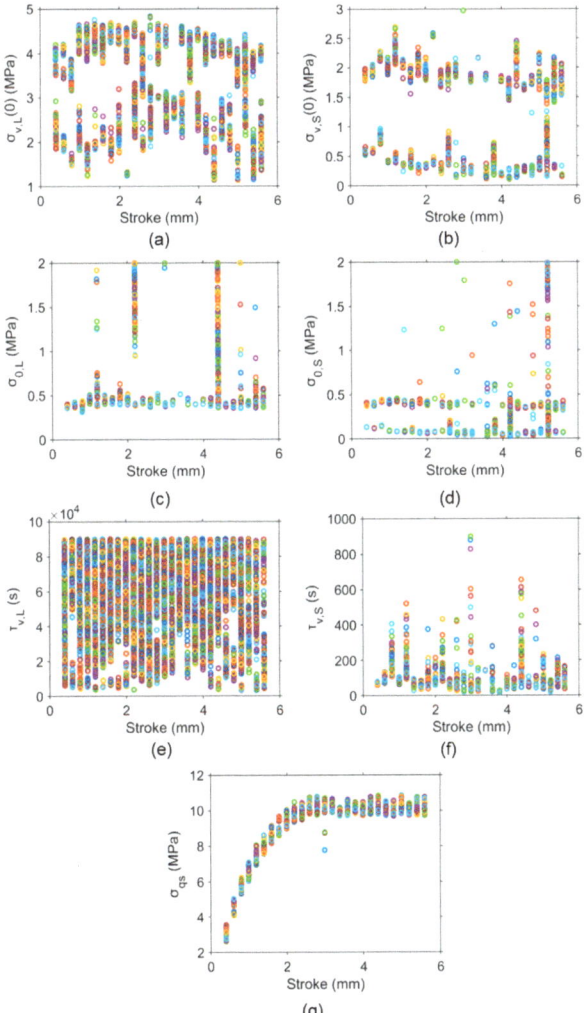

Figure 4. The 1000 sets of parameter values for simulation at the relaxation stages of different deformation levels in one RR test of HDPE-b: (**a**) $\sigma_{v,L}(0)$, (**b**) $\sigma_{v,S}(0)$, (**c**) $\sigma_{0,L}$, (**d**) $\sigma_{0,S}$, (**e**) $\tau_{v,L}$, (**f**) $\tau_{v,S}$, and (**g**) σ_{qs}. Different colors at one stroke are used to indicate the 1000 sets of parameter values.

For $\tau_{v,L}$ and $\tau_{v,S}$ values, as shown in Figure 4e,f, their values are scattered across the deformation levels considered in the RR test, indicating that variations in $\tau_{v,L}$ and $\tau_{v,S}$ values may not affect the two-path pattern for the fitting parameters $\sigma_{v,L}(0)$, $\sigma_{v,S}(0)$, and $\sigma_{0,S}$. These results confirm the previous suggestion that inaccurate values for the characteristic relaxation time have a minor influence on the simulation [167]. In the literature, the characteristic relaxation time was often fixed as a constant for different deformation levels and materials [51,174,181]. Although Izraylit et al. [80] determined the values of characteristic relaxation time at different deformation levels, they did not clearly present the curve-fitting process used in their study. Jar [179] obtained values of characteristic relaxation time as functions of deformation levels but only provided one set of fitting parameters.

Even with a two-path distribution for some of the fitting parameters, a single trend of variation with stroke could be established for σ_{qs}, as shown in Figure 4g. The band of variation for σ_{qs} is quite small, suggesting that the σ_{qs} values are not sensitive to the variation in the fitting parameter values. These findings suggest that the determination of σ_{qs} does not require a unique set of values for the fitting parameters, as long as the fitting parameter can provide a reasonable simulation of the test results.

Figure 5 summarizes the $\sigma_{v,L}(0)$ values of the pipe specimens, which clearly shows that a two-path pattern also exists for PE-Xa, PE2708, PE4710-yellow, and PE4710-black pipes, suggesting that the presence of two distinct paths for variations in $\sigma_{v,L}(0)$ with deformation is a common phenomenon. Figure 5 also shows that $\sigma_{v,L}(0)$ increases significantly at the early stage of the RR test, which is consistent with the observations reported in the literature [92]. Note that in the literature, Liu et al. [182], and Moore et al. [183–185] also compared modeling and experimental testing for the stress response of HDPE pipes, but they did not provide the viscous stress component of the stress response. Zhang and Jar [181] determined the viscous stress in the pipes but with the assumption that the characteristic relaxation time should be kept constant.

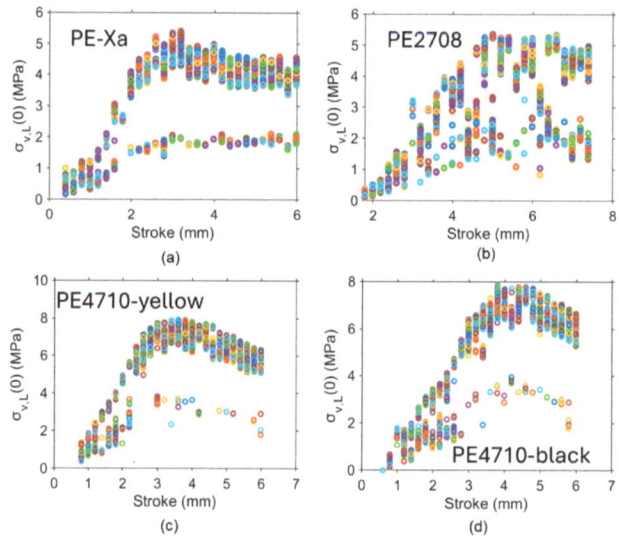

Figure 5. A two-path pattern of $\sigma_{v,L}(0)$ as a function of stroke for NPR specimens based on 1000 sets of parameter values: (**a**) PE-Xa, (**b**) PE2708, (**c**) PE4710-yellow, and (**d**) PE4710-black pipes. Different colors at one stroke are used to indicate the 1000 sets of parameter values.

The above findings suggest that it is possible to improve the uniqueness of model parameter values which could be used to characterize the mechanical performance of SCPs. However, a further study would be needed to confirm this possibility.

4.2. Best Five Fits

One of the main problems addressed in the literature about the deformation of SCPs is the evolution of the crystalline phase with an increase in deformation [3]. Therefore, if the fitting parameters are to be used to characterize a material's performance, the change in the fitting parameter values should reflect the evolution of SCP microstructures.

Many researchers [82,85] minimized the difference between the model and experiments to determine the model parameter values. Although the 1000 sets of parameter values are equally valid solutions of the model, because the uniqueness of the parameter values is absent among the 1000 sets, the five best sets of fitting parameter values were considered. Using the procedure depicted in Figure 3, five sets of fitting parameter values were identified which provided the closest simulation of the stress variation at the relaxation stages. These fitting parameter values for HDPE-b, along with its σ_{qs}, are summarized in

Figure 6 as functions of stroke. Note that some outliers exist, especially for $\sigma_{v,S}(0)$ and $\sigma_{0,S}$, but apart from these outliers, a general trend for $\sigma_{v,L}(0)$, $\sigma_{v,S}(0)$, $\sigma_{0,L}$, and $\sigma_{0,S}$ values is clearly given with the increase in stroke.

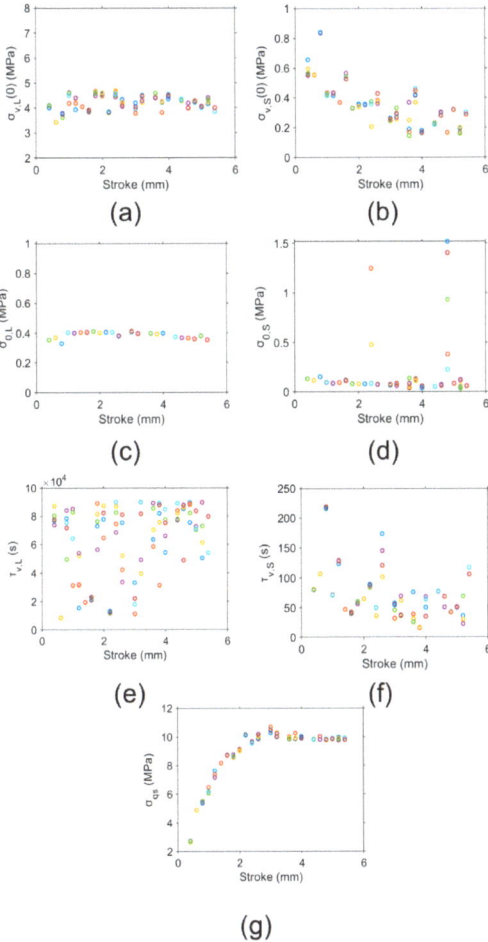

Figure 6. Best five sets of parameter values (in open red circles) selected from 1000 sets for the simulation of stress variation at the relaxation stages of HDPE-b and the corresponding σ_{qs}: (**a**) $\sigma_{v,L}(0)$, (**b**) $\sigma_{v,S}(0)$, (**c**) $\sigma_{0,L}$, (**d**) $\sigma_{0,S}$, (**e**) $\tau_{v,L}$, (**f**) $\tau_{v,S}$, and (**g**) σ_{qs}. Different colors at one stroke are used to indicate the five best sets of parameter values.

It should be pointed out that the five sets of parameter values shown in Figure 6 gave the maximum difference in stress response to deformation between the simulation and the experimental data of less than 0.04 MPa at the relaxation stages. In view of the fact that these values are significantly smaller than the resolution of the experimental measurements of 0.0682 MPa, as shown in Table 2, a further study using a test set-up that gives a better resolution than that in the current study would be needed to verify the validity of the five sets of parameter values. Nevertheless, Figure 6 clearly shows that fitting parameter values with a clear trend of dependence with deformation could be determined using the proposed approach for the data analysis.

It should also be noted that the $\sigma_{v,L}(0)$ values in Figure 6a were located along the upper path in Figure 4a, which indicates that the best five sets improved the uniqueness of the values of the long-term viscous stress at the beginning of relaxation. In the literature, Sweeney et al. [85] also described the long-term relaxation behavior using a Maxwell model, but the uniqueness of the model parameter values was not considered.

Figure 6b shows $\sigma_{v,S}(0)$ values for the five best sets of fitting parameters. These $\sigma_{v,S}(0)$ values are much smaller than their $\sigma_{v,L}(0)$ counterpart in Figure 6a, which is consistent with the values determined before by the manual curve fitting [179].

The five sets of $\sigma_{0,L}$ values shown in Figure 6c indicate that the five values at a given stroke are very consistent, and are in the value range consistent with those obtained previously [51] using a different test method (MR test). Figure 6d presents the $\sigma_{0,S}$ values, showing that apart from these outliers, their values are smaller than the corresponding $\sigma_{0,L}$ values at the same stroke, consistent with the previous observations [179]. The values of $\sigma_{0,L}$ and $\sigma_{0,S}$ in Figure 6c,d are consistent with the values reported in the literature [51,92,174].

The $\tau_{v,L}$ and $\tau_{v,S}$ values shown in Figure 6e,f show significant scattering with the increase in stroke, though the $\tau_{v,S}$ values are smaller than the $\tau_{v,L}$ values. This implies that the $\tau_{v,L}$ values and $\tau_{v,S}$ values exhibited high variability. However, the scattering $\tau_{v,L}$ and $\tau_{v,S}$ values did not affect the consistency of the corresponding fitting parameters $\sigma_{v,L}(0)$, $\sigma_{v,S}(0)$, $\sigma_{0,L}$, and $\sigma_{0,S}$. This aligns with the findings in the literature that the values of the characteristic relaxation time play a minor role in the simulation [167].

Figure 6g shows the σ_{qs} values as a function of stroke, which are consistent with values reported previously based on a different curve-fitting approach [81]. The figure suggests that σ_{qs} values increase initially and then reach a plateau, consistent with the trend observed previously [51]. As expected, even with the significant scattering of $\tau_{v,L}$ and $\tau_{v,S}$ values in Figure 6e,f, some outliers for $\sigma_{0,S}$ in Figure 6d, and some scattering for $\sigma_{v,L}(0)$ and $\sigma_{v,S}(0)$ in Figure 6a and 6b, respectively, the five sets of σ_{qs} values are still very consistent. In view of the measurement resolution shown in Table 2, this suggests that the σ_{qs} values determined from the current method have high consistency, not much affected by variations in fitting parameter values determined by the inverse approach.

It should be noted that, although some scattering is present, the consistency of the fitting parameter values shown in Figure 6 is much better than that reported in the literature. For example, Xu et al. [88] developed a generalized reduced gradient optimization algorithm, and used the algorithm to determine the parameter values for a three-branch model. Their results showed a much more significant scattering than those shown in Figure 6. Therefore, the proposed analysis method can capture a much more appropriate set of parameter values for the characterization of SCPs than the approaches currently available in the literature.

Table 3 lists the best five sets of fitting parameters for HDPE-b at the relaxation stage around the yield point. It was found that the five sets of $\sigma_{0,L}$ values are nearly identical to each other. In the literature, Xu et al. [88] determined three sets of model parameter values and the coefficient of variation was more than 50%. This indicates that the best-five-fits method in this study could provide better model parameter values than theirs. It was also found that the $\sigma_{v,L}(0)$ values are higher than the $\sigma_{v,S}(0)$ values, which is consistent with the results reported in the literature [179].

Figure 7 shows the simulation of the stress change in relaxation stages at different strokes using the fitting parameter values from the best five sets in Figure 6. The symbols in Figure 7 represent the experimental data and the lines represent the simulation data. This indicates that the parameter values determined from the current method can provide a quite accurate description of the relaxation behavior.

Figure 8 shows the $K_{v,L}$ and $K_{v,S}$ values of HDPE-b as functions of stroke. Figure 8 suggests that most of the $K_{v,L}$ values are higher than the $K_{v,S}$ at the same stroke. Note that the difference between $K_{v,L}$ and $K_{v,S}$ values has been an open question in the literature, as the works reported indicate that $K_{v,L}$ values could be either larger or smaller than $K_{v,S}$ [83,88,179]. This uncertainty could be explained by the results presented in Figure 4, as $K_{v,L}$ values are influenced by the choice of $\sigma_{v,L}(0)$ values from the two paths in Figure 5a. When the lower path in Figure 4a is used to determine $K_{v,L}$, in view of the fact that the corresponding $\sigma_{v,S}(0)$ values belong to the upper path in Figure 4b, $K_{v,S}$ must be larger than $K_{v,L}$. Conversely, the $K_{v,L}$ values are larger than $K_{v,S}$. As shown in Figure 8, for the best five sets of fitting parameter values, the $\sigma_{v,L}(0)$ values belong to the upper path. Therefore, the $K_{v,L}$ values for HDPE-b should be larger than the $K_{v,S}$ values. The above explanation is based on the identification of the two-path pattern for $\sigma_{v,L}(0)$ and $\sigma_{v,S}(0)$, which would not be possible without the collection of a large number of fitting parameter values (1000 sets). Similarly, it was found that the $K_{v,L}$ values are higher than the $K_{v,S}$ for pipes in this study.

Table 3. Best five sets of parameter values around the yield point for HDPE-b.

Model Parameters	Set 1	Set 2	Set 3	Set 4	Set 5
$\sigma_{v,S}(0)$ (MPa)	0.26	0.27	0.33	0.29	0.30
$\sigma_{0,S}$ (MPa)	0.06	0.06	0.09	0.08	0.08
$\tau_{v,S}$ (s)	62.51	69.78	36.70	37.80	37.88
$\sigma_{v,L}(0)$ (MPa)	4.21	4.28	4.47	4.51	4.47
$\sigma_{0,L}$ (MPa)	0.40	0.40	0.40	0.40	0.40
$\tau_{v,L}$ (s)	39,881.24	49,320.51	89,792.94	89,999.71	81,756.19

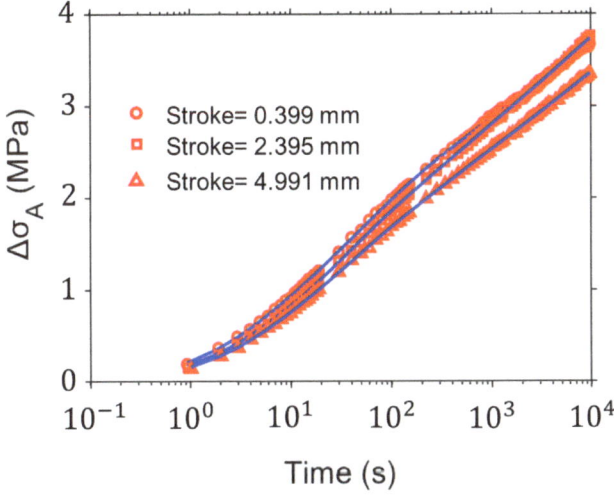

Figure 7. Simulation of stress change at relaxation stages of different strokes for HDPE-b using the fitting parameter values in Figure 6.

In addition, we believe that the $K_{v,L}$ and $K_{v,S}$ values could represent the microstructural changes in PE during the deformation process [179]. The accurate determination of $K_{v,L}$ and $K_{v,S}$ values is essential for examining the possible relationship between microstructural changes and mechanical performance of SCPs. This study provides an approach that could clearly distinguish the difference between $K_{v,L}$ and $K_{v,S}$ values, which has not been possible using other approaches reported in the literature.

Figure 9 compares $\sigma_A(0)$ and σ_{qs} for NPR specimens from the four pipes in Table 2. Markers in Figure 9b represent the average of the five σ_{qs} values that were determined based on the five sets of the best-fitting parameter values using the procedure described in Figure 3. The error bars in Figure 9b depict the standard deviation of the five σ_{qs} values. It was found that although the $\sigma_A(0)$ values for PE-Xa and PE2708 are lower than those for

PE4710-yellow and PE4710-black at the same stroke, their σ_{qs} values are much closer to each other.

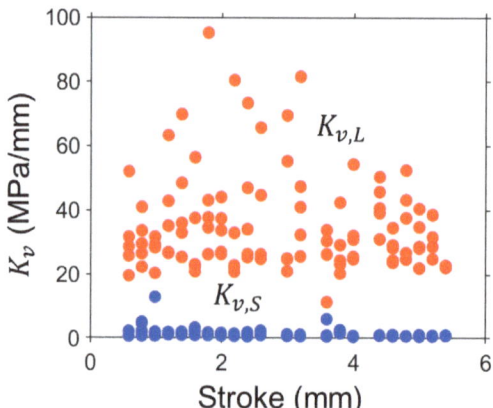

Figure 8. $K_{v,L}$ and $K_{v,S}$ as a function of stroke of HDPE-b.

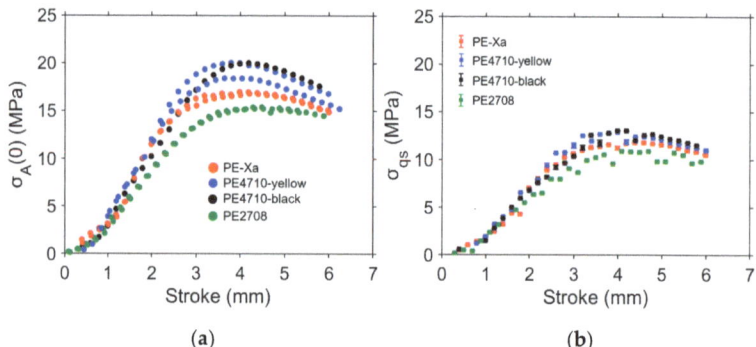

Figure 9. Summary RR test results for NPR specimens: (a) applied stress at the onset of relaxation, $\sigma_A(0)$, and (b) σ_{qs}.

5. Conclusions

This paper presents a new analysis method based on global and local optimization for the simulation of the relaxation, recovery, and loading behaviors of PE and its pipes in RR tests on cylindrical and NPR specimens, respectively. The results from the RR tests can be accurately mimicked using the three-branch model with the parameter values determined using the proposed analysis approach, and the maximum difference between the stress measured experimentally and that determined from the model is much smaller than the values reported in the literature.

Based on the proposed analysis method, 1000 sets of fitting parameter values were determined to simulate stress variations at the relaxation stages at different deformation levels, with the discrepancy between the experimental data and simulation results below 0.08 MPa. The $\sigma_{v,L}(0)$ values show two distinct paths with the increase in the stroke. The best five sets selected from the 1000 sets of parameter values provide a closer simulation of the relaxation behavior with a maximum difference in the stress response of less than 0.04 MPa. The results from the best five fits show that the proposed method can determine consistent values and a clear trend for $\sigma_{v,L}(0)$, $\sigma_{v,S}(0)$, $\sigma_{0,L}$, and $\sigma_{0,S}$. The results also indicate that the analysis method is better than any of the methods reported in the literature on the parameter identification of spring–dashpot models. The results from this study suggest that it is possible to improve the uniqueness of parameter values, which can then be used to characterize the viscous component of mechanical behavior for SCPs. This study also confirms that the $K_{v,L}$ values for PE should be larger than the $K_{v,S}$ at the same stroke, which solved the problem of uncertainty of the relationship between $K_{v,L}$ and $K_{v,S}$ in the literature. The results from the simulation suggest that variations in the characteristic relaxation time do not have much influence on the variations in other fitting parameter values, which confirms the previous finding that the effect of characteristic relaxation time has little influence on the variations in other fitting parameter values. However, further study is needed to improve the resolution of the measured results, so that the accuracy of the values based on the best five sets of fitting parameter values can be verified.

The overall conclusions of this study are as follows. The uniqueness of parameter values can be improved, except for the characteristic relaxation time, as the characteristic relaxation time has a minor influence on the modeling. On the other hand, if the experimental data have a sufficiently high resolution to reduce the uncertainty of the test results, it is then possible to explore the relationship between these parameters and microstructural changes in polyethylene during the deformation using the proposed method. This study provides a tool to improve the uniqueness of the model parameter values in a three-branch model.

Author Contributions: Conceptualization, P.-Y.B.J. and F.S.; methodology, P.-Y.B.J. and F.S.; software, P.-Y.B.J. and F.S.; validation, P.-Y.B.J. and F.S.; formal analysis, P.-Y.B.J. and F.S.; investigation, P.-Y.B.J. and F.S.; resources, P.-Y.B.J. and F.S.; data curation, P.-Y.B.J. and F.S.; writing—original draft preparation, P.-Y.B.J. and F.S.; writing—review and editing, P.-Y.B.J. and F.S.; visualization, P.-Y.B.J. and F.S.; supervision, P.-Y.B.J.; project administration, P.-Y.B.J.; funding acquisition, P.-Y.B.J. All authors have read and agreed to the published version of the manuscript.

Funding: This research was funded by Natural Sciences and Engineering Research Council of Canada (grant number RGPIN-2022-03588 Jar) and China Scholarship Council (grant number 201906450012). The APC was funded by China Scholarship Council (grant number 201906450012).

Institutional Review Board Statement: Not applicable.

Data Availability Statement: The data supporting the findings described in this manuscript are available from the corresponding authors upon request.

Acknowledgments: Sincere appreciation is given to the machine shop at the University of Alberta for the specimen preparation, and the Natural Sciences and Engineering Research Council of Canada and China Scholarship Council for the financial support.

Conflicts of Interest: The authors declare no conflicts of interest.

References

1. Sparks, T.D.; Banerjee, D. Materials Informatics and Polymer Science: Pushing the Frontiers of Our Understanding. *Matter* **2021**, *4*, 1454–1456. [CrossRef]
2. Stepto, R.; Horie, K.; Kitayama, T.; Abe, A. Mission and challenges of polymer science and technology. *Pure Appl. Chem.* **2003**, *75*, 1359–1369. [CrossRef]
3. Men, Y. Critical Strains Determine the Tensile Deformation Mechanism in Semicrystalline Polymers. *Macromolecules* **2020**, *53*, 9155–9157. [CrossRef]
4. Felder, S.; Holthusen, H.; Hesseler, S.; Pohlkemper, F.; Gries, T.; Simon, J.-W.; Reese, S. Incorporating Crystallinity Distributions into a Thermo-Mechanically Coupled Constitutive Model for Semi-Crystalline Polymers. *Int. J. Plast.* **2020**, *135*, 102751. [CrossRef]
5. Farge, L.; Boisse, J.; Dillet, J.; André, S.; Albouy, P.-A.; Meneau, F. Wide-Angle X-Ray Scattering Study of the Lamellar/Fibrillar Transition for a Semi-Crystalline Polymer Deformed in Tension in Relation with the Evolution of Volume Strain. *J. Polym. Sci. Part B Polym. Phys.* **2015**, *53*, 1470–1480. [CrossRef]
6. Uchida, M.; Tada, N. Micro-, Meso- to Macroscopic Modeling of Deformation Behavior of Semi-Crystalline Polymer. *Int. J. Plast.* **2013**, *49*, 164–184. [CrossRef]
7. Garcia-Gonzalez, D.; Zaera, R.; Arias, A. A Hyperelastic-Thermoviscoplastic Constitutive Model for Semi-Crystalline Polymers: Application to PEEK under Dynamic Loading Conditions. *Int. J. Plast.* **2017**, *88*, 27–52. [CrossRef]
8. Hao, P.; Laheri, V.; Dai, Z.; Gilabert, F.A. A Rate-Dependent Constitutive Model Predicting the Double Yield Phenomenon, Self-Heating and Thermal Softening in Semi-Crystalline Polymers. *Int. J. Plast.* **2022**, *153*, 103233. [CrossRef]
9. Li, C.Y. The Rise of Semicrystalline Polymers and Why Are They Still Interesting. *Polymer* **2020**, *211*, 123150. [CrossRef]

10. Regrain, C.; Laiarinandrasana, L.; Toillon, S.; Saï, K. Multi-Mechanism Models for Semi-Crystalline Polymer: Constitutive Relations and Finite Element Implementation. *Int. J. Plast.* **2009**, *25*, 1253–1279. [CrossRef]
11. Dusunceli, N.; Colak, O.U. Modelling Effects of Degree of Crystallinity on Mechanical Behavior of Semicrystalline Polymers. *Int. J. Plast.* **2008**, *24*, 1224–1242. [CrossRef]
12. Yeh, I.-C.; Andzelm, J.W.; Rutledge, G.C. Mechanical and Structural Characterization of Semicrystalline Polyethylene under Tensile Deformation by Molecular Dynamics Simulations. *Macromolecules* **2015**, *48*, 4228–4239. [CrossRef]
13. Yang, X.; Sun, Y.; Liao, T.; Men, Y. Strain Dependent Evolution of Structure and Stress in Propylene-Based Elastomer during Stress Relaxation. *Polymer* **2020**, *201*, 122612. [CrossRef]
14. Khan, F.; Yeakle, C. Experimental Investigation and Modeling of Non-Monotonic Creep Behavior in Polymers. *Int. J. Plast.* **2011**, *27*, 512–521. [CrossRef]
15. Lendlein, A.; Langer, R. Biodegradable, Elastic Shape-Memory Polymers for Potential Biomedical Applications. *Science* **2002**, *296*, 1673–1676. [CrossRef]
16. Lendlein, A.; Gould, O.E. Reprogrammable Recovery and Actuation Behaviour of Shape-Memory Polymers. *Nat. Rev. Mater.* **2019**, *4*, 116–133. [CrossRef]
17. Leterrier, Y. Durability of Nanosized Oxygen-Barrier Coatings on Polymers. *Prog. Mater. Sci.* **2003**, *48*, 1–55. [CrossRef]
18. Hutař, P.; Ševčík, M.; Frank, A.; Náhlík, L.; Kučera, J.; Pinter, G. The Effect of Residual Stress on Polymer Pipe Lifetime. *Eng. Fract. Mech.* **2013**, *108*, 98–108. [CrossRef]
19. Kartalis, C.N.; Papaspyrides, C.D.; Pfaendner, R. Recycling of Post-Used PE Packaging Film Using the Restabilization Technique. *Polym. Degrad. Stab.* **2000**, *70*, 189–197. [CrossRef]
20. Hou, S.; Qi, S.; Hutt, D.A.; Tyrer, J.R.; Mu, M.; Zhou, Z. Three Dimensional Printed Electronic Devices Realised by Selective Laser Melting of Copper/High-Density-Polyethylene Powder Mixtures. *J. Mater. Process. Technol.* **2018**, *254*, 310–324. [CrossRef]
21. Sobhan, K.; Mashnad, M. Tensile Strength and Toughness of Soil–Cement–Fly-Ash Composite Reinforced with Recycled High-Density Polyethylene Strips. *J. Mater. Civ. Eng.* **2002**, *14*, 177–184. [CrossRef]
22. Cha, J.-H.; Kim, Y.; Sathish Kumar, S.K.; Choi, C.; Kim, C.-G. Ultra-High-Molecular-Weight Polyethylene as a Hypervelocity Impact Shielding Material for Space Structures. *Acta Astronaut.* **2020**, *168*, 182–190. [CrossRef]
23. Zaribaf, F.P. Medical-Grade Ultra-High Molecular Weight Polyethylene: Past, Current and Future. *Mater. Sci. Technol.* **2018**, *34*, 1940–1953. [CrossRef]
24. Patil, A.; Patel, A.; Purohit, R. An Overview of Polymeric Materials for Automotive Applications. *Mater. Today Proc.* **2017**, *4*, 3807–3815. [CrossRef]
25. Barba, D.; Arias, A.; Garcia-Gonzalez, D. Temperature and Strain Rate Dependences on Hardening and Softening Behaviours in Semi-Crystalline Polymers: Application to PEEK. *Int. J. Solids Struct.* **2020**, *182–183*, 205–217. [CrossRef]
26. Ayoub, G.; Zaïri, F.; Naït-Abdelaziz, M.; Gloaguen, J.M. Modelling Large Deformation Behaviour under Loading–Unloading of Semicrystalline Polymers: Application to a High Density Polyethylene. *Int. J. Plast.* **2010**, *26*, 329–347. [CrossRef]
27. Atiq, O.; Ricci, E.; Baschetti, M.G.; De Angelis, M.G. Modelling Solubility in Semi-Crystalline Polymers: A Critical Comparative Review. *Fluid Phase Equilibria* **2022**, *556*, 113412. [CrossRef]
28. Nunes dos Santos, W.; Augusto Marcondes Agnelli, J.; Mummery, P.; Wallwork, A. Effect of Recycling on the Thermal Properties of Polymers. *Polym. Test.* **2007**, *26*, 216–221. [CrossRef]
29. Miaudet, P.; Derre, A.; Maugey, M.; Zakri, C.; Piccione, P.M.; Inoubli, R.; Poulin, P. Shape and Temperature Memory of Nanocomposites with Broadened Glass Transition. *Science* **2007**, *318*, 1294–1296. [CrossRef]
30. Li, Y.; He, Y.; Liu, Z. A Viscoelastic Constitutive Model for Shape Memory Polymers Based on Multiplicative Decompositions of the Deformation Gradient. *Int. J. Plast.* **2017**, *91*, 300–317. [CrossRef]
31. Liu, P.; Peng, L.; Chen, J.; Yang, B.; Chen, Y.; Luo, Z.; Han, C.C.; Huang, X.; Men, Y. Tensile Creep Failure of Isotactic Polypropylene under the Strain Criterion. *Macromolecules* **2022**, *55*, 9663–9670. [CrossRef]
32. Xu, S.; Zhou, J.; Pan, P. Strain-Induced Multiscale Structural Evolutions of Crystallized Polymers: From Fundamental Studies to Recent Progresses. *Prog. Polym. Sci.* **2023**, *140*, 101676. [CrossRef]
33. Sedighiamiri, A.; Govaert, L.E.; Kanters, M.J.W.; van Dommelen, J.A.W. Micromechanics of Semicrystalline Polymers: Yield Kinetics and Long-Term Failure. *J. Polym. Sci. B Polym. Phys.* **2012**, *50*, 1664–1679. [CrossRef]
34. Lim, S.D.; Rhee, J.M.; Nah, C. Predicting the Long-Term Creep Behavior of Plastics Using the Short-Term Creep Test. *Int. Polym. Process.* **2004**, *7*, 313–319. [CrossRef]
35. Kubát, J. Stress Relaxation in Solids. *Nature* **1965**, *205*, 378–379. [CrossRef]
36. Janssen, R. Deformation and Failure in Semi-Crystalline Polymer Systems. Master's Thesis, Eindhoven University of Technology, Eindhoven, The Netherlands, 2002.
37. Malpass, V.E. Prediction of Long-Term ABS Relaxation Behavior. *J. Appl. Polym. Sci.* **1968**, *12*, 771–788. [CrossRef]
38. Moser, A.P.; Folkman, S.L. *Buried Pipe Design*, 3rd ed.; McGraw-Hill: New York, NY, USA, 2008; ISBN 978-0-07-147689-8.
39. Moser, A.P. Structural Performance of Buried Profile-Wall High-Density Polyethylene Pipe and Influence of Pipe Wall Geometry. *Transp. Res. Rec.* **1998**, *1624*, 206–213. [CrossRef]
40. Frank, A.; Pinter, G.; Lang, R.W. Prediction of the Remaining Lifetime of Polyethylene Pipes after up to 30 Years in Use. *Polym. Test.* **2009**, *28*, 737–745. [CrossRef]

41. Frank, A.; Berger, I.J.; Arbeiter, F.; Hutař, P.; Pinter, G. Lifetime Prediction of PE100 and PE100-RC Pipes Based on Slow Crack Growth Resistance. In Proceedings of the 18th Plastic Pipes Conference PPXVIII, Berlin, Germany, 12–14 September 2016. [CrossRef]
42. Hoàng, E.M.; Lowe, D. Lifetime Prediction of a Blue PE100 Water Pipe. *Polym. Degrad. Stab.* **2008**, *93*, 1496–1503. [CrossRef]
43. Brown, N. Intrinsic Lifetime of Polyethylene Pipelines. *Polym. Eng. Sci.* **2007**, *47*, 477–480. [CrossRef]
44. Zha, S.; Lan, H.; Huang, H. Review on Lifetime Predictions of Polyethylene Pipes: Limitations and Trends. *Int. J. Press. Vessel. Pip.* **2022**, *198*, 104663. [CrossRef]
45. Frank, A.; Arbeiter, F.J.; Berger, I.J.; Hutař, P.; Náhlík, L.; Pinter, G. Fracture Mechanics Lifetime Prediction of Polyethylene Pipes. *J. Pipeline Syst. Eng. Pract.* **2019**, *10*, 04018030. [CrossRef]
46. Zhang, Y.; Jar, P.-Y.B. Time-Strain Rate Superposition for Relaxation Behavior of Polyethylene Pressure Pipes. *Polym. Test.* **2016**, *50*, 292–296. [CrossRef]
47. Zhang, Y.; Jar, P.-Y.B.; Xue, S.; Han, L.; Li, L. Measurement of Environmental Stress Cracking Resistance of Polyethylene Pipe: A Review. In Proceedings of the ASME 2019 Asia Pacific Pipeline Conference, Qingdao, China, 15–19 May 2019; American Society of Mechanical Engineers: New York City, NY, USA; p. V001T10A001.
48. Zhang, Y.; Jar, P.-Y.B.; Xue, S.; Li, L. Numerical Simulation of Ductile Fracture in Polyethylene Pipe with Continuum Damage Mechanics and Gurson-Tvergaard-Needleman Damage Models. *Proc. IMechE* **2019**, *233*, 2455–2468. [CrossRef]
49. Zhang, Y.; Qiao, L.; Fan, J.; Xue, S.; Jar, P.B. Molecular Dynamics Simulation of Plastic Deformation in Polyethylene under Uniaxial and Biaxial Tension. *Proc. Inst. Mech. Eng. Part L J. Mater. Des. Appl.* **2021**, *236*, 146442072110458. [CrossRef]
50. Fancey, K.S. A Mechanical Model for Creep, Recovery and Stress Relaxation in Polymeric Materials. *J. Mater. Sci.* **2005**, *40*, 4827–4831. [CrossRef]
51. Tan, N.; Jar, P.-Y.B. Determining Deformation Transition in Polyethylene under Tensile Loading. *Polymers* **2019**, *11*, 1415. [CrossRef]
52. Hong, K.; Rastogi, A.; Strobl, G. Model Treatment of Tensile Deformation of Semicrystalline Polymers: Static Elastic Moduli and Creep Parameters Derived for a Sample of Polyethylene. *Macromolecules* **2004**, *37*, 10174–10179. [CrossRef]
53. Castagnet, S. High-Temperature Mechanical Behavior of Semi-Crystalline Polymers and Relationship to a Rubber-like "Relaxed" State. *Mech. Mater.* **2009**, *41*, 75–86. [CrossRef]
54. Koerner, H.; Price, G.; Pearce, N.A.; Alexander, M.; Vaia, R.A. Remotely Actuated Polymer Nanocomposites—Stress-Recovery of Carbon-Nanotube-Filled Thermoplastic Elastomers. *Nat. Mater.* **2004**, *3*, 115–120. [CrossRef]
55. Shi, F. Studies on the Time-Dependent Behavior of Semi-Crystalline Polymers. *Res. Dev. Polym. Sci.* **2023**, *2*, 1–2. [CrossRef]
56. Yakimets, I.; Lai, D.; Guigon, M. Model to Predict the Viscoelastic Response of a Semi-Crystalline Polymer under Complex Cyclic Mechanical Loading and Unloading Conditions. *Mech. Time-Depend Mater.* **2007**, *11*, 47–60. [CrossRef]
57. Wilding, M.A.; Ward, I.M. Creep and Recovery of Ultra High Modulus Polyethylene. *Polymer* **1981**, *22*, 870–876. [CrossRef]
58. Wilding, M.A.; Ward, I.M. Tensile Creep and Recovery in Ultra-High Modulus Linear Polyethylenes. *Polymer* **1978**, *19*, 969–976. [CrossRef]
59. Sweeney, J.; Ward, I.M. A Unified Model of Stress Relaxation and Creep Applied to Oriented Polyethylene. *J. Mater. Sci.* **1990**, *25*, 697–705. [CrossRef]
60. Okereke, M.I.; Akpoyomare, A.I. Two-Process Constitutive Model for Semicrystalline Polymers across a Wide Range of Strain Rates. *Polymer* **2019**, *183*, 121818. [CrossRef]
61. Detrez, F.; Cantournet, S.; Séguéla, R. A Constitutive Model for Semi-Crystalline Polymer Deformation Involving Lamellar Fragmentation. *Comptes Rendus Mécanique* **2010**, *338*, 681–687. [CrossRef]
62. Olley, P.; Sweeney, J. A Multiprocess Eyring Model for Large Strain Plastic Deformation. *J. Appl. Polym. Sci.* **2011**, *119*, 2246–2260. [CrossRef]
63. Johnsen, J.; Clausen, A.H.; Grytten, F.; Benallal, A.; Hopperstad, O.S. A Thermo-Elasto-Viscoplastic Constitutive Model for Polymers. *J. Mech. Phys. Solids* **2019**, *124*, 681–701. [CrossRef]
64. Brusselle-Dupend, N.; Lai, D.; Feaugas, X.; Guigon, M.; Clavel, M. Mechanical Behavior of a Semicrystalline Polymer before Necking. Part II: Modeling of Uniaxial Behavior. *Polym. Eng. Sci.* **2003**, *43*, 501–518. [CrossRef]
65. DeMaio, A.; Patterson, T. Rheological Modeling of the Tensile Creep Behavior of Paper: Tensile Creep Behavior of Paper. *J. Appl. Polym. Sci.* **2007**, *106*, 3543–3554. [CrossRef]
66. Duxbury, J.; Ward, I.M. The Creep Behaviour of Ultra-High Modulus Polypropylene. *J. Mater. Sci.* **1987**, *22*, 1215–1222. [CrossRef]
67. Guedes, R.M.; Singh, A.; Pinto, V. Viscoelastic Modelling of Creep and Stress Relaxation Behaviour in PLA-PCL Fibres. *Fibers Polym.* **2017**, *18*, 2443–2453. [CrossRef]
68. Daneshyar, A.; Ghaemian, M.; Du, C. A Fracture Energy–Based Viscoelastic–Viscoplastic–Anisotropic Damage Model for Rate-Dependent Cracking of Concrete. *Int. J. Fract.* **2023**, *241*, 1–26. [CrossRef]
69. Jordan, B.; Gorji, M.B.; Mohr, D. Neural Network Model Describing the Temperature- and Rate-Dependent Stress-Strain Response of Polypropylene. *Int. J. Plast.* **2020**, *135*, 102811. [CrossRef]
70. Nechad, H.; Helmstetter, A.; El Guerjouma, R.; Sornette, D. Andrade and Critical Time-to-Failure Laws in Fiber-Matrix Composites: Experiments and Model. *J. Mech. Phys. Solids* **2005**, *53*, 1099–1127. [CrossRef]
71. Naraghi, M.; Kolluru, P.V.; Chasiotis, I. Time and Strain Rate Dependent Mechanical Behavior of Individual Polymeric Nanofibers. *J. Mech. Phys. Solids* **2014**, *62*, 257–275. [CrossRef]

72. Xu, P.; Zhou, Z.; Liu, T.; Pan, S.; Tan, X.; Chen, Z. The Investigation of Viscoelastic Mechanical Behaviors of Bolted GLARE Joints: Modeling and Experiments. *Int. J. Mech. Sci.* **2020**, *175*, 105538. [CrossRef]
73. Agbossou, A.; Cohen, I.; Muller, D. Effects of Interphase and Impact Strain Rates on Tensile Off-Axis Behaviour of Unidirectional Glass Fibre Composite: Experimental Results. *Eng. Fract. Mech.* **1995**, *52*, 923–935. [CrossRef]
74. Arruda, E.M.; Boyce, M.C. Evolution of Plastic Anisotropy in Amorphous Polymers during Finite Straining. *Int. J. Plast.* **1993**, *9*, 697–720. [CrossRef]
75. Du, Y.; Pei, P.; Suo, T.; Gao, G. Large Deformation Mechanical Behavior and Constitutive Modeling of Oriented PMMA. *Int. J. Mech. Sci.* **2023**, *257*, 108520. [CrossRef]
76. Zhao, W.; Liu, L.; Lan, X.; Leng, J.; Liu, Y. Thermomechanical Constitutive Models of Shape Memory Polymers and Their Composites. *Appl. Mech. Rev.* **2023**, *75*, 020802. [CrossRef]
77. Wang, Z.; Guo, J.; Seppala, J.E.; Nguyen, T.D. Extending the Effective Temperature Model to the Large Strain Hardening Behavior of Glassy Polymers. *J. Mech. Phys. Solids* **2021**, *146*, 104175. [CrossRef]
78. Hong, K.; Strobl, G. Characterizing and Modeling the Tensile Deformation of Polyethylene: The Temperature and Crystallinity Dependences. *Polym. Sci. Ser. A* **2008**, *50*, 483–493. [CrossRef]
79. Na, B.; Zhang, Q.; Fu, Q.; Men, Y.; Hong, K.; Strobl, G. Viscous-Force-Dominated Tensile Deformation Behavior of Oriented Polyethylene. *Macromolecules* **2006**, *39*, 2584–2591. [CrossRef]
80. Izraylit, V.; Heuchel, M.; Gould, O.E.C.; Kratz, K.; Lendlein, A. Strain Recovery and Stress Relaxation Behaviour of Multiblock Copolymer Blends Physically Cross-Linked with PLA Stereocomplexation. *Polymer* **2020**, *209*, 122984. [CrossRef]
81. Shi, F.; Ben Jar, P.-Y. Characterization of Polyethylene Using a New Test Method Based on Stress Response to Relaxation and Recovery. *Polymers* **2022**, *14*, 2763. [CrossRef]
82. Haario, H.; von Hertzen, R.; Karttunen, A.T.; Jorkama, M. Identification of the Viscoelastic Parameters of a Polymer Model by the Aid of a MCMC Method. *Mech. Res. Commun.* **2014**, *61*, 1–6. [CrossRef]
83. Johnson, T.P.M.; Socrate, S.; Boyce, M.C. A Viscoelastic, Viscoplastic Model of Cortical Bone Valid at Low and High Strain Rates. *Acta Biomater.* **2010**, *6*, 4073–4080. [CrossRef]
84. Christöfl, P.; Czibula, C.; Berer, M.; Oreski, G.; Teichert, C.; Pinter, G. Comprehensive Investigation of the Viscoelastic Properties of PMMA by Nanoindentation. *Polym. Test.* **2021**, *93*, 106978. [CrossRef]
85. Sweeney, J.; Bonner, M.; Ward, I.M. Modelling of Loading, Stress Relaxation and Stress Recovery in a Shape Memory Polymer. *J. Mech. Behav. Biomed. Mater.* **2014**, *37*, 12–23. [CrossRef] [PubMed]
86. Wayne Chen, W.; Jane Wang, Q.; Huan, Z.; Luo, X. Semi-Analytical Viscoelastic Contact Modeling of Polymer-Based Materials. *J. Tribol.* **2011**, *133*, 041404. [CrossRef]
87. Shahin, A.; Barsoum, I.; Islam, M.D. Constitutive Model Calibration of the Time and Temperature-Dependent Behavior of High Density Polyethylene. *Polym. Test.* **2020**, *91*, 106800. [CrossRef]
88. Xu, Q.; Engquist, B.; Solaimanian, M.; Yan, K. A New Nonlinear Viscoelastic Model and Mathematical Solution of Solids for Improving Prediction Accuracy. *Sci. Rep.* **2020**, *10*, 2202. [CrossRef] [PubMed]
89. Mierke, C.T. Viscoelasticity Acts as a Marker for Tumor Extracellular Matrix Characteristics. *Front. Cell Dev. Biol.* **2021**, *9*, 785138. [CrossRef]
90. Muliana, A. A Fractional Model of Nonlinear Multiaxial Viscoelastic Behaviors. *Mech. Time-Depend Mater.* **2022**, *27*, 1187–1207. [CrossRef]
91. ASTM D2837-22; Test Method for Obtaining Hydrostatic Design Basis for Thermoplastic Pipe Materials or Pressure Design Basis for Thermoplastic Pipe Products. ASTM International: West Conshohocken, PA, USA, 2022.
92. Tan, N.; Jar, P.B. Multi-Relaxation Test to Characterize PE Pipe Performance. *Plast. Eng.* **2019**, *75*, 40–45. [CrossRef]
93. Zhang, Y.; Ben Jar, P.-Y. Quantitative Assessment of Deformation-Induced Damage in Polyethylene Pressure Pipe. *Polym. Test.* **2015**, *47*, 42–50. [CrossRef]
94. Zhang, Y.; Jar, P.-Y.B. Phenomenological Modelling of Tensile Fracture in PE Pipe by Considering Damage Evolution. *Mater. Des.* **2015**, *77*, 72–82. [CrossRef]
95. Zhang, Y.; Ben Jar, P.-Y. Effects of Crosshead Speed on the Quasi-Static Stress–Strain Relationship of Polyethylene Pipes. *J. Press. Vessel. Technol.* **2017**, *139*, 021402. [CrossRef]
96. Shi, F.; Jar, P.-Y.B. Characterization of Loading, Relaxation, and Recovery Behaviors of High-Density Polyethylene Using a Three-Branch Spring-Dashpot Model. *Polym. Eng. Sci.* **2024**, *64*, 4920–4934. [CrossRef]
97. Kumar, S.; Liu, G.; Schloerb, D.; Srinivasan, M. Viscoelastic Characterization of the Primate Finger Pad In Vivo by Microstep Indentation and Three-Dimensional Finite Element Models for Tactile Sensation Studies. *J. Biomech. Eng.* **2015**, *137*. [CrossRef] [PubMed]
98. Blake, Y. *Review of Viscoelastic Models Applied to Cortical Bone*; Trinity College Dublin: Dublin, Ireland, 2021.
99. Heuchel, M.; Cui, J.; Kratz, K.; Kosmella, H.; Lendlein, A. Relaxation Based Modeling of Tunable Shape Recovery Kinetics Observed under Isothermal Conditions for Amorphous Shape-Memory Polymers. *Polymer* **2010**, *51*, 6212–6218. [CrossRef]
100. López-Guerra, E.A.; Solares, S.D. Modeling Viscoelasticity through Spring–Dashpot Models in Intermittent-Contact Atomic Force Microscopy. *Beilstein J. Nanotechnol.* **2014**, *5*, 2149–2163. [CrossRef] [PubMed]
101. Xu, S.; Odaira, T.; Sato, S.; Xu, X.; Omori, T.; Harjo, S.; Kawasaki, T.; Seiner, H.; Zoubková, K.; Murakami, Y.; et al. Non-Hookean Large Elastic Deformation in Bulk Crystalline Metals. *Nat. Commun.* **2022**, *13*, 5307. [CrossRef]

102. Yang, J.-L.; Zhang, Z.; Schlarb, A.K.; Friedrich, K. On the Characterization of Tensile Creep Resistance of Polyamide 66 Nanocomposites. Part II: Modeling and Prediction of Long-Term Performance. *Polymer* **2006**, *47*, 6745–6758. [CrossRef]
103. Alves, A.F.C.; Ferreira, B.P.; Andrade Pires, F.M. Constitutive Modeling of Amorphous Thermoplastics from Low to High Strain Rates: Formulation and Critical Comparison Employing an Optimization-Based Parameter Identification. *Int. J. Solids Struct.* **2023**, *273*, 112258. [CrossRef]
104. Laheri, V.; Hao, P.; Gilabert, F.A. Constitutive Recasting of Macromolecular-Based Thermoviscoplasticity as Yield Function-Based Formulation. *Int. J. Mech. Sci.* **2023**, *250*, 108278. [CrossRef]
105. Teoh, S.H. Effect of Saline Solution on Creep Fracture of Delrin®. *Biomaterials* **1993**, *14*, 132–136. [CrossRef]
106. Jadhao, V.; Robbins, M.O. Rheological Properties of Liquids under Conditions of Elastohydrodynamic Lubrication. *Tribol. Lett.* **2019**, *67*, 66. [CrossRef]
107. Hong, K. A Model Treating Tensile Deformation of Semi-Crystalline Polymers. Ph.D. Thesis, University of Freiburg Institute of Mathematics, Breisgau, Germany, 2005.
108. Tan, N.; Ben Jar, P.-Y. Reanalysis of the Creep Test Data and Failure Behavior of Polyethylene and Its Copolymers. *J. Mater. Eng. Perform.* **2022**, *31*, 2182–2192. [CrossRef]
109. Halsey, G.; White, H.J.; Eyring, H. Mechanical Properties of Textiles, I. *Text. Res.* **1945**, *15*, 295–311. [CrossRef]
110. Lee, H.-N.; Paeng, K.; Swallen, S.F.; Ediger, M.D. Direct Measurement of Molecular Mobility in Actively Deformed Polymer Glasses. *Science* **2009**, *323*, 231–234. [CrossRef] [PubMed]
111. Ghorbel, E.; Hadriche, I.; Casalino, G.; Masmoudi, N. Characterization of Thermo-Mechanical and Fracture Behaviors of Thermoplastic Polymers. *Materials* **2014**, *7*, 375–398. [CrossRef] [PubMed]
112. Men, Y.; Rieger, J.; Strobl, G. Role of the Entangled Amorphous Network in Tensile Deformation of Semicrystalline Polymers. *Phys. Rev. Lett.* **2003**, *91*, 095502. [CrossRef]
113. Nitta, K. On a Thermodynamic Foundation of Eyring Rate Theory for Plastic Deformation of Polymer Solids. *Philos. Mag. Lett.* **2023**, *103*, 2186190. [CrossRef]
114. Srikanth, K.; Sreejith, P.; Arvind, K.; Kannan, K.; Pandey, M. An Efficient Mode-of-Deformation Dependent Rate-Type Constitutive Relation for Multi-Modal Cyclic Loading of Elastomers. *Int. J. Plast.* **2023**, *163*, 103517. [CrossRef]
115. Kakaletsis, S.; Lejeune, E.; Rausch, M.K. Can Machine Learning Accelerate Soft Material Parameter Identification from Complex Mechanical Test Data? *Biomech. Model. Mechanobiol.* **2023**, *22*, 57–70. [CrossRef]
116. Klinge, S.; Steinmann, P. Inverse Analysis for Heterogeneous Materials and Its Application to Viscoelastic Curing Polymers. *Comput. Mech.* **2015**, *55*, 603–615. [CrossRef]
117. Polanco-Loria, M.; Daiyan, H.; Grytten, F. Material Parameters Identification: An Inverse Modeling Methodology Applicable for Thermoplastic Materials. *Polym. Eng. Sci.* **2012**, *52*, 438–448. [CrossRef]
118. Chen, Z.; Diebels, S.; Peter, N.J.; Schneider, A.S. Identification of Finite Viscoelasticity and Adhesion Effects in Nanoindentation of a Soft Polymer by Inverse Method. *Comput. Mater. Sci.* **2013**, *72*, 127–139. [CrossRef]
119. Yun, G.J.; Shang, S. A Self-Optimizing Inverse Analysis Method for Estimation of Cyclic Elasto-Plasticity Model Parameters. *Int. J. Plast.* **2011**, *27*, 576–595. [CrossRef]
120. Van Der Vossen, B.C.; Makeev, A.V. Mechanical Properties Characterization of Fiber Reinforced Composites by Nonlinear Constitutive Parameter Optimization in Short Beam Shear Specimens. *J. Compos. Mater.* **2021**, *55*, 2985–2997. [CrossRef]
121. Saleeb, A.F.; Gendy, A.S.; Wilt, T.E. Parameter-Estimation Algorithms for Characterizing a Class of Isotropic and Anisotropic Viscoplastic Material Models. *Mech. Time-Depend. Mater.* **2002**, *6*, 323–361. [CrossRef]
122. Maier, G.; Bocciarelli, M.; Bolzon, G.; Fedele, R. Inverse Analyses in Fracture Mechanics. *Int. J. Fract.* **2006**, *138*, 47–73. [CrossRef]
123. Lyu, Y.; Pathirage, M.; Ramyar, E.; Liu, W.K.; Cusatis, G. Machine Learning Meta-Models for Fast Parameter Identification of the Lattice Discrete Particle Model. *Comput. Mech.* **2023**, *72*, 593–612. [CrossRef]
124. Hoerig, C.; Ghaboussi, J.; Wang, Y.; Insana, M.F. Machine Learning in Model-Free Mechanical Property Imaging: Novel Integration of Physics with the Constrained Optimization Process. *Front. Phys.* **2021**, *9*, 600718. [CrossRef]
125. Andrade-Campos, A.; Thuillier, S.; Pilvin, P.; Teixeira-Dias, F. On the Determination of Material Parameters for Internal Variable Thermoelastic–Viscoplastic Constitutive Models. *Int. J. Plast.* **2007**, *23*, 1349–1379. [CrossRef]
126. Unger, J.F.; Könke, C. An Inverse Parameter Identification Procedure Assessing the Quality of the Estimates Using Bayesian Neural Networks. *Appl. Soft Comput.* **2011**, *11*, 3357–3367. [CrossRef]
127. Xu, H.; Jiang, X. Creep Constitutive Models for Viscoelastic Materials Based on Fractional Derivatives. *Comput. Math. Appl.* **2017**, *73*, 1377–1384. [CrossRef]
128. Van Den Bos, A. Nonlinear Least-Absolute-Values and Minimax Model Fitting. *IFAC Proc. Vol.* **1985**, *18*, 173–177. [CrossRef]
129. Powell, M.J.D. *Approximation Theory and Methods*; Cambridge University Press: Cambridge, UK; New York, NY, USA, 1981; ISBN 978-0-521-22472-7.
130. Setiyoko, A.; Basaruddin, T.; Arymurthy, A.M. Minimax Approach for Semivariogram Fitting in Ordinary Kriging. *IEEE Access* **2020**, *8*, 82054–82065. [CrossRef]
131. Boutaleb, S.; Zaïri, F.; Mesbah, A.; Naït-Abdelaziz, M.; Gloaguen, J.M.; Boukharouba, T.; Lefebvre, J.M. Micromechanics-Based Modelling of Stiffness and Yield Stress for Silica/Polymer Nanocomposites. *Int. J. Solids Struct.* **2009**, *46*, 1716–1726. [CrossRef]
132. Khan, A.S.; Lopez-Pamies, O.; Kazmi, R. Thermo-Mechanical Large Deformation Response and Constitutive Modeling of Viscoelastic Polymers over a Wide Range of Strain Rates and Temperatures. *Int. J. Plast.* **2006**, *22*, 581–601. [CrossRef]

133. Kemmer, G.; Keller, S. Nonlinear Least-Squares Data Fitting in Excel Spreadsheets. *Nat. Protoc.* **2010**, *5*, 267–281. [CrossRef]
134. Paetzold, M.; Andert, T.P.; Asmar, S.W.; Anderson, J.D.; Barriot, J.-P.; Bird, M.K.; Haeusler, B.; Hahn, M.; Tellmann, S.; Sierks, H.; et al. Asteroid 21 Lutetia: Low Mass, High Density. *Science* **2011**, *334*, 491–492. [CrossRef]
135. Messager, M.L.; Lehner, B.; Grill, G.; Nedeva, I.; Schmitt, O. Estimating the Volume and Age of Water Stored in Global Lakes Using a Geo-Statistical Approach. *Nat. Commun.* **2016**, *7*, 13603. [CrossRef]
136. Sprave, L.; Menzel, A. A Large Strain Gradient-Enhanced Ductile Damage Model: Finite Element Formulation, Experiment and Parameter Identification. *Acta Mech.* **2020**, *231*, 5159–5192. [CrossRef]
137. Sweeney, J. A Comparison of Three Polymer Network Models in Current Use. *Comput. Theor. Polym. Sci.* **1999**, *9*, 27–33. [CrossRef]
138. Mahnken, R.; Stein, E. Parameter Identification for Viscoplastic Models Based on Analytical Derivatives of a Least-Squares Functional and Stability Investigations. *Int. J. Plast.* **1996**, *12*, 451–479. [CrossRef]
139. Khan, A.; Zhang, H. Finite Deformation of a Polymer: Experiments and Modeling. *Int. J. Plast.* **2001**, *17*, 1167–1188. [CrossRef]
140. Zhang, C.; Cai, L.-H.; Guo, B.-H.; Miao, B.; Xu, J. New Kinetics Equation for Stress Relaxation of Semi-Crystalline Polymers below Glass Transition Temperature. *Chin. J. Polym. Sci.* **2022**, *40*, 1662–1669. [CrossRef]
141. Ayoub, G.; Zaïri, F.; Fréderix, C.; Gloaguen, J.M.; Naït-Abdelaziz, M.; Seguela, R.; Lefebvre, J.M. Effects of Crystal Content on the Mechanical Behaviour of Polyethylene under Finite Strains: Experiments and Constitutive Modelling. *Int. J. Plast.* **2011**, *27*, 492–511. [CrossRef]
142. Drozdov, A.D.; Klitkou, R.; Christiansen, J. deC. Multi-Cycle Deformation of Semicrystalline Polymers: Observations and Constitutive Modeling. *Mech. Res. Commun.* **2013**, *48*, 70–75. [CrossRef]
143. Kubát, J.; Seldén, R. The Stress Dependence of Activation Volumes in Creep and Stress Relaxation. *Mater. Sci. Eng.* **1978**, *36*, 65–69. [CrossRef]
144. Le, T.M.; Fatahi, B. Trust-Region Reflective Optimisation to Obtain Soil Visco-Plastic Properties. *Eng. Comput.* **2016**, *33*. [CrossRef]
145. Schmidt, U.; Mergheim, J.; Steinmann, P. Multiscale Parameter Identification. *Int. J. Mult. Comp. Eng.* **2012**, *10*, 327–342. [CrossRef]
146. Ramzanpour, M.; Hosseini-Farid, M.; Ziejewski, M.; Karami, G. Particle Swarm Optimization Method for Hyperelastic Characterization of Soft Tissues. In Proceedings of the ASME 2019 International Mechanical Engineering Congress and Exposition, Salt Lake City, UT, USA, 11–14 November 2019.
147. Pereira, J.O.; Farias, T.M.; Castro, A.M.; (Al-Baldawi), A.A.; Secchi, A.R.; Cardozo, N.S.M. Estimation of the Nonlinear Parameters of Viscoelastic Constitutive Models Using CFD and Multipass Rheometer Data. *J. Non-Newton. Fluid Mech.* **2020**, *281*, 104284. [CrossRef]
148. Syrjakow, M.; Szczerbicka, H. Efficient Parameter Optimization Based on Combination of Direct Global and Local Search Methods. In *Evolutionary Algorithms*; The IMA Volumes in Mathematics and its Applications; Davis, L.D., De Jong, K., Vose, M.D., Whitley, L.D., Eds.; Springer: New York, NY, USA, 1999; Volume 111, pp. 227–249. ISBN 978-1-4612-7185-7.
149. Renders, J.-M.; Flasse, S.P. Hybrid Methods Using Genetic Algorithms for Global Optimization. *IEEE Trans. Syst. Man Cybern. Part B* **1996**, *26*, 243–258. [CrossRef]
150. Mahinthakumar, G.; Sayeed, M. Hybrid Genetic Algorithm—Local Search Methods for Solving Groundwater Source Identification Inverse Problems. *J. Water Resour. Plann. Manag.* **2005**, *131*, 45–57. [CrossRef]
151. Maaranen, H.; Miettinen, K.; Penttinen, A. On Initial Populations of a Genetic Algorithm for Continuous Optimization Problems. *J. Glob. Optim.* **2007**, *37*, 405–436. [CrossRef]
152. Yen, J.; Liao, J.C.; Lee, B.; Randolph, D. A Hybrid Approach to Modeling Metabolic Systems Using a Genetic Algorithm and Simplex Method. *IEEE Trans. Syst. Man Cybern. Part B* **1998**, *28*, 173–191. [CrossRef] [PubMed]
153. Ahn, C.W.; Ramakrishna, R.S. A Genetic Algorithm for Shortest Path Routing Problem and the Sizing of Populations. *IEEE Trans. Evol. Comput.* **2002**, *6*, 566–579. [CrossRef]
154. Pál, K.F. Genetic Algorithm with Local Optimization. *Biol. Cybern.* **1995**, *73*, 335–341. [CrossRef]
155. Okamoto, M.; Nonaka, T.; Ochiai, S.; Tominaga, D. Nonlinear Numerical Optimization with Use of a Hybrid Genetic Algorithm Incorporating the Modified Powell Method. *Appl. Math. Comput.* **1998**, *91*, 63–72. [CrossRef]
156. Crain, T.; Bishop, R.H.; Fowler, W.; Rock, K. Interplanetary Flyby Mission Optimization Using a Hybrid Global-Local Search Method. *J. Spacecr. Rocket.* **2012**, *37*, 468–474. [CrossRef]
157. Attaviriyanupap, P.; Kita, H.; Tanaka, E.; Hasegawa, J. A Hybrid EP and SQP for Dynamic Economic Dispatch with Nonsmooth Fuel Cost Function. *IEEE Trans. Power Syst.* **2002**, *17*, 411–416. [CrossRef]
158. Henz, B.J.; Mohan, R.V.; Shires, D.R. A Hybrid Global–Local Approach for Optimization of Injection Gate Locations in Liquid Composite Molding Process Simulations. *Compos. Part A Appl. Sci. Manuf.* **2007**, *38*, 1932–1946. [CrossRef]
159. Wu, L.; Chen, Z.; Long, C.; Cheng, S.; Lin, P.; Chen, Y.; Chen, H. Parameter Extraction of Photovoltaic Models from Measured I-V Characteristics Curves Using a Hybrid Trust-Region Reflective Algorithm. *Appl. Energy* **2018**, *232*, 36–53. [CrossRef]
160. Nenov, H.B.; Dimitrov, B.H.; Marinov, A.S. Algorithms for Computational Procedure Acceleration for Systems Differential Equations in Matlab. In Proceedings of the 2013 36th International Convention on Information and Communication Technology, Electronics and Microelectronics (MIPRO), Opatija, Croatia, 20–24 May 2013; pp. 238–242.
161. Rivard, S.R.; Mailloux, J.-G.; Beguenane, R.; Bui, H.T. Design of High-Performance Parallelized Gene Predictors in MATLAB. *BMC Res. Notes* **2012**, *5*, 183. [CrossRef]
162. Tan, N. Deformation Transitions and Their Effects on the Long-Term Performance of Polyethylene and Its Pressure Pipe. Ph.D. Thesis, Department of Mechanical Engineering University of Alberta, Alberta, AL, Canada, 2021; p. 172.

163. Zheng, J.; Li, H.; Hogan, J.D. Strain-Rate-Dependent Tensile Response of an Alumina Ceramic: Experiments and Modeling. *Int. J. Impact Eng.* **2023**, *173*, 104487. [CrossRef]
164. Zheng, J.; Li, H.; Hogan, J.D. Advanced Tensile Fracture Analysis of Alumina Ceramics: Integrating Hybrid Finite-Discrete Element Modeling with Experimental Insights. *Eng. Fract. Mech.* **2024**, *302*, 110075. [CrossRef]
165. Jar, P.-Y.B. Effect of Tensile Loading History on Mechanical Properties for Polyethylene. *Polym. Eng. Sci.* **2015**, *55*, 2002–2010. [CrossRef]
166. Zheng, J.; Ji, M.; Zaiemyekeh, Z.; Li, H.; Hogan, J.D. Strain-Rate-Dependent Compressive and Compression-Shear Response of an Alumina Ceramic. *J. Eur. Ceram. Soc.* **2022**, *42*, 7516–7527. [CrossRef]
167. Fritsch, J.; Hiermaier, S.; Strobl, G. Characterizing and Modeling the Non-Linear Viscoelastic Tensile Deformation of a Glass Fiber Reinforced Polypropylene. *Compos. Sci. Technol.* **2009**, *69*, 2460–2466. [CrossRef]
168. Rafiee, R.; Mazhari, B. Simulation of the Long-Term Hydrostatic Tests on Glass Fiber Reinforced Plastic Pipes. *Compos. Struct.* **2016**, *136*, 56–63. [CrossRef]
169. Ebert, C.; Hufenbach, W.; Langkamp, A.; Gude, M. Modelling of Strain Rate Dependent Deformation Behaviour of Polypropylene. *Polym. Test.* **2011**, *30*, 183–187. [CrossRef]
170. Graba, M. Evaluation of Measurement Uncertainty in a Static Tensile Test. *Open Eng.* **2021**, *11*, 709–722. [CrossRef]
171. Mulliken, A.D.; Boyce, M.C. Mechanics of the Rate-Dependent Elastic–Plastic Deformation of Glassy Polymers from Low to High Strain Rates. *Int. J. Solids Struct.* **2006**, *43*, 1331–1356. [CrossRef]
172. Jar, P.B. Revisiting Creep Test on Polyethylene Pipe—Data Analysis and Deformation Mechanisms. *Polym. Eng. Sci.* **2021**, *61*, 586–599. [CrossRef]
173. Piyal Aravinna, A.G. *Estimation of Measurement Uncertainty in Determination of Tensile Strength of Reinforcement Steel*; Technical Report–2021/QC1; Central Engineering Consultancy Bureau: Colombo, Sri Lanka, 2021. [CrossRef]
174. Hong, K.; Rastogi, A.; Strobl, G. A Model Treating Tensile Deformation of Semicrystalline Polymers: Quasi-Static Stress- Strain Relationship and Viscous Stress Determined for a Sample of Polyethylene. *Macromolecules* **2004**, *37*, 10165–10173. [CrossRef]
175. Richeton, J.; Ahzi, S.; Daridon, L.; Rémond, Y. A Formulation of the Cooperative Model for the Yield Stress of Amorphous Polymers for a Wide Range of Strain Rates and Temperatures. *Polymer* **2005**, *46*, 6035–6043. [CrossRef]
176. Bergstrom, J.; Boyce, M.C. Constitutive Modeling of the Large Strain Time-Dependent Behavior of Elastomers. *J. Mech. Phys. Solids* **1998**, *46*, 931–954. [CrossRef]
177. Natarajan, V.D. Constitutive Behavior of a Twaron® Fabric/Natural Rubber Composite: Experiments and Modeling. Ph.D. Thesis, Texas A&M University, College Station, TX, USA, 2009.
178. Popa, C.M.; Fleischhauer, R.; Schneider, K.; Kaliske, M. Formulation and Implementation of a Constitutive Model for Semicrystalline Polymers. *Int. J. Plast.* **2014**, *61*, 128–156. [CrossRef]
179. Jar, P.-Y.B. Analysis of Time-dependent Mechanical Behavior of Polyethylene. *SPE Polym.* **2024**, *5*, 426–443. [CrossRef]
180. Pyrz, M.; Zairi, F. Identification of Viscoplastic Parameters of Phenomenological Constitutive Equations for Polymers by Deterministic and Evolutionary Approach. *Model. Simul. Mater. Sci. Eng.* **2007**, *15*, 85–103. [CrossRef]
181. Zhang, Y.; Jar, P.-Y.B. Comparison of Mechanical Properties Between PE80 and PE100 Pipe Materials. *J. Mater. Eng Perform* **2016**, *25*, 4326–4332. [CrossRef]
182. Liu, X.; Zhang, H.; Xia, M.; Wu, K.; Chen, Y.; Zheng, Q.; Li, J. Mechanical Response of Buried Polyethylene Pipelines under Excavation Load during Pavement Construction. *Eng. Fail. Anal.* **2018**, *90*, 355–370. [CrossRef]
183. Zhang, C.; Moore, I.D. Nonlinear Mechanical Response of High Density Polyethylene. Part II: Uniaxial Constitutive Modeling. *Polym. Eng. Sci.* **1997**, *37*, 414–420. [CrossRef]
184. Zhang, C.; Moore, I.D. Nonlinear Mechanical Response of High Density Polyethylene. Part I: Experimental Investigation and Model Evaluation. *Polym. Eng. Sci.* **1997**, *37*, 404–413. [CrossRef]
185. Moore, I. Profiled HDPE Pipe Response to Parallel Plate Loading. In *Buried Plastic Pipe Technology: 2nd Volume*; Eckstein, D., Ed.; ASTM International: West Conshohocken, PA, USA, 1994; p. 25. ISBN 978-0-8031-1992-5.

Disclaimer/Publisher's Note: The statements, opinions and data contained in all publications are solely those of the individual author(s) and contributor(s) and not of MDPI and/or the editor(s). MDPI and/or the editor(s) disclaim responsibility for any injury to people or property resulting from any ideas, methods, instructions or products referred to in the content.

Communication

Key Factors in Enhancing Pseudocapacitive Properties of PANI-InO$_x$ Hybrid Thin Films Prepared by Sequential Infiltration Synthesis

Jiwoong Ham [1], Hyeong-U Kim [2] and Nari Jeon [1,*]

[1] Department of Materials Science and Engineering, Chungnam National University, Daejeon 34134, Republic of Korea; maxbasic55@o.cnu.ac.kr
[2] Department of Plasma Engineering, Korea Institute of Machinery & Materials (KIMM), Daejeon 34103, Republic of Korea; guddn418@kimm.re.kr
* Correspondence: njeon@cnu.ac.kr

Abstract: Sequential infiltration synthesis (SIS) is an emerging vapor-phase synthetic route for the preparation of organic–inorganic composites. Previously, we investigated the potential of polyaniline (PANI)-InO$_x$ composite thin films prepared using SIS for application in electrochemical energy storage. In this study, we investigated the effects of the number of InO$_x$ SIS cycles on the chemical and electrochemical properties of PANI-InO$_x$ thin films via combined characterization using X-ray photoelectron spectroscopy, ultraviolet–visible spectroscopy, Raman spectroscopy, Fourier transform infrared spectroscopy, and cyclic voltammetry. The area-specific capacitance values of PANI-InO$_x$ samples prepared with 10, 20, 50, and 100 SIS cycles were 1.1, 0.8, 1.4, and 0.96 mF/cm^2, respectively. Our result shows that the formation of an enlarged PANI-InO$_x$ mixed region directly exposed to the electrolyte is key to enhancing the pseudocapacitive properties of the composite films.

Keywords: indium oxide; polyaniline; cyclic voltammetry; sequential infiltration synthesis (SIS); conducting polymer

Citation: Ham, J.; Kim, H.-U.; Jeon, N. Key Factors in Enhancing Pseudocapacitive Properties of PANI-InO$_x$ Hybrid Thin Films Prepared by Sequential Infiltration Synthesis. *Polymers* **2023**, *15*, 2616. https://doi.org/10.3390/polym15122616

Academic Editors: Célio Pinto Fernandes, Luís L. Ferrás and Alexandre M. Afonso

Received: 11 May 2023
Revised: 1 June 2023
Accepted: 6 June 2023
Published: 8 June 2023

Copyright: © 2023 by the authors. Licensee MDPI, Basel, Switzerland. This article is an open access article distributed under the terms and conditions of the Creative Commons Attribution (CC BY) license (https://creativecommons.org/licenses/by/4.0/).

1. Introduction

Organic–inorganic hybrid materials have received continued attention owing to their novel functionalities, which are not demonstrated in single-phase materials, whether organic or inorganic. In particular, electrically conductive organic–inorganic composite films have demonstrated enhanced properties for application in electrochemistry-related areas. Some recently reported examples of the benefits of using PANI–metal oxide composite films include enhanced efficiency in photovoltaic cells [1], sensitivities and linearities in sensors [2], photocatalytic efficiencies in catalysts [3], and retention properties in supercapacitors [4]. The different properties of hybrid thin films and their performances in different areas depend primarily on the synthesis routes of the thin films. Therefore, the development of novel techniques for preparing organic–inorganic hybrid thin films has received attention in various areas.

Sequential infiltration synthesis (SIS) is a new vacuum-based technique for preparing organic–inorganic composites and is considered a variant of atomic layer deposition (ALD). Although ALD exploits self-limiting chemical reactions on the surfaces, SIS utilizes chemical reactions within the bulk polymer phase [5]. For SIS reactions to occur readily, the precursors designated to be used must first infiltrate the polymer phases. Typically, the entrapment of infiltrated SIS precursors by specific functional groups of the polymer is preferred. Poly(methylmethacrylate) (PMMA) [6] and poly(2-vinylpyridine) (P2VP) [7] are typical types of polymers that have been widely used for SIS because the carbonyl groups in PMMA and pyridine groups in P2VP undergo Lewis acid–base reactions with typical SIS precursors such as trimethylaluminum and titanium tetrachloride (TiCl$_4$). However, the potential applications of hybrid thin films based on PMMA and P2VP are limited to areas where electrical conductivity is not required. Only several studies have been conducted on SIS with conducting polymers and their applications in electrochemistry.

Polyaniline (PANI) is a representative conducting polymer with controllable electrical conductivity owing to doping. PANI can exist in three different chemical states, namely, leucoemeraldine, pernigraniline, and emeraldine, depending on the degrees of oxidation and reduction [8,9]. PANI with an emeraldine base can be transformed into an emeraldine salt, which exhibits high electrical conductivity (~10^2 S/cm) when acid-doped [10]. Wang et al. reported the SIS process of doping PANI with $SnCl_4$ and $MoCl_5$ vapors [11], which exhibit the Lewis acidic nature; the doped PANI exhibited a moderate conductivity of ~9.8×10^{-5} S/cm. PANI-ZnO (~18.42 S/cm) [12] and PANI-InO_x (4–9 S/cm) [4] composite thin films prepared via SIS showed electrically conductive properties. Previously, we demonstrated the significant potential of PANI-InO_x composite films prepared using SIS for electrochemical energy storage, which warrants follow-up studies on the same system [4].

The aim of this study was to investigate the influence of metal oxides on the electrochemical properties of polyaniline–metal oxide composites as energy storage materials utilizing the SIS. Research related to SIS focusing on conducting polymers is limited to several papers, and studies specifically investigating their electrochemical properties are scarce. In this study, we investigated the variations in the chemical and electrochemical properties of PANI-InO_x films prepared via SIS as a function of the SIS cycle number. PANI-InO_x films exhibit a graded concentration of InO_x along the direction of the film thickness, where their structure is determined by the number of cycles. A combination of ultraviolet–visible (UV-vis) spectroscopy, Raman spectroscopy, and attenuated total reflectance–Fourier transform infrared (ATR-FTIR) spectroscopy was performed to understand the variation in the chemical structure of PANI in response to alloying with InO_x. The superior pseudocapacitive properties of the sample with the optimized cycle number (50 cy) are attributable to the increased volume of the PANI-InO_x mixed region, which is exposed to the electrolyte.

2. Materials and Methods

2.1. Sample Preparation

PANI with an emeraldine base powder (M_w ~10,000, Sigma–Aldrich, Saint Louis, MI, USA) was dissolved in methyl-2-pyrrolidone ($\geq$99%, Sigma–Aldrich, Saint Louis, MI, USA) with a concentration of 30 mg/mL. The solution was stirred for 24 h at 80 °C and 850 rpm. The solution was spun onto prepared substrates at 2000 rpm, and the as-spun substrates were baked at 70 °C in air. The thickness of the prepared PANI thin films was approximately 37 nm. Subsequently, SIS was performed using a thermal ALD reactor (Daeki HighTech, Daejeon, Republic of Korea) with a cross-flow design. The precursors used for the SIS were trimethylindium (TMIn, 99.999%, EasyChem) and H_2O (99.999%, Sigma Aldrich, Saint Louis, MI, USA). Ar carrier gas (99.999%) was continuously flowed at 5 sccm during the entire SIS. Both the TMIn and H_2O half-cycles involved a 1 s dose, 120 s of exposure, and 120 s of purging. The reactor chamber was isolated from the pump during the exposure step to facilitate the infiltration of the precursors into the polymer matrix. The SIS-prepared substrates were annealed at 270 °C for 1 h in the forming gas of H_2–N_2 (~3.9% H_2 in N_2).

Different types of substrates were used for different characterization methods: an Si substrate (n type, 1–10 Ohm·cm, iTASCO) with a 500-nm-thick SiO_2 layer was used for X-ray photoelectron spectroscopy (XPS) and Raman spectroscopy. Au-coated Si substrate (Au thickness: ~90 nm) was used for ATR-FTIR spectroscopy. A fused silica (iNexus, Inc., Seongnam, Republic of Korea), which had a transmittance of ~90% or higher in the wavelength range > 250 nm, was utilized for UV-vis spectroscopy. Electrochemical experiments were performed using glass substrates with a fluorine-doped tin oxide (FTO) layer measuring ~600 nm thick (NSG TEC 7, Pilkington, Lathom, UK).

2.2. Sample Characterization

HRXPS depth profiling was performed using an X-ray photoelectron spectrometer (K-alpha, Thermo Scientific, Waltham, MA, USA) with Ar^+ ion beams at 1 kV and an etching time of 10 s. The X-ray source used was monochromatic Al $K\alpha$ (1487 eV). The surfaces of the annealed samples were partly scratched using stainless-steel tweezers to create a surface step on the sample, and the thickness of the PANI-InO_x thin film was measured using a stylus profiler (Alpha-Step® D-500, KLA, Milpitas, CA, USA). The cleanliness of the scratched surface was confirmed based on spectroscopic ellipsometry data (FS-1, Film Sense, Lincoln, NE, USA) obtained from the surface and a comparison with those of a bare Si substrate. Raman spectroscopy was performed using a Raman spectrometer (LabRAM HR-800, Horiba, Japan) under the following conditions: 514 nm laser source measuring 0.7 μm, 1800 gr/mm grating, 10 s acquisition time, and 10 specular accumulations. UV-vis spectroscopy was performed using a UV-vis spectrometer (UV-2600, Shimadzu, Japan), where an FTIR spectrometer (Vertex 80v, Bruker, Billerica, MA, USA) with a mercury–cadmium–telluride detector and a diamond attenuated total reflection (ATR) crystal were used to obtain the ATR-FTIR spectroscopy data at a spectral resolution of 4 cm^{-1}. An electrochemical analyzer (CHI602E, CH Instruments, Bee Cave, TX, USA) was used to perform cyclic voltammetry (CV) experiments using a three-electrode setup comprising a Ag/AgCl reference electrode, a Pt wire counter electrode, and a PANI-InO_x FTO/glass working electrode. CV data were obtained using a pH 7 buffer solution as the electrolyte.

3. Results and Discussion

We analyzed PANI-InO_x composite thin films, which were prepared under different SIS cycles (10, 20, 50, and 100 cy) and annealed in a reducing atmosphere. The sample structure could be summarized as follows: (1) an InO_x-rich region, (2) a PANI-InO_x mixed region, and (3) a PANI-rich region, as shown in Figure 1a. The thicknesses and chemical compositions of the three regions differed depending on the number of SIS cycles (Figure 1b). The samples with 10 and 20 SIS cycles did not contain an InO_x surface layer and only presented a PANI-InO_x mixed region and a PANI bulk region. Owing to the repetition of the SIS cycle, the InO_x-rich region near the surface of the PANI at times prevented the additional infiltration of the TMIn precursor, thereby resulting in the formation of a thicker InO_x layer, which further developed via a mechanism similar to ALD. However, TMIn infiltration in later SIS cycles may not have been completely hindered because the concentration in the PANI-rich region of the 100 cy sample decreased gradually from ~40 to 0 at%.

Figure 1c shows the oxygen (O) and nitrogen (N) HRXPS data obtained at different locations on the PANI-InO_x film, as shown in Figure 1b. The O 1s HRXPS data were deconvoluted into three or four peaks originating from the following components: lattice oxygen (In-O) from InO_x at ~529.9 eV, oxygen vacancy ($V_\ddot{o}$) at ~531.0 eV, indium hydroxide (In-OH) at ~532.1 eV, and lattice oxygen (Si-O) from SiO_2 at ~533.1 eV [13,14]. In all four samples, the In-O component was the most dominant in the topmost region, which was either an InO_x-PANI mixed region (samples with 10 and 20 SIS cy) or InO_x-rich regions (Samples with 50 and 100 SIS cy). The components of higher BEs (i.e., −OH and $V_\ddot{o}$) became more dominant compared to the In-O component as the HRXPS analysis region shifted toward the substrate (i.e., PANI-rich region). This is consistent with the previous SIS results, which indicated that the oxidation of the SIS precursors within the polymer matrix was less complete than that on the polymer surface [6,15]. The average stoichiometries of the four samples were $InO_{0.85}$, InO, $InO_{1.38}$ and $InO_{1.44}$ for 10, 20, 50, and 100 cycles, respectively. The stoichiometry trend was reasonable, considering that the proportion of surface-like InO_x compared with that of bulk-like InO_x (i.e., synthesized within the polymer matrix) enhanced as the SIS cycle number increased.

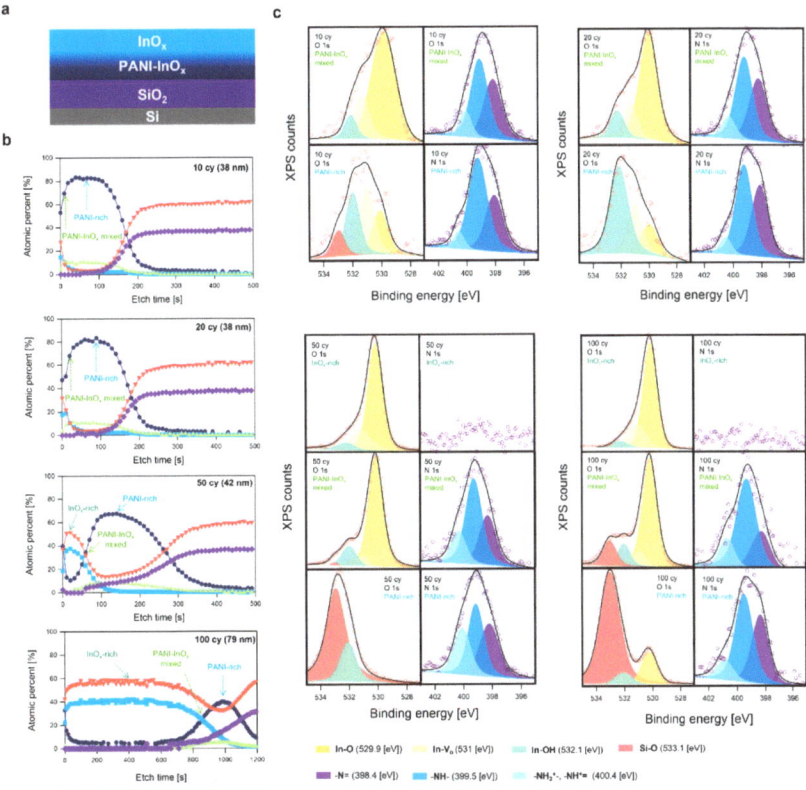

Figure 1. (a) Schematic illustration of structure of PANI-InO$_x$/SiO$_2$/Si samples. InO$_x$ content shows graded concentration along film thickness direction. (b) XPS depth profiles showing C, In, N, O, and Si atomic concentrations for four SIS samples with different cycle numbers: 10, 20, 50, and 100 cy. (c) O 1s and N 1s HRXPS data obtained at different locations (i.e., InO$_x$-rich region, PANI-InO$_x$ mixed region, and PANI-rich region) in four samples. Each location at which HRXPS data were captured are shown as arrows in (b).

The N 1s HRXPS data of the four samples were deconvoluted into three components: quinonoid imine (−N=) at ~398.4 eV, benzenoid amine (−NH−) at ~399.5 eV, and protonated amine/imine state (−NH$_2^+$−, =NH$^+$−) at ~400.4 eV [16]. In all the HRXPS spectra, the amine component was more dominant than those of the other components. Meanwhile, the PANI with an emeraldine base usually contained equal amounts of imine and amine components. The presence of protonated species along with a decrease in the number of imine units suggested that the protonated species may have originated from the imine units. PANI doped with HCl contains protonated species transformed from the imine components [17]. However, no clear correlation was indicated between the percentage of protonated species and the InO$_x$ content in any sample. Therefore, further studies are necessary to determine the potential chemical reactions contributing to the formation of protonated species during InO$_x$ alloying.

Figure 2a shows the UV-vis transmittance spectra of the four samples. The samples with 10, 20, and 50 cy showed a weak absorption band at ~610 nm, which was assigned to n–π* transition between the benzenoid and quinonoid rings. The increase in the absorption below ~400 nm from the three sample was likely related to the π–π* transition of the benzenoid ring, which is known to be located at ~320 nm [18]. The 100 cy sample showed stronger absorption in the UV region (<400 nm), whereas the n–π* transition was primarily suppressed. The absorption in the UV region was likely due to absorption by the thick InO_x layer. The Tauc plot of the 100 cy SIS shows that the bandgap of InO_x was ~3.5 eV, which was slightly lower level than that of the dense InO_x thin films reported in the literature [19,20]. The smaller bandgap of InO_x along with the presence of tail states was reasonable, considering that a significant portion of the InO_x phase was present within the polymer matrix along with a high concentration of oxygen vacancies. The optical bandgap of PANI-InO_x samples tends to decrease as the number of SIS cycles decreases (Figure S1). The ATR-FTIR spectra of the four samples (Figure 2b) showed IR bands related to the PANI phase: (1) stretching of the quinonoid ring at ~1600 cm^{-1}, (2) stretching of the benzenoid ring at ~1512 cm^{-1}, (3) stretching of C–N of the secondary aromatic amine at 1300 cm^{-1}, and (4) out-of-plane C–H deformation of the 1,4-distributed aromatic ring at 823 cm^{-1} [21–23]. The 100 cy samples showed significant IR bands, with lower intensities compared with the other samples owing to the presence of the InO_x surface layer. Similarly, the Raman spectra of the four samples (Figure 2c) exhibited significant Raman bands associated with the PANI phase, as summarized in Table 1. The presence of the C–N$^{+\bullet}$ stretching (radical cations) band at ~1350 cm^{-1} in all samples is consistent with the presence of protonated amine/imine species indicated in the HRXPS analysis. Radical cation bands are typically observed in acid-doped PANI, which suggests that alloying with InO_x may offer similar effects on the acid doping of PANI.

Table 1. Major Raman bands identified in the PANI-InO_x and PANI-only samples.

Raman Shift (cm^{-1})	Assignment
1610	C=C stretching vibration of a quinonoid ring
1560	N–H bending
1405	C–C stretching vibrations in a quinonoid ring
1350	C–N$^{+\bullet}$ Radical cation
1240	C–N stretching in a benzenoid ring
1170	C–H in-plane C–H bending quinonoid ring
817	out-of-plane C–H vibration

The CV results for the samples, measured at a scan rate of 10 mV/s, are shown in Figure 3. All the samples showed a pair of redox peaks at similar potentials (i.e., ~0.2 V vs. Ag/AgCl and ~−0.05 V vs. Ag/AgCl). The 50 cy sample indicated better-defined redox peaks with a higher current compared with the other samples. The area-specific capacitance values of PANI-InO_x samples prepared with 10, 20, 50, and 100 SIS cycles were 1.1, 0.8, 1.4, and 0.96 mF/cm^2, respectively. A detailed explanation of this calculation is reported in the literature [4]. The CV measurements were performed multiple times using different samples prepared under the same SIS conditions. The redox peak positions of the CV curves varied slightly within a ~100 mV range. Therefore, the subtle variation in the peak position observed for the different SIS samples (10, 20, 50, and 100 cy) is considered to be within the experimental error. CV curves collected at different scan rates are provided in Supplementary Materials. In order to investigate the capacitance stability of the sample after prolonged exposure to the electrolyte, we conducted a CV experiment consisting of 1000 cycles (Figure S2). This experiment assessed the evolution of the capacitance over time in response to extended electrolyte exposure. In the electrochemical impedance spectroscopy (EIS) test conducted on the PANI-InO_x 50 cycle, pure PANI, and pure InO_x samples in a previous study, semicircles were observed at high frequencies and straight lines were observed at low frequencies [4]. Among the three samples, the composite sample exhibited the smallest semicircle, indicating a lower charge transfer resistance.

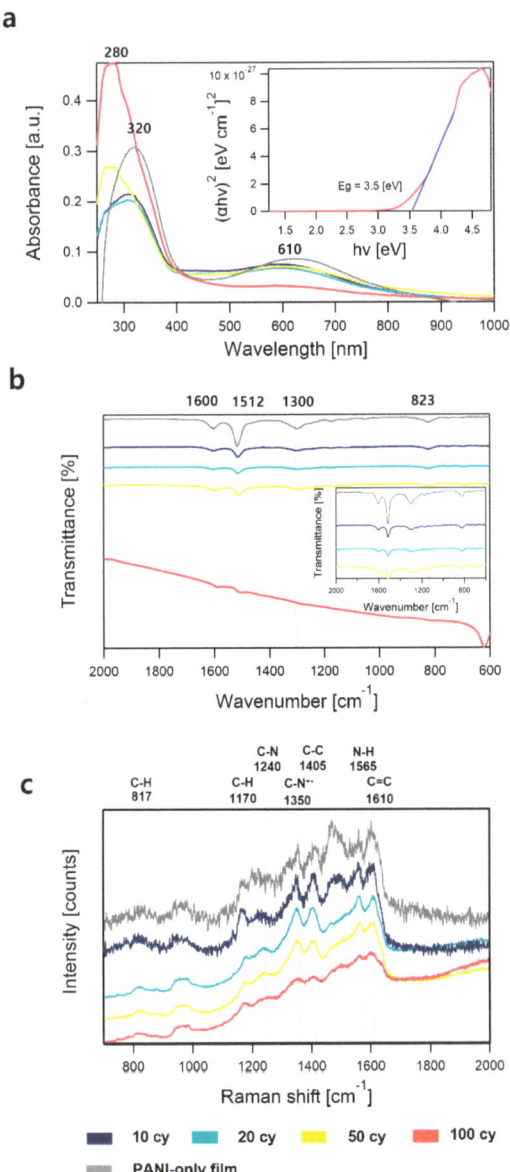

Figure 2. (a) UV-vis transmittance spectra, (b) ATR-FTIR absorbance spectra, and (c) Raman spectra of PANI-InO$_x$ samples at different number of cycles (10, 20, 50, and 100 cy) and PANI-only sample. The PANI-only sample was annealed under the same conditions as the PANI-InO$_x$ samples. Inset of (a) shows Tauc plot of 100 cy PANI-InO$_x$ sample. The dashed lines in panel (c) mark the location of Raman bands observed in the PANI-InO$_x$ film, which are also summarized in Table 1.

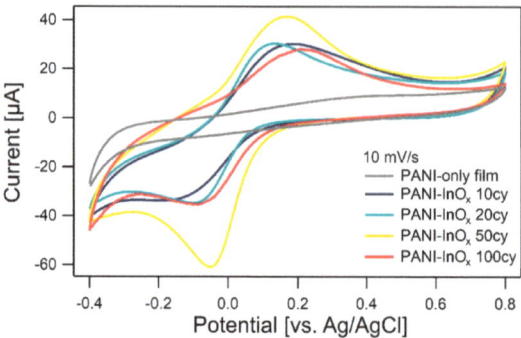

Figure 3. CV curves of PANI-InO$_x$ films prepared with different SIS cycles (10, 20, 50, and 100 cy) and PANI-only film. The CVs were collected at a scan rate of 10 mV/s in an aqueous electrolyte of neutral pH.

Further analysis is necessary to identify the redox reactions contributing to the observed CV peaks. However, the conversion of the emeraldine and pernigraniline states was speculated to be the primary redox reaction in the PANI-InO$_x$ samples in our previous study. The enhanced peak current of the 50 cy sample might have been related to the larger thickness of the PANI-InO$_x$ mixed region (Figure 1b) compared with those of the 10 and 20 cy samples. The 100 cy sample exhibited a sufficiently thick PANI-InO$_x$ mixed region; however, the InO$_x$ surface layer likely prevented direct contact between the mixed region and the electrolyte [4]. Furthermore, the 50 cy sample had a larger proportion of protonated amine/imine structures (Figure 1c), a fact which is likely related to the PANI-InO$_x$ mixed region.

4. Conclusions

It is only recently that SIS was proved to be a promising route for the preparation of organic–inorganic hybrid films for electrochemistry and electrochemical energy storage. The goal of this study was to provide a more comprehensive understanding of the effects of the number of SIS cycles (10, 20, 50, and 100 cycles) on the chemical and electrochemical properties of PANI-InO$_x$ composite films. The PANI-InO$_x$ films showed a graded composition of InO$_x$ within the PANI matrix, with ample concentrations of oxygen vacancies and hydroxide components. The entire film structure was composed of two or three components, including an InO$_x$-rich region, a PANI-InO$_x$ mixed region, and a PANI-rich region, and its structure depended on the number of cycles. Combined characterization using HRXPS, UV-vis, Raman, and FTIR spectroscopy consistently revealed the presence of cationic radicals, which might have been related to the transition from the quinonoid structure to the benzenoid structure. The 50 cy samples showed the highest pseudocapacitance among the tested samples, which was likely due to the relatively thick electrochemically active PANI-InO$_x$ region and the exposure of the PANI-InO$_x$ region to the electrolyte.

Supplementary Materials: The following supporting information can be downloaded at: https://www.mdpi.com/article/10.3390/polym15122616/s1, Figure S1: Tauc plots of PANI-InO$_x$ sample prepared with 10, 20, and 50 SIS cycles.; Figure S2: Capacity retention test of PANI-InO$_x$ samples prepared with 10, 20, 50, and 100 SIS cycles.; Figure S3: CV curves of 50 cy PANI-InO$_x$ sample collected at different scan rates. Reference [24] is cited in the supplementary materials.

Author Contributions: Conceptualization, J.H., H.-U.K. and N.J.; formal analysis, J.H.; investigation, J.H.; Data Curation, H.-U.K.; writing—original draft preparation, J.H.; writing—review and editing, N.J.; visualization, J.H.; supervision, N.J.; project administration, N.J. All authors have read and agreed to the published version of the manuscript.

Funding: This work was supported by the National Research Foundation of Korea (NRF) grant funded by the Korea government (MSIT) (No. RS-2023-00214170). This research was supported by Chungnam National University (2021–2022). This work was supported by KIMM institutional program (NK242F) and NST/KIMM.

Data Availability Statement: The data presented in this study are available in the article.

Conflicts of Interest: The authors declare no conflict of interest.

References

1. Nemade, K.; Dudhe, P.; Tekade, P. Enhancement of photovoltaic performance of polyaniline/graphene composite-based dye-sensitized solar cells by adding TiO_2 nanoparticles. *Solid State Sci.* **2018**, *83*, 99–106. [CrossRef]
2. Murugesan, T.; Kumar, R.R.; Anbalagan, A.K.; Lee, C.-H.; Lin, H.-N. Interlinked Polyaniline/ZnO Nanorod Composite for Selective NO_2 Gas Sensing at Room Temperature. *ACS Appl. Nano Mater.* **2022**, *5*, 4921–4930. [CrossRef]
3. Chen, S.; Huang, D.; Zeng, G.; Xue, W.; Lei, L.; Xu, P.; Deng, R.; Li, J.; Cheng, M. In-situ synthesis of facet-dependent $BiVO_4/Ag_3PO_4$/PANI photocatalyst with enhanced visible-light-induced photocatalytic degradation performance: Synergism of interfacial coupling and hole-transfer. *Chem. Eng. J.* **2020**, *382*, 122840. [CrossRef]
4. Ham, J.; Park, S.; Jeon, N. Conductive Polyaniline–Indium Oxide Composite Films Prepared by Sequential Infiltration Synthesis for Electrochemical Energy Storage. *ACS Omega* **2023**, *8*, 946–953. [CrossRef] [PubMed]
5. Leng, C.Z.; Losego, M.D. A physiochemical processing kinetics model for the vapor phase infiltration of polymers: Measuring the energetics of precursor-polymer sorption, diffusion, and reaction. *Phys. Chem. Chem. Phys.* **2018**, *20*, 21506–21514. [CrossRef]
6. Ham, J.; Ko, M.; Choi, B.; Kim, H.-U.; Jeon, N. Understanding Physicochemical Mechanisms of Sequential Infiltration Synthesis toward Rational Process Design for Uniform Incorporation of Metal Oxides. *Sensors* **2022**, *22*, 6132. [CrossRef]
7. Biswas, M.; Libera, J.A.; Darling, S.B.; Elam, J.W. Polycaprolactone: A Promising Addition to the Sequential Infiltration Synthesis Polymer Family Identified through in Situ Infrared Spectroscopy. *ACS Appl. Polym. Mater.* **2020**, *2*, 5501–5510. [CrossRef]
8. Babel, V.; Hiran, B.L. A review on polyaniline composites: Synthesis, characterization, and applications. *Polym. Compos.* **2021**, *42*, 3142–3157. [CrossRef]
9. Zare, E.N.; Makvandi, P.; Ashtari, B.; Rossi, F.; Motahari, A.; Perale, G. Progress in Conductive Polyaniline-Based Nanocomposites for Biomedical Applications: A Review. *J. Med. Chem.* **2020**, *63*, 1–22. [CrossRef]
10. Stejskal, J.; Trchová, M.; Bober, P.; Humpolíček, P.; Kašpárková, V.; Sapurina, I.; Shishov, M.A.; Varga, M. Conducting Polymers: Polyaniline. In *Encyclopedia of Polymer Science and Technology*; Wiley: Hoboken, NJ, USA, 2015; pp. 1–44. [CrossRef]
11. Wang, W.; Yang, F.; Chen, C.; Zhang, L.; Qin, Y.; Knez, M. Tuning the Conductivity of Polyaniline through Doping by Means of Single Precursor Vapor Phase Infiltration. *Adv. Mater. Interfaces* **2017**, *4*, 1600806. [CrossRef]
12. Wang, W.; Chen, C.; Tollan, C.; Yang, F.; Beltrán, M.; Qin, Y.; Knez, M. Conductive Polymer–Inorganic Hybrid Materials through Synergistic Mutual Doping of the Constituents. *ACS Appl. Mater. Interfaces* **2017**, *9*, 27964–27971. [CrossRef] [PubMed]
13. Hoch, L.B.; He, L.; Qiao, Q.; Liao, K.; Reyes, L.M.; Zhu, Y.; Ozin, G.A. Effect of Precursor Selection on the Photocatalytic Performance of Indium Oxide Nanomaterials for Gas-Phase CO_2 Reduction. *Chem. Mater.* **2016**, *28*, 4160–4168. [CrossRef]
14. Cañón, J.; Teplyakov, A.V. XPS characterization of cobalt impregnated SiO_2 and γ-Al_2O_3. *Surf. Interface Anal.* **2021**, *53*, 475–481. [CrossRef]
15. Ko, M.; Kirakosyan, A.; Kim, H. U.; Seok, H.; Choi, J.; Jeon, N. A new nanoparticle heterostructure strategy with highly tunable morphology via sequential infiltration synthesis. *Appl. Surf. Sci.* **2022**, *593*, 153387. [CrossRef]
16. Tantawy, H.R.; Kengne, B.-A.F.; McIlroy, D.N.; Nguyen, T.; Heo, D.; Qiang, Y.; Aston, D.E. X-ray photoelectron spectroscopy analysis for the chemical impact of solvent addition rate on electromagnetic shielding effectiveness of HCl-doped polyaniline nanopowders. *J. Appl. Phys.* **2015**, *118*, 175501. [CrossRef]
17. Song, E.; Choi, J.-W. Conducting Polyaniline Nanowire and Its Applications in Chemiresistive Sensing. *Nanomaterials* **2013**, *3*, 498–523. [CrossRef]
18. Andriianova, A.N.; Biglova, Y.N.; Mustafin, A.G. Effect of structural factors on the physicochemical properties of functionalized polyanilines. *RSC Adv.* **2020**, *10*, 7468–7491. [CrossRef]
19. Agbenyeke, R.E.; Jung, E.A.; Park, B.K.; Chung, T.-M.; Kim, C.G.; Han, J.H. Thermal atomic layer deposition of In_2O_3 thin films using dimethyl(N-ethoxy-2,2-dimethylcarboxylicpropanamide)indium and H_2O. *Appl. Surf. Sci.* **2017**, *419*, 758–763. [CrossRef]
20. Maeng, W.J.; Choi, D.-W.; Park, J.; Park, J.-S. Indium oxide thin film prepared by low temperature atomic layer deposition using liquid precursors and ozone oxidant. *J. Alloys Compd.* **2015**, *649*, 216–221. [CrossRef]
21. Su, N. Polyaniline-Doped Spherical Polyelectrolyte Brush Nanocomposites with Enhanced Electrical Conductivity, Thermal Stability, and Solubility Property. *Polymers* **2015**, *7*, 1599–1616. [CrossRef]

22. Patra, B.N.; Majhi, D. Removal of Anionic Dyes from Water by Potash Alum Doped Polyaniline: Investigation of Kinetics and Thermodynamic Parameters of Adsorption. *J. Phys. Chem. B* **2015**, *119*, 8154–8164. [CrossRef] [PubMed]
23. Wong, P.-Y.; Phang, S.-W.; Baharum, A. Effects of synthesised polyaniline (PAni) contents on the anti-static properties of PAni-based polylactic acid (PLA) films. *RSC Adv.* **2020**, *10*, 39693–39699. [CrossRef] [PubMed]
24. Yue, J.; Lin, L.; Jiang, L.; Zhang, Q.; Tong, Y.; Suo, L.; Hu, Y.-s.; Li, H.; Huang, X.; Chen, L. Interface Concentrated-Confinement Suppressing Cathode Dissolution in Water-in-Salt Electrolyte. *Adv. Energy Mater.* **2020**, *10*, 2000665. [CrossRef]

Disclaimer/Publisher's Note: The statements, opinions and data contained in all publications are solely those of the individual author(s) and contributor(s) and not of MDPI and/or the editor(s). MDPI and/or the editor(s) disclaim responsibility for any injury to people or property resulting from any ideas, methods, instructions or products referred to in the content.

Article

Numerical Simulation of Three-Dimensional Free Surface Flows Using the K–BKZ–PSM Integral Constitutive Equation †

Juliana Bertoco [1,*], Antonio Castelo [2], Luís L. Ferrás [3,4] and Célio Fernandes [4,5]

1. Center for Mathematics, Computing and Cognition — CMCC, Federal University of ABC — UFABC, Santo André 09210-580, Brazil
2. Department of Applied Mathematics and Statistics, University of São Paulo — USP, São Carlos 13566-590, Brazil; castelo@icmc.usp.br
3. Department of Mechanical Engineering (Section of Mathematics), Faculty of Engineering of the University of Porto — FEUP, 4200-465 Porto, Portugal; lferras@fe.up.pt
4. Center of Mathematics — CMAT, University of Minho, 4800-058 Guimarães, Portugal; cbpf@fe.up.pt
5. Transport Phenomena Research Centre — CEFT, Faculty of Engineering at University of Porto — FEUP, 4200-465 Porto, Portugal
* Correspondence: jubertoco@alumni.usp.br
† Dedicated to the Memory of Our Dear Friend and Colleague, Late Professor Murilo Francisco Tomé.

Abstract: This work introduces a novel numerical method designed to address three-dimensional unsteady free surface flows incorporating integral viscoelastic constitutive equations, specifically the K–BKZ–PSM (Kaye–Bernstein, Kearsley, Zapas–Papanastasiou, Scriven, Macosko) model. The new proposed methodology employs a second-order finite difference approach along with the deformation fields method to solve the integral constitutive equation and the marker particle method (known as marker-and-cell) to accurately capture the evolution of the fluid's free surface. The newly developed numerical method has proven its effectiveness in handling complex fluid flow scenarios, including confined flows and extrudate swell simulations of Boger fluids. Furthermore, a new semi-analytical solution for velocity and stress fields is derived, considering fully developed flows of a K–BKZ–PSM fluid in a pipe.

Keywords: K–BKZ; PSM; free surface; Boger fluids; finite difference

Citation: Bertoco, J.; Castelo, A.; Ferrás, L.L.; Fernandes, C. Numerical Simulation of Three-Dimensional Free Surface Flows Using the K–BKZ–PSM Integral Constitutive Equation. *Polymers* 2023, 15, 3705. https://doi.org/10.3390/polym15183705

Academic Editor: Brian J. Edwards

Received: 1 August 2023
Revised: 4 September 2023
Accepted: 6 September 2023
Published: 8 September 2023

Copyright: © 2023 by the authors. Licensee MDPI, Basel, Switzerland. This article is an open access article distributed under the terms and conditions of the Creative Commons Attribution (CC BY) license (https://creativecommons.org/licenses/by/4.0/).

1. Introduction

Discovered and developed in the late twentieth century, viscoelastic materials have been used in a number of different applications (polymer industry, biomedicine, automotive industry, food, paints, etc.). Their use is often based on trial and error procedures, resulting in wasting raw material and time before a good end product is achieved. To mitigate this problem, numerical simulations are often used to predict the best material processing conditions. Usually, the simulations are based on the finite element, finite volume and finite difference methods, and the constitutive equations are, in most cases, defined by rheological differential models, such as Oldroyd-B [1], UCM (Upper-convected Maxwell) [2,3], PTT (Phan–Thien–Tanner) [4,5], FENE-P (Finite Extensible Nonlinear Elastic Peterlin) [6,7] and Giesekus models [7]. Simulations of three-dimensional real-world applications require a great deal of computational effort, making the convergence of the algorithms in a reasonable amount of time a difficult task [4]. However, recent technological advances in scientific computing and the development of faster computers have led researchers to perform simulations in more complex geometries and use more sophisticated rheological models (that use integral equations instead of partial differential equations).

It is known that integral constitutive equations can better model various viscoelastic fluids, such as high-density polyethylene (HDPE) [8,9] and low-density polyethylene (LDPE) [10] (used in the injection molding industry), and one of the most successful integral models is the K–BKZ–PSM [11–13] (see also [14,15]). Therefore, there is significant interest among research groups worldwide in developing numerical methods to deal with the K–BKZ model, particularly with an emphasis on its application to polymer flows. Many studies have focused on simulating data and phenomena associated with polymer melt flows in rheology and polymer processing; however, there is still a need for further efforts to tackle numerical solutions of the K–BKZ–PSM for three-dimensional, time-dependent, free surface flows.

The vast majority of problems studied in the literature (considering integral models) are about confined flows, such as entry flows [9,16,17] and flows in abrupt contractions [18–22]. Regarding free surface flows, Mitsoulis and Malamataris [20] extended the implementation of the free boundary condition (FBC) method to viscoelastic fluids governed by integral constitutive equations. Specifically, they focused on the K–BKZ–PSM model. To validate their numerical approach, they used the finite element method (FEM) to obtain results in simple test cases, including planar flow at an angle and Poiseuille flow in a tube, where analytical solutions are available for comparison. Furthermore, they have applied the FBC method to the K–BKZ–PSM model using data from a benchmark polymer melt, specifically the IUPAC-LDPE melt. Some other researchers have also considered flows with free surfaces [8,14,23–27]. Ganvir et al. [25] developed a novel approach for simulating extrudate swell using an Arbitrary Lagrangian Eulerian (ALE) technique in conjunction with a finite element formulation. The constitutive behavior of the melt was modeled using a differential exponential Phan–Thien–Tanner (PTT) viscoelastic model. With the proposed method, they have conducted simulations of extrudate swell in both planar and axisymmetric extrusion scenarios, which involve an abrupt contraction ahead of the die exit. Regarding three-dimensional (3D) flows, Rasmussen [28] developed a Galerkin finite element method for simulating three-dimensional transient viscoelastic flows. The method used a Lagrangian kinematic description and integral constitutive models. The numerical implementation was validated with the calculation of various transient and steady drag correction factors for the motion of a sphere in a cylinder containing an upper convected Maxwell fluid. Later, Marín and Rasmussen [29] extended the Galerkin finite element method for simulating three-dimensional transient and non-isothermal flows of K–BKZ type fluids. Tomé et al. [27] proposed a novel numerical approach to tackle the simulation of 3D viscoelastic unsteady free surface flows governed by the Oldroyd-B differential constitutive equation. The numerical method involves solving the governing equations using a finite difference approach on a 3D-staggered grid. To validate the accuracy and reliability of the proposed technique, an exact solution of the flow of an Oldroyd-B fluid inside a 3D-pipe was employed. The results obtained through numerical simulations included the analysis of transient extrudate swell and jet buckling.

Summarizing, previous studies in the field of free surface flows have predominantly centered around two-dimensional (2D) scenarios and used the finite element method. These investigations primarily revolve around the extrudate swell problem, considering both steady and unsteady flows; however, it is worth noting that these studies relied on differential viscoelastic constitutive equations. Therefore, this work introduces a novel numerical method specifically designed to address 3D unsteady free surface flows incorporating integral viscoelastic constitutive equations, specifically, the K–BKZ–PSM model. The key innovation lies in the development of a robust numerical method for integral models using the finite difference method on a staggered grid, which enables accurate predictions of extrudate swell phenomena. We also derive a new semi-analytical solution for the fully developed flow of a K–BKZ–PSM viscoelastic fluid, which can serve for other authors to verify their own numerical implementations of the K–BKZ–PSM integral viscoelastic model.

It is worth noting that the FEM prominently features in the limited body of work concerning this subject. Nevertheless, both FEM and FDM stand as extensively used numerical approaches for tackling partial differential equations (PDEs). When properly employed within suitable conditions, both techniques exhibit stability. Within our research group, a longstanding tradition exists regarding leveraging the finite difference method [2,4,27,30–32], resulting in a profound mastery of its implementation. Furthermore, the group has made innovative strides in enhancing the fundamental finite difference methodology. This progression equips us to adeptly handle different grid structures and a range of discretization choices. Consequently, this method takes precedence in our current work.

The paper is structured as follows. In Section 2, we introduce the governing equations for isothermal and incompressible viscoelastic flows modelled by the K–BKZ–PSM constitutive equation. Section 3 is devoted to the numerical method, where the variant of the marker particle method that employs the finite difference method on a staggered grid is described for 3D flows using the K–BKZ–PSM viscoelastic integral model. In Section 4, we derive a new semi-analytical solution for the fully developed flow of a K–BKZ–PSM viscoelastic fluid. For validation of the newly developed numerical method, two case studies are analyzed in Section 5, the confined pipe flow and the extrudate swell free surface flow of a Boger fluid. The paper ends with the conclusions in Section 6.

2. Governing Equations

The isothermal and incompressible fluid flow considered in this work is governed by the dimensionless continuity and linear momentum equations [27],

$$\nabla \cdot \mathbf{v} = 0, \tag{1}$$

$$\frac{\partial \mathbf{v}}{\partial t} + \nabla \cdot (\mathbf{v}\mathbf{v}) = -\nabla p + \frac{1}{Re}\nabla^2 \mathbf{v} + \nabla \cdot \mathbf{\Phi} + \frac{1}{Fr^2}\mathbf{g}, \tag{2}$$

together with a constitutive equation for the stress. $\mathbf{\Phi}$ is a stress tensor given by

$$\mathbf{\Phi} = \boldsymbol{\tau} - \frac{1}{Re}\dot{\boldsymbol{\gamma}}, \text{ with } \dot{\boldsymbol{\gamma}} = \nabla \mathbf{v} + (\nabla \mathbf{v})^{\mathrm{T}}, \tag{3}$$

where $\boldsymbol{\tau}$ is a non-Newtonian stress tensor, $\mathbf{v}(u,v,w)$ is the velocity field, p is the kinematic pressure, $\mathbf{g}$ is the gravity acceleration vector and t is the time. In these equations, $Fr = U/\sqrt{Lg}$ is the Froude number, $Re = \rho_0 U L/\eta_0$ is the Reynolds number, η_0 is the zero-shear-rate viscosity, ρ_0 is the fluid density, g the magnitude of the gravity acceleration vector and U and L are the characteristic velocity and length scales, respectively. Note that all variables are dimensionless, with: $\mathbf{x} = \bar{\mathbf{x}}/L$, $\mathbf{v} = \bar{\mathbf{v}}/U$, $t = \bar{t}U/L$, $p = \bar{p}/(\rho U^2)$ and $\mathbf{\Phi} = \bar{\mathbf{\Phi}}/(\rho U^2)$.

The constitutive equation for the non-Newtonian stress tensor is given by the K–BKZ–PSM model [11],

$$\boldsymbol{\tau}(t) = \int_{-\infty}^{t} M(t-t') H(I_1, I_2) \mathbf{B}_{t'}(t) dt', \tag{4}$$

where

$$M(t-t') = \sum_{k=1}^{m_1} \frac{a_k}{\lambda_k Wi} e^{-\frac{t-t'}{\lambda_k Wi}}, \tag{5}$$

is the memory function, λ_k is the relaxation time of the fluid, a_k is a model parameter and m_1 is the number of modes. $H(I_1, I_2)$ is the Papanastasiou–Scriven–Macosko decay function, being given by

$$H(I_1, I_2) = \frac{\alpha}{\alpha - 3 + \beta I_1 + (1-\beta) I_2}. \tag{6}$$

$\mathbf{B}_{t'}(t)$ is the Finger tensor, and $I_1 = tr[\mathbf{B}_{t'}(t)]$, $I_2 = \frac{1}{2}((I_1)^2 - tr[\mathbf{B}_{t'}^2(t)])$ are the first and second invariants of $\mathbf{B}_{t'}(t)$, respectively. The parameters a_k, λ_k, α and β are obtained from a fit to rheological data. $Wi = \lambda_{ref} U/L$ is the Weissenberg number, the viscosity is given by $\eta_0 = \sum_{k=1}^{m_1} a_k \lambda_k$ and $\lambda_{ref} = \sum_{k=1}^{m_1} \frac{a_k \lambda_k^2}{a_k \lambda_k}$ is the mean relaxation time [14].

In this work, the method of *deformation fields* [33] is used to update the Finger tensor as the fluid flows. In this methodology, $(N+1)$-deformation instants (t') are defined in the interval $[0, t]$ where the history of deformation is stored. This deformation is updated by solving the transport equation for $\mathbf{B}_{t'}(t)$,

$$\frac{\partial}{\partial t}\mathbf{B}_{t'}(\mathbf{x},t) + \mathbf{v}(\mathbf{x},t)\cdot\nabla\mathbf{B}_{t'}(\mathbf{x},t) = [\nabla\mathbf{v}(\mathbf{x},t)]^T\cdot\mathbf{B}_{t'}(\mathbf{x},t) + \mathbf{B}_{t'}(\mathbf{x},t)\cdot\nabla\mathbf{v}(\mathbf{x},t). \tag{7}$$

The governing equations are solved in a Cartesian 3D system (x, y, z, t) where

$$p = p(x, y, z, t),$$
$$\mathbf{v} = (u(x,y,z,t), v(x,y,z,t), w(x,y,z,t))^T,$$
$$\boldsymbol{\tau}(x,y,z,t) = \begin{bmatrix} \tau^{xx} & \tau^{xy} & \tau^{xz} \\ \tau^{xy} & \tau^{yy} & \tau^{yz} \\ \tau^{xz} & \tau^{yz} & \tau^{zz} \end{bmatrix} \text{ and } \mathbf{B}_{t'(t)}(x,y,z,t) = \begin{bmatrix} B^{xx} & B^{xy} & B^{xz} \\ B^{xy} & B^{yy} & B^{yz} \\ B^{xz} & B^{yz} & B^{zz} \end{bmatrix}.$$

This results in the following system of equations that need to be solved for the pressure, velocity and stress:

continuity equation:

$$\frac{\partial u}{\partial x} + \frac{\partial v}{\partial y} + \frac{\partial w}{\partial z} = 0. \tag{8}$$

linear momentum equations:

$$\begin{aligned}
\frac{\partial u}{\partial t} + \frac{\partial(uu)}{\partial x} + \frac{\partial(vu)}{\partial y} + \frac{\partial(wu)}{\partial z} &= -\frac{\partial p}{\partial x} + \frac{1}{Re}\left[\frac{\partial^2 u}{\partial x^2} + \frac{\partial^2 u}{\partial y^2} + \frac{\partial^2 u}{\partial z^2}\right] \\
&\quad + \frac{\partial\Phi^{xx}}{\partial x} + \frac{\partial\Phi^{xy}}{\partial y} + \frac{\partial\Phi^{xz}}{\partial z} + \frac{1}{Fr^2}g_x, \\
\frac{\partial v}{\partial t} + \frac{\partial(uv)}{\partial x} + \frac{\partial(vv)}{\partial y} + \frac{\partial(wv)}{\partial z} &= -\frac{\partial p}{\partial y} + \frac{1}{Re}\left[\frac{\partial^2 v}{\partial x^2} + \frac{\partial^2 v}{\partial y^2} + \frac{\partial^2 v}{\partial z^2}\right] \\
&\quad + \frac{\partial\Phi^{xy}}{\partial x} + \frac{\partial\Phi^{yy}}{\partial y} + \frac{\partial\Phi^{yz}}{\partial z} + \frac{1}{Fr^2}g_y, \\
\frac{\partial w}{\partial t} + \frac{\partial(uw)}{\partial x} + \frac{\partial(vw)}{\partial y} + \frac{\partial(ww)}{\partial z} &= -\frac{\partial p}{\partial z} + \frac{1}{Re}\left[\frac{\partial^2 w}{\partial x^2} + \frac{\partial^2 w}{\partial y^2} + \frac{\partial^2 w}{\partial z^2}\right] \\
&\quad + \frac{\partial\Phi^{xz}}{\partial x} + \frac{\partial\Phi^{yz}}{\partial y} + \frac{\partial\Phi^{zz}}{\partial z} + \frac{1}{Fr^2}g_z,
\end{aligned} \tag{9}$$

where g_x, g_y, g_z are the Cartesian components of the gravity vector.

stress tensor $\boldsymbol{\Phi}$:

$$\begin{aligned}
\Phi^{xx} &= \tau^{xx} - \frac{2}{Re}\frac{\partial u}{\partial x}, \\
\Phi^{xy} &= \tau^{xy} - \frac{1}{Re}\left[\frac{\partial u}{\partial y} + \frac{\partial v}{\partial x}\right], \\
\Phi^{xz} &= \tau^{xz} - \frac{1}{Re}\left[\frac{\partial u}{\partial z} + \frac{\partial w}{\partial x}\right], \\
\Phi^{yy} &= \tau^{yy} - \frac{2}{Re}\frac{\partial v}{\partial y}, \\
\Phi^{yz} &= \tau^{yz} - \frac{1}{Re}\left[\frac{\partial v}{\partial z} + \frac{\partial w}{\partial y}\right], \\
\Phi^{zz} &= \tau^{zz} - \frac{2}{Re}\frac{\partial w}{\partial z}.
\end{aligned} \tag{10}$$

stress tensor $\boldsymbol{\tau}$:

$$\tau^{xx}(t) = \int_{-\infty}^{t} \sum_{k=1}^{m_1} \frac{a_k}{Wi\lambda_k} e^{-\frac{(t-t')}{Wi\lambda_k}} \frac{\alpha}{\alpha - 3 + \beta I_1 + (1-\beta)I_2} B_{t'}^{xx}(t) dt',$$

$$\tau^{xy}(t) = \int_{-\infty}^{t} \sum_{k=1}^{m_1} \frac{a_k}{Wi\lambda_k} e^{-\frac{(t-t')}{Wi\lambda_k}} \frac{\alpha}{\alpha - 3 + \beta I_1 + (1-\beta)I_2} B_{t'}^{xy}(t) dt',$$

$$\tau^{xz}(t) = \int_{-\infty}^{t} \sum_{k=1}^{m_1} \frac{a_k}{Wi\lambda_k} e^{-\frac{(t-t')}{Wi\lambda_k}} \frac{\alpha}{\alpha - 3 + \beta I_1 + (1-\beta)I_2} B_{t'}^{xz}(t) dt', \quad (11)$$

$$\tau^{yy}(t) = \int_{-\infty}^{t} \sum_{k=1}^{m_1} \frac{a_k}{Wi\lambda_k} e^{-\frac{(t-t')}{Wi\lambda_k}} \frac{\alpha}{\alpha - 3 + \beta I_1 + (1-\beta)I_2} B_{t'}^{yy}(t) dt',$$

$$\tau^{yz}(t) = \int_{-\infty}^{t} \sum_{k=1}^{m_1} \frac{a_k}{Wi\lambda_k} e^{-\frac{(t-t')}{Wi\lambda_k}} \frac{\alpha}{\alpha - 3 + \beta I_1 + (1-\beta)I_2} B_{t'}^{yz}(t) dt',$$

$$\tau^{zz}(t) = \int_{-\infty}^{t} \sum_{k=1}^{m_1} \frac{a_k}{Wi\lambda_k} e^{-\frac{(t-t')}{Wi\lambda_k}} \frac{\alpha}{\alpha - 3 + \beta I_1 + (1-\beta)I_2} B_{t'}^{zz}(t) dt'.$$

Finger tensor $\boldsymbol{B}$:

$$\frac{\partial B^{xx}}{\partial t} + \frac{\partial (uB^{xx})}{\partial x} + \frac{\partial (vB^{xx})}{\partial y} + \frac{\partial (wB^{xx})}{\partial z} = 2\left[\frac{\partial u}{\partial x}B^{xx} + \frac{\partial u}{\partial y}B^{xy} + \frac{\partial u}{\partial z}B^{xz}\right],$$

$$\frac{\partial B^{xy}}{\partial t} + \frac{\partial (uB^{xy})}{\partial x} + \frac{\partial (vB^{xy})}{\partial y} + \frac{\partial (wB^{xy})}{\partial z} = \frac{\partial v}{\partial x}B^{xx} + \left[\frac{\partial u}{\partial x} + \frac{\partial v}{\partial y}\right]B^{xy}$$
$$+ \frac{\partial v}{\partial z}B^{xz} + \frac{\partial u}{\partial y}B^{yy} + \frac{\partial u}{\partial z}B^{yz},$$

$$\frac{\partial B^{xz}}{\partial t} + \frac{\partial (uB^{xz})}{\partial x} + \frac{\partial (vB^{xz})}{\partial y} + \frac{\partial (wB^{xz})}{\partial z} = \frac{\partial w}{\partial x}B^{xx} + \frac{\partial w}{\partial y}B^{xy} + \left[\frac{\partial u}{\partial x} + \frac{\partial w}{\partial z}\right]B^{xz}$$
$$+ \frac{\partial u}{\partial y}B^{yz} + \frac{\partial u}{\partial z}B^{zz}, \quad (12)$$

$$\frac{\partial B^{yy}}{\partial t} + \frac{\partial (uB^{yy})}{\partial x} + \frac{\partial (vB^{yy})}{\partial y} + \frac{\partial (wB^{yy})}{\partial z} = 2\left[\frac{\partial v}{\partial x}B^{xy} + \frac{\partial v}{\partial y}B^{yy} + \frac{\partial v}{\partial z}B^{yz}\right],$$

$$\frac{\partial B^{yz}}{\partial t} + \frac{\partial (uB^{yz})}{\partial x} + \frac{\partial (vB^{yz})}{\partial y} + \frac{\partial (wB^{yz})}{\partial z} = \frac{\partial w}{\partial x}B^{xy} + \frac{\partial v}{\partial x}B^{xz} + \frac{\partial w}{\partial y}B^{yy}$$
$$+ \left[\frac{\partial v}{\partial y} + \frac{\partial w}{\partial z}\right]B^{yz} + \frac{\partial v}{\partial z}B^{zz},$$

$$\frac{\partial B^{zz}}{\partial t} + \frac{\partial (uB^{zz})}{\partial x} + \frac{\partial (vB^{zz})}{\partial y} + \frac{\partial (wB^{zz})}{\partial z} = 2\left[\frac{\partial w}{\partial x}B^{xz} + \frac{\partial w}{\partial y}B^{yz} + \frac{\partial w}{\partial z}B^{zz}\right].$$

3. Numerical Method

The governing equations are solved by a variant of the marker particle method [4,27], which employs the finite difference method on a staggered grid. This methodology is implemented in the FREEFLOW-3D code developed by researchers from the Institute of Mathematical and Computing Sciences (ICMC) at the University of São Paulo (USP) in Brazil. Code details can be found in [4,27,30,31] considering 2D, 3D and radial symmetry flows. The precision of the numerical technique and its validation for three-dimensional viscoelastic flows with a free surface is presented in the works of Tomé et al. [27,30] (which only uses differential constitutive equations). The use of integral models in three-dimensional flows (considering free surface problems) has not yet been tested on this system (see Tomé et al. [4,27,30,31]). The novelty of this work is to incorporate equations for viscoelastic fluids using integral models (more complex than the differential type models) in the FREEFLOW-3D system.

In this methodology, the velocity field is approximated in the face of the cells, and the other variables, denoted by ζ, are evaluated in the center of the computational cells (see Figure 1a). The technique adopted here is presented by Tomé et al. [30,31] for differential models, where the free surface is defined by marker particles that move with the local fluid velocity. In addition, the computational cells are defined as (see also Figure 1b):

- Fluid entrance: *Inflow — I*,
- Fluid exit: *Outflow — O*,
- Rigid boundaries: *Boundary — B*,
- Empty cells: *Empty — E*,
- Free surface cells: *Surface — S*,
- Full cells: *Full — F*.

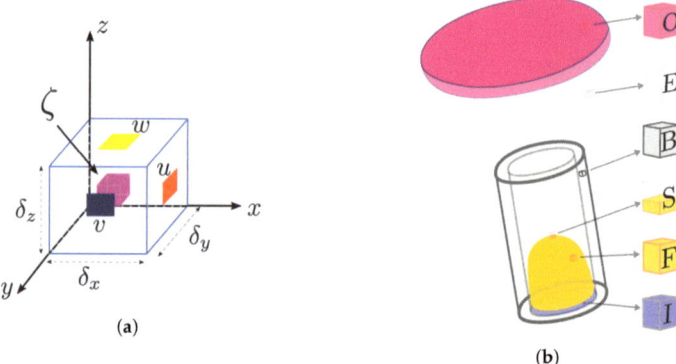

Figure 1. (a) Typical three-dimensional staggered cell and (b) illustration of cell type classification used.

To solve Equations (1) and (2), one must specify boundary conditions for the velocity field. A velocity field (V_{inf}) is prescribed in the fluid inlet cells (*inflows*), and a fully developed flow is assumed in the *outflow* (a homogeneous Neumann boundary condition $\partial v/\partial n = 0$ is assumed, where n is the normal direction to the contour). In the fluid inlet cells (*inflows*), we assume that $\frac{\partial p}{\partial n} = 0$ and the Finger tensor $B_{t'}$ is the identity matrix. In outflow regions, we assume Neumann conditions for Finger tensor $\left(\frac{\partial B}{\partial n} = 0\right)$, and we assume $p = 0$. We also take $v = 0$ in rigid boundaries. Details of the boundary conditions adopted in this work can be found in Tomé et al. [30] or Castelo et al. [31].

The solutions $v(x, t_{n+1})$, $p(x, t_{n+1})$ and $\tau(x, t_{n+1})$ at time step $t_{n+1} = t + \Delta t$ are obtained in the following way: first, using the values of $\tau(x, t_n)$, the velocity and pressure fields at time t_{n+1} are calculated. Then, $v(x, t_{n+1})$ is used to calculate the tensor $\tau(x, t_{n+1})$ by the method of deformation fields, and, lastly, the free surface is updated. Specifically, the following steps are performed:

Step 1 — *Calculation of* $v(x, t_{n+1})$ *and* $p(x, t_{n+1})$

It is assumed that, at time t, the variables $v(x, t) = v^{(n)}$, $p(x, t) = p^{(n)}$, $\tau(x, t) = \tau^{(n)}$ and the marker's positions $x(t) = x^{(n)}$ are known. Then, $v(x, t_{n+1})$ and $p(x, t_{n+1})$ are obtained as follows:

1. Calculate $\dot{\gamma}^{(n)} = \left[\nabla v^{(n)} + \left(\nabla v^{(n)}\right)^T\right]$, and, from the EVSS [34] transformation, obtain $\Phi = \tau^{(n)} - \frac{1}{Re}\dot{\gamma}^{(n)}$;

2. Calculate an intermediate velocity field $\tilde{v}^{(n+1)}$ using the ideas of the projection method [30,31] to uncouple the conservation of mass and momentum equations. An intermediate velocity field $\tilde{v}^{(n+1)}$ is obtained from Equation (2) using explicit Euler Methods, where $p^{(n)}$ is an approximation to $p^{(n+1)}$. The boundary conditions for $\tilde{v}^{(n+1)}$ are the same as those for the final velocity $v^{(n+1)}$. Details of boundary conditions for full cells (F),

outflow cells (O) and free surface cells (S) are provided in detail in Tomé et al. [4,30] and will not be presented here for the sake of conciseness. It can be shown that $\tilde{\mathbf{v}}^{(n+1)}$ possesses the correct vorticity at time t_{n+1}, but it does not conserve mass in general. Therefore, there is a potential function $\psi^{(n+1)}$ so that,

$$\mathbf{v}^{(n+1)} = \tilde{\mathbf{v}}^{(n+1)} - \nabla \psi^{(n+1)}. \tag{13}$$

3. Solve the Poisson equation for the potential function ψ for every F-cell in the domain,

$$\nabla^2 \psi^{(n+1)} = \nabla \cdot \tilde{\mathbf{v}}^{(n+1)}. \tag{14}$$

The boundary conditions required for solving this Poisson equation are the homogeneous Neumann conditions for rigid walls and inflows, while homogeneous Dirichlet conditions are used at outflows.

4. Compute the final velocity field from Equation (13);
5. Compute the final pressure field (see [4]) by

$$p^{(n+1)} = p^{(n)} + \frac{\psi^{(n+1)}}{\Delta t}. \tag{15}$$

Details of the discretization of the equations (temporal and spatial) considering all types of cells (see Figure 1b) are given in [4,27,30].

Step 2 — *Calculation of the extra stress tensor $\tau(\mathbf{x}, t_{n+1})$ and free surface update*

To calculate the extra stress tensor $\tau(\mathbf{x}, t_{n+1})$, initially, the Finger tensor is updated at t_{n+1} for every full cell (F) and surface cell (S) for every intermediate time t' using Equation (7). Details of the calculation of the Finger B tensor (in two dimensions) can be found in [32] and will not be presented here because the extension to three dimensions is straightforward. Note that, for each computational cell (F and S), Equation (7) is solved N times (for each t'), considering each of the components of the Finger tensor. Thus, considering three-dimensional flows, the computational cost to obtain the deformation history in each cell is high, demanding a great deal of memory and simulation time (since, for each cell, it is necessary to calculate the Finger tensor N times for each component of the deformation matrix). For inflow cells (I), boundary cells (B) and empty cells (E), the Finger tensor is the identity tensor. In the outflow, the Neumann condition is assumed.

The definition of the points t' for the calculation of the components of the Finger tensor and the tensor τ are given as follows. Let t'_j, $j = 0, 1, \cdots, N$, be $(N+1)$-points in the interval $[0, t_{n+1}]$. Then, the constitutive equation can be written in the form

$$\tau(t_{n+1}) = \int_{-\infty}^{0} M(t_{n+1} - t') H(I_1, I_2) \mathbf{B}_{t'}(t_{n+1}) dt' \\ + \sum_{j=0}^{\frac{N-2}{2}} \int_{t'_{2j}}^{t'_{2j+2}} M(t_{n+1} - t') H(I_1, I_2) \mathbf{B}_{t'}(t_{n+1}) dt', \tag{16}$$

where an even N is assumed. For $t' < 0$, $\mathbf{B}_{t'}(t_{n+1}) = \mathbf{B}_0(t_{n+1})$, and, therefore, the first integral becomes

$$\int_{-\infty}^{0} M(t_{n+1}) H(I_1(\mathbf{B}_0(t_{n+1})), I_2(\mathbf{B}_0(t_{n+1}))) \mathbf{B}_0(t_{n+1}) dt', \tag{17}$$

and can be solved exactly.

Each integral under the summation operator $\int_{t'_{2j}}^{t'_{2j+2}} M(t_{n+1} - t')H(I_1, I_2)\mathbf{B}_{t'}(t_{n+1})dt'$ is approximated by the 3-points quadrature formula

$$\begin{aligned}I_3 =& A_0 \times H\left(I_1\left(\mathbf{B}_{t'_{2j}}(t_{n+1})\right), I_2\left(\mathbf{B}_{t'_{2j}}(t_{n+1})\right)\right)\mathbf{B}_{t'_{2j}}(t_{n+1}) \\&+ A_1 \times H\left(I_1\left(\mathbf{B}_{t'_{2j+1}}(t_{n+1})\right), I_2\left(\mathbf{B}_{t'_{2j+1}}(t_{n+1})\right)\right)\mathbf{B}_{t'_{2j+1}}(t_{n+1}) \\&+ A_2 \times H\left(I_1\left(\mathbf{B}_{t'_{2j+2}}(t_{n+1})\right), I_2\left(\mathbf{B}_{t'_{2j+2}}(t_{n+1})\right)\right)\mathbf{B}_{t'_{2j+2}}(t_{n+1}).\end{aligned} \quad (18)$$

The coefficients A_0, A_1, A_2 are obtained by solving the (3×3) linear system

$$\begin{aligned}A_0 + A_1 + A_2 &= b_0 = \int_{t'_{2j}}^{t'_{2j+2}} M(t_{n+1} - t')dt', \\A_0 \times t'_{2j} + A_1 \times t'_{2j+1} + A_2 \times t'_{2j+2} &= b_1 = \int_{t'_{2j}}^{t'_{2j+2}} M(t_{n+1} - t')t'dt', \\A_0 \times \left(t'_{2j}\right)^2 + A_1 \times \left(t'_{2j+1}\right)^2 + A_2 \times \left(t'_{2j+2}\right)^2 &= b_2 = \int_{t'_{2j}}^{t'_{2j+2}} M(t_{n+1} - t')(t')^2 dt',\end{aligned} \quad (19)$$

and are found to be

$$\begin{aligned}A_2 &= \frac{t'_{2j+1}t'_{2j}b_0 + b_2 - t'_{2j}b_1 - t'_{2j+1}b_1}{\left(t'_{2j+2}\right)^2 - t'_{2j+1}t'_{2j+2} + t'_{2j+1}t'_{2j} - t'_{2j}t'_{2j+2}}, \\A_1 &= \frac{-t'_{2j}b_1 + t'_{2j}b_0 t'_{2j+2} + b_2 - t'_{2j+2}b_1}{\left(t'_{2j+1} - t'_{2j}\right)\left(t'_{2j+1} - t'_{2j+2}\right)}, \\A_0 &= -\frac{-t'_{2j+1}b_1 + t'_{2j+1}b_0 t'_{2j+2} + b_2 - t'_{2j+2}b_1}{\left(-t'_{2j+2} + t'_{2j}\right)\left(t'_{2j+1} - t'_{2j}\right)}.\end{aligned} \quad (20)$$

One of the key issues of the deformation fields method is how the integration nodes $0 = t'_0 < t'_1 < \cdots < t'_N = t_{n+1}$ are distributed over the interval $[0, t_{n+1}]$ because such distribution can affect the accuracy of the results when solving complex flows. In this work, we used an ad hoc methodology for the discretization of the interval $[0, t_{n+1}]$, where a geometric progression is employed to calculate the integration nodes. Note that we consider time-dependent flows, and, therefore, the integration nodes are calculated at time t_{n+1} as follows:

(a) Set $t'_0 = 0$ and $t'_N = t_{n+1}$;
(b) $t'_{N-j} = t'_N - q^j, j = 1, 2, \cdots, N-1$, where $q = (t_{n+1}/\Delta t)^{1/N}$.

The last step in the calculation is to update the position of the moving free surface (the S-cell in Figure 1b). The fluid surface is represented by a piecewise linear surface composed of triangles and quadrilaterals having marker particles on their vertices (see [30]). The particle coordinates, stored at each time step, are updated, solving the equation

$$\frac{d\mathbf{x}}{dt} = \mathbf{v}, \quad (21)$$

by Euler's method. With the new coordinates of each marker particle, a reclassification of the free surface cells is performed. A free surface cell can become an empty cell (E-cell in Figure 1b) or a full cell (F-cell in Figure 1b) or remain an S-type cell. Details on the marker particles that define the free surface and the steps for inserting and removing particles will not be shown here, but the reader can consult Tomé et al. [30] or Castelo et al. [31].

4. Semi-Analytical Solution

We will now derive a semi-analytical solution for a fully developed three-dimensional tube flow of a K–BKZ–PSM fluid to validate the numerical implementation. Due to the complexity of the integral model, some simplifications need to be assumed to develop the analytical solution, which is only possible for some types of domains. We consider cylindrical coordinates (see Figure 2) for simplicity and assume pure shear flow. After finding the semi-analytical solution, we present the change in variables to obtain the solution in Cartesian coordinates.

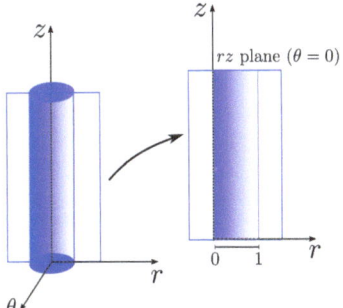

Figure 2. Representation of the pipe and a section in the rz plane.

We assume that $r \in [0, 1], u = 0, w^{in} = w(r), \dot{\gamma} = \dfrac{\partial w}{\partial r}$, and

$$\mathbf{B}(r, \theta, z) = \begin{bmatrix} 1 & 0 & \dot{\gamma}(t-t') \\ 0 & 1 & 0 \\ \dot{\gamma}(t-t') & 0 & 1 + \dot{\gamma}^2(t-t')^2 \end{bmatrix}. \qquad (22)$$

The invariants I_1 and I_2 that are required in the Papanastasiou function $H(I_1, I_2)$ take the form

$$I_1 = I_2 = 3 + \dot{\gamma}^2(t-t')^2, \qquad (23)$$

and the tensor components are given by

$$\tau^{rr} = \tau^{\theta\theta} = \frac{a_1 \alpha}{Wi} \int_{-\infty}^{t} \frac{e^{-(t-t')/Wi}}{\alpha + \dot{\gamma}^2(t-t')^2} dt',$$

$$\tau^{rz} = \frac{a_1 \alpha}{Wi} \int_{-\infty}^{t} \frac{\dot{\gamma}(t-t') e^{-(t-t')/Wi}}{\alpha + \dot{\gamma}^2(t-t')^2} dt', \qquad (24)$$

$$\tau^{zz} = \frac{a_1 \alpha}{Wi} \int_{-\infty}^{t} \frac{[1 + \dot{\gamma}^2(t-t')^2] e^{-(t-t')/Wi}}{\alpha + \dot{\gamma}^2(t-t')^2} dt'.$$

Taking the change in variables $s = t - t'$, these equations are rewritten as

$$\tau^{rr} = \tau^{\theta\theta} = \frac{a_1 \alpha}{Wi} \int_0^\infty \frac{1}{\alpha + \dot{\gamma}^2 s^2} e^{-s/Wi} ds,$$

$$\tau^{rz} = \frac{a_1 \alpha}{Wi} \int_0^\infty \frac{\dot{\gamma} s e^{-s/Wi}}{\alpha + \dot{\gamma}^2 s^2} ds, \qquad (25)$$

$$\tau^{zz} = \frac{a_1 \alpha}{Wi} \int_0^\infty \frac{[1 + \dot{\gamma}^2 s^2] e^{-s/Wi}}{\alpha + \dot{\gamma}^2 s^2} ds.$$

Thus, the equations of continuity and balance of momentum become

$$-\frac{\partial p}{\partial r} + \frac{1}{r}\frac{\partial}{\partial r}(r\tau^{rr}) - \frac{\tau^{\theta\theta}}{r} = 0, \qquad (26)$$

$$-\frac{\partial p}{\partial z} + \frac{1}{r}\frac{\partial}{\partial r}(r\tau^{rz}) = 0. \tag{27}$$

Integrating Equation (26), we obtain

$$p(r,z) = \int \frac{1}{r}\frac{\partial}{\partial r}(r\tau^{rr})dr - \int \frac{\tau^{\theta\theta}}{r}dr + F(z). \tag{28}$$

Thus,

$$\frac{\partial p}{\partial z} = F'(z), \tag{29}$$

and Equation (27) is rewritten as

$$\frac{1}{r}\frac{\partial}{\partial r}(r\tau^{rz}) = F'(z). \tag{30}$$

The left hand side of Equation (30) is just a function of r, so F' must be constant, let us say C, and therefore $C = dp/dz$. In this way, it follows that

$$\tau^{rz}(r) = \frac{1}{2}Cr + \frac{h(z)}{r}, \tag{31}$$

where $\tau^{rz}(r=0)$ should be finite and therefore $h(z) = 0$, leading to

$$\tau^{rz}(r) = \frac{1}{2}Cr. \tag{32}$$

The second equation in Equation (25) can then be rewritten as

$$\frac{1}{2}Cr = \frac{a_1\alpha}{Wi}\int_0^\infty \frac{\dot{\gamma}se^{-s/Wi}}{\alpha + \dot{\gamma}^2 s^2}ds. \tag{33}$$

The inlet boundary condition allows one to determine the constant C. The inflow velocity $w^{in}(r)$ is given by

$$w^{in}(r) = 1 - r^2, \quad \text{where } w^{in}(0) = 1 \text{ and } w^{in}(1) = 0, \tag{34}$$

thus,

$$\int_0^1 rw^{in}dr = \int_0^1 r(1-r^2)dr = \frac{1}{4}. \tag{35}$$

By mass conservation,

$$\int_0^1 rw^{in}(r)dr = \frac{1}{4}, \tag{36}$$

and integrating by parts leads to

$$\int_0^1 rw^{in}(r)dr = \left[\frac{1}{2}r^2 w^{in}(r)\right]_0^1 - \frac{1}{2}\int_0^1 r^2 \dot{\gamma}dr, \tag{37}$$

with

$$\int_0^1 r^2 \dot{\gamma}dr = -\frac{1}{2}. \tag{38}$$

Thus, to determine C, we must obtain $\dot{\gamma}(r)$ from Equation (33) and verify that

$$F(C) = \int_0^1 r^2 \dot{\gamma}dr + \frac{1}{2} = 0 \text{ is satisfied}. \tag{39}$$

The steps to calculate semi-analytical solutions are as follows:

Step 1: Set an interval $[C_0, C_1]$ such that $F(C_0) \times F(C_1) < 0$.

Step 2: Determine the zero for $|F(C)|$ taking $|F(C)| < \epsilon$, where ϵ is a small value (ϵ is the tolerance for the error). We carefully selected the value of ϵ to ensure the attainment of a semi-analytical solution accurate to six significant digits. Using Gauss–Laguerre quadrature in Equation (33), obtain $\hat{\gamma}(r)$. Using Equation (39), obtain the value of $F(C)$ using Simpson 1/3 quadrature.

Step 3: Lastly, determine $\tau^{rr}(r)$, $\tau^{zz}(r)$ and $\tau^{rz}(r)$ using the first and third equations in Equations (25) and (32), respectively.

The solution in three dimensions is obtained by making the change in coordinates as follows:

$$\begin{pmatrix} \tau^{xx} & \tau^{xy} & \tau^{xz} \\ \tau^{xy} & \tau^{yy} & \tau^{yz} \\ \tau^{xz} & \tau^{yz} & \tau^{zz} \end{pmatrix} = XYX^T$$

where $X = \begin{pmatrix} \cos(\theta) & -\sin(\theta) & 0 \\ \sin(\theta) & \cos(\theta) & 0 \\ 0 & 0 & 1 \end{pmatrix}$, $Y = \begin{pmatrix} \tau^{rr} & 0 & \tau^{rz} \\ 0 & \tau^{\theta\theta} & 0 \\ \tau^{rz} & 0 & \tau^{zz} \end{pmatrix}$ and

$$\begin{aligned} \tau^{xx} &= [\cos(\theta)^2 + \sin(\theta)^2]\tau^{rr}, \\ \tau^{xy} &= [\cos(\theta)\sin(\theta)]\tau^{rr} - [\cos(\theta)\sin(\theta)]\tau^{rr}, \\ \tau^{xz} &= \cos(\theta)\tau^{rz}, \\ \tau^{yy} &= [\cos(\theta)^2 + \sin(\theta)^2]\tau^{rr}, \\ \tau^{yz} &= \sin(\theta)\tau^{rz}, \\ \tau^{zz} &= \tau^{zz}. \end{aligned} \quad (40)$$

Thus, in three-dimensional Cartesian coordinates, we have that

$$\begin{aligned} \tau^{xx} &= \tau^{rr}, \\ \tau^{xy} &= 0, \\ \tau^{xz} &= \frac{x}{\sqrt{x^2+y^2}}\tau^{rz}, \\ \tau^{yy} &= \tau^{rr}, \\ \tau^{yz} &= \frac{y}{\sqrt{x^2+y^2}}\tau^{rz}, \\ \tau^{zz} &= \tau^{zz}. \end{aligned} \quad (41)$$

5. Results

The numerical code will now be used to solve confined (see Section 5.1) and free surface flows (see Section 5.2).

5.1. Confined Pipe Flows

In this confined pipe flow, the fluid is assumed to have only one relaxation mode. Therefore, it is possible to compare the simulation results with the semi-analytical solution presented before. The parameters used in this simulation are (see Tomé et al. [32] and Quinzani et al. [35]):

- Diameter $L = 0.01$ m, $U = 0.025$ m.s^{-1}, $\rho = 801.5$ Kg.m^{-3};
- $\lambda_{ref} = 0.1396$ s, $a_1 = 1.6648$ Pa, $\eta_0 = 0.2324$ Pa.s;
- Number of deformation fields $N = 50$;
- $Re = \frac{\rho L U}{\eta_0} = 0.8621$, $Wi = \lambda_{ref}\frac{U}{L} = 0.349$;
- Geometry: 0.01 m × 0.01 m × 0.05 m;
- Meshes (number of cells in the x, y and z directions): M1 = 12 × 12 × 60 ($\delta x = \delta y = \frac{0.01}{12}$), M2 = 16 × 16 × 80 ($\delta x = \delta y = \frac{0.01}{16}$), M3 = 20 × 20 × 100

($\delta x = \delta y = \frac{0.01}{20}$), M4 = 24 × 24 × 120 ($\delta x = \delta y = \frac{0.01}{24}$) and M5 = 28 × 28 × 140 ($\delta x = \delta y = \frac{0.01}{28}$).

Figure 3a,b show the velocity profiles w and u, respectively, with corresponding cross-section velocity distributions in the plane xz (obtained mesh M3). The velocity profiles are fully developed at $\bar{t} = 20$ s, suggesting that they have reached a steady-state condition. The velocity profile w shows a parabolic shape in the cross-section represented in Figure 3a, while the influence of the inflow can be observed in the cross-section depicted in Figure 3b, but the solution for u exhibits the expected physical behavior outside this region. Notice that, in all full cells, the initial velocity vector is defined as $\mathbf{v} = (u, v, w) = (0, 0, 0)$.

Similar to the velocity vector, initial conditions for pressure and tensors are set to zero in full cells. As shown in Figure 4a,b, the pressure and τ^{xz} tensor profiles are in agreement with the physically expected profiles, i.e., linear profiles across the longitudinal direction (flow direction) and transverse direction (perpendicular to the flow), respectively. In addition, the values obtained for τ^{xz} are comparable with the behavior of the analytical solution (see Figure 5b). In the cross-section represented in Figure 4b, there is an influence of the inflow (tensors are defined as zero in the inflow, and the Finger tensor is defined as the identity matrix), but, outside this region, the expected linear profile is obtained as previously stated (for further details, refer to Figure 5b).

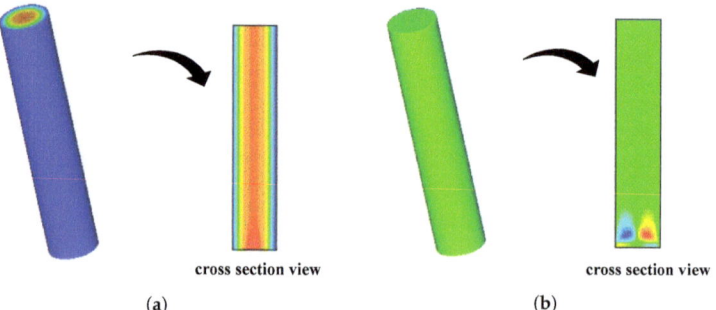

Figure 3. Velocity components u and w of $\mathbf{v}(u, v, w)$ along the plane xz ($y = 0$) at $\bar{t} = 20$ s. (**a**) Visualization of the velocity profile w. (**b**) Visualization of the velocity profile u.

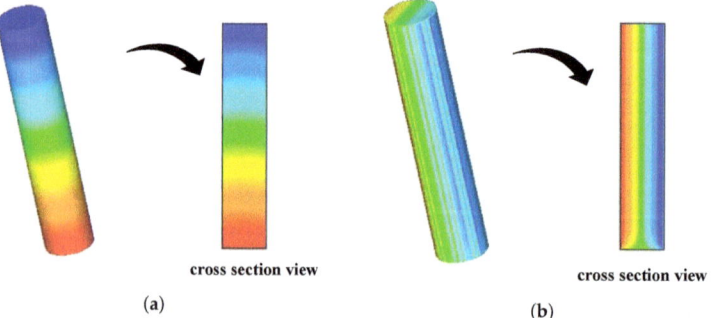

Figure 4. Pressure and τ^{xz} distribution along the xz plane ($y = 0$). (**a**) Visualization of the pressure p. (**b**) Visualization of the τ^{xz} tensor component.

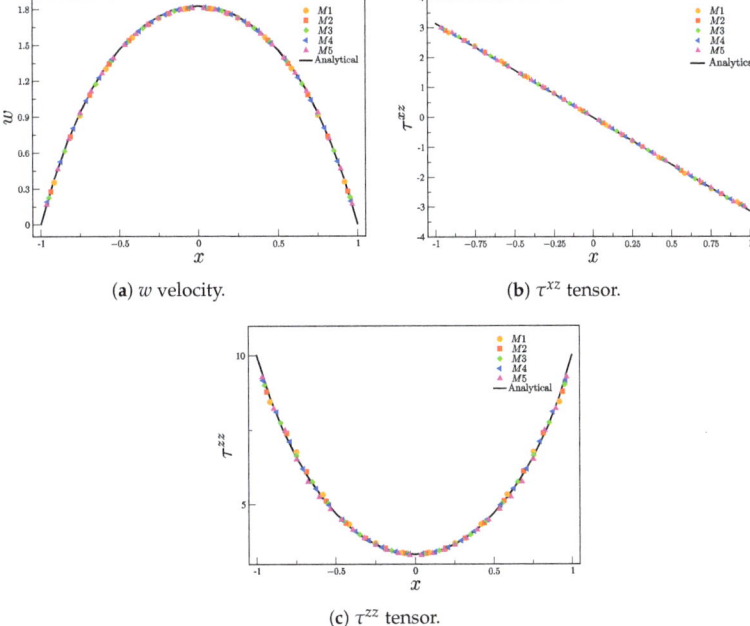

Figure 5. Comparison between the analytical and numerical solutions for (**a**) w velocity component and tensors components (**b**) τ^{xz} and (**c**) τ^{zz}.

Figure 5a–c show, respectively, the solution for the velocity w and stress components τ^{xz} and τ^{zz} using meshes M1–M5. The profiles were obtained at $y = 0$ and were taken in the center of the pipe, $z = z_{max}/2$ for $t = 10$ (or $\bar{t} = 25$ s). The results were compared with the semi-analytical solution, with good agreement between the numerical results (M1–M5) and the semi-analytical results. The instant $t = 10$ ($\bar{t} = 25$ s) was chosen because the velocity and stresses already show a steady state behavior (the velocity residual is small (see Equation (42)). The residual (for the velocities) is calculated as

$$R_{es} = \frac{\sqrt{\sum_{i=1}^{N_c}\{(u_i^t - u_i^{t-\Delta t})^2 + (v_i^t - v_i^{t-\Delta t})^2 + (w_i^t - w_i^{t-\Delta t})^2\}}}{N_c}, \quad (42)$$

where t is the simulation time, Δt is the time step and N_c is the number of computational cells. The slight variances between the analytical and numerical solutions can be attributed to approximations made in the numerical simulations, which differ from the precise analytical solution. Although the tube's length appears adequate for the complete development of velocity and shear stress profiles, this completeness is not reflected in the τ^{zz} tensor. Consequently, these disparities remain minor. It is worth noting that, across all meshes, the average relative error remains below 5%.

Figure 6 shows the calculation of the residuals R_{es} in meshes M1–M5 up to time $t = 10$ (or, equivalently, $\bar{t} = 25$ s). It can be observed that the residuals in the five meshes M1–M5 decrease and show convergence towards a steady-state solution, thus proving the robustness of the numerical method. As expected, we also observe a smaller residual for the most refined meshes.

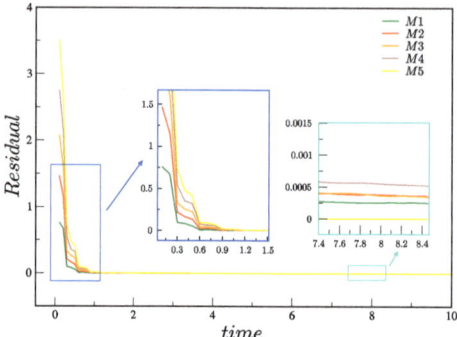

Figure 6. Total residual in meshes M1, M2, M3, M4 and M5 up to $t = 10$ (or $\bar{t} = 25$ s).

5.2. Free Surface Flows

In this subsection, we test the numerical method's robustness by simulating the extrudate swell phenomenon of Boger fluids. Please note that our aim is not to conduct an in-depth study of this type of flow in Boger fluids; instead, we focus on assessing the reliability of the numerical approach.

The phenomenon known as extrudate swell is very present in various industrial processes. In this phenomenon, the fluid flows over a *pipe/die* and swells outside the free surface region (the cross-sectional area of the extrudate—the material being extruded—is larger after exiting the die compared to the die orifice). This behavior is mainly due to the elastic recovery of the polymer chains after being subjected to high pressures and shear forces during the extrusion process. Figure 7a shows the domain used in the simulation, and Figure 7b illustrates the phenomenon of extrudate swell (with the contour lines representing the trajectory of the fluid's free surface).

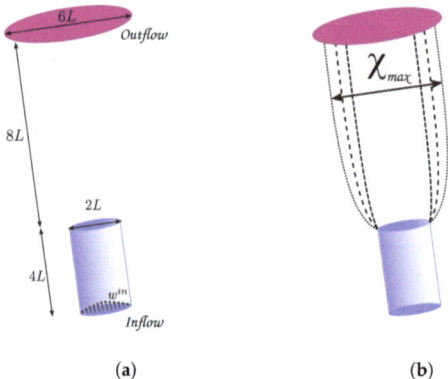

(a) (b)

Figure 7. Schematic of a free surface simulation in the FREEFLOW-3D software. (**a**) Schematic representation of the domain; (**b**) illustration of the extrudate swell. The fluid exits the tube and starts to swell.

It should be remarked that simulating extrudate swell can be challenging due to several numerical difficulties, which are now outlined: the extrusion process involves highly non-uniform and complex flow patterns, especially near the die exit. These flows experience rapid changes in pressure and velocity, making it difficult to model accurately; simulating extrusion processes requires discretizing the computational domain into smaller elements or cells, and the geometry of the die can be quite intricate. Moreover, the simulation must maintain numerical stability, which can be problematic in high-pressure and high-shear regions; simulating extrudate swell is computationally intensive, especially for large-scale industrial extrusion processes and using integral models; extrudate swell is a time-dependent phenomenon as the material continuously deforms and recovers during extrusion. Capturing this transient behavior accurately in numerical simulations requires precise time-stepping algorithms and may increase computational complexity. To address these challenges, researchers often resort to simplifications and assumptions to reduce computational complexity. However, in the context of this work, we take a different approach by considering the complete system of equations and accounting for the full 3D geometry.

To test the robustness of the new numerical procedure, two different Weissenberg numbers were considered (case C1 – $Wi = 0.43$ and case C2 – $Wi = 0.64$), both using non-shear-thinning highly elastic polymer solutions (Boger fluids – see Table 1). Boger fluids are a type of dilute polymer solution known for their remarkable elasticity, particularly at low apparent shear rates, and this unique characteristic gives rise to a significant extrudate swell during the extrusion process [14,36–38]. This makes Boger fluids ideal to test the numerical implementation. Numerical simulation of extruded swelling in two dimensions using the data used here (see Table 1) is presented in Mitsoulis [14].

Table 1. Parameters of the fluid used in the extrudate swell problem (see Mitsoulis [14]).

	$\rho_0 = 868$ kg/m^3, $\alpha = 34{,}214$		
	$\lambda_{ref} = 0.081$ s, $\beta = 0.1$, $\eta_0 = 2.4$ Pa.s		
k	λ_k [s]	a_k [Pa]	η_k [Pa.s]
1	0.4887×10^{-3} s	3.1295×10^3 Pa	1.5294 Pa.s
2	0.4464×10^{-1} s	5.0917×10^0 Pa	0.2273 Pa.s
3	2.8384×10^{-1} s	2.2783×10^0 Pa	0.6457 Pa.s

The following parameters were used in the simulations (see Mitsoulis [14] and Tomé et al. [32]):

- Pipe dimension: 0.04 m × 0.04 m × 0.08 m; $\delta x = \delta y = \frac{0.04}{16}$ (see Figure 7a);
- Pipe diameter $2L = 0.04$ m;
- Number of deformation fields $N = 50$;
- C1 – $U = 0.1067$ m.s^{-1}, $Re = 0.7$, $Wi = 0.43$;
- C2 – $U = 0.1600$ m.s^{-1}, $Re = 1.1$, $Wi = 0.64$.

Figure 8 shows the flow development of a Boger fluid for two different Weissenberg number values (cases C1 and C2). We conducted flow measurements at six different time points to analyze the behavior of the fluid in the system. The first four time instants were identical for both C1 and C2 cases, while the last two time points differed. Specifically, for the lower inlet velocity case, we considered time points $\bar{t} = 5$ and 6 s, and, for the other case, the time points were $\bar{t} = 4$ and 4.4 s. During the initial stages of both cases, the fluid exhibited a smooth flow with a parabolic profile as it exited the tube. However, as the process continued, swell occurred, causing the cross-sectional area of the extrudate to increase after leaving the die. This swelling behavior significantly affects the flow dynamics and needs to be carefully considered in the analysis. The simulation results show the development of the fluid front as it reaches the wall. Notably, there are distinct differences in swelling between two specific time instances: $\bar{t} = 6$ s (C1) and $\bar{t} = 4.4$ s (C2).

To characterize the extrudate swell phenomenon, an important parameter is the dimensionless swelling rate $\chi = \chi_{max}/(2r)$, where χ_{max} is the maximum swelling value and r is the pipe radius. For case C1, the maximum swelling value is found to be $\chi = 1.78$, while, for case C2, the swelling rate increases to $\chi = 1.95$. This was expected since the Weissenberg number represents the ratio of the characteristic time scale of the elastic forces acting on a fluid to the characteristic time scale of viscous forces, and, when a polymer melt is subjected to shear flow (for example, in an extruder), the long polymer chains experience deformation due to the flow-induced stretching and alignment. A higher Weissenberg number indicates a more elastic behavior of the polymer melt, leading to more significant elastic recovery and increased extrudate swell.

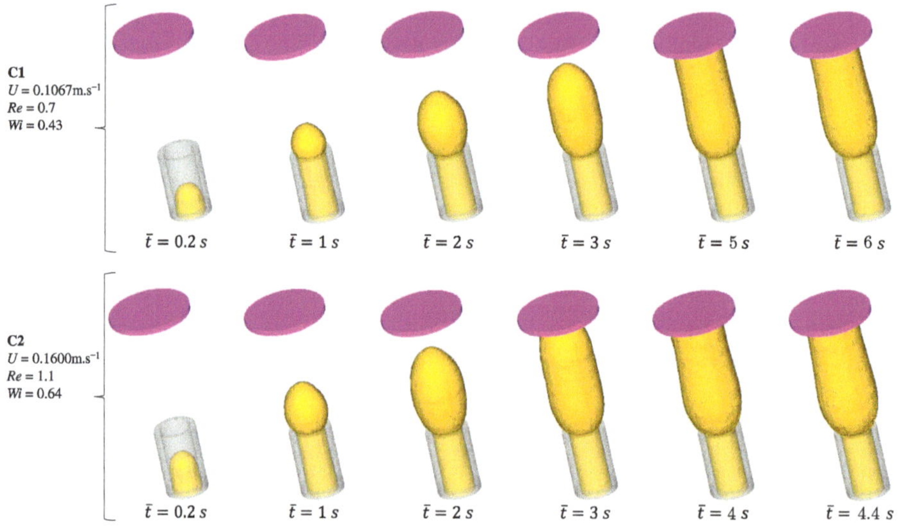

C1
$U = 0.1067 \text{m.s}^{-1}$
$Re = 0.7$
$Wi = 0.43$

$\bar{t} = 0.2\ s$ $\bar{t} = 1\ s$ $\bar{t} = 2\ s$ $\bar{t} = 3\ s$ $\bar{t} = 5\ s$ $\bar{t} = 6\ s$

C2
$U = 0.1600 \text{m.s}^{-1}$
$Re = 1.1$
$Wi = 0.64$

$\bar{t} = 0.2\ s$ $\bar{t} = 1\ s$ $\bar{t} = 2\ s$ $\bar{t} = 3\ s$ $\bar{t} = 4\ s$ $\bar{t} = 4.4\ s$

Figure 8. Flow development of a Boger fluid for two different inlet velocities (cases C1 and C2).

We may therefore conclude that the numerical method employed in the simulations demonstrates its capability to capture the transient physics of the extrudate swell problem in detail, even for a small difference in the Weissenberg number (cases C1 and C2). It accurately predicts the swelling behavior and allows for a better understanding of the process dynamics. By accounting for the material properties and flow conditions, the simulation provides valuable insights into the extrusion process and contributes to the optimization of extrusion operations.

6. Conclusions

In this work, a novel numerical method was developed to address three-dimensional unsteady free surface flows incorporating integral viscoelastic constitutive equations, specifically the K–BKZ–PSM (Kaye–Bernstein, Kearsley, Zapas–Papanastasiou, Scriven, Macosko) model. To implement this new approach, we integrated it into the FREEFLOW-3D code [27], enhancing its capabilities for handling viscoelastic fluid behavior.

To validate the numerical methodology, we conducted simulations of K–BKZ–PSM fluids in a pipe. The results were compared with a newly derived semi-analytical solution, and we found that the simulations performed on five different meshes yielded excellent agreement with the analytical solution. Furthermore, we applied our methodology to tackle flows with free surfaces. One notable example was the simulation of the classic extrudate swell problem, which involved a highly elastic polymeric solution known as the Boger fluid.

The significance of this work lies in the scarcity of literature concerning the simulation of unsteady three-dimensional flows of K–BKZ–PSM fluids (and integral viscoelastic models in general) using finite differences, especially when considering problems with moving free surfaces. Therefore, we hope that our contributions will inspire and encourage other researchers to further develop and explore the numerical methods we have presented here.

In conclusion, our newly developed numerical method has proven its effectiveness in handling complex fluid flow scenarios, including free surface flows such as extrudate swell simulations. The successful validation against analytical solutions reinforces the reliability of our approach and opens up opportunities for broader applications in the field of viscoelastic fluid dynamics.

Author Contributions: Conceptualization, J.B., A.C., L.L.F. and C.F.; Methodology, J.B., A.C., L.L.F. and C.F.; Software, J.B. and A.C.; Validation, A.C., L.L.F. and C.F.; Supervision, A.C., L.L.F. and C.F. All authors have read and agreed to the published version of the manuscript.

Funding: This research was funded by São Paulo Research Foundation (FAPESP) grants 2013/07375-0, 2019/07316-0; the National Council for Scientific and Technological Development (CNPq) grant 307483/2017-7; FCT (Fundação para a Ciência e a Tecnologia) and CMAT (Centre of Mathematics of the University of Minho) projects UIDB/00013/2020 and UIDP/00013/2020 and FCT funding, project 2022.06672.PTDC and contract 2022.00753.CEECIND.

Institutional Review Board Statement: Not applicable

Data Availability Statement: Not applicable

Acknowledgments: J. Bertoco acknowledges the support by CMCC — Center for Mathematics, Computing and Cognition, UFABC. A. Castelo thanks the financial support from the São Paulo Research Foundation (FAPESP) grants 2013/07375-0, 2019/07316-0 and the National Council for Scientific and Technological Development (CNPq) grant 307483/2017-7. A. Castelo and J. Bertoco acknowledge the support by ICMC — Institute of Mathematics and Computational Sciences — University of Sao Paulo (USP). L.L. Ferrás would like to thank FCT (Fundação para a Ciência e a Tecnologia) for financial support through CMAT (Centre of Mathematics of the University of Minho) projects UIDB/00013/2020 and UIDP/00013/2020 and would also like to thank FCT for the funding of project 2022.06672.PTDC. C. Fernandes acknowledges the support by FEDER funds through the COMPETE 2020 Programme and National Funds through FCT (Portuguese Foundation for Science and Technology) under the contract FCT/2022.00753.CEECIND. Research carried out using the computational resources of the Center for Mathematical Sciences Applied to Industry (CeMEAI) funded by FAPESP grant 2013/07375-0.

Conflicts of Interest: The authors declare no conflicts of interest.

Abbreviations

The following abbreviations are used in this manuscript:

K–BKZ	Kaye–Bernstein, Kearsley, Zapas
PSM	Papanastasiou, Scriven, Macosko
EVSS	Elastic–Viscous Split Stress
UCM	Upper-convected Maxwell
PTT	Phan–Thien–Tanner
FENE-P	Finite Extensible Nonlinear Elastic Peterlin
HDPE	High-Density Polyethylene
LDPE	Low-Density Polyethylene
FEM	Finite element Method
FDM	Finite Difference Method
PDE	Partial Differential Equation
ALE	Arbitrary Lagrangian Eulerian
FBC	Free Boundary Condition

References

1. Clermont, J.-R.; Normandin, M. Numerical simulation of extrudate swell for Oldroyd-B fluids using the stream-tube analysis and a streamline approximation. *J. Non-Newton. Fluid Mech.* **1993**, *50*, 193–215. [CrossRef]
2. Tomé, M.F.; Castelo, A.; Afonso, A.M.; Alves, M.A.; Pinho, F.T. Application of the log-conformation tensor to three-dimensional time-dependent free surface flows. *J. Non-Newton. Fluid Mech.* **2012**, *175–176*, 44–54. [CrossRef]
3. Mompean, G.; Thais, L.; Tomé, M.F.; Castelo, A. Numerical prediction of three-dimensional time-dependent viscoelastic extrudate swell using differential and algebraic models. *Comput. Fluids* **2011**, *44*, 68–78. [CrossRef]
4. Tomé, M.F.; Paulo, G.S.; Pinho, F.T.; Alves, M.A. Numerical solution of the PTT constitutive equation for unsteady three-dimensional free surface flows. *J. Non-Newton. Fluid Mech.* **2010**, *165*, 247–262. [CrossRef]
5. Béraudo, C.; Fortin, A.; Coupez, T.; Demay, Y.; Vergnes, B.; Agassant, J.F. A finite element method for computing the flow of multi-mode viscoelastic fluids: Comparison with experiments. *J. Non-Newton. Fluid Mech.* **1998**, *75*, 1–23. [CrossRef]
6. Mu, Y.; Zhao, G.; Wu, X.; Zhai, J. Modeling and simulation of three-dimensional planar contraction flow of viscoelastic fluids with PTT, Giesekus and FENE-P constitutive models. *Appl. Math. Comput.* **2012**, *218*, 8429–8443. [CrossRef]

7. Hulsen, M.A.; van der Zanden, J. Numerical simulation of contraction flows using a multi-mode Giesekus model. *J. Non-Newton. Fluid Mech.* **1991**, *38*, 183–221. [CrossRef]
8. Goublomme, A.; Crochet, M.J. Numerical prediction of extrudate swell of a high-density polyethylene: Further results. *J. Non-Newton. Fluid Mech.* **1993**, *47*, 281–287. [CrossRef]
9. Park, H.J.; Mitsoulis, E. Numerical simulation of circular entry flows of fluid M1 using an integral constitutive equation. *J. Non-Newton. Fluid Mech.* **1992**, *42*, 301–314. [CrossRef]
10. Dupont, S.; Crochet, M.J. The vortex growth of a K.B.K.Z. fluid in an abrupt contraction. *J. Non-Newton. Fluid Mech.* **1988**, *29*, 81–91. [CrossRef]
11. Papanastasiou, A.C.; Scriven, L.E.; Macosko, C.W. An Integral Constitutive Equation for Mixed Flows: Viscoelastic Characterization. *J. Rheol.* **1983**, *27*, 387–410. [CrossRef]
12. Bernstein, B.; Kearsley, E.A.; Zapas, L.J. A Study of Stress Relaxation with Finite Strain. *Trans. Soc. Rheol.* **1963**, *7*, 391–410. [CrossRef]
13. Kaye, A. Non-Newtonian Flow in Incompressible Fluids. *Aerosp. Eng. Rep.* **1963**, *134*.
14. Mitsoulis, E. Extrudate swell of Boger fluids. *J. Non-Newton. Fluid Mech.* **2010**, *165*, 812–824. [CrossRef]
15. Mitsoulis, E. 50 Years of the K-BKZ Constitutive Relation for Polymers. *ISRN Polym. Sci.* **2013**, *2013*, 1–22. [CrossRef]
16. Luo, X.-L.; Mitsoulis, E. An efficient algorithm for strain history tracking in finite element computations of non-Newtonian fluids with integral constitutive equations. *Int. J. Numer. Methods Fluids* **1990**, *11*, 1015–1031. [CrossRef]
17. Ansari, M.; Alabbas, A.; Hatzikiriakos, S.G.; Mitsoulis, E. Entry Flow of Polyethylene Melts in Tapered Dies. *Int. Polym. Process.* **2010**, *25*, 287–296. [CrossRef]
18. Chai, M.S.; Yeow, Y.L. Modelling of fluid M1 using multiple-relaxation-time constitutive equations. *J. Non-Newton. Fluid Mech.* **1990**, *35*, 459–470. [CrossRef]
19. Luo, X.-L.; Mitsoulis, E. A numerical study of the effect of elongational viscosity on vortex growth in contraction flows of polyethylene melts. *J. Rheol.* **1990**, *34*, 309–342. [CrossRef]
20. Mitsoulis, E.; Malamataris, N.A. The free (open) boundary condition with integral constitutive equations. *J. Non-Newton. Fluid Mech.* **2012**, *177–178*, 97–108. [CrossRef]
21. Olley, P.; Spares, R.; Coates, P.D. A method for implementing time-integral constitutive equations in commercial CFD packages. *J. Non-Newton. Fluid Mech.* **1999**, *86*, 337–357. [CrossRef]
22. Araújo, M.S.B.; Fernandes, C.; Ferrás, L.L.; Tuković, Ž.; Jasak, H.; Nóbrega, J.M. A stable numerical implementation of integral viscoelastic models in the OpenFOAM® computational library. *Comput. Fluids* **2018**, *172*, 728–740. [CrossRef]
23. Luo, X.-L.; Mitsoulis, E. Memory Phenomena in Extrudate Swell Simulations for Annular Dies. *J. Rheol.* **1989**, *33*, 1307–1327. [CrossRef]
24. Goublomme, A.; Draily, B.; Crochet, M.J. Numerical prediction of extrudate swell of a high-density polyethylene. *J. Non-Newton. Fluid Mech.* **1992**, *44*, 171–195. [CrossRef]
25. Ganvir, V.; Lele, A.; Thaokar, R.; Gautham, B.P. Prediction of extrudate swell in polymer melt extrusion using an Arbitrary Lagrangian Eulerian (ALE) based finite element method. *J. Non-Newton. Fluid Mech.* **2009**, *156*, 21–28. [CrossRef]
26. Ahmed, R.; Liang, R.F.; Mackley, M.R. The experimental observation and numerical prediction of planar entry flow and die swell for molten polyethylenes. *J. Non-Newton. Fluid Mech.* **1995**, *59*, 129–153. [CrossRef]
27. Tomé, M.F.; Castelo, A.; Ferreira, V.G.; McKee, S. A finite difference technique for solving the Oldroyd-B model for 3D-unsteady free surface flows. *J. Non-Newton. Fluid Mech.* **2008**, *154*, 179–206. [CrossRef]
28. Rasmussen, H.K. Time-dependent finite-element method for the simulation of three-dimensional viscoelastic flow with integral models. *J. Non-Newton. Fluid Mech.* **1999**, *84*, 217–232. [CrossRef]
29. Marín, J.M.R.; Rasmussen, H.K. Lagrangian finite element method for 3D time-dependent non-isothermal flow of K-BKZ fluids. *J. Non-Newton. Fluid Mech.* **2009**, *162*, 45–53. [CrossRef]
30. Tomé, M.F.; Filho, A.C.; Cuminato, J.A.; Mangiavacchi, N.; Mckee, S. GENSMAC3D: A numerical method for solving unsteady three-dimensional free surface flows. *Int. J. Numer. Methods Fluids* **2001**, *37*, 747–796. [CrossRef]
31. A. Castello F.; Tomé, M.F.; César, C.N.L.; McKee, S.; Cuminato, J.A. Freeflow: An integrated simulation system for three-dimensional free surface flows. *Comput. Vis. Sci.* **2000**, *2*, 199–210. [CrossRef]
32. Tomé, M.F.; Bertoco, J.; Oishi, C.M.; Araujo, M.S.B.; Cruz, D.; Pinho, F.T.; Vynnycky, M. A finite difference technique for solving a time strain separable K-BKZ constitutive equation for two-dimensional moving free surface flows. *J. Comput. Phys.* **2016**, *311*, 114–141. [CrossRef]
33. Hulsen, M.A.; Peters, E.A.J.F.; van den Brule, B.H.A.A. A new approach to the deformation fields method for solving complex flows using integral constitutive equations. *J. Non-Newton. Fluid Mech.* **2001**, *98*, 201–221. [CrossRef]
34. Rajagopalan, D.; Armstrong, R.C.; Brown, R.A. Finite element methods for calculation of steady, viscoelastic flow using constitutive equations with a Newtonian viscosity. *J. Non-Newton. Fluid Mech.* **1990**, *36*, 159–192. [CrossRef]
35. Quinzani, L.M.; McKinley, G.H.; Brown, R.A.; Armstrong, R.C. Modeling the rheology of polyisobutylene solutions. *J. Rheol.* **1990**, *34*, 705–748. [CrossRef]

36. Mitsoulis, E. The numerical simulation of Boger fluids: A viscometric approximation approach. *Polym. Eng. Sci.* **1986**, *26*, 1552–1562. [CrossRef]
37. López-Aguilar, J.E.; Tamaddon-Jahromi, H.R. Computational Predictions for Boger Fluids and Circular Contraction Flow under Various Aspect Ratios. *Fluids* **2020**, *5*, 85. [CrossRef]
38. Satrape, J.V.; Crochet, M.J. Numerical simulation of the motion of a sphere in a Boger fluid. *J. Non-Newton. Fluid Mech.* **1994**, *55*, 91–111. [CrossRef]

Disclaimer/Publisher's Note: The statements, opinions and data contained in all publications are solely those of the individual author(s) and contributor(s) and not of MDPI and/or the editor(s). MDPI and/or the editor(s) disclaim responsibility for any injury to people or property resulting from any ideas, methods, instructions or products referred to in the content.

Article

A Phenomenological Model for Enthalpy Recovery in Polystyrene Using Dynamic Mechanical Spectra

Koh-hei Nitta *, Kota Ito and Asae Ito

Division of Material Sciences, Graduate School of Natural Science and Technology, Kanazawa University, Kakuma Campus, Kanazawa 920-1192, Japan; asae@se.kanazawa-u.ac.jp (A.I.)
* Correspondence: nitta@se.kanazawa-u.ac.jp; Tel.: +81-76-234-4818; Fax: +81-76-264-6220

Abstract: This paper studies the effects of annealing time on the specific heat enthalpy of polystyrene above the glass transition temperature. We extend the Tool–Narayanaswamy–Moynihan (TNM) model to describe the endothermic overshoot peaks through the dynamic mechanical spectra. In this work, we accept the viewpoint that the enthalpy recovery behavior of glassy polystyrene (PS) has a common structural relaxation mode with linear viscoelastic behavior. As a consequence, the retardation spectrum evaluated from the dynamic mechanical spectra around the primary T_g peak is used as the recovery function of the endothermic overshoot of specific heat. In addition, the sub-T_g shoulder peak around the T_g peak is found to be related to the structural relaxation estimated from light scattering measurements. The enthalpy recovery of annealed PS is quantitatively described using retardation spectra of the primary T_g, as well as the kinetic process of the sub-T_g relaxation process.

Keywords: enthalpy relaxation; dynamic mechanical spectra; creep compliance; polystyrene

Citation: Nitta, K.-h.; Ito, K.; Ito, A. A Phenomenological Model for Enthalpy Recovery in Polystyrene Using Dynamic Mechanical Spectra. *Polymers* 2023, 15, 3590. https://doi.org/10.3390/polym15173590

Academic Editor: Javier González-Benito

Received: 23 July 2023
Revised: 25 August 2023
Accepted: 26 August 2023
Published: 29 August 2023

Copyright: © 2023 by the authors. Licensee MDPI, Basel, Switzerland. This article is an open access article distributed under the terms and conditions of the Creative Commons Attribution (CC BY) license (https://creativecommons.org/licenses/by/4.0/).

1. Introduction

When an amorphous polymer is cooled from well above glass transition temperature T_g to some lower temperature, rotational mobility around the main-chain bonds is frozen, and the polymer has no time to attain conformational equilibrium within the time scale of the given experiment. As a result, the glassy polymer solidified below T_g usually shows time-dependent physical properties such as heat capacity C_p, modulus, and density. This is a feature known as "physical aging" [1–3]. Then, a series of atomic rearrangements of the main chain towards the new equilibrium state proceeds in order to lose the excess configurational energy, and the subsequent time-dependent change is referred to as "structural relaxation" [4–7].

Figure 1 illustrates schematical plots of enthalpy H and heat capacity C_p of a glassy polymer under cooling and subsequent reheating. The glass transition temperature T_g has been conventionally defined as a temperature that is the intersection of the slopes of the liquid state and the glassy one. The maintenance at temperature T_a below T_g after cooling from a liquid state induces enthalpy relaxation toward the equilibrium value. This phenomenon is called enthalpy relaxation or physical aging. The equilibrium state is represented by the extension of the slope of the liquid state (the dashed line in Figure 1). A cross point of the dashed line and the H–T curve gives a limiting fictive temperature T_f^0. The enthalpy H overshoots and deviates downward from the equilibrium line during heating after enthalpy relaxation. The enthalpy recovery behavior can be detected as a peak around the glass transition during reheating on the C_p–T curve.

The experimental technique most frequently used for the characterization of structural relaxation is the measurement of enthalpy changes by differential scanning calorimetry (DSC). The internal rotation of the main chain toward reducing configurational energy, i.e., structural relaxation, emerges as an overshoot in the C_p curve under a constant heating rate after annealing at a temperature below T_g. The overshoot peak area resulting from the difference in both corresponding C_p–temperature lines is in accordance with the enthalpy difference ΔH_{DSC} between the extrapolated enthalpy of the liquid and that of the glass. The overshoot in the C_p curve shows a positive dependence on the annealing temperature and time. This thermodynamic feature is known as "enthalpy relaxation" [8–10]. Consequently, the changes in the enthalpy difference ΔH with annealing time at a fixed annealing temperature can be expressed using a decay function $\phi(t)$ of the annealing time t as follows:

$$\Delta H_{DSC} = \Delta H_e (1 - \phi(t)) \tag{1}$$

and here we introduce the normalized recovery function $\Lambda_T(t)$ as:

$$\Lambda_T(t) = 1 - \phi(t) \tag{2}$$

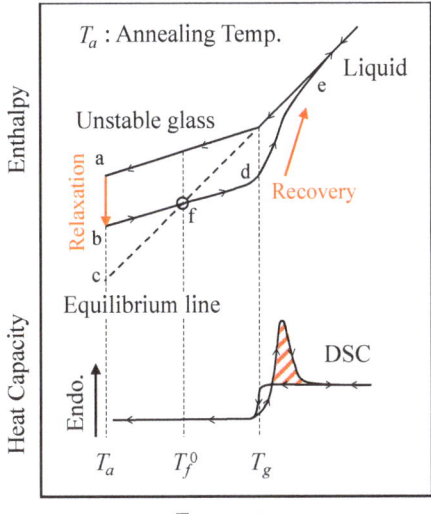

Figure 1. Schematic plots of enthalpy and heat capacity during cooling and subsequent heating at a fixed rate for a glassy polymer.

The enthalpy loss $\Delta H_{DSC} = H_0 - H(t)$, where H_0 is the initial enthalpy at the annealing time $t = 0$, and $H(t)$ is the enthalpy at any annealing time t, and the final loss $\Delta H_e = H_0 - H_\infty$, where H_∞ is the enthalpy at equilibrium ($t = \infty$), and $\phi(t)$ is an exponential decay function. It has been identified that the decay function $\phi(t)$, describing the kinetics of enthalpy recovery, is characterized by two essential features: non-exponentiality and non-linearity. The non-exponential character is brought about by the distribution of relaxation times, which can be usually described by the Kohlaush–Williams–Watts (KWW) equation [11–13] as follows:

$$\phi(t) = e^{-(t/\tau_0)^\beta} \tag{3}$$

where $\beta (0 < \beta < 1)$ is a parameter associated with the distribution of characteristic time τ_0. The small value of β implies a broad distribution, and $\beta = 1$ implies a single characteristic time. Additionally, the non-linearity is caused by the discrepancy of the approach from below and from above an equilibrium point. This discrepancy is revealed by the double dependence of τ_0 on both the actual temperature T and the fictive temperature T_f in Equation (3) [14,15], which is defined as the temperature at which the glass would be in the equilibrium structural state if it was instantaneously brought to the temperature. In this view, a non-linearity parameter $x(0 < x \leq 1)$ is introduced to partition the activation energy into the actual temperature and the fictive temperature [16,17] as follows:

$$\tau_0 = A \exp\left[x\Delta h/RT + (1-x)\Delta h/RT_f\right] \quad (4)$$

where A is an arbitrary relaxation time, Δh is activation energy, and R is the gas constant.

So far, much effort has been made to develop mathematical models that permit the comparison of experimental C_p data obtained after various thermal histories with theoretical predictions. Some fitting parameters in these phenomenological models are used to predict the overshooting behavior of the experimental C_p curve during a constant rate of heating [16–25]. The Tool–Narayanaswamy–Moynihan (TNM) model [15,26,27] has been most widely used to describe the enthalpy relaxation behavior. Here, we outline the enthalpy relaxation in the TMN model under a constant heating rate test.

A DSC experiment under continuous heating/cooling can be treated as a series of infinitesimal temperature jumps (T-jumps). In general, then, the response of the fictive temperature to the departure of the enthalpy from its equilibrium value follows the recovery function of the enthalpy loss:

$$T_f = T_0 + \Delta T \Lambda_T(t) \quad (5)$$

where T_0 is the initial temperature at $t = 0$, and ΔT is the magnitude of the T-jump. Taking the relaxation function to be linear with respect to T-jumps, the net response of the system can be formulated as the superposition of the response to the series of T-jumps that constitute their thermal histories. Combination of Equations (1) and (3) with the Boltzmann superposition principle leads to the evolution of fictive temperature as a function of the actual temperature:

$$T_f(T) = T_0 + \int_0^t dt' \left(\frac{dT}{dt}\right)_{t=t'} \left(1 - \exp\left[-\left(\int_{t'}^t dt'' \frac{1}{\tau_0}\right)^\beta\right]\right) \quad (6)$$

where t' and t'' are dummy integral variables of time. Differentiation of Equation (6) yields the normalized specific heat capacity as:

$$C_p^N(T) = \frac{dT_f}{dT} \quad (7)$$

This mathematical process has been used to give a description of the specific heat as a function of actual temperature. TNM-based models simulate the enthalpy relaxation in a series of various polymeric materials by controlling fitting parameters such as A, x, Δh, and β.

The recovery function $\Lambda_T(t)$, describing the enthalpy changes, is a retardation function rather than a relaxation one. Therefore, the characteristic time τ_0 in Equation (3) should be referred to as "retardation time" rather than "relaxation time", and, in many endothermic experiments described as "enthalpy relaxation", the "enthalpy recovery" is recorded as a function of temperature or time, as suggested by Hodge [8]. This implies that it is not correct to directly compare DSC and dynamic mechanical analysis (DMA) [28].

In this work, we propose a novel phenomenological model equation as a natural extension of the linear viscoelasticity to simulate the overshoot of C_p under constant heating rate as a function of annealing time t at an annealing temperature below sub-T_g for atactic polystyrene. Struik also pointed out the experimental fact that physical aging occurs in the temperature range of T_g to sub-T_g [1]. The retardation time function can be straightforwardly applied to the enthalpy recovery to give the enthalpy H or specific heat C_p curves as a function of the annealing time. The present work is fundamentally based on the concept that the time-dependent behavior of enthalpy recovery and dynamic mechanical spectra is associated with a common structural relaxation mode in the glassy state.

The relationship between the time scale of enthalpy relaxation and that of some accompanying mechanical responses such as creep, stress–relaxation, and stress–strain behavior has been widely studied by a number of workers [29–36]. However, there does not seem to have been any universal behavior found between thermo and mechanical dynamics, as suggested by Simon et al. [31] and Robertson et al. [33]. This is because much effort has been made to interrelate both time-dependent behaviors via the decay function $\phi(t)$. Yoshida [30] demonstrated that the enthalpy recovery rate after annealing corresponds to the change in dynamic storage modulus with annealing for amorphous and semicrystalline engineering plastics. A series of configurational rearrangements of the main chain reflecting the strain evolution towards the new equilibrium state under a creep test in the linear viscoelastic region corresponds to the structural rearrangement reflecting the enthalpy recovery during heating after a fixed annealing process to lose the excess configurational energy. In this work, on the basis of this view, we expand the TNM model to reproduce the C_p overshooting peak due to the enthalpy recovery process as determined from DSC experiments where we used retardation spectra and kinetic data of relaxation peaks from linear dynamic mechanical data in place of some fitting parameters of the TNM model.

This paper is organized as follows. In Section 2, we propose a phenomenological theory for expressing the evolution of fictive temperature in which the retardation time spectrum of viscoelastic properties is incorporated in the recovery response in the TNM model. Section 3 shows the experimental procedure for monitoring the annealing in PS by DSC and DMA. The enthalpy recovery data and DMA spectra are summarized in Section 4. Section 5 compares the experimental DSC curves for the enthalpy recovery behavior of various annealed PSs with the curves predicted from their DMA spectra. We give concluding remarks in Section 6.

2. Theory

In the case of the TNM model, the Boltzmann superposition principle is applied to determine the response to thermal histories. Consider first the fictive temperature T_f response to a sequence of T-jumps ΔT_j applied at time t_j. Then, the T_f response at any time t becomes the sum of the recovery response $\Lambda_T(t)$ given by Equation (2) for each T-jump:

$$T_f(t) = T_0 + \Delta T_1 \Lambda_T(t - t_1) + \Delta T_2 \Lambda_T(t - t_2) + \Delta T_3 \Lambda_T(t - t_3) + \cdots \quad (8)$$

The Riemann integral form of Equation (8) corresponds to Equation (6).

An analogous derivation can be made for the response of strain to a sequence of stress steps on the basis of a corresponding principle; thus, each stress step makes an independent contribution to the final strain in which incremental stresses $\Delta \sigma_1, \Delta \sigma_2, \Delta \sigma_3$, etc., are added at times t_1, t_2, t_3, etc., respectively. The total strain $\varepsilon(t)$ is obtained by the addition of all the contributions:

$$\varepsilon(t) = \Delta \sigma_1 D(t - t_1) + \Delta \sigma_2 D(t - t_2) + \Delta \sigma_3 D(t - t_3) + \cdots \quad (9)$$

where $D(t) = \varepsilon(t)/\sigma$ is the tensile compliance as a function of time. It should be noted here that the compliance $D(t)$ is independent of applied stress. As a result, the formulation of the Boltzmann superposition principle for a multistep creep of linear viscoelastic solids generalizes the summation of Equation (9) to be the Riemann integral form.

$$\varepsilon(t) = \int_{-\infty}^{t} dt' \left(\frac{d\sigma}{dt}\right)_{t=t'} D(t - t') \quad (10)$$

Here, we assume that a stress history initiates at $t = 0$. In the general case where there exists a distribution in retardation time τ, the compliance is then given by:

$$D(t - t') = \int_{-\infty}^{\infty} d\ln \tau L(\ln \tau) \{1 - \exp[-(t - t')/\tau']\} \quad (11)$$

215

where τ' is the retardation time at t', and $L(\ln \tau)$ is the retardation time spectrum. Here, we assume that the enthalpy recovery behavior has a common structural relaxation mode with linear viscoelastic behavior. It follows that the recovery function $\Lambda_T(t)$ in enthalpy corresponds to the retardation function in creep strain. Consequently, we have a phenomenological equation for expressing the evolution of fictive temperature as the Riemann integral form:

$$T_f(t) = T_0 + \int_0^t dt' \left(\frac{dT}{dt}\right)_{t=t'} \int_{-\infty}^{\infty} d\ln\tau L^N(\ln\tau)\{1 - \exp[-(t-t')/\tau']\} \quad (12)$$

where $L^N(\ln \tau)$ is a normalized retardation spectrum defined by:

$$L^N(\ln \tau) = L(\ln\tau) / \int_{-\infty}^{\infty} d\ln\tau L(\ln\tau)$$

Thus, the enthalpy recovery curves can be obtained only from the viscoelastic data without any fitting parameters.

3. Experimental

3.1. Sample Preparation

Pellets of atactic polystyrene (PS) with a molecular weight $M_w = 28 \times 10^4$ and $M_n = 11 \times 10^4$ were comp-molded at 230 °C under 40 MPa to films about 0.2 mm thick for DSC and DMA experiments. Then, the comp-molded samples were cooled to 25 °C at around -40 °C/min. The cooled specimens were annealed for 1, 5, 10, and 24 h in a vacuum oven at 80 °C for DMA measurements.

3.2. DSC Measurements

The DSC measurements were performed using a Perkin Elmer Diamond DSC with a controlled cooling accessory (Intracooler 2P) under nitrogen gas flux (flow rate of about 4.5 psi and 20.0 mL/min). The temperature and heat capacity data were calibrated with standard indium and sapphire, respectively. The samples of about 3 mg, cut out from the slow-cooled films, were sealed in aluminum pans. Figure 2 shows the thermal history for the measurements of the dependence of enthalpy relaxation on the annealing time. The sample was kept at 230 °C for 10 min and cooled to an annealing temperature of 80 °C ($= T_g - 20$ °C) at -40 °C/min, kept for annealing time of 1, 5, 10, and 24 h, further cooled to 50 °C, and then heated to 230 °C at 20 °C/min for monitoring of enthalpy recovery.

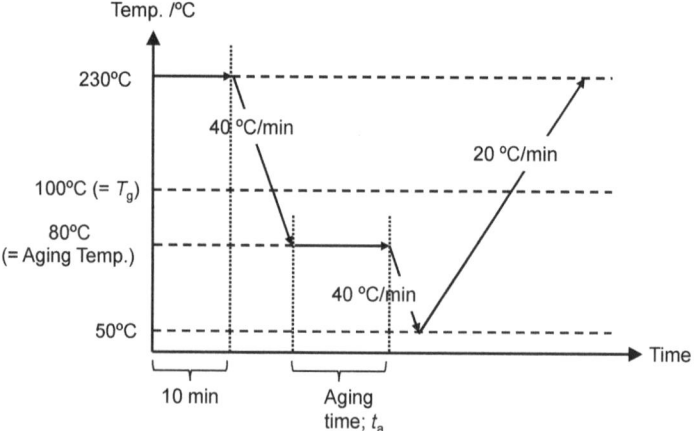

Figure 2. Schematic illustration of temperature scanning in aging experiment.

3.3. DMA Measurements

The temperature dependence of storage and loss moduli of the film specimens were examined in the temperature range from 50 °C to 150 °C at a heating rate of 2 °C/min under fixed frequencies of 10, 30, 100, and 200 Hz under a nitrogen gas atmosphere using a dynamic mechanical analyzer (UBM DVE-V4). The rectangular sample specimens with 3 mm width, 0.2 mm thickness, and 20 mm length were cut out from the slow-cooled and annealed films. The grip-to-grip distance was 10 mm, and the dynamic amplitude was set to be 1 µm for the dynamic mechanical analysis (DMA) measurements.

In addition, the dynamic mechanical data over a wide range of frequencies were measured. The dynamic mechanical spectrum in the glass transition region was measured by the elongation mode in the temperature range from 70 °C to 110 °C and the frequency range from 0.5 to 300 Hz using DVE-V4 using the rectangular sample specimens with 3 mm width, 0.2 mm thickness, and 20 mm length. The viscoelastic properties from the glass transition region to the flow region were measured using the strain-controlled rotary rheometer; the measurements were made using a strain-controlled rotational rheometer (UBM G300NT) with a parallel plate of 25 mm diameter and a gap distance of 0.5 mm. A temperature range of 110 to 230 °C, a frequency range of 0.01 to 20 Hz, and a displacement of 0.5° were applied for the measurements. It should be noted that the displacement was set as 0.1° because the sample showed a glassy state. The shear modulus G values obtained directly by the rheometer were converted to the tensile modulus E values using $E = 3G$.

3.4. Light Scattering Measurements

To quantitatively examine the effects of annealing on heterogeneous molecular aggregation, we carried out small-angle light scattering (SALS) measurement with a light scattering photometer (IST Planning SALS-100S) using a He–Ne laser with stabilized power supply (NF EC1000S). The detector used was a photomultiplier (H10721-20) that can rotate horizontally around a film specimen to scan scattering angles from 30° to 60°. The transmitted intensity was amplified by IST photosensor and recorded by a transient computer memory every 0.2° as a function of angle.

The well-collimated incident beam was polarized in any direction in a plane perpendicular to the incident beam by using a polarization rotator and monochromatized to have a wavelength of 632.8 ± 2.4 nm using a band-pass filter. An analyzer was placed between the film specimen and the photomultiplier whose polarization direction was also rotatable in a plane perpendicular to the incident beam, thus, achieving any combination of polarized scatterings, such as V_V (vertical polarizer, vertical analyzer) and H_V (vertical polarizer, horizontal analyzer) polarizations.

4. Results

The set of DSC second run curves after an isothermal annealing stage at 80 °C was examined. All of the curves measured after annealing showed an overshoot peak at a temperature higher than T_g. The main feature of these experiments was that the area and the position of the overshoot increased with increasing annealing time. From the original DSC second run curves, it was possible to evaluate the normalized specific heat $C_p^N(T)$ from the specific enthalpy lines corresponding to the liquid and glassy states as:

$$C_p^N(T) = \frac{\bar{C}_p - \bar{C}_p^{glass}}{\bar{C}_p^{liquid} - \bar{C}_p^{glass}} \tag{13}$$

The limiting fictive temperature T_f^0 [27] was determined as the extrapolated intersection of the pre-transition (or glass region) and post-transition (or liquid region) in enthalpy units, as indicated by point (f) of Figure 1. The method for determining T_f^0 of PS annealed for 24 h at 80 °C is exemplified in Figure S1. The fictive temperature T_f, as shown in Figure 3, was obtained by adding T_f^0 to the integration $C_p^N(T)$ with time (or temperature) using Equation (7). In addition, the normalized enthalpy loss ΔH^N could be calculated from $C_p^N(T)$ data as a function of the annealing time. It is confirmed in Figure 4 that the enthalpy recovery proceeded toward an equilibrium value as the annealing time increased.

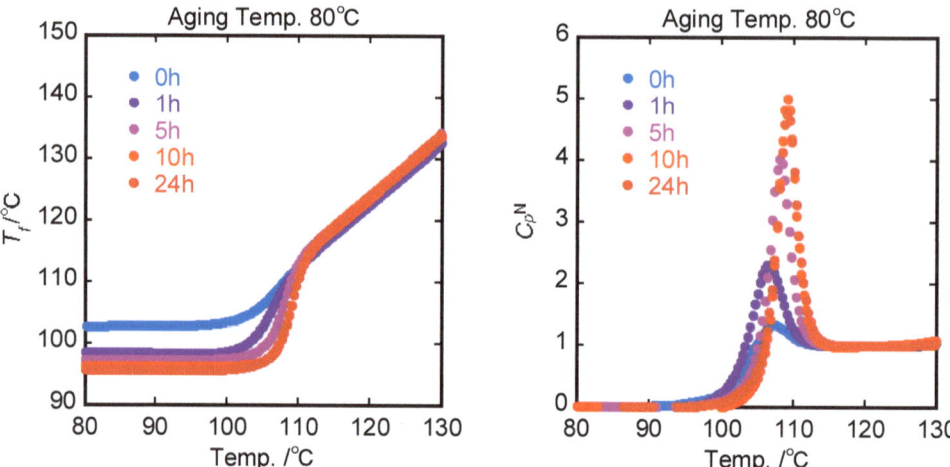

Figure 3. Annealing time dependence of fictive temperature and normalized heat capacity curves at an annealing temperature of 80 °C.

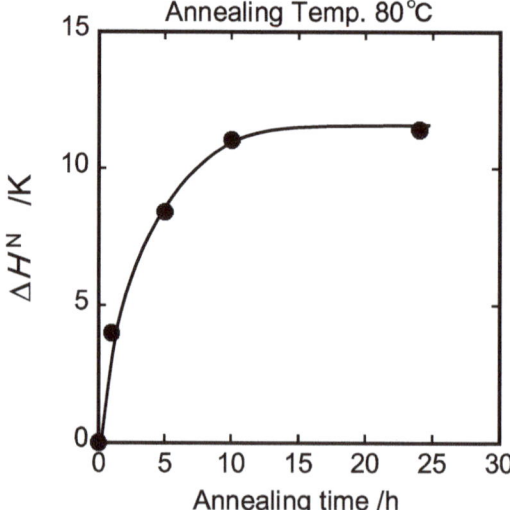

Figure 4. Annealing time dependence of the normalized change of enthalpy relaxation.

Temperature dependences of dynamic storage modulus E' and loss modulus E'' measured at 10 Hz are shown in Figure 5. We observed a shoulder peak (sub-T_g) at some lower temperatures of primary peak (T_g). It should be noted here that the primary T_g relaxation peak was almost insensitive to the annealing time, whereas the magnitude of the sub-T_g shoulder changed with the annealing time. The E'' values for other frequencies plotted against the inverse of temperature are exemplified for PS annealed for 1 h in Figure 6.

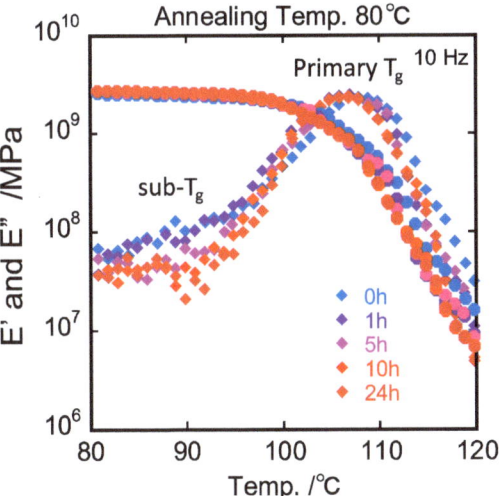

Figure 5. Annealing time dependence of dynamic mechanical spectra measured at 10 Hz for PS annealed at 80 °C.

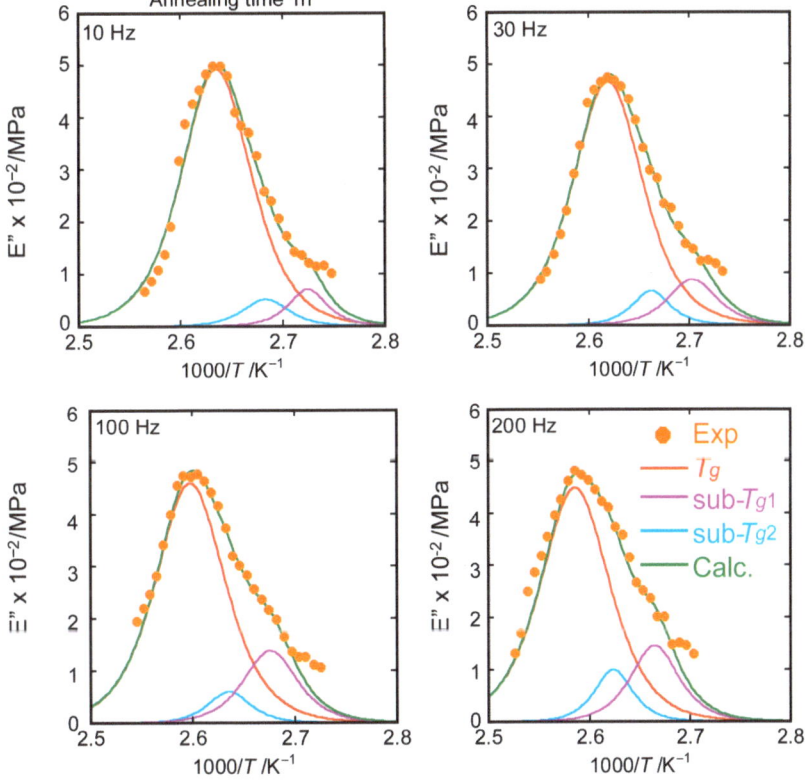

Figure 6. Curve fitting for dynamic mechanical spectra of loss modulus E'' around glass relaxation region at 1, 30, 100, and 200 Hz for PS annealed for 1 h at 80 °C.

The appearance of similar sub-T_g peaks in various polymers on the endothermic curves below T_g was reported in several articles [16,37–40]. At longer annealing time or aging, sub-T_g endotherm peaks approach limiting values as the annealed or aged glass approaches the equilibrium state, and the sub-T_g peaks evolve into overshoots [8]. In addition, such sub-T_g endotherms and mechanical spectra have been observed in glassy polymers subjected to varied pressure and mechanical stress [41–44]. The secondary or subglass relaxation has been considered to reflect the transformation of the non-thermal perturbation such as the non-homogeneous (non-equilibrium) structure resulting from quenching from T_g into the homogeneous (equilibrium) structure. Wypych et al. [45] found from low-frequency Raman scattering data of a glassy PMMA that the aging makes the nano-scale structural aggregation more homogeneous, leading to an equilibrium structure with lower energy.

To characterize the position, the shape, and the relaxation strength of sub-T_g, we analyzed the E'' peak in the glass transition region by peak decomposition using the Cole–Cole function given by [46]:

$$E''(\omega, \tau) = (E_U - E_R) \frac{(\omega\tau)^\beta \sin(\beta\pi/2)}{1 + 2(\omega\tau)^\beta \cos(\beta\pi/2) + (\omega\tau)^{2\beta}} \quad (14)$$

where E_U and E_R represent the limiting values of the storage modulus at infinite and zero frequency, respectively. In this function, the broadening of E'' is reflected by a decrease in β to values lower than unity. One example of the fitting to the dynamic glass transition is given in Figure 6. All other fitting data are shown in Figure S2 of the supporting information.

The decomposition was performed on the dynamic loss modulus as a function of T^{-1} and possibly into one primary T_g and two individual peaks, sub-T_{g1} and sub-T_{g2}, appearing in the range from 90 °C to 100 °C. The frequency dependences of the primary peak and two sub-T_g peaks give their activation enthalpy values under the Arrhenius activation law. All the data complied with the Arrhenius relation and gave rise to the apparent activation enthalpies shown in Figure 7. It is interesting to note that the activation enthalpies of both sub-T_gs exponentially decayed with increasing annealing time, whereas that of primary T_g was almost constant (about 500 kJ/mol). The Arrhenius plots are shown in Figure S3.

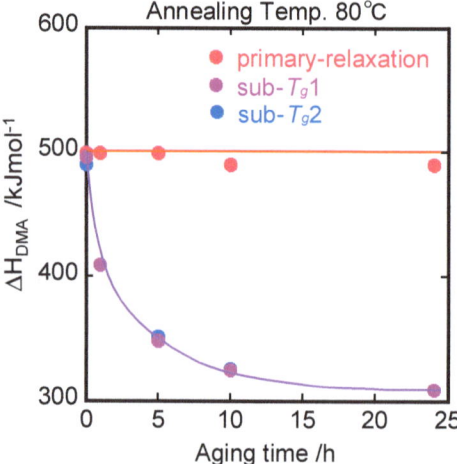

Figure 7. Annealing time dependence of the activation energies of T_g and sub-T_g.

The master curve of frequency dependences with the reference temperature (T_r) of 100 °C is exemplified for pristine PS in Figure 8. The thermos-rheological simplicity was confirmed to be established for the PS sample for a wide range of frequency. The starting numeral of E' and E'' denotes the measurement temperature in °C.

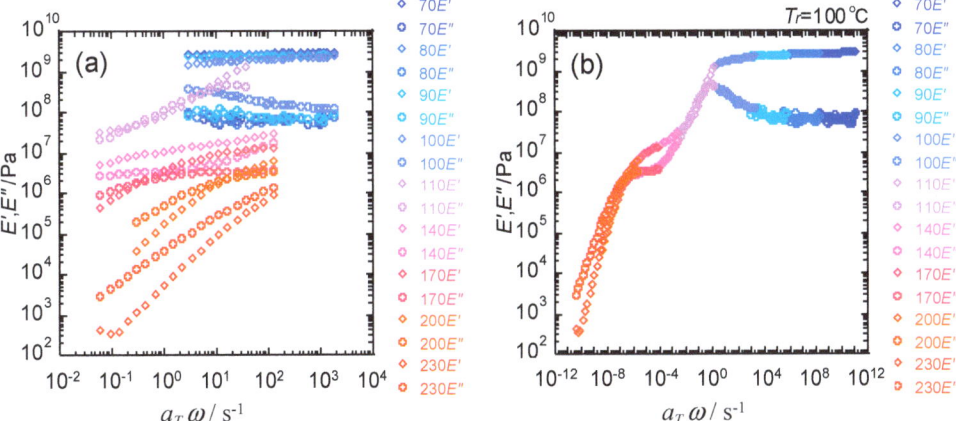

Figure 8. Frequency dependencies of (**a**) storage elastic modulus (E') and loss elastic modulus (E'') and (**b**) their master curves at $T_r = 100$ °C.

V_V and H_V light scattering intensities are plotted against the scattering angle θ for annealed samples in Figure 9. The H_V scattering intensities are almost independent of angle θ, but the V_V scattering intensity negatively depends on θ, and its magnitude decreases with increasing annealing time, suggesting the existence of a non-thermal ordered structure in the glassy PS matrix. In addition, the size is postulated to be enough to cause observable angle dependence over the accessible range of the light scattering vector $|s| = 2\sin(\theta/2)$. The V_V intensity composed of the angle-dependent "excess" $V_V^{ex}(\theta)$ term superimposed on angle-independent "background" V_V^0 and H_V scattering terms is as follows [47]:

$$V_V = V_V^{ex}(\theta) + V_V^0 + \frac{3}{4}H_V \quad (15)$$

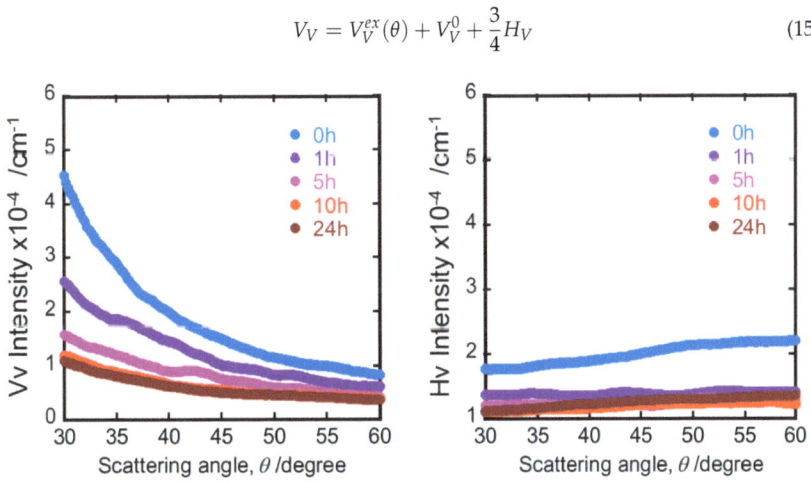

Figure 9. Scattering angle depends of V_V and H_V scattering intensities.

The angle-dependent $V_V^{ex}(\theta)$ resulting from thermally induced density fluctuations is given by [48]:

$$V_V^{ex}(\theta) = \frac{\pi^2}{9n^4\lambda^4}\left(n^2-1\right)^2\left(n^2+2\right)^2 kT\beta_T \qquad (16)$$

where n is the refractive index of PS, k is the Boltzmann constant, and β_T is the isothermal compressibility at T_g. The excess angle-dependent $V_V^{ex}(\theta)$ results from the light scattering between dipoles induced by intersegmental correlations due to the ordering of segmental units. We can obtain the theoretical $V_V^{ex}(\theta)$ based on the exponential correlation function introduced by Debye–Bueche [49] as:

$$V_V^{ex}(\theta) = \frac{8\pi^2 \langle \eta^2 \rangle a^3}{\varepsilon_0^2 \lambda^4 (1 + \nu^2 a^2 s^2)} \qquad (17)$$

where $\nu = 2\pi/\lambda$, ε_0 is the permittivity of glassy PS, $\langle \eta^2 \rangle$ is the mean square of density fluctuations, and a is the correlation length. Debye–Bueche plots for a series of PSs with different annealing times give the correlation length a value (see Figure 10).

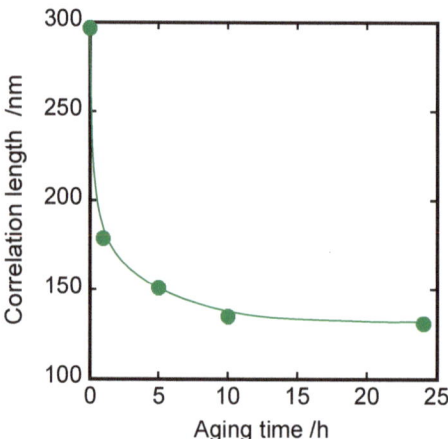

Figure 10. Annealing time dependence of correlation length.

The a value approximately decayed from 300 nm to 100 nm. This is consistent with the results of light and X-ray scattering data by Fujiki et al. [50], where a glassy PS was shown to possess heterogeneities with 100–400 nm. Consequently, the non-thermal heterogeneous structure evolved in initial PS glass is considered to reduce toward a homogeneous structure as the annealing time increases. This is related to the structural relaxation during the annealing process. In addition, the structural relaxation is suggested to be associated not with the primary T_g but with the sub-T_g relaxation by comparison to Figure 7.

5. Discussion

The dynamic moduli $E'(\omega)$ and $E''(\omega)$ data in the glass transition zone from each master curve can be converted to the retardation spectrum $L(\ln \tau)$ using Shwarzl–Staverman two-order approximation [51,52] given by:

$$L(1/\omega) = \frac{2}{\pi}\left[D''(\omega) - \frac{d^2 D''(\omega)}{d(\ln \omega)^2}\right]_{\tau=1/\omega} \qquad (18)$$

and

$$D''(\omega) = \frac{E''(\omega)}{E'(\omega)^2 + E''(\omega)^2} \qquad (19)$$

Substituting the experimental $L^N(\ln \tau)$ data into Equation (12) provides us with the theoretical normalized $C_p^N(T)$ curves based on dynamic viscoelasticity. The experimental $C_p^N(T)$ curves measured by DSC (Figure 4) are shown in Figure 11, together with the theoretical ones determined from DMA spectra.

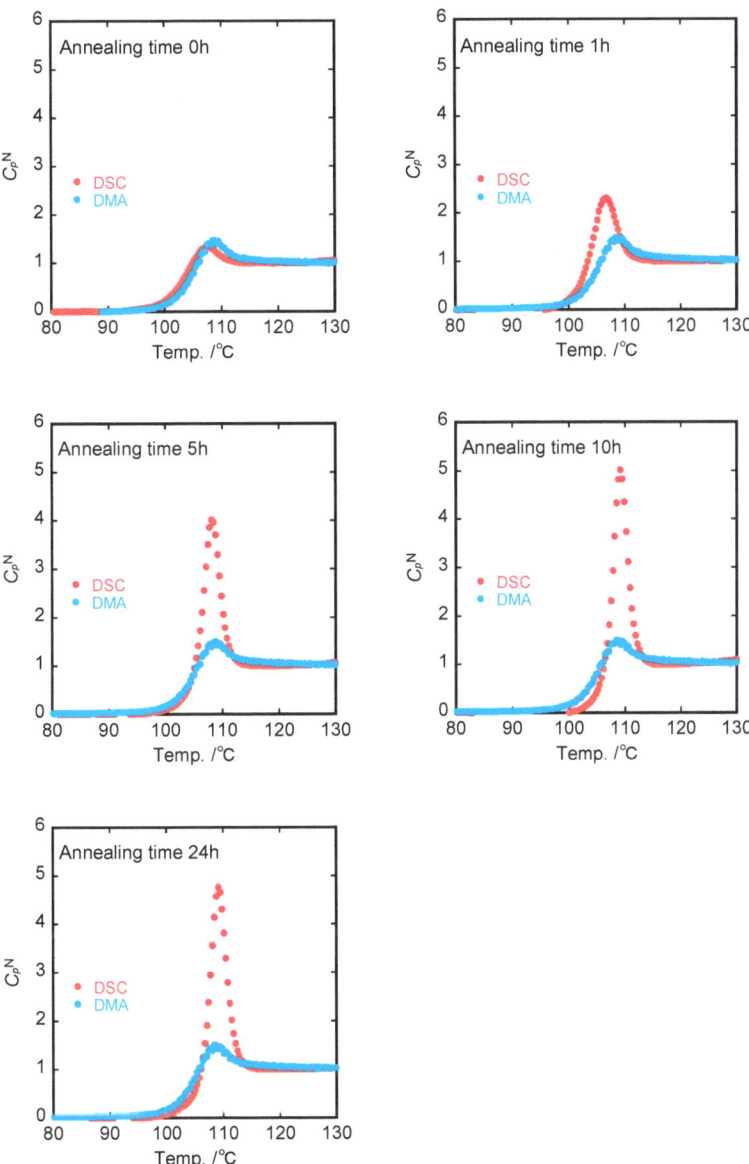

Figure 11. Comparison of normalized heat capacity evaluated from DSC and DMA (primary T_g).

The peak intensity of $C_p^N(T)$ by DSC measurements increases with increasing annealing time, whereas $C_p^N(T)$ estimated from DMA is independent of the annealing time which is because the activation energy of the primary E'' peak (T_g) is almost constant. As a result, although the $C_p^N(T)$ curve by DSC was in accordance with the $C_p^N(T)$ curve calculated from DMA spectra only for the non-annealed PS sample, the difference between experimental and calculated heights of $C_p^N(T)$ peaks was enhanced more and more with the annealing time. This suggests that the annealing effect of sub-T_g on the E'' spectra dominantly contributes to the overshoot in the specific heat in the heating process.

When the original TNM phenomenological model quantitatively fits experimental endothermic data, the non-exponentiality plays a central role in describing the experimental specific heat curves [16]. The non-exponentiality is likely related to the sub-T_g process which broadens the T_g retardation peak. The evidence of the existence of this distribution was demonstrated by Kovac et al. [38,53] with memory effects which showed that the response of the glass is a function of its overall previous thermal and mechanical history. The total loss of the activation enthalpy ΔH_{subT_g} of sub-T_g is given by the sum of activation energy loss of both sub-T_{g1} and sub-T_{g2}.

$$\Delta H_{subT_g} = |H_1(t) - H_1(0)|_{subT_{g1}} + |H_2(t) - H_2(0)|_{subT_{g2}} \qquad (20)$$

The relationship between enthalpy loss ΔH_{DSC} of DSC endotherms and ΔH_{subT_g} is compared in Figure 12a.

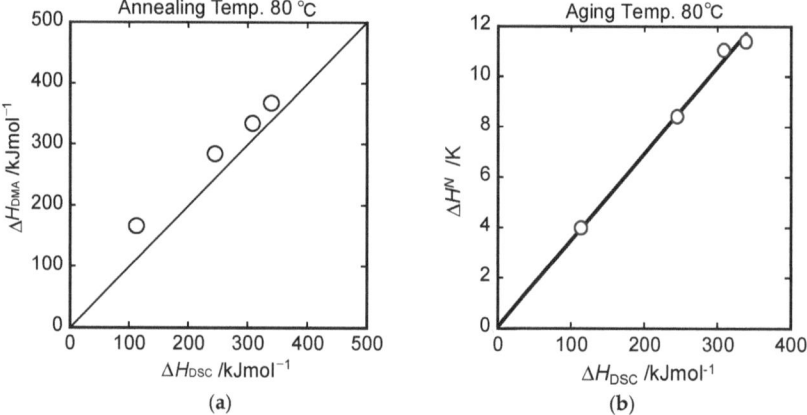

Figure 12. (a) Comparison of (a) enthalpy loss evaluated from DSC and activation energy of sub-T_g and (b) enthalpy loss ΔH_{DSC}/kJmol^{-1} and normalized enthalpy loss ΔH^N/K.

The unit of ΔH_{DSC} was converted from J/g to J/mol using the M_n value. The normalized values of enthalpy loss of DSC endotherms were previously plotted against the annealing time, as shown in Figure 4. The linear relation between the enthalpy loss ΔH_{DSC} and ΔH_{subT_g} with slope equal to unity passed almost through the original point, indicating that the enthalpy relaxation process was well correlated to the sub-T_g relaxation process on the dynamic mechanical spectra. Furthermore, the normalized enthalpy loss ΔH^N, as shown in Figure 4, was found to be proportional to the enthalpy loss ΔH_{DSC} of DSC (see Figure 12b). The conversion factor obtained from its slope was 29.0 kJ/mol K. Consequently, the activation energy changes in dynamic mechanical spectra can be converted to the normalized enthalpy changes in DSC curves.

It follows that sub-T_g endotherms are superimposed on the primary glass transition heat capacity 'background', observed at the same cooling and heating rates [54]. In addition, we assumed that the specific heat curve is expressed by a Gaussian distribution curve using normalized values of enthalpy loss of sub-T_g. The calculated $C_p^N(T)$ curves involving the primary and sub-T_g relaxation process on dynamic mechanical spectra are shown in Figure 13, together with the experimental normalized specific heat curves. It is confirmed that the curves calculated by the present model were almost close to the experimental ones for any annealed sample. This implies that the enthalpy recovery under the annealing process is dominated by the structural relaxation of the non-equilibrium aggregation state toward the equilibrium state.

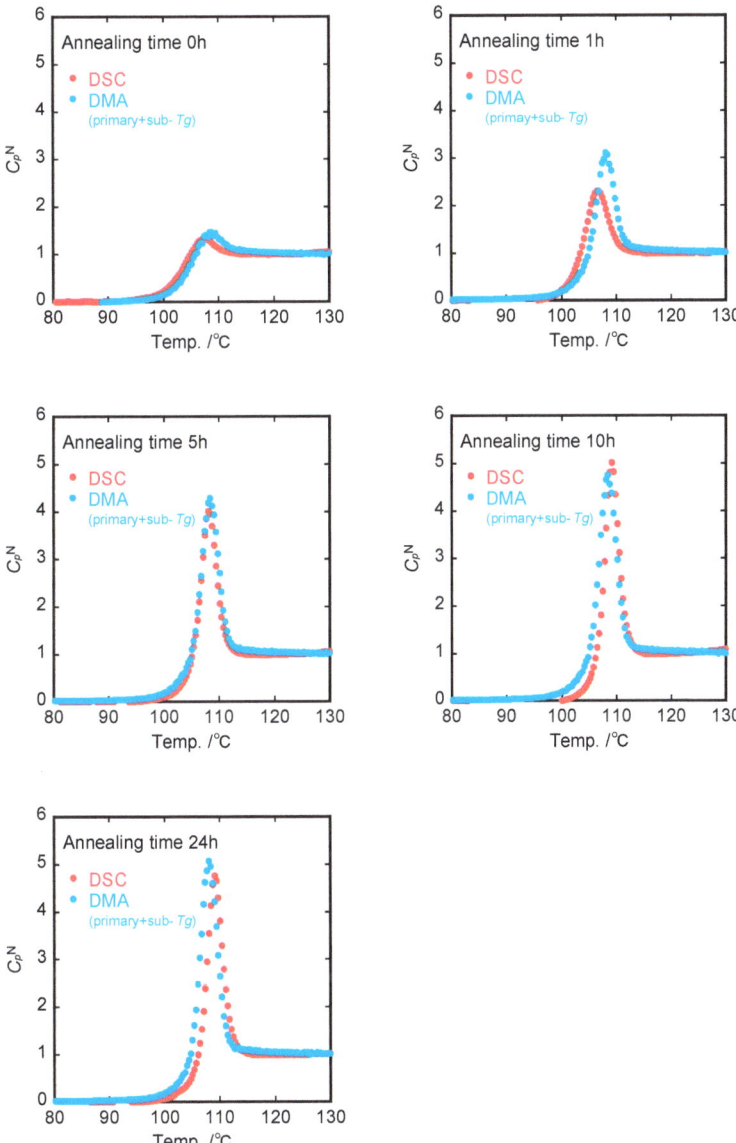

Figure 13. Comparison of normalized heat capacity evaluated from DSC and DMA (both sub-T_g and primary T_g).

So far, a correlation between enthalpy relaxation and mechanical behavior [33,54–57], such as viscoelastic, stress relaxation, creep compliance, and stress–strain properties, has been widely studied for various glassy polymers. Although the absolute values of enthalpy relaxation time, which are obtained from a KWW equation, and mechanical characteristic time differ, there exists an empirical correlation between calorimetric and viscoelastic changes during structural relaxation under annealing processes. Sasaki et al. [58] found that enthalpy relaxation and dielectric relaxation evaluated by the same stretched exponential function (KWW equation) are almost identical. However, there exists no consensus with regards to aging or annealing effects for enthalpy or density and mechanical properties despite their same molecular origin. Pye et al. [59] and Cangialosi et al. [60] demonstrated that glass transition for PS occurs on multiple time scales and occurs via a two-step process. In this work, the use of the viscoelastic retardation spectra instead of the KWW function made it possible to relate the enthalpy relaxation behavior to the viscoelastic behavior for annealed PS samples. The annealing effects for sub-T_g processes dominated the enthalpy recovery process for annealed PS systems, implying that the segmental motions locally frozen in the non-equilibrium conditions are released under annealing or aging. Thus, it is likely that enthalpy recovering is dominated mainly by the secondary dispersion, i.e., sub-T_g, while the overall mechanical properties are dominated mainly by the primary dispersion, i.e., T_g. Consequently, the differences in C_p^N between DSC and DMA, including only the primary T_g relaxation, become larger as the annealing time increases, as shown in Figure 11.

According to the light scattering results (see Figure 10), the non-thermal heterogeneous structure with a size of a few 100 nm evolved in initial PS glass, which reduced toward homogeneous structure as the annealing time increased. This structural relaxation process governs the sub-T_g dispersion. The fact that the relaxation of the inhomogeneous structure toward the equilibrium is caused by the annealing process lowers the relaxation strength of sub-T_g with the annealing time. Hence, the reduction of the activation energy of sub-T_g due to the annealing reflects the relaxation of the non-equilibrium structure, resulting in a decrease in the activation enthalpy of sub-T_g corresponding to the enthalpy recovery in endotherms. The molecular understanding of the enthalpy recovery phenomena is illustrated in Figure 14.

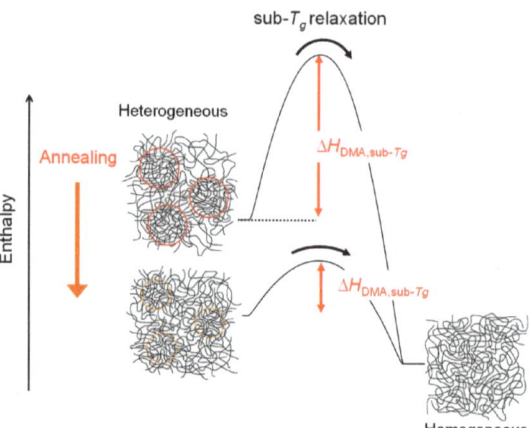

Figure 14. Schematic illustration of sub-T_g relaxation process.

6. Conclusions

In this work, the annealing time dependence of enthalpy recovery of polystyrene (PS) annealed at 80 °C was studied by the modified Tool–Narayanaswamy–Moynihan (TNM) model with creep retardation parameters determined from dynamic mechanical spectra. In this work, we accepted the viewpoint that the enthalpy recovery behavior of glassy PS has a common structural relaxation mode with linear viscoelastic behavior. As a consequence, the retardation spectrum evaluated from the dynamic mechanical spectra around the primary T_g peak was used as the recovery function of the endotherms overshoot of specific heat. In addition, we identified shoulder peaks at some lower temperatures than the T_g peak, and the sub-T_g was suggested to be ascribed to the non-thermal perturbed structures observed from light scattering measurements. It was found that the activation enthalpies of primary T_g are independent of annealing time, whereas that of sub-T_g exponentially decreased with increasing annealing time. The activation process of sub-T_g is dominated by the structural relaxation of the perturbed structures toward the homogeneous state reduction. Consequently, the enthalpy recovery of annealed PS can be quantitatively described by the use of kinetic data of the sub-T_g relaxation process in addition to the retardation spectra of the primary T_g in place of the use of some fitting parameters of the TNM model. This implies that the enthalpy recovery under annealing process is dominated by the structural relaxation of the non-equilibrium aggregation state toward the equilibrium state.

The use of the characteristic time distribution obtained via the viscoelastic retardation spectra, including sub-T_g, allowed us to predict the enthalpy recovery behavior of annealed PS systems. We believe that this makes it possible to give a unified description of the thermodynamical behavior of glassy polymer solids. Moreover, the present work takes an important step towards a structural understanding of the changes in mechanical properties during physical aging encountered in practice.

Supplementary Materials: The following supporting information can be downloaded at https://www.mdpi.com/article/10.3390/polym15173590/s1, Figure S1: Enthalpy change estimated from heat flow of DSC data for PS annealed for 24 h at 80 °C. The solid line corresponds to the extension (equilibrium) line of the liquid state; Figure S2: Curve fitting for dynamic mechanical spectra of loss modulus E'' around glass relaxation region at 1, 30, 100, and 200 Hz, for PS annealed for (a) 0, (b) 5 h, (c) 10 h, and (d) 24 h at 80 °C; Figure S3: Arrhenius plots for primary and dual sub-glass transitions for PS annealed at 80 °C.

Author Contributions: Conceptualization, theoretical consideration, K.-h.N.; methodology, K.-h.N. and K.I.; formal analysis, K.I.; data curation, K.I. and A.I.; writing—original draft preparation, K.-h.N. and K.I.; writing—review and editing, A.I. All authors have read and agreed to the published version of the manuscript.

Funding: This research received no external funding.

Institutional Review Board Statement: Not applicable.

Data Availability Statement: The data that support the findings of the study are available on request from the corresponding author.

Acknowledgments: The authors would like to deeply thank M. Yamaguchi (Japan Advanced Institute of Science and Technology) for experimental help with rheometer measurements.

Conflicts of Interest: The authors declare no conflict of interest.

References

1. Struik, L.C.E. *Physical Aging in Amorphous Polymers and Other Materials*; Van Krevelen, D.W., Ed.; Elsevier Scientific Pub. Co.: Amsterdam, NY, USA, 1977.
2. Struik, L.C.E. Physical aging in plastics and other glassy materials. *Polym. Eng. Sci.* **1977**, *17*, 165–173. [CrossRef]
3. Grigoriadi, K.; Westrik, J.; Vogiatzis, G.G.; van Breemen, L.C.A.; Anderson, P.D.; Hutter, M. Physical Ageing of Polystyrene: Does Tacticity Play a Role? *Macromolecules* **2019**, *52*, 5948–5954. [CrossRef] [PubMed]
4. Moynihan, C.T.; Macedo, P.B.; Montrose, C.J.; Montrose, C.J.; Gupta, P.K.; DeBolt, M.A.; Dill, J.F.; Dom, B.E.; Drake, P.W.; Easteal, A.J.; et al. Structural Relaxation in Vitreous Materials. *Ann. N. Y. Acad. Sci.* **1976**, *279*, 15–35. [CrossRef]
5. Angell, C.A.; Ngai, K.L.; McKenna, G.B.; McMillan, P.F.; Martin, S.W. Relaxation in glassforming liquids and amorphous solids. *J. Appl. Phys.* **2000**, *88*, 3113–3157. [CrossRef]
6. Weeks, E.R.; Crocker, J.C.; Levitt, A.C.; Schofield, A.; Weitz, D.A. Three-dimensional direct imaging of structural relaxation near the colloidal glass transition. *Science* **2000**, *287*, 627–631. [CrossRef]
7. Torre, R.; Bartolini, P.; Righini, R. Structural relaxation in supercooled water by time-resolved spectroscopy. *Nature* **2004**, *428*, 296–299. [CrossRef] [PubMed]
8. Hodge, I.M. Enthalpy relaxation and recovery in amorphous materials. *J. Non-Cryst. Solids* **1994**, *169*, 211–266. [CrossRef]
9. Sakatusji, W.; Konishi, T.; Miyamoto, Y. Enthalpy relaxation and annealing effect in polystyrene. *Phys. Rev. E* **2013**, *88*, 012605. [CrossRef] [PubMed]
10. Zheng, Q.; Zhang, Y.; Montazerian, M.; Gulbiten, O.; Mauro, J.C.; Zanotto, E.D.; Yue, Y. Understanding Glass through Differential Scanning Calorimetry. *Chem. Rev.* **2019**, *119*, 7848–7939. [CrossRef]

11. Kohlrausch, R. Theorie des elektrischen Rückstandes in der Leidener Flasche. *Ann. Phys. Chem.* **1854**, *167*, 179–214. [CrossRef]
12. Williams, G.; Watts, D.C. Non-symmetrical dielectric relaxation behaviour arising from a simple empirical decay function. *Trans. Faraday Soc.* **1970**, *66*, 80–85. [CrossRef]
13. Kohlrausch, F. Ueber die elastische Nachwirkung bei der Torsion. *Ann. Phys. Chem.* **1863**, *195*, 337–368. [CrossRef]
14. Tool, A.Q.; Eicitlin, C.G. Variations Caused in the Heating Curves of Glass by Heat Treatment1. *J. Am. Ceram. Soc.* **1931**, *14*, 276–308. [CrossRef]
15. Tool, A.Q. Relation between Inelastic Deformability and Thermal Expansion of Glass in Its Annealing Range. *J. Am. Ceram. Soc.* **1946**, *29*, 240–253. [CrossRef]
16. Hodge, I.M.; Berens, A.R. Effects of annealing and prior history on enthalpy relaxation in glassy polymers. 2. Mathematical modelingrelaxation in glassy polymers. 2. Mathematical modeling. *Macromolecules* **1982**, *15*, 762–770. [CrossRef]
17. Hodge, I.M.; Huvard, G.S. Effects of annealing and prior history on enthalpy relaxation in glassy polymers. 3. Experimental and modeling studies of polystyrene. *Macromolecules* **1983**, *16*, 371–375. [CrossRef]
18. Scherer, G.W. Use of the Adam-Gibbs Equation in the Analysis of Structural Relaxation. *J. Am. Ceram. Soc.* **1984**, *67*, 504–511. [CrossRef]
19. Tribone, J.J.; O'Reilly, J.M.; Greener, J. Analysis of enthalpy relaxation in poly(methyl methacrylate): Effects of tacticity, deuteration, and thermal history. *Macromolecules* **1986**, *19*, 1732–1739. [CrossRef]
20. Privalko, V.P.; Demchenko, S.S.; Lipatov, Y.S. Structure-dependent enthalpy relaxation at the glass transition of polystyrenes. *Macromolecules* **1986**, *19*, 901–904. [CrossRef]
21. O'Reilly, J.M.; Hodge, I.M. Effects of heating rate on enthalpy recovery in polystyrene. *J. Non-Cryst. Solids* **1991**, *131–133*, 451–456. [CrossRef]
22. Gomez Ribelles, J.L.; Monleon Pradas, M. Structural Relaxation of Glass-Forming Polymers Based on an Equation for Configurational Entropy. 1. DSC Experiments on Polycarbonate. *Macromolecules* **1995**, *28*, 5867–5877. [CrossRef]
23. Brunacci, A.; Cowie, J.M.G.; Ferguson, R.; Gómez Ribelles, J.L.; Vidaurre Garayo, A. Structural Relaxation in Polystyrene and Some Polystyrene Derivatives. *Macromolecules* **1996**, *29*, 7976–7988. [CrossRef]
24. Hutchinson, J.M.; Kumar, P. Enthalpy relaxation in polyvinyl acetate. *Thermochim. Acta* **2002**, *391*, 197–217. [CrossRef]
25. Boucher, V.M.; Cangialosi, D.; Alegría, A.; Colmenero, J. Enthalpy Recovery of Glassy Polymers: Dramatic Deviations from the Extrapolated Liquidlike Behavior. *Macromolecules* **2011**, *44*, 8333–8342. [CrossRef]
26. Narayanaswamy, O.S. A Model of Structural Relaxation in Glass. *J. Am. Ceram. Soc.* **1971**, *54*, 491–498. [CrossRef]
27. Moynihan, C.T.; Easteal, A.J.; Bolt, M.A.; Tucker, J. Dependence of the Fictive Temperature of Glass on Cooling Rate. *J. Am. Ceram. Soc.* **1976**, *59*, 12–16. [CrossRef]
28. Robertson, C.G.; Santangelo, P.G.; Roland, C.M. Comparison of glass formation kinetics and segmental relaxation in polymers. *J. Non-Cryst. Solids* **2000**, *275*, 153–159. [CrossRef]
29. Mijović, J.; Ho, T. Proposed correlation between enthalpic and viscoelastic measurements of structural relaxation in glassy polymers. *Polymer* **1993**, *34*, 3865–3869. [CrossRef]
30. Yoshida, H. Relationship between enthalpy relaxation and dynamic mechanical relaxation of engineering plastics. *Thermochim. Acta* **1995**, *266*, 119–127. [CrossRef]
31. Simon, S.L.; Plazek, D.J.; Sobieski, J.W.; McGregor, E.T. Physical aging of a polyetherimide: Volume recovery and its comparison to creep and enthalpy measurements. *J. Polym. Sci. Part B Polym. Phys.* **1997**, *35*, 929–936. [CrossRef]
32. Cowie, J.M.G. An ion conducted tour through some polymer electrolytes. *Polym. Int.* **1998**, *47*, 20–27. [CrossRef]
33. Robertson, C.G.; Monat, J.E.; Wilkes, G.L. Physical aging of an amorphous polyimide: Enthalpy relaxation and mechanical property changes. *J. Polym. Sci. Part B Polym. Phys.* **1999**, *37*, 1931–1946. [CrossRef]
34. Alves, N.M.; Mano, J.F.; Balaguer, E.; Meseguer Dueñas, J.M.; Gómez Ribelles, J.L. Glass transition and structural relaxation in semi-crystalline poly(ethylene terephthalate): A DSC study. *Polymer* **2002**, *43*, 4111–4122. [CrossRef]
35. Dionísio, M.; Alves, N.M.; Mano, J.F. Molecular dynamics in polymeric systems. *e-Polymers* **2004**, *4*, 044. [CrossRef]
36. Pan, P.; Zhu, B.; Inoue, Y. Enthalpy Relaxation and Embrittlement of Poly(l-lactide) during Physical Aging. *Macromolecules* **2007**, *40*, 9664–9671. [CrossRef]
37. Illers, V.K.H. Einfluß der thermischen vorgeschichte auf die eigenschaften von polyvinylchlorid. *Die Makromol. Chem.* **1969**, *127*, 1–33. [CrossRef]
38. Kovacs, A.J.; Aklonis, J.J.; Hutchinson, J.M.; Ramos, A.R. Isobaric volume and enthalpy recovery of glasses. II. A transparent multiparameter theory. *J. Polym. Sci. Polym. Phys. Ed.* **1979**, *17*, 1097–1162. [CrossRef]
39. Tant, M.R.; Wilkes, G.L. An overview of the nonequilibrium behavior of polymer glasses. *Polym. Eng. Sci.* **1981**, *21*, 874–895. [CrossRef]
40. Greaves, G.N.; Sen, S. Inorganic glasses, glass-forming liquids and amorphizing solids. *Adv. Phys.* **2007**, *56*, 1–166. [CrossRef]
41. Weitz, A.; Wunderlich, B. Thermal analysis and dilatometry of glasses formed under elevated pressure. *J. Polym. Sci. Polym. Phys. Ed.* **1974**, *12*, 2473–2491. [CrossRef]
42. Prest, W.M.; Roberts, F.J. Enthalpy Recovery in Pressure-Vitrified and Mechanically Stressed Polymeric Glasses. *Ann. N. Y. Acad. Sci.* **1981**, *371*, 67–86. [CrossRef]
43. Tanaka, A.; Jono, Y.; Wakabayashi, N.; Nitta, K.-H.; Onogi, S. A Novel Mechanical Dispersion and Molecular Ordering in Styrene-Butadiene-Styrene Triblock Copolymer Films. *Polym. J.* **1991**, *23*, 1091–1097. [CrossRef]

44. Descamps, M.; Aumelas, A.; Desprez, S.; Willart, J.F. The amorphous state of pharmaceuticals obtained or transformed by milling: Sub-Tg features and rejuvenation. *J. Non-Cryst. Solids* **2015**, *407*, 72–80. [CrossRef]
45. Wypych, A.; Duval, E.; Boiteux, G.; Ulanski, J.; David, L.; Mermet, A. Effect of physical aging on nano- and macroscopic properties of poly(methyl methacrylate) glass. *Polymer* **2005**, *46*, 12523–12531. [CrossRef]
46. Ferri, D.; Laus, M. Activation Law Parameters of Viscoelastic Subglass Relaxations from Dynamic-Mechanical Moduli Measurements as a Function of Temperature. *Macromolecules* **1997**, *30*, 6007–6010. [CrossRef]
47. Claiborne, C.; Crist, B. Light and X-ray scattering by polystyrene glasses. *Colloid Polym. Sci.* **1979**, *257*, 457–466. [CrossRef]
48. Einstein, A. Theorie der Opaleszenz von homogenen Flüssigkeiten und Flüssigkeitsgemischen in der Nähe des kritischen Zustandes. *Ann. Phys.* **1910**, *33*, 1275–1295. [CrossRef]
49. Debye, P.; Bueche, A.M. Scattering by an Inhomogeneous Solid. *J. Appl. Phys.* **1949**, *20*, 518–525. [CrossRef]
50. Fujiki, M.; Oikawa, S. Light Scattering Study on the Structure of Pure Polystyrene. *Polym. J.* **1984**, *16*, 609–617. [CrossRef]
51. Schwarzl, F.; Staverman, A.J. Higher approximation methods for the relaxation spectrum from static and dynamic measurements of visco-elastic materials. *Appl. Sci. Res.* **1953**, *4*, 127–141. [CrossRef]
52. Tschoegl, N.W. *The Phenomenological Theory of Linear Viscoelastic Behavior an Introduction*; Springer: Berlin/Heidelberg, Germany, 1989.
53. Hutchinson, J.M.; Kovacs, A.J. A simple phenomenological approach to the thermal behavior of glasses during uniform heating or cooling. *J. Polym. Sci. Polym. Phys. Ed.* **1976**, *14*, 1575–1590. [CrossRef]
54. Chen, H.S.; Wang, T.T. Sub-subTgstructural relaxation in glassy polymers. *J. Appl. Phys.* **1981**, *52*, 5898–5902. [CrossRef]
55. Hutchinson, J.M.; Smith, S.; Horne, B.; Gourlay, G.M. Physical Aging of Polycarbonate: Enthalpy Relaxation, Creep Response, and Yielding Behavior. *Macromolecules* **1999**, *32*, 5046–5061. [CrossRef]
56. Adolf, D.B.; Chambers, R.S.; Stavig, M.E.; Kawaguchi, S.T. Critical Tractions for Initiating Adhesion Failure at Interfaces in Encapsulated Components. *J. Adhes.* **2006**, *82*, 63–92. [CrossRef]
57. Ohara, A.; Kodama, H. Correlation between enthalpy relaxation and mechanical response on physical aging of polycarbonate in relation to the effect of molecular weight on ductile-brittle transition. *Polymer* **2019**, *181*, 121720. [CrossRef]
58. Sasaki, K.; Takatsuka, M.; Kita, R.; Shinyashiki, N.; Yagihara, S. Enthalpy and Dielectric Relaxation of Poly(vinyl methyl ether). *Macromolecules* **2018**, *51*, 5806–5811. [CrossRef]
59. Pye, J.E.; Roth, C.B. Two simultaneous mechanisms causing glass transition temperature reductions in high molecular weight freestanding polymer films as measured by transmission ellipsometry. *Phys. Rev. Lett.* **2011**, *107*, 235701. [CrossRef] [PubMed]
60. Cangialosi, D.; Boucher, V.M.; Alegria, A.; Colmenero, J. Direct evidence of two equilibration mechanisms in glassy polymers. *Phys. Rev. Lett.* **2013**, *111*, 095701. [CrossRef]

Disclaimer/Publisher's Note: The statements, opinions and data contained in all publications are solely those of the individual author(s) and contributor(s) and not of MDPI and/or the editor(s). MDPI and/or the editor(s) disclaim responsibility for any injury to people or property resulting from any ideas, methods, instructions or products referred to in the content.

Article

Analyzing Homogeneity of Highly Viscous Polymer Suspensions in Change Can Mixers

Michael Roland Larsen [1,2,*], Erik Tomas Holmen Olofsson [1,3] and Jon Spangenberg [1]

1. Department of Civil and Mechanical Engineering, Technical University of Denmark, 2800 Kongens Lyngby, Denmark; ethol@dtu.dk (E.T.H.O.); josp@dtu.dk (J.S.)
2. Dansac A/S, 3480 Fredensborg, Denmark
3. Haldor Topsoe A/S, 2800 Kongens Lyngby, Denmark
* Correspondence: mirola@dtu.dk

Citation: Larsen, M.R.; Holmen Olofsson, E.T.; Spangenberg, J. Analyzing Homogeneity of Highly Viscous Polymer Suspensions in Change Can Mixers. *Polymers* **2024**, *16*, 2675. https://doi.org/10.3390/polym16182675

Academic Editor: Maria Graça Rasteiro

Received: 2 October 2023
Revised: 24 August 2024
Accepted: 8 September 2024
Published: 23 September 2024

Copyright: © 2024 by the authors. Licensee MDPI, Basel, Switzerland. This article is an open access article distributed under the terms and conditions of the Creative Commons Attribution (CC BY) license (https://creativecommons.org/licenses/by/4.0/).

Abstract: The mixing of highly viscous non-Newtonian suspensions is a critical process in various industrial applications. This computational fluid dynamics (CFD) study presents an in-depth analysis of non-isothermal mixing performance in change can mixers. The aim of the study was to identify parameters that significantly influence both distributive and dispersive mixing in these mixers, which are essential for optimizing industrial mixing processes. The study employed a numerical design of experiments (DOE) approach to identify the parameters that most significantly influence both distributive and dispersive mixing, as measured by the Kramer mixing index (M_{Kramer}) and the Ica Manas-Zloczower mixing index ($\overline{\lambda_{MZ}}$). The investigated parameters included mixing time, number of arms, arm size ratio, revolutions per minute (RPM), z-axis rotation, z-axis movement, and initial and mixing temperatures. The methodology involved employing the bootstrap forest algorithm for predicting the mixing indices, achieving an R^2 of 0.949 for M_{Kramer} and an R^2 of 0.836 for $\overline{\lambda_{MZ}}$. The results indicate that the z-axis rotation has the greatest impact on both distributive and dispersive mixing. An increased number of arms negatively impacted λ_{MZ}, but had a small positive effect on M_{Kramer}. Surprisingly, in this study, neither the initial temperature of the material nor the mixing temperature significantly impacted the mixing performance. These findings highlight the relative importance of operational parameters over traditional temperature factors and provide a new perspective on mixing science.

Keywords: mixing; suspensions; homogeneity; computational fluid dynamics; parameter investigation

1. Introduction

Mixing fluids and powders is a fundamental process that has been present for centuries in various industries. Achieving uniformity in such mixtures, also known as homogeneity, is critical to ensure consistent quality and performance in food, concrete, medical device, and catalyst industries.

Homogeneity in particle-based slurries can be quantified using a distributive and dispersive mixing index [1]. Distributive mixing refers to the uniform component distribution within the matrix, ensuring a consistent spatial arrangement. In parallel, dispersive mixing aims to intentionally reduce the size of cohesive components, particularly clusters of solid particles, to enhance homogeneity.

Mixing indices have been developed to quantitatively evaluate the level of mixing in fluid-powder mixtures. Among them, the distributive mixing index is one of the earliest, originating with Lacey in the mid-19th century, who proposed using standard deviation from a concentration to estimate mixing quality [2,3]. This idea was later expanded upon by Kramer [4] and Ashton-Valentin [5]. Despite originating in a different era, the Kramer mixing index remains a vital tool in contemporary research; it has, e.g., recently been used to in cooperation with computational fluid dynamics (CFD) [6]. With the advent of CFD, newer approaches, such as the cluster distribution index, have been developed to evaluate the distribution of Lagrangian particles in a mixer [7,8]. Another particle-based approach is the "Scale of Segregation" developed by P.V. Danckwerts [9] and used in CFD simulations by Connelly and Kokini in 2007 [10].

The dispersive mixing index quantifies a mixer's ability to break up agglomerates, which is a critical factor in ensuring product quality. The efficiency of agglomerate breakage for highly viscous fluids depends on the type of flow, with elongation flow being the most effective [11]. In 1992, Ica Manas-Zloczower introduced the widely used mixing index λ_{MZ}, which describes the relationship between simple vorticity flow $(\lambda_{MZ} \approx 0)$ and simple elongation flow $(\lambda_{MZ} = 1)$ [12]. This mixing index has been extensively used in the literature [13,14].

Viscosity reduction via heat is a well-established phenomenon in the mixing literature [15,16]. This phenomenon is important for increasing the Reynolds number and, thereby, the distributive mixing efficiency [17]. Fluid friction, also known as viscous dissipation, tends to be neglected despite being a notable factor when dealing with highly viscous fluids [17]. The impact of viscous dissipation should, therefore, not be ignored, as demonstrated in past studies [15,18].

In the context of evaluating numerous parameters, machine learning techniques can be highly effective. Bootstrap Forest, a commercialized adaptation based on the random forest algorithm, excels in evaluating multiple parameters and their interactions. For instance, Duan and Takemi [19] demonstrated its utility in predicting urban surface roughness aerodynamic parameters, showing its robustness in handling complex, nonlinear relationships. Similarly, Ganesh et al. [20] applied random forest regression to accurately estimate fluid flow characteristics in curved pipes, highlighting its capability to reduce computational costs while maintaining high accuracy in predictions. Both studies utilized the random forest algorithm to achieve significant insights.

This study aims to expand the knowledge of dispersive and distributive mixing of highly viscous non-Newtonian suspensions in "change can mixers". In this regard, a CFD model was developed and exploited to perform a numerical design of experiments (DOE). The evaluation of the homogeneity of the suspension was performed by Kramer's and Ica Manas-Zloczower's mixing indexes, and seven process parameters were included in the DOE that the Taguchi and definitive screening design techniques inspired. The results of the DOE were analyzed by a bootstrap forest algorithm to evaluate the significance of each process parameter, leading to the development of a predictive model for the mixing indices. Following this introduction, the paper is organized as follows: Section 2 elucidates the methodology setup, encompassing comprehensive information about the numerical model and DOE. Section 3 presents and analyzes the study's findings, emphasizing the influence of the investigated parameters on the mixing performance. Ultimately, Section 4 summarizes the conclusions drawn from the study, underscoring the implications and significance of the findings.

2. Methodology

2.1. Material and Mixer Information

The suspension that the CFD model simulated has been characterized experimentally in terms of density, heat capacity and conductivity, as well as rheology. The latter is presented in the next subsection. The suspension was a combination of a highly viscous polymer fluid and a powder blend consisting of both colloidal and microscale particles. The specific fluid and powder are protected intellectual property and can, therefore, not be disclosed. In this regard, it is important to note that the primary focus is on the mixing process dynamics of the highly viscous polymer suspension rather than on a detailed analysis of the material composition itself. Table 1 shows the physical data of the mixed material. The density was measured with a "Sartorius YDK03 Density Kit for Analytical Balances" using the Archimedes principle. The heat capacity and conductivity were measured on a "TCi-3-A" from C-Therm Technologies Ltd. (Fredericton, NB, Canada) using the modified transient plane source (MTPS) method.

Table 1. Physical data for the mixed fluid.

Name of the Property	Values	Unit
Density	1133	$\frac{kg}{m^3}$
Heat capacity	1389	$\frac{J}{kg \cdot °C}$
Heat conductivity	$k(T \leq 23\,°C) = 0.52$ $k(23\,°C < T < 71\,°C) = -1.88 \cdot 10^{-3}\,T + 0.56$ $k(T \geq 71\,°C) = 0.43$	$\frac{J}{m \cdot K}$

Details about the change can mixer that CFD model simulates are illustrated in Figure 1. Heat is introduced into the system from the wall and bottom of the mixer. The top does not apply heat due to the physical setup where the rotor arms are mounted. The change can mixer operates under vacuum conditions to prevent air from becoming trapped in the mixing material. During operation, the mixer's top lid rotates clockwise and spins around the z-axis fulcrum, while the arms themselves also rotate around their own z-axes. For clarity throughout this study, the top lid's z-axis revolutions per minute is denoted as RPM and the rotation around the arm's z-axis per minute is referred to as the z-axis rotation.

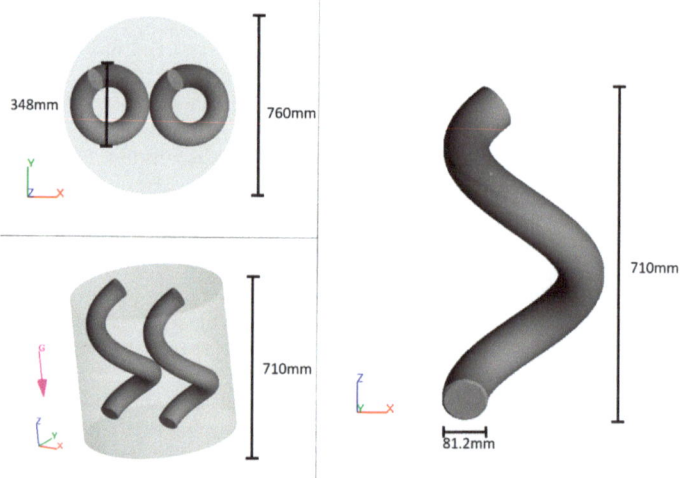

Figure 1. Illustration of the change can mixer (**left**) and mixing arm (**right**).

2.2. Numerical Model

The CFD model simulates the mixing of a suspension in an industrial-scale mixer using the commercial software, FLOW-3D (2022R1).

This software, known for its accurate fluid dynamics simulations, was chosen for its advanced TruVOF technique, which is particularly effective in modeling the free surface flows that are applied for these simulations. The software has previously been used to simulate other processes, such as 3D printing [21] and concrete casting [22]. Figure 2 shows an example of the simulation setup at $t = 0$. The red domain represents the powder component, while the blue/turquoise part represents the suspending fluid. The rotation begins at $t > 0$ and remains constant throughout the simulation, and a no-slip boundary condition is applied on all solid surfaces, including the rotational arms. The computational domain is meshed with a uniform grid consisting of ~134,000 cells. The number of cells was determined through a mesh sensitivity analysis. To ensure the reliability of our simulations, all models achieved satisfactory residual and convergence levels, indicating that the results can be trusted for higher accuracy.

Figure 2. CFD model at $t = 0$. The blue/turquoise mass represents the fluid, while the red mass represents the powder.

The simulation is computed as a transient non-isothermal flow because the viscosity is temperature-dependent. The material model substance is considered incompressible, hence, the density ρ is assumed constant. Thus, the flow is computed by considering mass, momentum, and energy conservation, as shown below in (1)–(4).

$$\nabla \cdot \mathbf{v} = 0 \tag{1}$$

$$\rho \frac{D\mathbf{v}}{Dt} = -\nabla p - [\nabla \cdot \boldsymbol{\tau}] + \rho \mathbf{G} \tag{2}$$

$$\rho C_p \frac{DT}{Dt} = -(\nabla \cdot \mathbf{q}) - (\boldsymbol{\tau} : \nabla \mathbf{v}) \tag{3}$$

$$\mathbf{q} = -k\nabla T \tag{4}$$

where the pressure, velocity vector, and gravitational vector are denoted as p, v, and $\mathbf{G}$, respectively. The gravitational force is defined as $(0, 0, -9.82) \frac{m}{s^2}$. The heat flux is represented as q, the specific heat capacity by C_p and k is the thermal conductivity. Additionally, the deviatoric stress tensor $\boldsymbol{\tau}$, is calculated as seen in (5) and (6).

$$\boldsymbol{\tau} = 2\mu(\dot{\gamma}, T)\mathbf{D} \tag{5}$$

$$\mathbf{D} = \frac{1}{2}(\nabla v) + \nabla v^T \tag{6}$$

where $\mathbf{D}$ is the deformation rate tensor and $\dot{\gamma}$ is the shear rate calculated from the trace of $\mathbf{D}$, $\dot{\gamma} = \sqrt{2\mathrm{tr}(\mathbf{D}^2)}$. To compute the equations from above, the software uses the finite volume method to discretize the governing equations. The free surface is calculated using the volume of fluid (VOF) technique [23]. The viscous stress and pressure are both solved implicitly, while the equation of advection is solved explicitly with 2nd-order accuracy.

The material was simulated with the properties described in the previous section (excl. rheology). The simulated material exhibits a viscoplastic behavior and is modeled using the Herschel–Bulkley model, with a slight modification to account for temperature dependency. Equation (7) describes the modified Herschel–Bulkley viscosity model, where τ_0 is the yield stress, k_{HB} is the consistency index, n is the flow index, and μ_0 as the initial viscosity. The temperature-dependent energy function, E in Equation (8) is represented by the empirically adjusted constants a and c. The reference temperature and fluid temperature are denoted by T^{ref} and T, respectively. In this context, the term "reference temperature" (T^{ref}) is an empirical fitting factor utilized within the Flow3D model.

$$(\dot{\gamma}, T) = \begin{cases} \mu_{max}, & for \ \dot{\gamma} \leq \dot{\gamma}_c \\ \mu_0 E(T)^n \dot{\gamma}^{1-n} + \frac{\tau_0}{\dot{\gamma}}, & for \ \dot{\gamma} > \dot{\gamma}_c \end{cases} \quad (7)$$

$$E(T) = \exp\left(a\left(\frac{T^{ref}}{T} - c\right)\right) \quad (8)$$

The applied values used for this model are shown in Table 2. The model was fitted to rheological measurements at 80 °C and 100 °C with shear rates between 1 s^{-1} and 215.7 s^{-1}. The rheological experiments were performed on a Dynisco LCR-7001 capillary rheometer with a 1 mm × 20 mm capillary unit. In Figure 3, the comparison between experimental data and model is seen. The viscosity of a suspension changes depending on the powder volume fraction, as demonstrated by Einstein in 1906 [24]. While this interaction affects the mixing (especially in the initial stages), this study will assume that the interaction has a negligible effect on the numerical results and is therefore not considered.

Table 2. Viscosity data of the simulated fluid.

Symbol	Value	Unit
τ_0	2	Pa
n	0.415	-
μ_0	29,307	Pa·s
a	0.853	-
c	0.00175	°C
T^{ref}	82.557	°C
$\dot{\gamma}_c$	0.14	s^{-1}

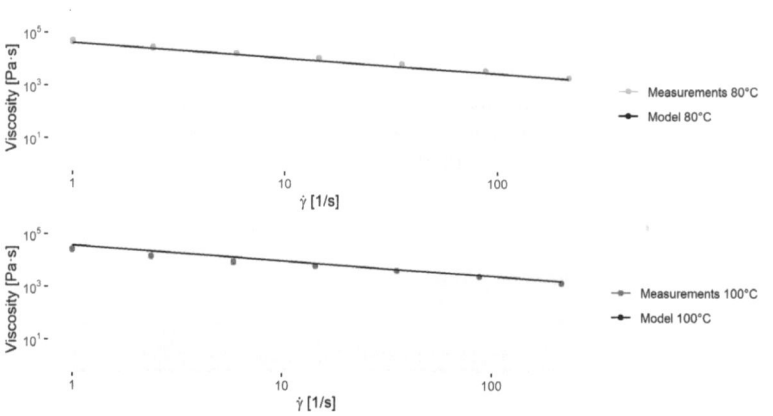

Figure 3. The model plotted to measurements at 80 and 100 °C.

The evaluation of the mixing was done using both a dispersive and distributive mixing indices. The Kramer index [4] was employed to measure the distributive mixing. The index requires an artificial dimensionless scalar concentration, c_i, with zero diffusion to ensure that the only factor propelling the mixing is the rotation itself. The powder and fluid were assigned a concentration of 1 and 0, respectively. This concentration did not affect the physical properties such as viscosity and density. The mean of the concentration $\bar{c}$, was 0.5. The dimensionless concentration $\hat{c}_i$ was calculated using Equation (9) from the scalar concentration. The dimensionless variance and M_{Kramer} can be found through Equations (9)–(11):

$$\hat{c}_i = \frac{c_i - \bar{c}}{\bar{c}} \tag{9}$$

$$S^2 = \frac{1}{(N_f - 1)} \sum_{i=1}^{N_f} (\hat{c}_i)^2 \tag{10}$$

$$M_{Kramer} = \frac{\sigma_0 - S}{\sigma_0 - \sigma_r} \tag{11}$$

S^2 is the dimensionless variance, and N_f represents the number of elements that contain fluid. The σ_0 and σ_r are defined as $\sigma_0 = (P \cdot (1 - P))^{\frac{1}{2}}$ and $\sigma_r = \frac{\sigma_0}{N_f}$, respectively, where P is the average concentration of the powder, which is 0.5. A large M_{Kramer} value indicates a homogeneous mixture.

The evaluation of the dispersive mixing was performed with the Manas–Zlacower mixing index λ_{MZ} shown in Equation (12). It was calculated by the shear rate and the vorticity, ω. The index indicates which type of flow is currently present. It is desirable to get as many elements as close to 1 as possible as it is proven that elongation flow breaks up agglomerates faster [25,26].

$$\lambda_{MZ} = \frac{|\dot{\gamma}|}{|\dot{\gamma}| + |\omega|} \tag{12}$$

2.3. Design of Experiments

DOE is a systematic method used to investigate process output by varying multiple parameters. This study used a definitive screening design to cover pre-set parameters and minimize simulation time. This choice was driven by the unique ability of Definitive Screening Design to provide a comprehensive yet efficient exploration of the process space, effectively assessing main effects and factor interactions. Furthermore, it is suitable for handling nonlinear effects [27]. Previous studies using DOE yielded satisfactory results [28,29]. The definitive screening design initially suggested 18 simulations. Some of the simulations were very computationally heavy, so while waiting for these simulations to finish, an additional 17 simulations were executed in order to cover even more of the parameter space. This brings the total number of simulations to 35, each identified by a unique simulation number (Sim. No.). The data treatment of the DOE was performed in SAS JMP®, employing the standard least squares method. The process parameters and their variation are presented below:

The initial and mixing temperature is known to reduce the viscosity of fluids [17,30], which will increase Reynolds number and potentially improve mass transfer during mixing [17]. The initial temperature is modified only for the suspending fluid, as the powder requires a constant initial temperature of 33 °C. The initial temperature of the fluid and the mixing temperature are varied between 20 °C and 80 °C. The RPM value and the z-axis rotation also affect the suspension mixing, as demonstrated by [31]. In this study, the RPM value was varied between 3 and 30, while the z-axis rotation was altered between 0 and 157.5. This study also explored the influence of moving the rotational arm 10 cm up and down along the z-axis. This z-axis movement had a frequency that was varied between 0 and 6 per minute. Finally, the number of arms in the mixer were varied from 1 to 3, and the size ratio was varied between 2/3 and 1.1. The size ratio represents that one arm in the mixer has a diameter that is given by the size ratio multiplied by the original diameter of 81.2 mm, cf. Figure 1. In Table 3, the numerical DOE is shown. Note that the mixing time is 300 s in all scenarios.

Table 3. Numerical DOE.

Sim. No.	Init. Temp	Mix Temp	RPM	Z-Axis Rotation	Z-Axis Movement	Arms	Size Ratio
1	50	80	22.5	157.5	6	3	1
2	50	20	3	0	0	2	0.667
3	80	50	30	0	0	2	1
4	20	50	3	21	6	3	0.667
5	80	20	16.5	0	6	3	1
6	20	80	16.5	115.5	0	2	0.667
7	80	80	30	150	6	2	0.667
8	20	20	3	15	0	3	1
9	80	20	3	21	6	2	0.75
10	20	80	30	0	0	3	0.75
11	80	80	3	0	3	3	0.667
12	20	20	22.5	157.5	3	2	1
13	80	80	3	21	0	2	1
14	20	20	30	0	6	3	0.667
15	80	20	22.5	157.5	0	3	0.667
16	20	80	3	0	6	2	1
17	50	50	16.5	82.5	3	2	0.75
18	50	50	16.5	82.5	3	3	0.75
19	80	20	10	10	0	2	0.75
20	80	20	10	10	0	3	0.75
21	80	20	10	30	0	2	1.1
22	80	20	10	30	0	3	1.1
23	80	20	20	20	0	2	1.1
24	80	20	20	20	0	3	1.1
25	80	20	20	60	0	2	0.75
26	80	20	20	60	0	3	0.75
27	80	20	20	20	0	1	1
28	20	80	16.5	115.5	0	1	1
29	80	20	22.5	157.5	0	1	1
30	80	20	10	30	0	1	1.1
31	80	80	3	0	3	1	0.667
32	50	50	16.5	82.5	3	1	0.75
33	80	20	10	30	0	2	0.75
34	80	20	10	50	0	2	0.75
35	20	20	3	0	0	1	1.1

2.4. Bootstrap Forest

A predictive model, specifically designed for distributive and dispersive mixing indices, was developed using the bootstrap forest algorithm in the JMP 16 Pro software, a type of random forest algorithm [32]. The algorithm is an ensemble learning method that combines multiple decision trees, $F_i(x)$, to create a more accurate and robust model. The bootstrap forest algorithm employs bootstrapping, a resampling technique that creates multiple subsets of the original dataset by repeatedly sampling with replacement. Each subset is used to train an individual decision tree within the forest, enhancing the model's robustness and accuracy. The predicted value, $\check{y}$, is obtained by averaging the predictions from all the individual trees in the forest, as seen in Equation (13).

$$\check{y} = \frac{1}{N} \sum_{i=1}^{N} F_i(x) \tag{13}$$

where N is the number of trees, and $F_i(x)$ is the prediction of the *i*th tree for the input vector row x that specifies the specific parameter found in Table 3.

The selected data to be evaluated by the algorithm was a balanced blend of systematic and random sampling. Specific time steps were selected at regular intervals and supplemented by additional points randomly selected via a random number generator. Each simulation yielded 26 data points, 11 chosen systematically and 15 chosen randomly. This method ensured computational efficiency and controlled memory usage, which was particularly important given the intensive computational requirements and memory capacity necessary for processing the 35 simulations. Note that the treatment of time-dependency in the analysis is described in Sections 3.5.1 and 3.5.2 for dispersive and distributive mixing, respectively.

The collected data was then partitioned into training and validation subsets, following a conventional 70-30 split in line with standard machine learning practice [33]. The coefficient of determination, R^2, was calculated to assess the performance of the fit of the predictive model:

$$R^2 = 1 - \frac{\sum_{i=1}^{N}(y_i - \check{y}_i)^2}{\sum_{i=1}^{N}(y_i - \bar{y})^2} \tag{14}$$

where y_i denotes a specific simulation observation and $\bar{y}$ represents the mean value of all observations for the output function; this calculation serves as a pivotal metric for assessing the adequacy of the predictive model in replicating the observed data.

Before splitting the data into training and validation, a predictor screening was carried out, which essentially is a bootstrap forest analysis without validation. This screening quantified each parameter's individual contribution. A parameter contribution threshold of 5% was set. Parameters with an average contribution below this threshold during the predictor screening were excluded, while those that made over a 5% contribution were included in the full bootstrap forest analysis. This strategy helped reduce the model's complexity and improved interpretability by eliminating less impactful variables. The 5% threshold was established based on the same criteria used in linear regression, where parameters with a p-value below 0.05 are excluded.

3. Results and Discussion

This section delineates the findings of the analysis concerning the mixing performance in change can mixers. Specifically, the velocity profile, temperature profile, dispersive mixing index, and distributive mixing index will be delved into. Given the intricacy of presenting 35 distinct simulations, the focus will be on two representative simulations for illustration purposes: Sim. No. 7 and Sim. No. 35. These two simulation were chosen as they are quite different.

Subsequently, the parameter study with all 35 simulations utilizing the bootstrap forest analysis is presented. Initially, the influence of parameters on the dispersive mixing index is scrutinized. Next, the impact of parameters on the distributive mixing index is outlined.

3.1. Velocity Profile

The velocity profiles for two representative simulations, Sim. No. 7 and 35, are depicted in Figure 4 at $t = 60$ s. The maximum velocity magnitude and velocity gradients are substantially lower in the simulation with one arm as compared to the simulation with two arms. This outcome was anticipated due to the lower RPM and z-axis rotation values utilized in Sim. No. 35. Both simulations illustrate that most flow takes place near the mixing arms, which is also expected. In addition, Sim. No. 7 displays the highest velocities around the large arm, which is due to the arms having the same angular velocity, resulting in a higher velocity for the large arm at the outer edge.

3.2. Temperature Profile

The temperature profile analysis compares the average fluid temperature for Sim. No. 7 and 35, as illustrated in Figure 5. Both simulations commence with different initial heat values and wall temperatures. Sim. No. 7 starts with an initial temperature of 80 °C, identical to the wall temperature, while Sim. No. 35 begins with an initial temperature of 20 °C, which also aligns with the wall temperature. In Sim. No. 7, the temperature surpasses the wall temperature, which is above 250 °C. This can be ascribed to viscous dissipation, a phenomenon where heat is generated due to fluid friction during the mixing process. It is important to note that temperatures reaching 250 °C or higher could potentially ruin the material being mixed. Such elevated temperatures may lead to degradation or other undesirable changes in the material properties. Conversely, in Sim. No. 35, the lower RPM and z-axis rotation values lead to minimal or no viscous dissipation. As a result, the temperature remains below 30 °C.

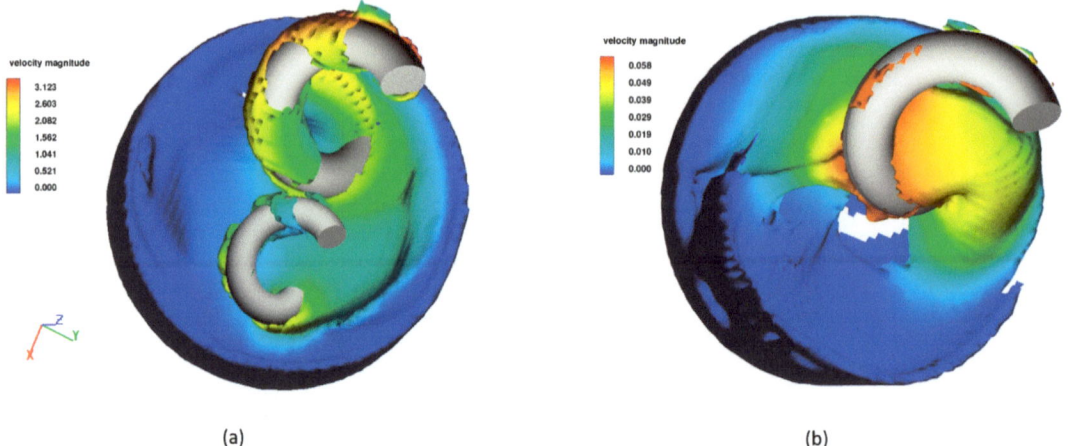

Figure 4. Velocity profiles at 60 s for (**a**) Sim. No. 7 and (**b**) Sim. No. 35. The velocity values are in m/s.

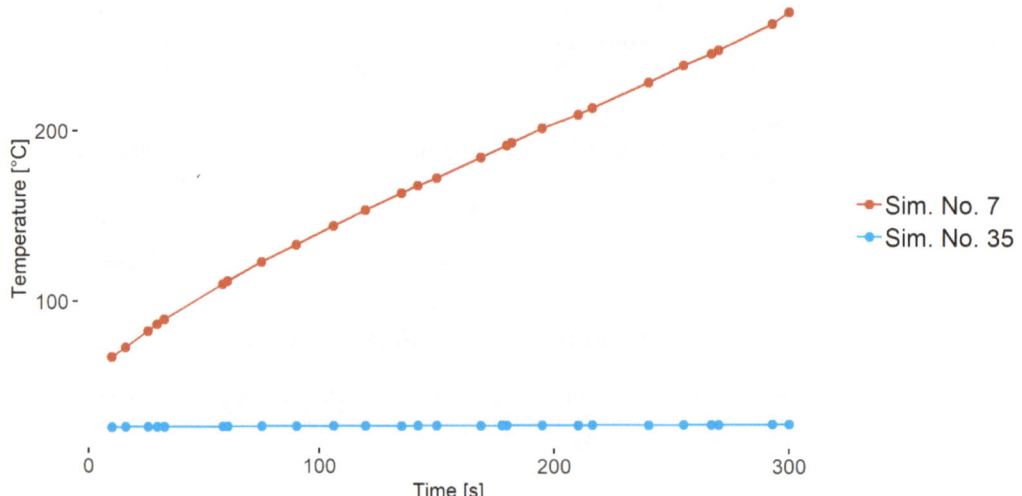

Figure 5. Average fluid temperature for Sim. No. 7 and Sim. No. 35 as a function of time.

3.3. Dispersive Mixing

Figure 6 presents the average dispersive mixing index $\overline{\lambda_{MZ}}$ for Sim. No. 7 and 35 at various time intervals. The dispersive mixing index values for Sim. No. 7 primarily ranges between 0.30 and 0.32, while for Sim. No. 35, they approximately span from 0.28 to 0.34. It is interesting to note that the mean value of the dispersive index is fairly similar for Sim. No. 7 and 35, even though they have quite different process parameters. However, that is not the case for all simulations, e.g., the mean value of Sim. No. 23 is 0.26.

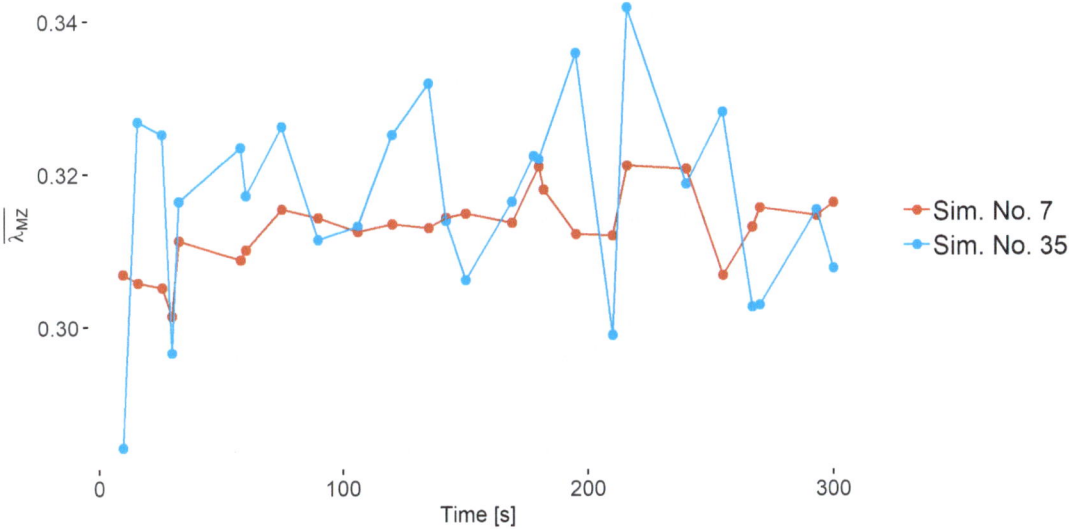

Figure 6. Dispersive mixing index for Sim. No. 7 and 35 as a function of time.

3.4. Distributive Mixing

In Figure 7, the M_{Kramer} values (i.e., the distributive mixing index) as a function of time is shown for Sim. No. 7 and 35. Both start below 0 due to the initial high standard deviation and subsequently, both simulations exhibit an increase in the M_{Kramer} values, which indicates that the mixes become more homogeneous. Sim. No. 7 undergoes a significantly faster mixing process compared to Sim. No. 35. After 300 s, Sim. No. 7 reaches around 0.75, while Sim. No. 35 only attains approximately −0.5. This observation aligns with the velocity profile analysis, where Sim. No. 7 exhibits a faster velocity compared to Sim. No. 35, and thus faster mass transfer. The results highlight that the mixing process is time-dependent but that other factors, such as the specific parameters, also play a crucial role in influencing the mixing behavior.

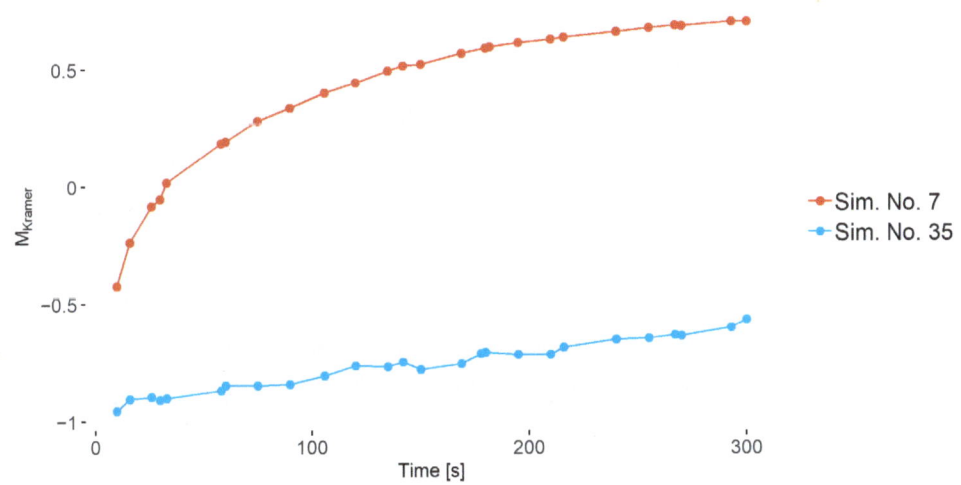

Figure 7. Distributive mixing index for Sim. No. 7 and Sim. No. 35 compared to time.

3.5. Bootstrap Forrest

3.5.1. Dispersive Modeling

Since time primarily introduced fluctuations rather than significant alterations in the modeling of the dispersive mixing index, as seen in Figure 6, it was excluded from this analysis. Nevertheless, when analyzing the dispersive mixing index, all 26 lambda values for each simulation were considered instead of using an average to better understand the estimated capabilities of the predictive model. The process parameters from Table 3 were considered in the analysis. An initial investigation was conducted using the predictor screening analysis to minimize the number of parameters. Parameters with a contribution of less than 5% were not selected for further analysis.

Figure 8 illustrates a bar chart of the predictor screening analysis, highlighting the influence of each parameter on dispersive mixing. It is evident that the z-axis rotation has the highest impact, followed by the number of arms, Size ratio, and RPM, all of which exceed the significance line. Conversely, z-axis movement, initial temperature, and mixing temperature did not exhibit significant contributions and were not considered further.

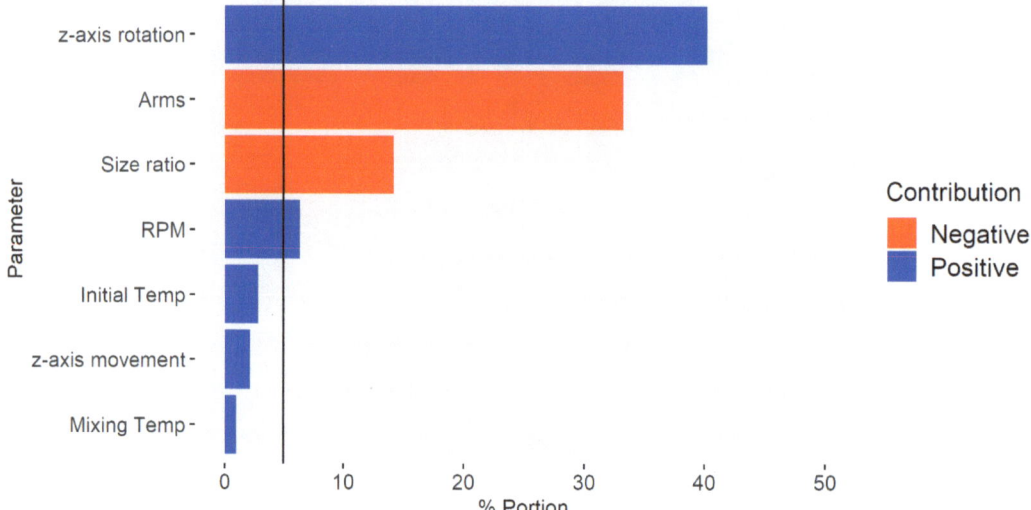

Figure 8. Predictor screening analysis of the influence of each parameter on dispersive mixing. Note the significance line at 5%.

Subsequently, the bootstrap forest algorithm was executed, focusing only on the four significant parameters, cf. Figure 9a. This second analysis aimed to evaluate these four parameters' influence and generate the predictive model. The impact sequence of the individual parameters on dispersive mixing remained unchanged, indicating that the threshold value was thoughtfully established. The R^2 value for the validation set was found to be 0.836. Figure 9b presents a comparison of predicted values and simulation outcomes, demonstrating commendable accuracy.

It is important to mention that the model's accuracy is challenged by the relatively large fluctuations observed for some of the simulations; e.g., Sim. No. 1 that had a z-axis rotation of 157.5, cf. Figure 10. This suggests that z-axis rotation positively contributes to the $\overline{\lambda_{MZ}}$ value but also introduces increased uncertainty. Another intriguing finding was the influence of the number of arms; fewer arms resulted in a higher dispersion. As observed in Figure 11 when comparing the average vorticity of Sim. No. 21, 22, and 27, all having identical parameters except for the number of arms, it is evident that the simulations with fewer arms exhibit lower vorticity, which explains the improvement in $\overline{\lambda_{MZ}}$.

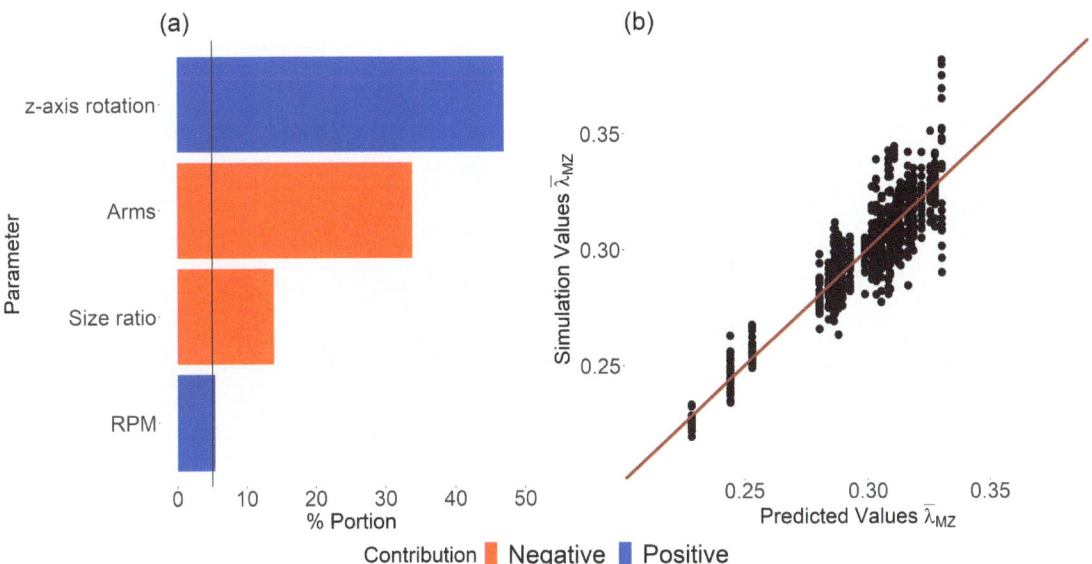

Figure 9. (a) Influence of each parameter on dispersive mixing, and (b) comparison between predicted and simulation values. Note the significance line at 5% in (a).

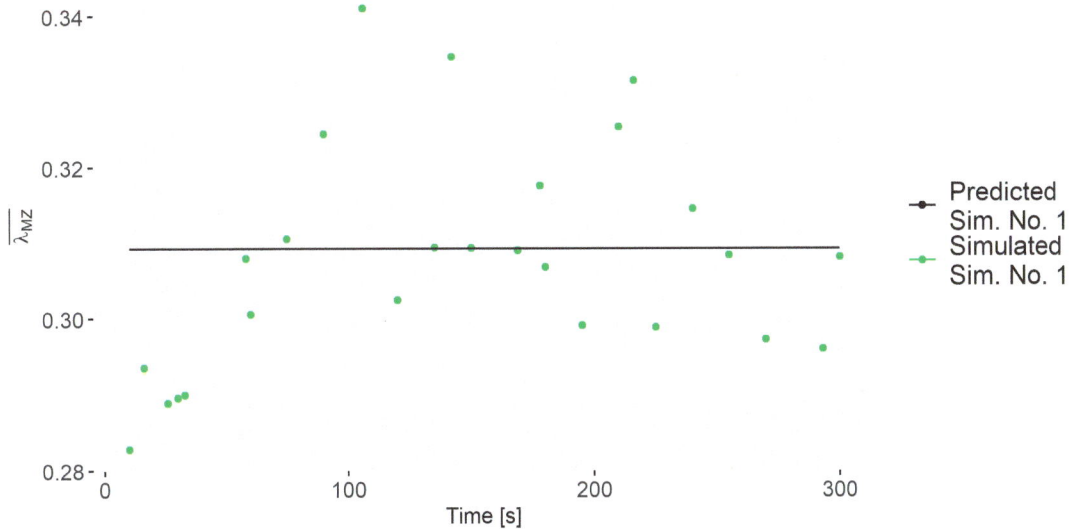

Figure 10. $\overline{\lambda_{MZ}}$ values for Batch 12 with predicted and simulation values.

3.5.2. Distributive Mixing Index

In contrast to dispersive mixing, time significantly affected the distributive mixing index, as evidenced in Figure 7. Thus, predictive screenings for M_{Kramer} were conducted at specific time steps across all simulations: first at 10 s, then at 30 s, followed by 30-s intervals. The aim was to identify which parameters, on average, had the most influence on M_{Kramer}. As seen from Figure 12, the z-axis rotation is the dominant factor after 10 s and RPM is the second dominating factor. The arms are 5.1%, which is above the significance line. Thus, the bootstrap forest parameters consisted of time, z-axis rotation, RPM, and arms. Figure 13a shows that all dominant parameters contribute positively to mixing when increased. In addition, there is good coherence between the predictive M_{Kramer} values and the simulated M_{Kramer} values, as seen in Figure 13b. The R^2 for the validation set was 0.949.

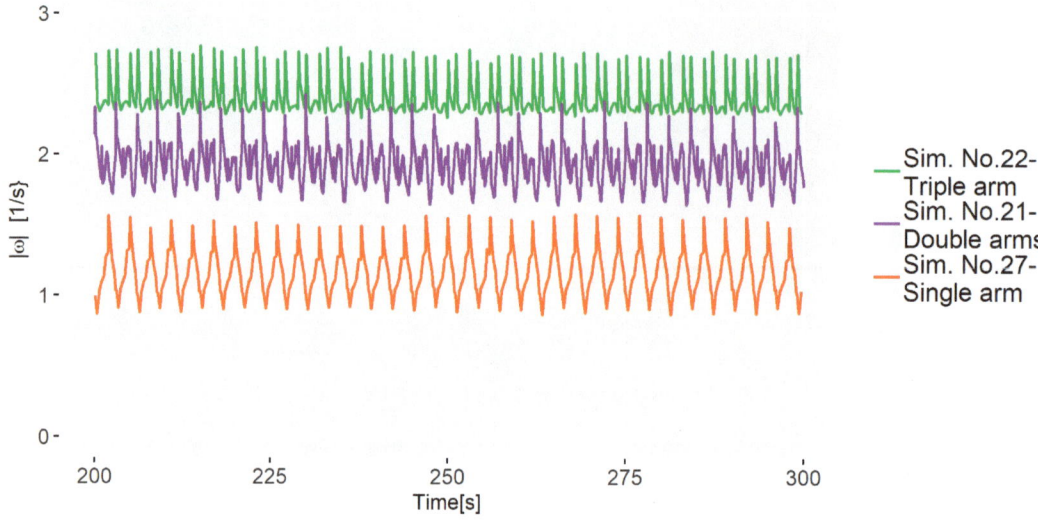

Figure 11. Vorticity comparison for three simulations with varying numbers of arms.

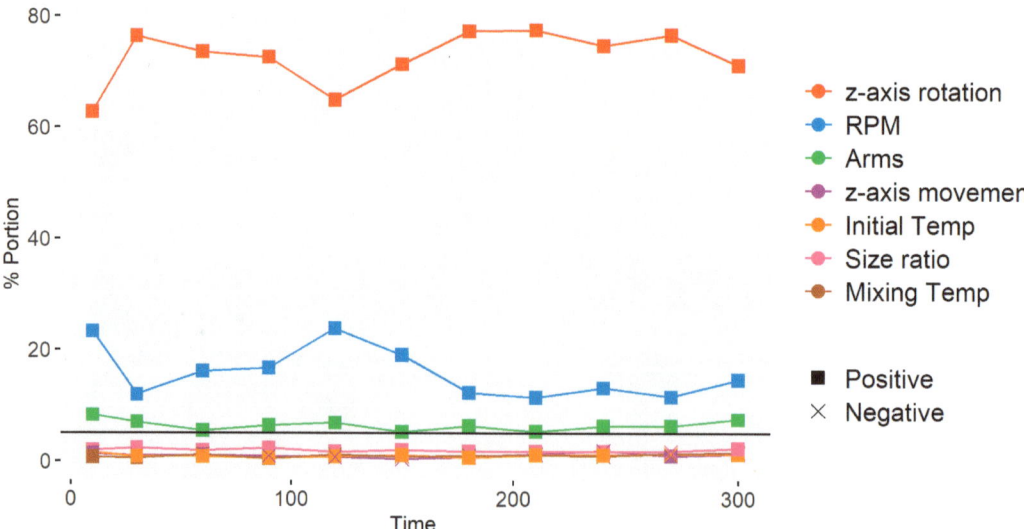

Figure 12. Screening of each parameter to $t = 10$ s and then $t = 30$ s and then every 30 s. Note the significance line at 5%. Positive and negative indicate the way in which the parameter affects the mixing when increased.

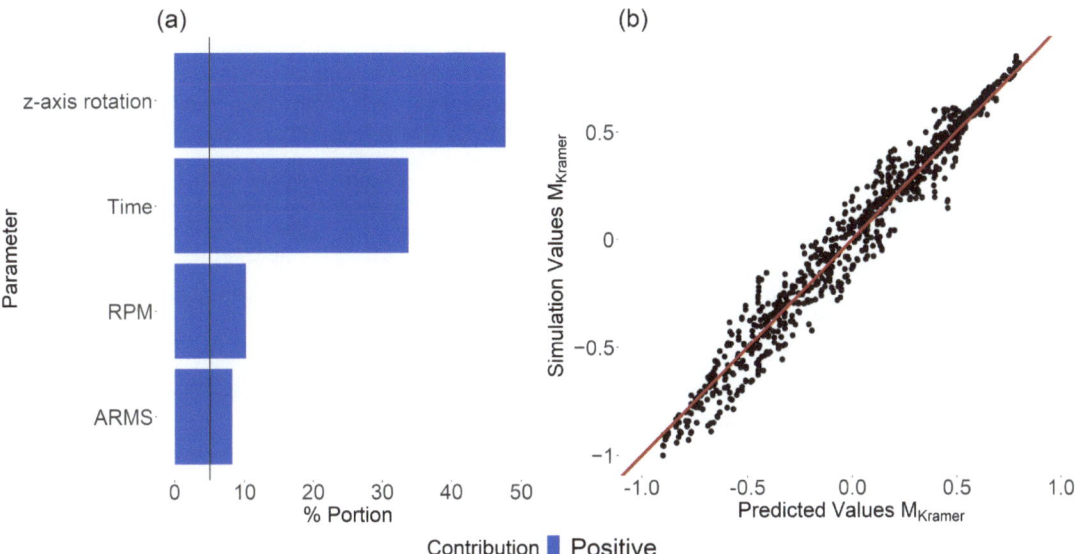

Figure 13. (a) Contribution for each parameter, and (b) the predicted vs the simulated values. Note the significance line at 5% in (a).

There were two interesting observations, first, z-axis rotation is more influential on mixing than time, and second, increasing the number of arms has a positive effect on the distributive mixing index, which is opposite as compared to the dispersion mixing index. This indicates that one needs to select the number of arms carefully in a change can mixer depending on whether the material at hand requires a focus on distributive or dispersive mixing. Additionally, even in small timestep intervals, neither the initial nor the mixing temperature shows to have any influence on the distributive mixing index, further highlighting the paramount importance of the mechanical parameters.

4. Conclusions

This study examined the impact of eight parameters on dispersive and distributive mixing in a change can mixer, utilizing both CFD models and the machine learning technique known as the bootstrap forest algorithm. A total of 35 simulations with each 26 time-dependent data points constituted the dataset for this research. The CFD models yielded results in accordance with theoretical expectations concerning fluid velocity and temperature; however, certain simulations revealed elevated average fluid temperatures compared to the set mixing temperatures, which can be attributed to viscous dissipation.

- Four key parameters significantly influenced dispersive mixing, represented by the average value of λ_{MZ}, $\overline{\lambda_{MZ}}$, achieving an R^2 value of 0.836 in the predictive model. Similarly, distributive mixing, denoted by M_{Kramer}, had an R^2 value of 0.949 in the predictive models.
- The z-axis rotation and RPM positively affected both mixing indexes, with z-axis rotation showing the greatest impact but also increased uncertainty.
- The number of arms negatively influenced dispersive mixing but positively impacted distributive mixing. The size ratio negatively affected dispersive mixing, while time significantly influenced distributive mixing.
- The z-axis movement, mixing temperature, and initial temperature demonstrated no significant effect in this study.

These findings provide essential direction for the refinement of mixing operations in industrial contexts. Future work could explore the validity of these simulations by adjusting the Z-axis rotation within the mixer, as such adjustments are anticipated to enhance both dispersive and distributive mixing. Subsequent research may benefit from validating these results in actual mixing scenarios, as well as assessing their relevance to different mixer types and materials. The investigative approach adopted herein offers a foundational framework for analogous future research in mixing science.

Author Contributions: M.R.L. has set up the experiments and simulations, and wrote the article; E.T.H.O. has assisted in optimizing and improving the simulation setup; and J.S. brought guidance to conceptualizing the study and analyzing the results, as well as assisted in writing the paper. All authors have read and agreed to the published version of the manuscript.

Funding: The support of this research was granted through the Innovation Fund Denmark (Grant no. 9065-00242B).

Institutional Review Board Statement: Not applicable.

Data Availability Statement: The original contributions presented in the study are included in the article, further inquiries can be directed to the corresponding author.

Acknowledgments: The authors wish to express their gratitude to Flow Science Inc. for providing licenses to FLOW-3D and to the Innovation Fund Denmark for their financial support. This article is based on the author's PhD thesis [34].

Conflicts of Interest: Author Michael Roland Larsen was employed by the company Dansac A/S. Author Erik Tomas Holmen Olofsson was employed by the company Haldor Topsoe A/S. The remaining authors declare that the research was conducted in the absence of any commercial or financial relationships that could be construed as a potential conflict of interest.

References

1. Manas-Zloczower, I. Analysis of mixing in polymer processing equipment. *Rheol. Bull.* **1997**, *66*, 5–8. Available online: http://www.rheology.org/sor/publications/rheology_b/Jan97/mixing.pdf (accessed on 22 July 2022). [CrossRef]
2. Lacey, P.M.C. The mixing of solid particles. *Chem. Eng. Res. Des.* **1943**, *75*, S49–S55. [CrossRef]
3. Lacey, P.M.C. Developments in the theory of particle mixing. *J. Appl. Chem.* **1954**, *4*, 257–268. [CrossRef]
4. Kramer, H.A. *Effect of Grain Velocity and Flow Rate upon the Performance of a Diverter-Type Sampler*; U.S. Department of Agriculture, Agricultural Research Service: Washington, DC, USA, 1968.
5. Ashton, M.D.; Valentin, F.H.H. The mixing of powders and particles in industrial mixers. *Trans. Inst. Chem. Eng.* **1966**, *44*, 166–188.
6. Larsen, M.R.; Ottsen, T.; Olofsson, E.T.H.; Spangenberg, J. Numerical Modeling of the Mixing of Highly Viscous Polymer Suspensions in Partially Filled Sigma Blade Mixers. *Polymers* **2023**, *15*, 1938. [CrossRef]
7. Ahmed, I.; Chandy, A.J. 3D numerical investigations of the effect of fill factor on dispersive and distributive mixing of rubber under non-isothermal conditions. *Polym. Eng. Sci.* **2019**, *59*, 535–546. [CrossRef]
8. Cheng, W.; Xin, S.; Chen, S.; Zhang, X.; Chen, W.; Wang, J.; Feng, L. Hydrodynamics and mixing process in a horizontal self-cleaning opposite-rotating twin-shaft kneader. *Chem. Eng. Sci.* **2021**, *241*, 116700. [CrossRef]
9. Danckwerts, P.V. The definition and measurement of some characteristics of mixtures. *Appl. Sci. Res. Sect. A* **1952**, *3*, 279–296. [CrossRef]
10. Connelly, R.K.; Kokini, J.L. Examination of the mixing ability of single and twin screw mixers using 2D finite element method simulation with particle tracking. *J. Food Eng.* **2007**, *79*, 956–969. [CrossRef]
11. Tolt, T.; Feke, D.L. Analysis and application of acoustics to suspension processing. *Proc. Intersoc. Energy Convers. Eng. Conf.* **1988**, *4*, 327–331.
12. Yang, H.-H.; Manas-Zloczower, I. Flow field analysis of the kneading disc region in a co-rotating twin screw extruder. *Polym. Eng. Sci.* **1992**, *32*, 1411–1417. [CrossRef]
13. Wang, J.; Tan, G.; Wang, J.; Feng, L.F. Numerical study on flow, heat transfer and mixing of highly viscous non-newtonian fluid in Sulzer mixer reactor. *Int. J. Heat Mass Transf.* **2022**, *183*, 122203. [CrossRef]
14. Pandey, V.; Maia, J.M. Comparative computational analysis of dispersive mixing in extension-dominated mixers for single-screw extruders. *Polym. Eng. Sci.* **2020**, *60*, 2390–2402. [CrossRef]
15. Zhu, X.Z.; Wang, G.; He, Y.D. Numerical Simulation of Temperature and Mixing Performances of Tri-screw Extruders with Non-isothermal Modeling. *Res. J. Appl. Sci. Eng. Technol.* **2013**, *5*, 3393–3401. [CrossRef]
16. Venczel, M.; Bognár, G.; Veress, Á. Temperature-Dependent Viscosity Model for Silicone Oil and Its Application in Viscous Dampers. *Processes* **2021**, *9*, 331. [CrossRef]
17. Bird, R.B.; Stewart, W.E.; Lightfoot, E.N. *Transport Phenomena*; John Wiley & Sons, Inc.: Hoboken, NJ, USA, 2007.
18. Tomar, A.S.; Harish, K.G.; Prakash, K.A. Numerical estimation of thermal load in a three blade vertically agitated mixer. *E3S Web Conf.* **2019**, *128*, 08004. [CrossRef]

19. Duan, G.; Takemi, T. Predicting Urban Surface Roughness Aerodynamic Parameters Using Random Forest. *J. Appl. Meteorol. Clim.* **2021**, *60*, 999–1018. [CrossRef]
20. N., G.; Jain, P.; Choudhury, A.; Dutta, P.; Kalita, K.; Barsocchi, P. Random Forest Regression-Based Machine Learning Model for Accurate Estimation of Fluid Flow in Curved Pipes. *Processes* **2021**, *9*, 2095. [CrossRef]
21. Comminal, R.; da Silva, W.R.L.; Andersen, T.J.; Stang, H.; Spangenberg, J. *Influence of Processing Parameters on the Layer Geometry in 3D Concrete Printing: Experiments and Modelling*; Springer International Publishing: Cham, Switzerland, 2020; pp. 852–862. [CrossRef]
22. Jacobsen, S.; Cepuritis, R.; Peng, Y.; Geiker, M.R.; Spangenberg, J. Visualizing and simulating flow conditions in concrete form filling using pigments. *Constr. Build. Mater.* **2013**, *49*, 328–342. [CrossRef]
23. Hirt, C.W.; Nichols, B.D. Volume of fluid (VOF) method for the dynamics of free boundaries. *J. Comput. Phys.* **1981**, *39*, 201–225. [CrossRef]
24. Einstein, A. Eine neue Bestimmung der Moleküldimensionen. *Ann. Phys.* **1906**, *324*, 289–306. [CrossRef]
25. Manas-Zloczower, I.; Feke, D.L. Analysis of Agglomerate Separation in Linear Flow Fields. *Int. Polym. Process.* **1988**, *2*, 185–190. [CrossRef]
26. Manas-Zloczower, I.; Feke, D.L. Analysis of Agglomerate Rupture in Linear Flow Fields. *Int. Polym. Process.* **1989**, *4*, 3–8. [CrossRef]
27. Takagaki, K.; Ito, T.; Arai, H.; Obata, Y.; Takayama, K.; Onuki, Y. The Usefulness of Definitive Screening Design for a Quality by Design Approach as Demonstrated by a Pharmaceutical Study of Orally Disintegrating Tablet. *Chem. Pharm. Bull.* **2019**, *67*, 1144–1151. [CrossRef]
28. Jung, U.-H.; Kim, J.-H.; Kim, J.-H.; Park, C.-H.; Jun, S.-O.; Choi, Y.-S. Optimum design of diffuser in a small high-speed centrifugal fan using CFD & DOE. *J. Mech. Sci. Technol.* **2016**, *30*, 1171–1184. [CrossRef]
29. Lira, J.O.B.; Riella, H.G.; Padoin, N.; Soares, C. CFD + DoE optimization of a flat plate photocatalytic reactor applied to NOx abatement. *Chem. Eng. Process.-Process Intensif.* **2020**, *154*, 107998. [CrossRef]
30. Ferry, J.D.; Parks, G.S. Viscous Properties of Polyisobutylene. *Physics* **1935**, *6*, 356–362. [CrossRef]
31. Rajavathsavai, D.; Khapre, A.; Munshi, B. Study of mixing behavior of cstr using CFD. *Braz. J. Chem. Eng.* **2014**, *31*, 119–129. [CrossRef]
32. Ho, T.K. Random decision forests. In Proceedings of the 3rd International Conference on Document Analysis and Recognition, Montreal, QC, Canada, 14–16 August 1995; pp. 278–282. [CrossRef]
33. Saha, S.; Roy, J.; Pradhan, B.; Hembram, T.K. Hybrid ensemble machine learning approaches for landslide susceptibility mapping using different sampling ratios at East Sikkim Himalayan, India. *Adv. Space Res.* **2021**, *68*, 2819–2840. [CrossRef]
34. Bendixen, M.R. *Experimental and Numerical Analysis of Mixing for Adhesive Barriers*; Technical University of Denmark: Lyngby, Denmark, 2023; ISBN 8774757806/9788774757801.

Disclaimer/Publisher's Note: The statements, opinions and data contained in all publications are solely those of the individual author(s) and contributor(s) and not of MDPI and/or the editor(s). MDPI and/or the editor(s) disclaim responsibility for any injury to people or property resulting from any ideas, methods, instructions or products referred to in the content.

Article

Numerical Modeling of the Mixing of Highly Viscous Polymer Suspensions in Partially Filled Sigma Blade Mixers

Michael Roland Larsen [1,2,*], Tobias Ottsen [3,*], Erik Tomas Holmen Olofsson [1,4] and Jon Spangenberg [1,*]

1 Department of Mechanical Engineering, Technical University of Denmark, 2800 Kgs. Lyngby, Denmark; etho@mek.dtu.dk
2 Dansac A/S, 3480 Fredensborg, Denmark
3 Manex ApS, 3500 Værløse, Denmark
4 Haldor Topsoe A/S, 2800 Kgs. Lyngby, Denmark
* Correspondence: mirola@dtu.dk (M.R.L.); tobias.ottsen@manex.dk (T.O.); josp@dtu.dk (J.S.)

Abstract: This paper presents a non-isothermal, non-Newtonian Computational Fluid Dynamics (CFD) model for the mixing of a highly viscous polymer suspension in a partially filled sigma blade mixer. The model accounts for viscous heating and the free surface of the suspension. The rheological model is found by calibration with experimental temperature measurements. Subsequently, the model is exploited to study the effect of applying heat both before and during mixing on the suspension's mixing quality. Two mixing indexes are used to evaluate the mixing condition, namely, the Ica Manas-Zlaczower dispersive index and Kramer's distributive index. Some fluctuations are observed in the predictions of the dispersive mixing index, which could be associated with the free surface of the suspension, thus indicating that this index might not be ideal for partially filled mixers. The Kramer index results are stable and indicate that the particles in the suspension can be well distributed. Interestingly, the results highlight that the speed at which the suspension becomes well distributed is almost independent of applying heat both before and during the process.

Keywords: mixing; suspensions; homogeneity; computational fluid dynamics; viscous dissipation; distributive- and dispersive-mixing

Citation: Larsen, M.R.; Ottsen, T.; Holmen Olofsson, E.T.; Spangenberg, J. Numerical Modeling of the Mixing of Highly Viscous Polymer Suspensions in Partially Filled Sigma Blade Mixers. *Polymers* 2023, 15, 1938. https://doi.org/10.3390/polym15081938

Academic Editor: Maria Graça Rasteiro

Received: 30 June 2022
Revised: 21 March 2023
Accepted: 10 April 2023
Published: 19 April 2023

Copyright: © 2023 by the authors. Licensee MDPI, Basel, Switzerland. This article is an open access article distributed under the terms and conditions of the Creative Commons Attribution (CC BY) license (https://creativecommons.org/licenses/by/4.0/).

1. Introduction

Attaining and ensuring homogeneity in particle-based slurries is not a trivial task—not least due to the complexity in quantifying homogeneity. The problem only gets more complicated when dealing with highly viscous non-Newtonian fluids. However, applications where homogeneity control is paramount are broad, extending to many industries such as food, pharmaceuticals, concrete and medical devices.

Some of the earliest attempts to obtain a quantification of homogeneity were made by Lacey in 1954 [1] and later by P. V. Danckwerts [2]. Both stated the importance of the sampling procedure, which is essential in determining whether particles in a population were evenly distributed. Other studies have used density as a method for the quantification of homogeneity [3,4]. These approaches have used standard statistics methods such as Coefficient of Variance [3] or Analysis of Variance (ANOVA) [4]. This works if you have appreciable unevenness in the distribution and a significant density discrepancy between components [5].

One of the pioneers within the field of mixing is Ica Manas-Zlaczower [6–11]. Her famous dispersive mixing index from 1992 is still used today [12–17], along with her distributive Cluster Distribute Index [12,17,18]. There have not been many alternatives to Zlaczower's dispersive mixing index. On the other hand, there are a number of alternatives to the Cluster Distribution Index—for example, the Scale of Segregation, a distributive mixing index that was developed back in 1952 [19] and later applied in Computational Fluid Dynamics (CFD) models by Connelly and Kokini in 2007 [20]. Lastly, there is the

Lacey mixing index [21], which has been modified over the years. Other indices that have been developed based on the Lacey mixing index include the Kramer mixing index [22] and the Ashton–Valentin mixing index [23].

Heating is an important aspect of mixing, since it can reduce viscosity drastically [12,17,24,25]. The advantage of reducing the viscosity is that it leads to an increased Reynolds Number, which improves mass transfer (i.e., mixing) [25]. The viscous dissipation of heat is a phenomenon that is often neglected in fluid dynamics [25]. However, when dealing with highly viscous fluids, viscous heating can be present if they are mixed with a certain force. Consequently, this leads to an impact on the energy balance and should therefore only cautiously be ignored, as shown in previous studies [12,26,27].

Sigma blades can often be used as the rotating mixing element in a mixer when dealing with highly viscous fluids [28]. They can be used either as a single-element [3] or, as often seen, as a twin-arm mixer [29]. Sigma blades have previously been simulated by Connelly [30,31] and evaluated with respect to the dispersive mixing index. However, to the best knowledge of the authors, no numerical models have been presented in the literature that account for a free surface, a non-Newtonian material behavior and viscous dissipation when mixing with Sigma blades.

This paper presents a new non-isothermal, non-Newtonian CFD model for the sigma blade mixing of an adhesive suspension, where both the free surface and viscous heating are accounted for. The suspension is modeled as a viscoplastic fluid, and model calibration is performed via optical temperature measurements. To evaluate the mixing quality, we have used Zlaczower's dispersive mixing index as well as Kramer's distributive mixing index. The model is used to investigate the effect of applying heat both before and during the process on the mixing quality. The rest of the paper is organized as follows: Section 2 introduces the experimental setup as well as the numerical model, while in Section 3, the results are presented and discussed. Section 4 summarizes the conclusions of the study.

2. Materials and Methods

2.1. Experiments

The mixing material consists of a highly viscous polymer fluid and a multi-component powder blend. The powder contains both colloidal and microscale particles. The specific fluid/powder suspension is a piece of intellectual property and cannot be disclosed. The viscosity of suspensions changes with the volume fraction of the powder [32,33], but this is not accounted for in the numerical model, as it is assumed to have a limited effect on the results. The physical data, besides viscosity, are shown in Table 1. The density measurements were performed with a Sartorius YDK03 Density Kit for Analytical Balances, which uses the Archimedes principle. The heat capacity and conductivity measurements were performed on a TCi-3-A from C-Therm Technologies Ltd., which uses the Modified Transient Plane Source (MTPS) method that has been used in textile research [34].

Table 1. Physical data for the fluid.

Name of the Property	Values	Unit
Density	1133	$\frac{kg}{m^3}$
Heat capacity	1389	$\frac{J}{kg \cdot °C}$
Heat conductivity	$k(t \leq 23\ °C) = 0.52$ $k(23\ °C < t < 71\ °C) = -1.88 \cdot 10^{-3}\ t + 0.56$ $k(t \geq 71\ °C) = 0.43$	$\frac{J}{m \cdot K}$

The experimental mixing was carried out on a Linden LK II 1 machine; see Figure 1. The machine has a mixing chamber with two sigma blades rotating in opposite directions. The front blade rotates at 60 rpm, and the one behind rotates at 19 rpm. Historically, this has been a typical setting for highly viscous mixing [28]. The mixer also has a control panel to switch the rotation direction. In addition, there is a digital display that indicates the temperature, as measured at the bottom of the mixer. Attached to the mixing equipment is a heat exchanger that adjusts the temperature for the mixing process. A vacuum pump is also mounted to ensure that air is not trapped inside the mixture. Two walls in the mixer can provide heat to the system. If this function is on, the temperature will be 80 °C.

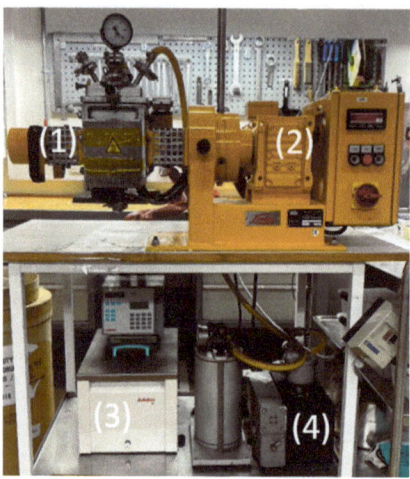

Figure 1. The mixing setup. (1) Chamber of the mixer, (2) control panel and digital display of the temperature, (3) heat exchanger, (4) vacuum pump.

2.2. Numerical Model

The CFD model simulates the mixing of the suspension in the mixing chamber and is developed in the commercial software FLOW-3D, which has successfully simulated other processes such as casting [35,36] and 3D printing [37,38] highly viscous fluids. An illustration of the mixer geometry is seen in Figure 2. At t = 0, the fluid is well distributed at the bottom, and the powder is placed on top of the fluid. At t > 0, the sigma blades start rotating. A no-slip boundary condition is applied on all solid surfaces. The computational domain is meshed with a uniform grid consisting of ~50,000 elements, which was arrived at after a mesh sensitivity analysis. The two side walls perpendicular to the sigma blades apply heat to the system; see Figure 2.

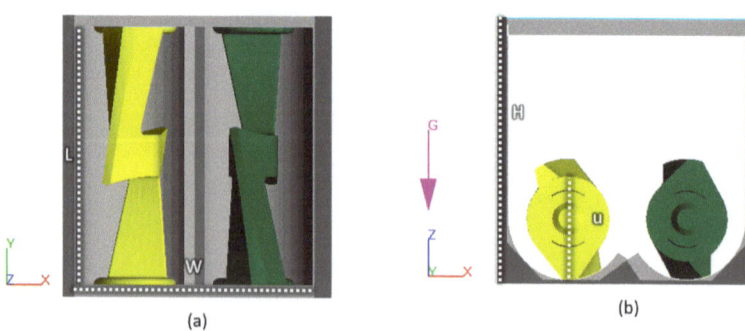

Figure 2. Illustration of the mixer geometry from (**a**) above and (**b**) the side. W and H have a distance of 120 mm, while L and u are 126.4 mm and 50.23 mm, respectively.

The flow is transient and non-isothermal since the viscosity is temperature dependent. The material is modeled as an incompressible substance, and thus, the density is approximated as constant. Hence, the flow is computed by considering the mass, momentum and energy conservation:

$$\frac{\partial \rho}{\partial t} + (\nabla \cdot \rho \mathbf{v}) = 0 \quad (1)$$

$$\rho \frac{D v}{Dt} = -\nabla p - [\nabla \cdot \tau] + \rho \mathbf{G} \quad (2)$$

$$\rho C_p \frac{DT}{Dt} = -(\nabla \cdot q) - (\tau : \nabla v) \quad (3)$$

$$q = -k\nabla T \quad (4)$$

where ρ is the density, p is the pressure, k is the thermal conductivity C_p, is the specific heat capacity and q is the heat flux vector. τ is the material deviatoric stress tensor and is calculated as $\tau = 2\mu(\dot{\gamma}, T)D$. D is the trace of the deformation rate tensor, and it is defined as $D = \frac{1}{2}(\nabla v) + \nabla v^T$. $\dot{\gamma}$ is the shear rate and is calculated by $\dot{\gamma} = \sqrt{2\mathrm{tr}(D^2)}$. The gravitational acceleration, G, is given by $\left((0,0,-9.82)\frac{m}{s^2}\right)$. The software used the finite volume method to discretize the governing equations. The equation of energy is calculated explicitly, while the viscous stress and pressure are solved implicitly. The advection is solved explicitly with first-order accuracy. The free surface is calculated with the volume of fluid technique [39].

The material behaves as a non-isothermal viscoplastic fluid and is simulated by a modified Carreu model:

$$\mu(\dot{\gamma}, T) = \begin{cases} \mu_{max}, & for \ \dot{\gamma} \leq \dot{\gamma}_c \\ \mu_\infty + \frac{\mu_0 E(T) - \mu_\infty}{\left(\dot{\gamma}^2(\lambda E(T))^2\right)^{\frac{1-n}{2}}}, & for \ \dot{\gamma} > \dot{\gamma}_c \end{cases} \quad (5)$$

$$E(T) = \exp\left(a\left(\frac{T^{ref}}{T-b}\right)\right) \quad (6)$$

where μ_0 is the zero-share-rate viscosity, μ_∞ is the infinity-share-rate viscosity, n is the power exponent, λ is the time constant, E is an energy function which is dependent on the fluid temperature, a and b are empirical constants and T^{ref} is the reference temperature. The applied values of the variables in Equations (5) and (6) are seen in Table 2. The shear rate-dependent variables are obtained via a rheological characterization of the fluid; see Figure 3. The temperature-dependent variables are obtained by a calibration, as the fluid was too viscous to perform measurements at low temperatures. The calibration is reported in Section 3.1.

Table 2. Viscosity data of the simulated fluid.

Symbol	Value	Unit
λ	1	s
n	−0.295	-
μ_0	4.35 10^8	Pa·s
μ_∞	2.9	Pa·s
a	−20	-
b	−95.2	°C
t^{ref}	20	°C
$\dot{\gamma}_c$	1	s^{-1}

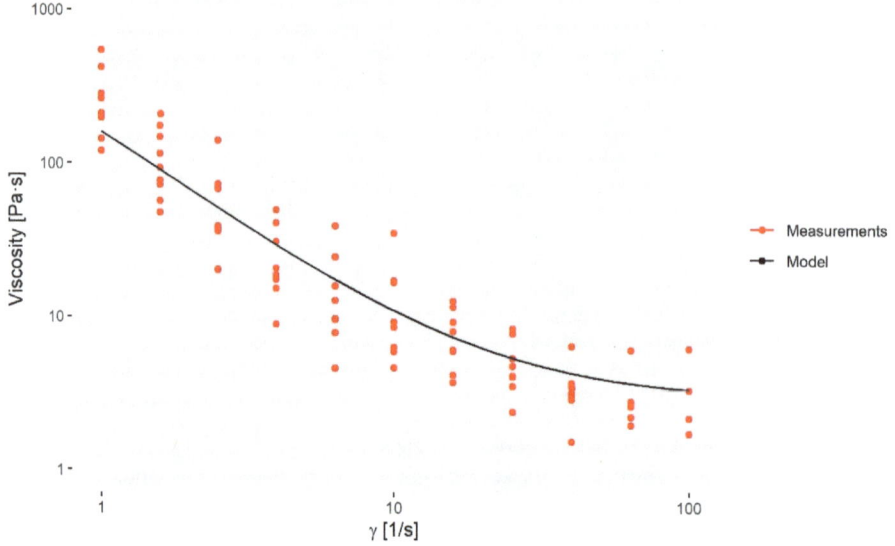

Figure 3. Viscosity measurements compared to the model.

The evaluation of the mixing is carried out by a dispersive- and distributive-mixing index. The dispersive mixing is quantified through the Manas–Zlacower mixing index, λ_{mz}; see Equation (8). ω is the vorticity. When λ_{mz} is close to 0, the mixing is purely rotational driven, while at 0.5 and 1, the mixing is shear-driven and purely elongation-driven, respectively. The latter leads to a desirable faster breakup of agglomerates [11,13,29].

$$\lambda_{mz} = \frac{|\dot{\gamma}|}{|\dot{\gamma}| + |\omega|} \tag{7}$$

The second approach is a distributive mixing index where an artificial concentration or material is used. The concentration is defined as a scalar with zero diffusion, which does not affect the physics (such as the viscosity and density). Thus, the only effect is pure mixing. The mean of the concentration, $\bar{c}$, is 0.5. The dimensionless concentration, $\hat{c}_i$, and the dimensionless variance across the domain that contains fluid, S^2, are found by Equations (8) and (9). From the variance, the Kramer dispersive mixing index, M_{Kramer}, can be found by Equation (10) [17], where $\sigma_r = \frac{\sigma_0}{N_f}$ and $\sigma_0 = (P \cdot (1-P))^{\frac{1}{2}}$, with P being the proportion of the component containing the concentration, which is set to 0.5 for this paper.

$$\hat{c}_i = \frac{c_i - \bar{c}}{\bar{c}} \tag{8}$$

$$S^2 = \frac{1}{(N_f - 1)} \sum_{i=1}^{N_f} (\hat{c}_i)^2 \tag{9}$$

$$M_{Kramer} = \frac{\sigma_0 - S}{\sigma_0 - \sigma_r} \tag{10}$$

Table 3 presents the process parameters for the simulations that are studied in this paper.

Table 3. Information about the simulations.

Simulation ID	Boundary Condition	Initial Condition	Simulation Time
CA23	Adiabatic	23 °C	900 s
A23	Adiabatic	23 °C	1800 s
A50	Adiabatic	50 °C	1800 s
A80	Adiabatic	80 °C	1800 s
NA23	80 °C	23 °C	1800 s
NA50	80 °C	50 °C	1800 s
NA80	80 °C	80 °C	1800 s

3. Results and Discussion

3.1. Calibration

In the experimental setup, an optical thermal probe was placed inside the mixing chamber to measure the temperature of the suspension. The probe logged a surface temperature every 10 s. The initial temperature of the fluid was measured to be 23 °C, and no heat was applied while mixing. The experimental findings and the results of the calibration simulation, $CA23$, are shown in Figure 4. The experimental measurements show that viscous heating is present and that the temperature increases most rapidly during the first 250 s. The fluctuation around 250 and 500 s is possibly due to the brushed steel occasionally being measured instead of the fluid. The parameters a and b in Table 2 are calibrated in order to make the CFD model predict the same temperature evolution as that seen in the experiment. By doing so, the experimental findings and the numerical results are in very good agreement. The absolute mean error between the two is 1.72 °C. The rheological model obtained by calibration is presented in Figure 5 for the temperatures 23, 50 and 80 °C.

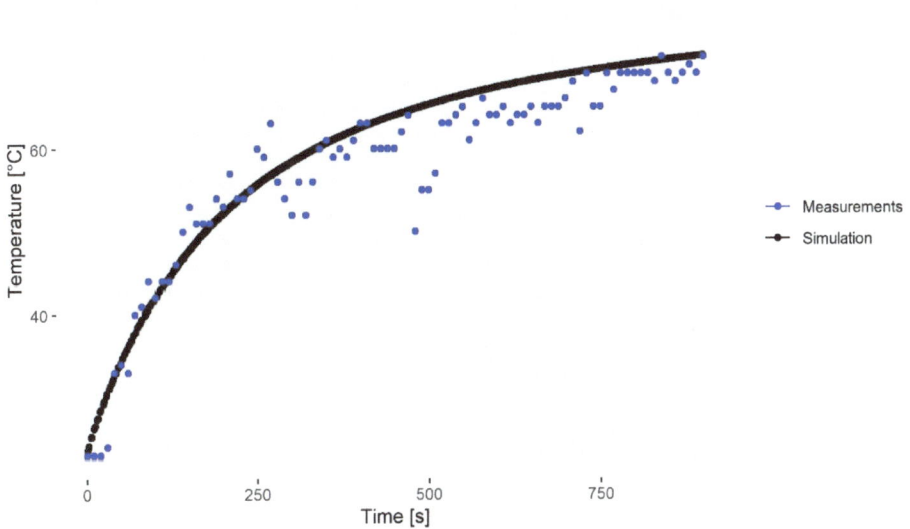

Figure 4. Experimental and simulation results of the surface temperature of the fluid as a function of time. The absolute mean error between the simulation and measurement is 1.72 °C.

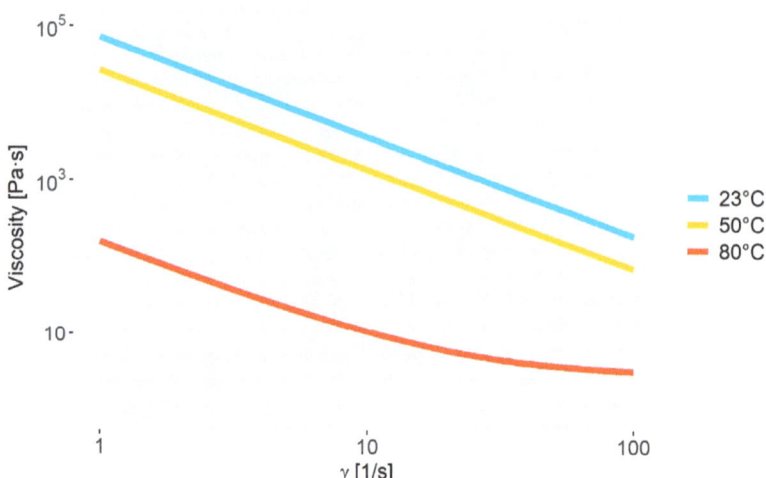

Figure 5. Viscosity curve as a function of the shear rate at different temperatures.

3.2. Temperature Distribution

The temperature distribution is close to uniform when mixed under adiabatic boundary conditions, but this is not the case when applying the non-adiabatic boundary condition, as seen in Figure 6, which illustrates the temperature profile for $NA50$ at $t = 15$ s. The profile shows that the temperature is highest near the heated walls. The temperature decrease along the x-axis is due to the fluid not being exposed to the fixed wall temperature. The average temperature increases with time due to the viscous heating and heated walls.

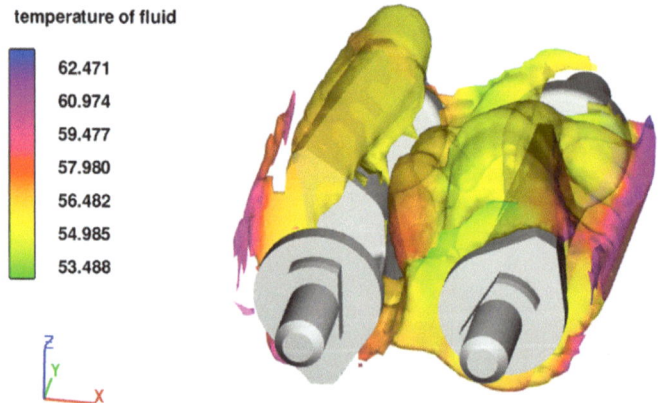

Figure 6. Temperature profile for $NA50$ at $t = 15$ s.

The average temperature of the fluid as a function of time is illustrated in Figure 7 for all simulations except $CA23$. The simulations with applied wall temperature increase faster in terms of temperature as compared to their counterpart with adiabatic boundary conditions, except for $NA80$, where the walls effectively cool the suspension. $A23$ and $A50$ do not have enough time to reach 80 °C, while $A80$ reaches a temperature slightly higher than 80 °C.

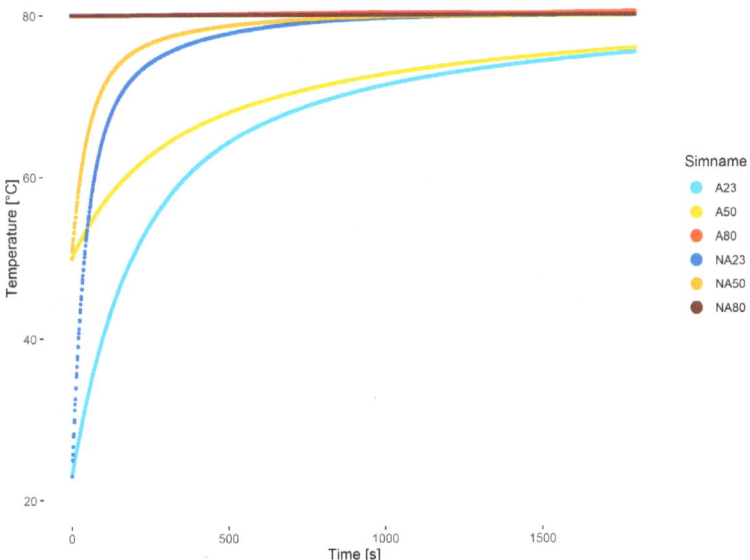

Figure 7. Average temperature of the fluid as a function of time.

3.3. Velocity Field

The velocity field gives a good indication of how the flow pattern is inside of the mixer. In Figure 8, the velocity field for *A*80 is presented at different times. As expected, the sigma blade at the lowest *x*-value is the one with the highest velocity, while the second sigma blade spins slower to ensure mixing in the whole system. Noticeably, some of the fluid sticks to the wall, which is due to the viscoplastic effect of the fluid and the no-slip condition on the walls.

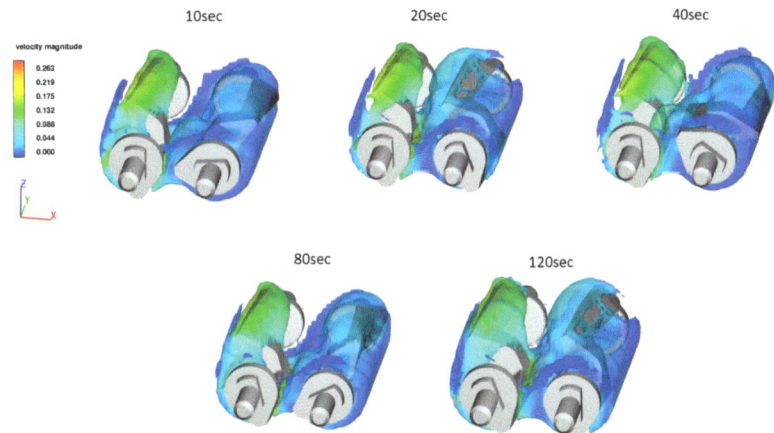

Figure 8. Velocity profile for an initial temperature of 80 °C with adiabatic boundary conditions at different time values. The velocity values are in m/s.

3.4. Dispersive Mixing

In Figure 9, the histogram results of λ_{mz} at $t = 1800$ s are shown for all simulations except $CA23$. The results are fairly similar for all simulations, which indicates that preheating and heating during the process do not have a big influence on the mixing quality after 30 min. The dispersive mixing index is primarily between 0 and 0.5, which is similar to the findings of Ahmed and Chandy [12], who simulated a two-wing rotors mixer. However, the λ_{mz} values are lower than the predictions by Connelly and Kokini [30], who studied a fully filled sigma blade mixer, which indicates that a partially filled mixer seems to have a decreasing effect on how well dispersed the particles will be in the suspension. Note that the presence of λ_{mz} values between 0 and 0.1 is mainly due to the viscoplastic effect that makes the suspension more prone to have zones with a shear rate of zero and therefore limited mixing.

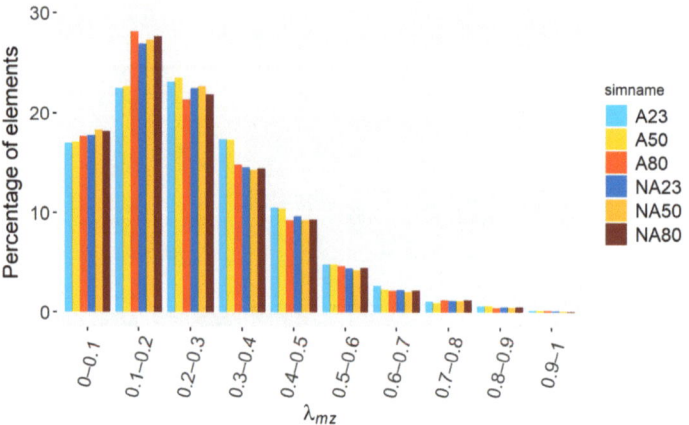

Figure 9. Histogram of the λ_{mz} at $t = 1800$ s.

Figure 10 shows the mean value, $\overline{\lambda_{mz}}$, illustrated at different times. It is seen that the mean value fluctuates by ~10%. The fluctuations are most likely due to the free surface of the mixture (i.e., the mixer is partially filled).

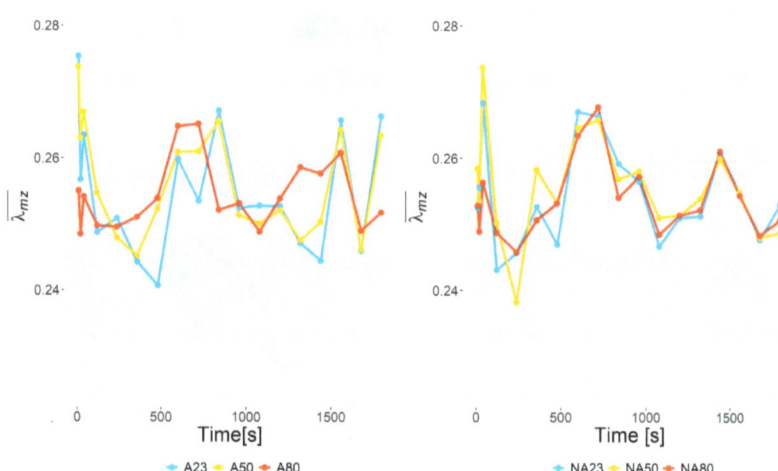

Figure 10. The mean value of λ_{mz} for different time step values.

3.5. Distributive Mixing

In Figure 11, the Kramer mixing index is illustrated at different time values for all simulations except $CA23$. The results show that all simulations obtain a high M_{Kramer} value (i.e., close to 1) after 2 min, which indicates that the particles in the suspension end up being well distributed. $A80$ and $NA80$ reach a high M_{Kramer} value the fastest (due to the lower viscosity at high temperatures), but this is marginally faster than the rest, thereby highlighting that energy potentially can be saved by not preheating the mixture without compromising the mixing of the suspension. Similarly, the results show that no substantial gain will be obtained regarding the mixing time by applying heat during the process. One should keep in mind that at a fixed mixing velocity, the Sigma blades require more energy at low temperatures due to the higher viscosity of the material. In addition, the Sigma blades will be exposed to larger forces at high viscosities, which can affect the blades' lifespan.

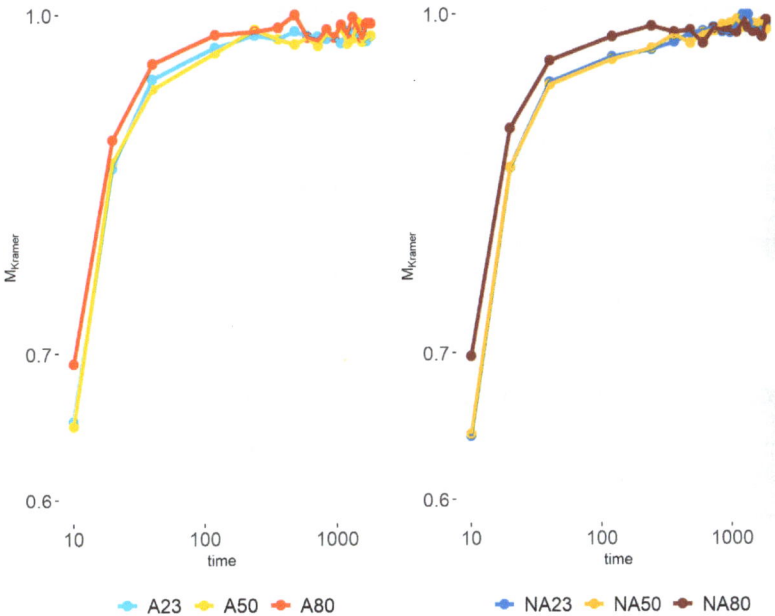

Figure 11. Kramer mixing index at different time values.

4. Conclusions

This work presents a non-isothermal, non-Newtonian CFD model that simulates a partially filled Sigma blade mixer by taking into account viscous heating and the free surface of the suspension. The model is calibrated against physical temperature measurements, and good agreement between the experimental and numerical results is obtained. The results also highlight the importance of accounting for viscous heating when modeling the mixing of the suspension at hand. A numerical parameter study (i.e., varying wall temperature and pre-heating) was made in order to improve the understanding of the mixing conditions. The results of the model showed that the dispersive mixing index, $\overline{\lambda_{mz}}$, was primarily between 0 and 0.5, and it was not affected much by changing the wall and pre-heating temperature. Fluctuations of approximately 10% were found in the dispersive mixing index for most simulations, which was associated with the partially filled mixer (i.e., free surface of the suspension). This can be seen as a limitation of the mixing index. The model predicted via the Kramer mixing index that the suspension in all simulations would be well distributed after 1800 s. In addition, the results illustrated that pre-heating as well as heating during the process did not provide a substantial gain when it came to how fast a high Kramer index would be obtained. This illustrated that one could potentially save energy and time by eliminating the heating. This was a very interesting finding, as, intuitively, one could have expected that heating would lead to lower viscosity and therefore substantially faster mixing. In future work, focus will be put on extending the model to account for the effect of powder concentration variations on the viscosity and thermal conductivity, as this can especially affect the initial phase of the mixing.

Author Contributions: M.R.L. has set up the experiments, simulations and wrote the article; T.O. has contributed to the calibration setup and data treatment; E.T.H.O. has helped out optimizing and improve the simulation setup and J.S. has come with guidance the paper setup, the numerical model and wrote the paper. All authors have read and agreed to the published version of the manuscript.

Funding: The support of this research were done through Innovation Fund Denmark (Grant no. 9065-00242B).

Institutional Review Board Statement: Not applicable.

Informed Consent Statement: Not applicable.

Data Availability Statement: Data is available on request.

Acknowledgments: The authors would like to acknowledge FLOW-3D for their support in regard to licenses.

Conflicts of Interest: The authors declare no conflict of interest.

References

1. Lacey, P.M.C. Developments in the theory of particle mixing. *J. Appl. Chem.* **2007**, *4*, 257–268. [CrossRef]
2. Danckwerts, P.V. Theory of mixtures and mixing. In *Insights Into Chemical Engineering*; Elsevier: Amsterdam, The Netherlands, 1981; pp. 262–268. [CrossRef]
3. Abdillah, L.H.; Winardi, S.; Sumarno, S.; Nurtono, T. Effect of Mixing Time to Homogeneity of Propellant Slurry. *IPTEK J. Proc. Ser.* **2018**, *4*, 94. [CrossRef]
4. Yang, M.; Li, X.; Shi, T.; Yang, S. Mixture homogeneity in a high-viscous flow mixer. *Nongye Gongcheng Xuebao/Trans. Chin. Soc. Agric. Eng.* **2011**, *27*, 137–142. [CrossRef]
5. Watano, S. *Powder Technology Handbook*, 4th ed.; Higashitani, K., Makino, H., Matsusaka, S., Eds.; CRC Press: Boca Raton, FL, USA, 2019; pp. 401–409. [CrossRef]
6. Manas-Zloczower, I.; Cheng, H. Analysis of mixing efficiency in polymer processing equipment. *Macromol. Symp.* **1996**, *112*, 77–84. [CrossRef]
7. Wang, W.; Manas-Zloczower, I. Temporal distributions: The basis for the development of mixing indexes for scale-up of polymer processing equipment. *Polym. Eng. Sci.* **2001**, *41*, 1068–1077. [CrossRef]
8. Cheng, H.; Manas-Zloczower, I. Chaotic Features of Flow in Polymer Processing Equipment-Relevance to Distributive Mixing. *Int. Polym. Process.* **1997**, *12*, 83–91. [CrossRef]
9. Manas-Zloczower, I.; Feke, D.L. Analysis of Agglomerate Rupture in Linear Flow Fields. *Int. Polym. Process.* **1989**, *4*, 3–8. [CrossRef]
10. Yang, H.-H.; Manas-Zloczower, I. Analysis of Mixing Performance in a VIC Mixer. *Int. Polym. Process.* **1994**, *9*, 291–302. [CrossRef]
11. Yang, H.-H.; Manas-Zloczower, I. Flow field analysis of the kneading disc region in a co-rotating twin screw extruder. *Polym. Eng. Sci.* **1992**, *32*, 1411–1417. [CrossRef]
12. Ahmed, I.; Chandy, A.J. 3D numerical investigations of the effect of fill factor on dispersive and distributive mixing of rubber under non-isothermal conditions. *Polym. Eng. Sci.* **2018**, *59*, 535–546. [CrossRef]
13. Wang, J.; Tan, G.; Wang, J.; Feng, L.-F. Numerical study on flow, heat transfer and mixing of highly viscous non-newtonian fluid in Sulzer mixer reactor. *Int. J. Heat Mass Transf.* **2021**, *183*, 122203. [CrossRef]
14. Pandey, V.; Maia, J.M. Comparative computational analysis of dispersive mixing in extension-dominated mixers for single-screw extruders. *Polym. Eng. Sci.* **2020**, *60*, 2390–2402. [CrossRef]
15. Marschik, C.; Osswald, T.A.; Roland, W.; Albrecht, H.; Skrabala, O.; Miethlinger, J. Numerical analysis of mixing in block-head mixing screws. *Polym. Eng. Sci.* **2018**, *59*, E88–E104. [CrossRef]
16. Danda, C.; Pandey, V.; Schneider, T.; Norman, R.; Maia, J.M. Enhanced Dispersion and Mechanical Behavior of Polypropylene Composites Compounded Using Extension-Dominated Extrusion. *Int. Polym. Process.* **2020**, *35*, 281–301. [CrossRef]
17. Ahmed, I.; Poudyal, H.; Chandy, A.J. Fill Factor Effects in Highly-Viscous Non-Isothermal Rubber Mixing Simulations. *Int. Polym. Process.* **2019**, *34*, 182–194. [CrossRef]
18. Cheng, W.; Xin, S.; Chen, S.; Zhang, X.; Chen, W.; Wang, J.; Feng, L. Hydrodynamics and mixing process in a horizontal self-cleaning opposite-rotating twin-shaft kneader. *Chem. Eng. Sci.* **2021**, *241*, 116700. [CrossRef]

19. Danckwerts, P.V. The definition and measurement of some characteristics of mixtures. *Appl. Sci. Res. Sect. A* **1952**, *3*, 279–296. [CrossRef]
20. Connelly, R.K.; Kokini, J.L. Examination of the mixing ability of single and twin screw mixers using 2D finite element method simulation with particle tracking. *J. Food Eng.* **2007**, *79*, 956–969. [CrossRef]
21. Lacey, P. The mixing of solid particles. *Chem. Eng. Res. Des.* **1997**, *75*, S49–S55. [CrossRef]
22. Kramer, H.A. *Effect of Grain Velocity and Flow Rate Upon the Performance of a Diverter-Type Sampler*; U.S. Dept. of Agriculture, Agricultural Research Service: Washington, DC, USA, 1968.
23. Ashton, M.D.; Valentin, F.H.H. The mixing of powders and particles in industrial mixers. *Trans. Inst. Chem. Eng.* **1966**, *44*, 166–188.
24. Ferry, J.D.; Parks, G.S. Viscous Properties of Polyisobutylene. *Physics* **1935**, *6*, 356–362. [CrossRef]
25. Bird, R.B.; Stewart, W.E.; Lightfoot, E.N. *Transport Phenomena*; John Wiley & Sons, Inc.: Hoboken, NJ, USA, 2007.
26. Wang, G.; He, Y.; Zhu, X. Numerical Simulation of Temperature and Mixing Performances of Tri-screw Extruders with Non-isothermal Modeling. *Res. J. Appl. Sci. Eng. Technol.* **2013**, *5*, 3393–3401. [CrossRef]
27. Tomar, A.S.; Harish, K.G.; Prakash, K.A. Numerical estimation of thermal load in a three blade vertically agitated mixer. *E3S Web Conf.* **2019**, *128*, 08004. [CrossRef]
28. Parker, N.H. How to select double-arm mixers. *Chem. Eng.* **1965**, *72*, 121–128.
29. Paul, E.L.; Atiemo-Obeng, V.A.; Kresta, S.M. 16. Mixing of Highly Viscous Fluids, Polymers, and Pastes. In *Handbook of Industrial Mixing-Science and Practice*; John Wiley & Sons: Hoboken, NJ, USA, 2004; pp. 987–1025. Available online: https://app.knovel.com/hotlink/khtml/id:kt007EO511/handbook-industrial-mixing/mixing-highly-viscous (accessed on 10 March 2023).
30. Connelly, R.K.; Kokini, J.L. 3D numerical simulation of the flow of viscous newtonian and shear thinning fluids in a twin sigma blade mixer. *Adv. Polym. Technol.* **2006**, *25*, 182–194. [CrossRef]
31. Connelly, R.K.; Kokini, J.L. Mixing simulation of a viscous Newtonian liquid in a twin sigma blade mixer. *AIChE J.* **2006**, *52*, 3383–3393. [CrossRef]
32. Spangenberg, J.; Scherer, G.W.; Hopkins, A.B.; Torquato, S. Viscosity of bimodal suspensions with hard spherical particles. *J. Appl. Phys.* **2014**, *116*, 184902. [CrossRef]
33. Jabbari, M.; Spangenberg, J.; Hattel, J.H. Particle migration using local variation of the viscosity (LVOV) model in flow of a non-Newtonian fluid for ceramic tape casting. *Chem. Eng. Res. Des.* **2016**, *109*, 226–233. [CrossRef]
34. Venkataraman, R.M.M.; Mishra, R.; Militky, J. Comparative Analysis of High Performance Thermal Insulation Materials. *J. Text. Eng. Fash. Technol.* **2017**, *2*, 1–10. [CrossRef]
35. Jacobsen, S.; Cepuritis, R.; Peng, Y.; Geiker, M.R.; Spangenberg, J. Visualizing and simulating flow conditions in concrete form filling using pigments. *Constr. Build. Mater.* **2013**, *49*, 328–342. [CrossRef]
36. Spangenberg, J.; Roussel, N.; Hattel, J.H.; Thorborg, J.; Geiker, M.R.; Stang, H.; Skocek, J. Prediction of the Impact of Flow-Induced Inhomogeneities in Self-Compacting Concrete (SCC). In *Design, Production and Placement of Self-Consolidating Concrete*; Springer: Dordrecht, The Netherlands, 2010; pp. 209–215. [CrossRef]
37. Comminal, R.; da Silva, W.R.L.; Andersen, T.J.; Stang, H.; Spangenberg, J. Influence of Processing Parameters on the Layer Geometry in 3D Concrete Printing: Experiments and Modelling; Springer: Cham, Switzerland, 2020; pp. 852–862. [CrossRef]
38. Mollah, T.; Comminal, R.; Serdeczny, M.P.; Pedersen, D.B.; Spangenberg, J. Stability and deformations of deposited layers in material extrusion additive manufacturing. *Addit. Manuf.* **2021**, *46*, 102193. [CrossRef]
39. Hirt, C.W.; Nichols, B.D. Volume of fluid (VOF) method for the dynamics of free boundaries. *J. Comput. Phys.* **1981**, *39*, 201–225. [CrossRef]

Disclaimer/Publisher's Note: The statements, opinions and data contained in all publications are solely those of the individual author(s) and contributor(s) and not of MDPI and/or the editor(s). MDPI and/or the editor(s) disclaim responsibility for any injury to people or property resulting from any ideas, methods, instructions or products referred to in the content.

Article

Knot Formation on DNA Pushed Inside Chiral Nanochannels

Renáta Rusková *[ ] and Dušan Račko *[ ]

Polymer Institute of the Slovak Academy of Sciences, Dúbravská cesta 9, 845 41 Bratislava, Slovakia
* Correspondence: renata.ruskova@savba.sk (R.R.); dusan.racko@savba.sk (D.R.)

Abstract: We performed coarse-grained molecular dynamics simulations of DNA polymers pushed inside infinite open chiral and achiral channels. We investigated the behavior of the polymer metrics in terms of span, monomer distributions and changes of topological state of the polymer in the channels. We also compared the regime of pushing a polymer inside the infinite channel to the case of polymer compression in finite channels of knot factories investigated in earlier works. We observed that the compression in the open channels affects the polymer metrics to different extents in chiral and achiral channels. We also observed that the chiral channels give rise to the formation of equichiral knots with the same handedness as the handedness of the chiral channels.

Keywords: polymer; DNA; molecular simulations; nanochannels; topology; knot; chirality; membranes

Citation: Rusková, R.; Račko, D. Knot Formation on DNA Pushed Inside Chiral Nanochannels. *Polymers* **2023**, *15*, 4185. https://doi.org/10.3390/polym15204185

Academic Editors: Alexandre M. Afonso, Luís L. Ferrás and Célio Pinto Fernandes

Received: 31 August 2023
Revised: 18 October 2023
Accepted: 19 October 2023
Published: 22 October 2023

Copyright: © 2023 by the authors. Licensee MDPI, Basel, Switzerland. This article is an open access article distributed under the terms and conditions of the Creative Commons Attribution (CC BY) license (https://creativecommons.org/licenses/by/4.0/).

1. Introduction

Polymers are long molecules consisting of many building units called monomers [1]. The units can be chemically identical, and still it is possible to think of an infinite number of polymers that could be constructed and yet differ by only how the monomers are connected. As such, the polymers represent a combinatorial problem that is suitable for studying by computer simulations.

The way the monomers are connected defines the polymer's topology. A relatively new topology is represented by polymer knots [2]. The knots are formed by winding a long polymer chain around itself. The polymer knots occur naturally both in synthetic and biological polymers, such as DNA and proteins. The first synthetic knots with controlled topology were made possible by the end of the 1980s [3]. Knotted polymers have been drawing increasing scientific attention given the progress in macromolecular synthesis, biology, mathematics, and molecular simulations.

It is difficult to pinpoint when and how this scientific interest in knotted molecules began, whether it was sparked by imagination or, as is often the case, by observing nature. But now, it is clear that the topological state of molecules has strong biological and technological effects and implications. While in biology, the knots can be very harmful on a genome [4–8] but very important on proteins [9–11], whose biological function is yet to be fully revealed, the distinct effects of polymer knotted topology currently pose problems that need to be engineered due to a current lack of polymers with a well-defined knotted topology and methods to synthesize them in sufficient amounts.

As pointed out above, polymers, especially with regard to their topology, represent a problem suitable to be studied using computers. While the progress in polymer synthesis enabled controlled preparations of knotted molecules of up to eight crossings [12], in computer simulations, knotted polymers with well-defined topology are easily prepared. Thus, computer simulations are currently a useful and indispensable approach in investigating knotted polymers.

As we also mentioned, the knots are formed naturally by biological processes [13,14], but are also inevitable in physical processes, when the probability of a polymer being knotted increases with its length [15]. The simplest and most practical way to produce knots would be using a diluted solution of long polymers and ligating the ends of the polymer. This would, however, lead to the production of a wide spectrum of knots. Computer simulations are useful to provide a prediction of how the length of polymers, such as DNA, and ionic environments would be used to control the topology of the polymer knots [16–18].

The next layer of complexity is added by investigating polymer knotting in confined spaces. The confinement state is the most typical state where polymers occur in nature [19,20]. Confinement is a state where the natural dimension of the polymer is larger than the size of the confinement geometry. Biopolymers, such as DNA, are naturally confined in such small spaces as viral capsids or cellular nuclei. The state of polymer confinement is also essential for nanotechnology, including nanocomposite or nanofluidic experiments involved in genomic studies. The confinement of polymers, yet while a very complex state, being a subject of experimental, theoretical, and simulations studies, is known to enhance and stabilize the knottedness of polymers [17,21,22]. Computer simulations provide insights beyond the possibilities of experimental imaging.

Consequently, the level of the complexity of the problem can be extended by adding external forces into consideration that induce compression of the confined polymer. The compression of polymer chains under confinement was studied by means of Monte Carlo (MC) [23–25] and molecular dynamics (MD) simulations [26–35] in channels with square [23,31–34], cylindrical [24,25,27–30,35], or helical [36] sections or in structured channels [26,37,38], where the compressive force was applied by pulling on distant ends of the confined polymer in a direction against each other [23,24,27], by using a piston compression similar to the gaskets used in the experiments described in [24,25,28,29,31–36] or by flow of media [26,37,38], while compression in a spherical confinement was also investigated [39,40]. While previous research has generally highlighted the enhanced effects of compression in generating entanglements within polymers, only a few studies have investigated the topology of compressed polymers [28,36,40]. These works quantified knot complexity and the populations of knots as a function of compressive force.

It has been demonstrated that the theoretical and computational insights into polymer compression within nanochannels can be experimentally validated by confining DNA within nanofluidic channels and inducing compression through specific experimental setups [41,42]. More recently, the earlier computational findings regarding knot formation in confined spaces, along with developed nanofluidic experiments, have led to the creation of the 'knot factory' nanofluidic device. This innovative device produces knots when polymer chains are compressed inside nanochannels [43]. Furthermore, computer simulations have been instrumental in exploring various aspects of this device, including the compression duration and the compressive force's impact on knot topology [28]. Our own computational work has also contributed by investigating the effects of nanofluidic channel sizes and the strength of confinement [36]. The computer simulations carried out by us [36] extrapolated the model [28] used previously to model the experimental setup in [43], and also devised an in silico experiment to test whether the chiral geometry of the channels can induce the handedness and influence the chirality of knotted structures.

Chirality is a prominent property of knots. As mentioned above, knots are formed by a substantially long polymer chain winding around itself. One of the parameters to characterize knots is the crossing number that quantifies how many times the polymer winds around itself. The direction of the polymer winding around itself defines the chirality of the knot. The crossing number is a combinatorial property, and the number of possible knot types that can be constructed with a given crossing number increases substantially. This is why it is feasible to create as many as 1.7 million prime knots with up to 16 crossings, out of which fewer than 2000 knots are achiral [44]. The chirality of polymer knots exhibits unique physical properties. However, due to the limited availability of polymers with well-defined knotted topologies, practical applications are currently largely theoretical. Nevertheless, experimental evidence suggests that chiral knots play a role in biology [45], can be employed to control optical properties [46], offer potential for new energy-harvesting sources at the nanoscale [47], and they may find applications in stereoselective chemosensing [48] and also in the progress in organized entanglements in chemistry [49].

In the context of knot formation, a recent computer simulation was devised to explore an intriguing scenario of whether and how knotting could be induced by simply pushing DNA inside nanochannels, without the need for the more complex lab-on-chip nanofluidic experiments [34]. This scenario holds practical significance because it offers a relatively straightforward way to induce knot formation in polymers, which has relevance in various applications where polymers are pushed through narrow channels, such as in chromatographic resins or membranes. Additionally, experimental efforts are underway to develop chiral membranes capable of separating chiral enantiomers based on their geometric properties, which includes the use of structures like helical nanochannels [50,51]. The emerging applications of chiral knots mentioned above together with chiral separation devices are part of the emerging field of chiral nanotechnology [52], where our computer simulations contribute by proposing methods for producing knotted structures while controlling chirality through physical means.

In our current study, we are exploring the formation of knots in polymers as they are pushed into open, infinitely long nanochannels with varying sizes and geometries. To simulate chiral environments, we designed these channels with a helical geometry and induced different chirality by altering the winding direction of the helical loops within the nanochannels. We developed a novel computational approach to identify the chiral properties of the knots that form in the DNA strands as they are pushed inside these channels. Our method utilizes Knoto-ID [53], a topological software, to determine the chirality of the knots in reference to the Rolfsen knot table [54]. Furthermore, it identifies handedness based on a new, biologically motivated knot table [55]. We compare and discuss the simulations also with respect to the compression in finite channels by pushing against an impenetrable wall developed and modeled in Ref. [36]. The structure of this manuscript is as follows: in Section 3.1, we present the results concerning polymer metrics; in Section 3.2, we discuss various aspects of monomer distribution; Section 3.3 explores knotting probabilities; Section 3.4 introduces the computational routine used to analyze knot handedness, and, finally, in Section 3.5, we discuss the possible mechanism through which the chiral environment impacts the handedness of knots. We believe that our findings significantly contribute to our current understanding of polymers in confined spaces, especially under compressive forces. These insights can have implications in the fields of DNA biology, polymer physics, and the emerging field of chiral nanotechnology.

2. Materials and Methods

2.1. Model of DNA

The dsDNA is modeled as a discretized beaded chain consisting of N = 300 beads representing DNA portions with a width of $1\,\sigma$ corresponding to 2.5 nm [56]. The beads interact via bonded and nonbonded interactions. The bonded interactions are represented by covalent bonds modeled by bond-stretching and angle-bending potentials. The bond-stretching is modeled by a harmonic potential in the form $U_S(r) = K_s(r-r_0)^2$, where r is the position vector of the bead, r_0 is the equilibrium value set to $r_0 \equiv \ell = 1\,\sigma$, and K_S is the penalty against bond stretching. In order to prevent artificial strand passages under compressive forces of confinement and piston compression, the force constant K_S was set to $80\,\varepsilon_0$, where $\varepsilon_0 = k_B T$ represents energy of thermal fluctuations. The angle-bending interaction is modeled by a harmonic potential in the form $U_b(\theta) = K_b(\theta - \theta_0)^2$, where θ is the angle between vectors of two consecutive monomers in the chain, θ_0 is the equilibrium angle set to $\theta_0 = \pi$, and K_b is the force constant of the interaction representing the energy penalty against bending of the angle. The force constant K_b relates to the persistence length of the DNA molecule and we set the value $K_b = 20\,\sigma/\varepsilon_0$, thus imposing on the beaded chain a common value for the persistence length of the DNA molecule [57]. The nonbonded terms of the potential involve excluded volume interaction. The excluded volume interaction describes the volume occupied by monomers of the chain and it was modeled by fully repulsive cut and shifted Lennard-Jones potential $U_{ex}(r_{ij}) = 4\varepsilon_0[(\sigma/|r_{ij}|)^{12} - (\sigma/|r_{ij}|)^6] + 1/4$ if $|r_{ij}| < 2^{1/6}\sigma$ and $U_{ex}(r_{ij}) = 0$ otherwise, where $|r_{ij}|$ is a distance between a pair of beads position vectors r_i and r_j, where $i \neq j$.

2.2. Model of Nano-Channels

The channels were modeled by using an implicit helical confinement with a helical geometry developed in the previous work [36]. The channel is modeled as a tube with radius R_{ch} and central axis described by the equation of a helix given as $r_0(t) = kt\hat{i} + R_H\cos(wt)\hat{j} + R_H\sin(wt)\hat{k}$, where t is a periodic parameter in radial space and w gives a subtended angle as t increases [58]. The parameter w also carries information on the handedness of the channels, where $w/|w|<0$ corresponds to left-handed channels with negative winding and $w/|w|>0$ is used to model right-handed channels with positive winding. The channels are defined by four parameters: sign(w), radius of the channel R_{ch}, radius of the helix, R_H, and the pitch, k. The pitch of the channel determines the distance between the helical loops of the channel, $d_H = 2\pi k\sigma$. If the radius of helix is set to $R_H = 0$, the geometry of the channel corresponds to a simple cylindrical channel regardless of the settings of k and w. The model for implicit helical confinement was implemented into the Extensible Simulation Package for Research on Soft Matter Systems (ESPResSo v. 4.2) software that was used in the simulations [59,60]. The model for the implicit helical confinement implemented an algorithm for calculation of the distance of a point from the helix [61]. Despite the algorithm involving an iterative step, we experienced that the simulations are very fast and stable for a wide range of settings tested in the current and the previous work [36]. Within the simulations, we simulated DNA chains in the channels with the radii of the channel corresponding to three confinement strengths given in terms of the ratio of the channel diameter to the polymer's persistence, $D/P = 2^i$, where $i = -1, 0$, and 1. The corresponding diameters of the channels in physical units correspond to 10, 20, and 40 nm. These sizes of nanochannels are relevant to genomic experiments [62] and are also achievable in the preparation of chiral membranes [50]. As for the particular parameter settings, the radius of the helix was set to $R_H = 1/3\ R_{ch}$, based on our previous work, where the effects of the chiral confinement were the strongest in the range of $R_H = 1/3 - \frac{1}{2}R_{ch}$ [63]. In the case of the simulations of cylindrical channels, the setting of R_H was 0. The pitch was set to $k = D/2\pi$ to allow for comparisons and discussion of differences between simulations of chains pushed into open infinite channels and those compressed in the blinded nanochannels of knot factories investigated recently [36]. Both of the possible chiral scenarios were simulated with the handedness of the channels being $w = -1$ and 1. The polymer was inserted into the channel as a stretched chain with beads placed along the major axis of symmetry of the channel corresponding to the x-axis, with initial coordinates $r_{i,x} = I$, where $i = 1\ldots N$, and $r_{i,y} = r_{i,z} = 0$.

2.3. Push by External Force and MD Simulation

The push of the DNA chains was intermediated by a piston modeled by an additional bead with a very large radius, i.e., a radius of the piston $R_P \gg R_{ch}$. The "piston bead" moved along the main axis of symmetry of the channel corresponding to the x-axis, while the movement in the y and z directions was constrained. The piston bead also interacted with the chain only by the excluded volume interaction in the form, as provided above, and with the setting of $\sigma_P = 100\ \sigma$. We applied external force on the piston bead with the range of values $F\sigma/\varepsilon_0 = 0.1, 0.5, 1, 2$, and 5. The region of interest for these forces was chosen based on previous studies [24,25,36] while omitting the special regime of very small forces in narrow channels. The pushing forces were transformed to velocities of pushing after performing the simulations and computing the piston velocity as the overall distance traveled by the piston over the period of simulation time, $v = d_T/\tau_{sim}$. The pushing of the DNA inside the channel was carried out as in the recent molecular simulation work by using Langevin dynamics [34]. We carried out Langevin molecular dynamics simulations by solving Langevin equations of motion for each bead

$m\ddot{\mathbf{r}} = -\nabla U(\mathbf{r}) - \gamma m \dot{\mathbf{r}} + R(t)\sqrt{2\varepsilon_0 m \gamma} + F_{ext}$, where each term in the equation represents the force acting on the bead, $\nabla U(\mathbf{r})$ is given by the molecular potential, $-\gamma m \dot{\mathbf{r}}$ represents the friction, and $R(t)\sqrt{2\varepsilon_0 m \gamma}$ is the random kicking force, where $R(t)$ is a delta-correlated stationary Gaussian process. The last term in the equation is the added external force and applies only to the piston bead with $F_{ext}\sigma/\varepsilon_0 = F\sigma/\varepsilon_0 = 0.1, 0.5, 1, 2,$ and 5. The Langevin equation does not consider hydrodynamic interactions between monomers. For simulations involving out-of-equilibrium systems, especially when intense interpenetrations and interactions among polymer segments are present, a more suitable thermostat, such as dissipative particle dynamics (DPD), is required. This is notably encountered in bulk polymer brushes [64]. However, results from the Langevin dynamics and DPD converge when simulating systems with a low extent of monomer interactions, typically at concentrations below the collective regime [65]. For instance, recent work employed Langevin dynamics to simulate the pushing of DNA inside nanochannels [34]. This choice was justified by the fact that the chain conformations reside in the transition region between the Odijk and deGennes regimes where the hydrodynamic interactions are effectively screened and the collective regime does not apply [66].

The equations of motion were integrated with the step size $\Delta \tau = 0.010$. The optimal size of the integration step $\Delta \tau$ was determined through previous simulations [36,63]. It is small enough to maintain the stability of the equations of motion during integration, while also maximizing the performance of computer simulations and the usage of computer time. The coarse-grained time is transformed to $\tau = 6\pi\eta\sigma^3/\varepsilon$ physical units [56], where η is the viscosity of media. After the chains were inserted into the channel with the initial coordinates $r_{i,x} = i$, where $i = 1\ldots N$, and $r_{i,y} = r_{i,z} = 0$, we performed an initial pre-equilibration run with 10^7 iterative steps. Afterwards, the main simulation started with 10 repeated runs for each parameter setting performing 10^9 integrations, while we were also collecting data for analyses of polymer metrics, monomer distributions in the channels, and topological analyses of polymer knottedness.

3. Results and Discussion

3.1. General Polymer Metrics

First of all, we characterized the simulated systems by evaluating basic polymer metrics. Polymer metrics provide first and important information about polymer conformation, its size, and polymer behavior in the presence of confinement and compressive forces, realized here by a pushing force mediated by a piston or gasket. Polymer metrics also allow the bridging of gaps in standing theoretical understanding and gaining of insights into how the confinement and presence of external forces alter the behavior of the polymer.

Based on our previous experience from our previous study [36] and also in context of existing works relevant to studying polymers under compression in nanochannels [24,25,30], we chose to evaluate the polymer metrics in terms of polymer chain span, although the polymer metrics in terms of end-to-end distance and gyration radius are provided in Figure S1. Given a certain configuration, the span is calculated as the distance separating the two farthest beads of the chain.

The span is calculated as the maximum distance between two monomers, represented by coarse-grained beads, that can be found on the chain. The distances are calculated by using position vectors of the beads. When computing the span, one can use all Cartesian coordinates or compute the span only using the coordinate along the major axis of inertia of the channel. We decided on the latter case, since it diminishes chain size effects. The span is defined as $S(x) = \max |r_{i,x}, r_{j,x}|$, where $i \neq j$ and $i \in 1\ldots N$. The computed span is shown in Figure 1. In Figure 1, we not only show and compare span as obtained on polymers pushed inside the open infinite channels with helical and cylindrical geometry, but we took also advantage of having studied the case of polymers compressed by a gasket in blinded channels with an impenetrable wall at their bottom, corresponding to the experimental setting of knot factories [36].

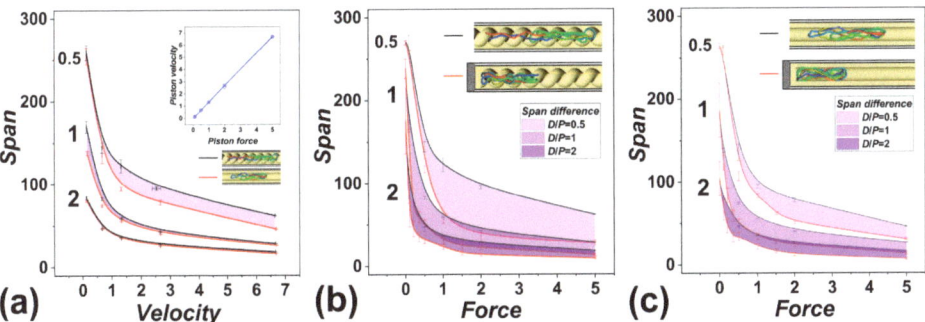

Figure 1. Polymer metrics in terms of the molecule's span during pushing of DNA molecule inside channels as a function of pushing speed and confinement strength expressed as D/P. (a) The panel compares the polymer span in units of σ/ℓ pushed with different velocities of the piston v in units of $1000\,\sigma/\tau$. The black lines indicate the span of a polymer pushed in helical channels, and red lines correspond to the simulations in a cylindrical channel. The areas in shades of purple show differences with the respective cases and a given confinement strength. The inset shows the velocities of the piston obtained for the external force employed on the piston. (b) The panel shows the span of polymer pushed in open infinite helical channels and compares the data with a previously investigated case of a polymer compressed in a blinded channel [36]. (c) The panel shows a comparison of the polymer's span in open infinite channels versus blinded channels with cylindrical geometry and as a function of force in the units of $F\sigma/\varepsilon_0$ and confinement strength expressed in terms of D/P. The insets show rendered snapshots obtained for $D/P = 0.5$ and $F = 5\varepsilon_0/\sigma$, where the polymer is shown in a rainbow color scheme, with one end of the polymer shown in blue and the other in red.

Figure 1a shows the dependence of the polymer's span, as obtained in channels in helical and cylindrical geometries, for three confinement strengths expressed in terms of the ratio of the diameter of the channel to polymer persistence length, $D/P = 0.5, 1,$ and 2. The values of the span are shown as black lines for helical channels and red lines for cylindrical channels, as also illustrated by inset snapshots from the simulations. The filled area indicates differences of span obtained for a given setting confinement strength D/P and between the two investigated geometries of the channels.

The span is shown as a function of velocity of the piston pushing the polymer along the infinite channels. The push by the piston is realized by applying an external force to the piston. Hence, the velocity of the piston was obtained from the simulated trajectories as a total distance traveled by the center of the piston from the beginning of the simulation to its final position at the end of the simulation over the simulation time τ, $v = |\mathbf{r}_P(0), \mathbf{r}_P(\tau)|/\tau$. The values of the velocities, the polymer spans, and other data evaluated later in the work were averaged over ten simulated trajectories. The obtained velocities are shown in the inset graph as a function of applied external force. The velocities v range from 1.2×10^{-4}–$6.6 \times 10^{-3}\,\sigma/\tau$, while the physical dimensions of the units are $[\sigma] = 2.5$ nm and $[\tau] = 74$ ns $\times\,(\eta/\eta_0)$ ns [56], where η_0 is the viscosity of pure water, $\eta_0 = 1$ cP, and η is the viscosity of the actual buffer used in the nanofluidic experiment. The buffers in nanofluidic applications often consist of a solution containing polymers, saccharose, agarose, etc., to increase the hydrodynamic drag of the media on the molecule [41,42], while viscosity can be increased to 10–80 cP [67]. The correction for the viscosity gives some space for variation to transformed value of the physical time units; nonetheless, the values of experimental velocities of pushing, in the order of μm/s, are accessible.

We would like to note that we opted for realizing the push intermediated by an applied external force instead of directly moving the piston by a constant distance at a time, as simulated in some existing works [34]. This allowed us to directly compare the metrics of the polymer pushed in open infinite channels to the previously obtained results on polymers compressed by the external forces inside channels that were blinded by an impenetrable wall [36].

The comparison of polymer metrics pushed in helical and cylindrical geometries on Figure 1a shows that there is a significant difference between the chain span observed in very narrow channels, with $D/P = 0.5$. The difference seems to be higher than in the compression in channels with an impenetrable wall. The difference in polymer span also seems to disappear at weak confinement strengths in terms of D/P, especially if strong compressive forces are applied.

In Figure 1b,c, we show the computed polymer spans as obtained in channels with helical and cylindrical geometry. Figure 1b compares the case of polymers pushed inside infinite open channels to the case of polymers compressed in channels with an impenetrable wall and helical geometry. Figure 1c, on the other hand, compares the pushing versus compression in channels with cylindrical geometry. The data for the current simulations are shown as black lines, and they are compared to the simulations in blinded channels shown as red lines. The insets show the simulation settings by snapshots taken from the simulations augmented by hatching schematics.

The comparison of the computed span shows, in general, smaller compaction of the polymer when compressed by pushing inside the open channels than in the case when the polymer is compressed against the impenetrable wall in the nanochannels. We understand the obtained results as follows. Given the form of the equations of motions in the Langevin dynamics, provided in the Methodology section (Section 2.3), the hydrodynamic drag force, represented by $-\gamma m \dot{r}$ in the Langevin equation, opposes the motion of the particle, leading to a damping effect. The strength of this damping is determined by the value of γ. As the external force acting on the particle increases, the hydrodynamic drag force remains proportional to the velocity of the particle ($-\gamma m \dot{r}$). The linear relationship between hydrodynamic drag and external force is confirmed by the computed velocities as a function of external forces, shown as inset in Figure 1a. When applying compressive force to a polymer confined in an open channel, some of the compressive energy is dissipated by the movement of the chain through the media. As the force-to-displacement ratio still follows the established relationship—$F \cdot D \propto S^{-9/4}$ (with F corresponding to force, D is the diameter of the channels, and S is the span) [30], this means that opening the channels acts like compressing the polymer with smaller force.

We also showed in the previous work that the helical confinement in narrow channels acted to a certain extent like cylindrical channels with a smaller diameter, i.e., channels with higher confinement strength. The conformation of the DNA molecule is determined by a balance of several ongoing forces: the confinement force, elastic force, and hydrodynamic force intermediated by the pushing force [38]. This is a complex relation, where we cannot directly compare the obtained data to a predictive model, but we may bridge the theoretical understanding with our computer experiments. In Figure 1b,c, the computed polymer metrics in terms of polymer span show not only that there are differences between compression in open infinite channels and compression in the nanochannels against an impenetrable wall, but the extent of the differences in DNA compaction is significantly influenced by the geometry of the channels.

As investigated in the existing studies, the compaction of the polymer under external forces is relevant and related to conformational changes [24,25,36], and it is important for devising a control mechanism for the topological state of the polymer for nanotechnological applications [28,34,36,43]. The data also indicate that compression by pushing the polymer into open channels, as compared to the case of compressing the polymer against an impenetrable wall inside finite nanochannels, is much less effective, especially in the case of the helical nanochannels, which will lead to lower degree of knotting, at least at the current setting of the helical geometry in terms of the pitch of the channels. In addition to the data in Figure 1b,c, it is noteworthy that at strong confinement forces in terms of D/P and small compressive forces, the existence of a special regime was discovered, forming a shoulder on the dependencies of the span versus compressive forces [24,25], that was not properly captured, as we did not probe the region in detail by applying a range of sufficiently small compressive forces. The reason is that there are only minor topological differences throughout this region, with the polymer being mostly unknotted; hence, the region is not in the focus of this work.

3.2. The Monomer Distributions upon Pushing

The polymer metrics provide one-dimensional information on the polymer behavior and the effects of confinement and compressive force. Another convenient property that is directly accessible from computer molecular simulations to represent the situation of a dynamically moving polymer molecule is distributions of monomers. Here, we analyze the radial and axial distributions of monomers across the major axes of inertia of the confining channels.

Figure 2 is a composite figure that shows the situation of the polymer's monomers within the channel, providing complementary information to the polymer metrics evaluated in the previous Section 3.1. We believe the figure, as presented, provides advantageous insight for readers when convening to a concise explanation of the figure. Figure 2 is divided into six panels. Each panel shows the heatmaps of the monomer distributions inside the nanochannel at the very left side of the panel. The heatmaps present valuable pictures of the overall distribution of monomers. Since the monomers travel through the channel, sometimes to very long distances, during pushing, we modified the method of calculating the heatmaps employed in earlier works [36]. The heatmaps show concentrations of monomers through the periodized distance of one helical turn, $d_H = 2\pi k\sigma$, where $k = D/2\pi$ in helical channels, and d_H simply equals D. There are five heatmaps corresponding to five settings of the external force, from the bottom up following the increasing velocity of pushing, as shown in the inset of Figure 1a.

The heatmaps indicate expulsion of monomers into the lateral sides of the channel with an increasing velocity of pushing. This effect is numerically captured in the graphs showing radial distribution functions that are displayed adjacent to the heatmaps in each of the panels. The radial distributions show a similar shape to those already obtained for cylindrical [36,68] and square channels [69], with a maximum of the number density of monomers in the middle of the channels. As the velocity of pushing increases, the radial distribution functions flatten, and the distributions show a drop in the monomer concentration in the center of the channel. The direction of the concentration drop with increasing velocity of pushing is indicated by a downwards arrow in the part of the distribution corresponding to the center of the channel. At the same time, the monomers are pushed and redistributed towards the lateral sides with increasing velocity of pushing. The increasing number concentration, expressed as a normalized frequency of occurrence, at the lateral sides of the channel in the direction of increasing velocity of pushing is indicated by an upwards arrow.

To the right of the radial distributions on each panel, we also show the axial distribution of monomers along the main axis of inertia of the channels. These distributions show the number density or concentration of monomers from the position of the piston. In the axial distributions, the position of the piston is always at the origin. The values on the x-axis also reflect the direction of push in the simulations, which went from right to left in Cartesian coordinates. The shape of the concentration profiles is very similar to what is observed experimentally in the dynamic nonequilibrium segmental concentration profile of a single nanochannel-confined DNA molecule [42,43]. In our case, the profiles correspond to equilibrium profiles obtained for different velocities of pushing. The axial distribution shows an evolution with an emerging peak in the number density of monomers near the surface of the piston. The maximum of the distribution increases with increasing velocity of pushing. At the same time, the distribution or occupancy of the channel along the major axis of inertia in the direction of pushing becomes narrower. This narrow region can be associated with a conformational transition with increased spooling [34,35,70], identified earlier to occur under large compressive forces. The insets in the graphs with axial distribution of monomers also show positioning of monomers along the channel in terms of their bead index as a function of their coordinates along the main axis of inertia (x-coordinate). This kind of projection was used in some existing works studying polymers in nanochannels by other authors [34,35]. The insets show the positionings for two limiting cases of the velocity settings used in the current simulations, i.e., $F\sigma/\varepsilon_0 = 0.1$ and 5. Each panel also shows a representative snapshot from the simulation showing polymer in the channel with a particular geometry obtained at the end of the simulation for the setting of $F\sigma/\varepsilon_0 = 1$.

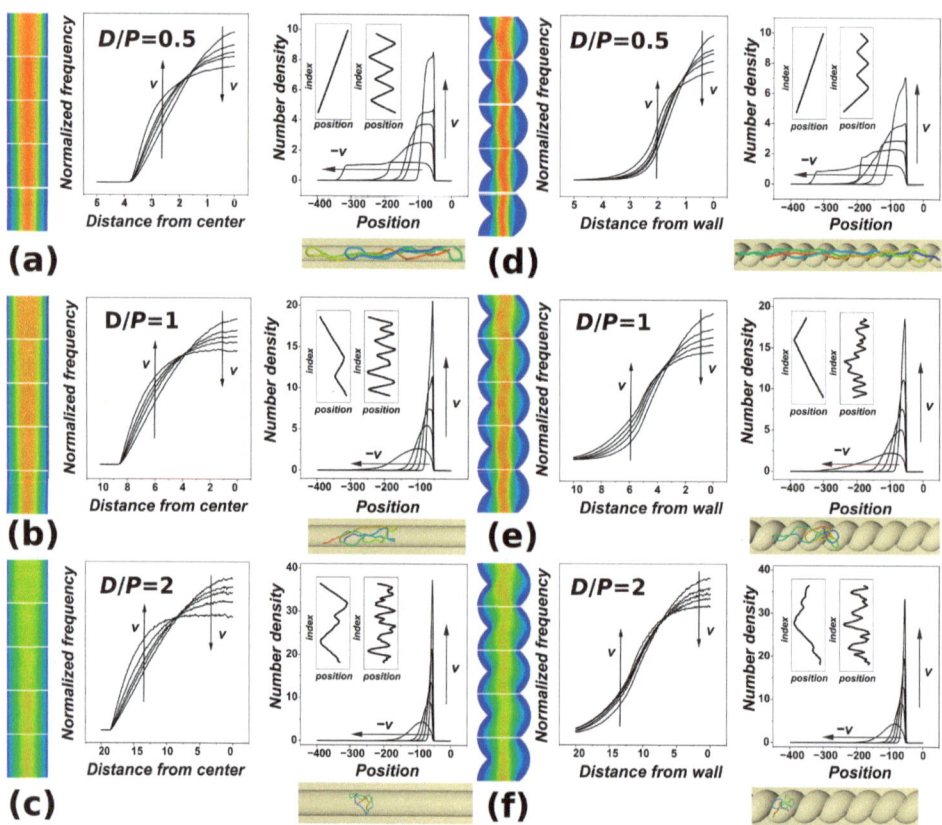

Figure 2. Monomer distributions of DNA polymer pushed inside channels with cylindrical and helical geometry. The panels show the distributions obtained in cylindrical (**a**–**c**) and helical channels (**d**–**f**). The distributions are also shown for different confinement strengths in terms of the D/P ratio, equal to 0.5 (**a**,**d**); $D/P = 1$ (**b**,**e**); and $D/P = 2$ in panels (**c**,**f**). The very left of each panel shows heatmaps of the monomer distributions with a color gradient ranging from blue to red, where the distribution's peak is indicated by red and zero occurrences are denoted by blue. The left graph in every panel shows monomer distribution along the channel from the position (in units of σ/ℓ) of the piston as a function of piston velocity, taking into regard also direction of pushing in the simulations. The insets of this graph also show index-to-bead projections of axial monomer distributions obtained for $F = 0.1\varepsilon_0/\sigma$ and $F = 5\varepsilon_0/\sigma$. The graphs to the right in the pair in every panel show radial distribution functions of monomers from the geometric center of the channel in units of σ/ℓ, in simulations represented by the $x = 0$ axis. The arrows indicate the direction of increasing or decreasing velocity. Each panel also shows a snapshot obtained for the particular geometry of the channel, confinement strength D/P, and force = 1 ε_0/σ. The polymer is depicted in rainbow coloring, with the first bead in blue and the last bead in the chain in red.

Figure 2a–c show the distribution of monomers for cylindrical channels, and Figure 2d–f show the information as obtained in the channels with helical geometry. The rows show computed distributions for different settings of confinement strength, indicated on the graphs as $D/P = 0.5$, 1, and 2. We can see that in the absolute numbers, a larger maximum on the axial distribution near the position of the piston surface is achieved in the case of cylindrical channels. Also, the observed flattening of the radial distributions is more extensive in the case of cylindrical channels. This indicates that when pushing the polymers inside open channels, the force is less effective in compressing the polymers, and the monomers do not fully explore the helical grooves of the helical channels.

Consistently with the polymer metrics shown in Figure 1, the data indicate that the extent of compaction is lower than that previously observed for finite channels with an impenetrable wall [36]. The previous studies showed that the level of compaction with higher compressive forces applied is directly related to the extent of knotting [28,34,43]. In our previous simulations, we also saw that helical geometry by itself enhances knotting as compared to simple cylindrical geometry. In the current simulations, the lower level of compaction and lower effectiveness of compression with given forces in open channels suggest that smaller levels of knotting will be expected, especially for very narrow channels and strong confinement.

3.3. Knotting Probabilities and Topology

In this section, we evaluate the topological state of DNA polymers under compression induced by pushing through open infinite channels. The knotting probability is evaluated as the frequency of finding knots in ten runs along the trajectories that contain 5000 structures each for topological analyses. The occurrence of knots was evaluated by Knoto-ID software [53] (v1.3.0), which uses Jones's polynomial, allowing us to obtain information on the handedness of the knots. The knotting probability analyses focused on obtaining information on the complexity of knots in terms of the crossing numbers, knot groups evaluating presence of amphichiral knots, twist knots, torus knots, unknots, and unidentified knots with a crossing number larger than 11. Finally, the analyses also investigated the chirality of knots in terms of the occurrence of right-handed and left-handed knots.

Figure 3 presents and summarizes the findings about the topology of the polymers. The first two columns compare the knotting probability as a function of the velocity of pushing or compressive force. The arrows above the columns indicate the direction of the increasing velocity of pushing. The frequencies of occurrence of knots with a given complexity characterized by their crossing numbers are shown as stacked areas distinguished by colors in a thermometer scale. The crossing numbers corresponding to particular colors at the employed scale are indicated at the bottom. The thermometer scale goes through a spectrum of colors from plain blue to plain red, where the plain blue corresponds to unknots and the plain red areas show presence of very complex knots with a crossing number larger than 11. The rows correspond to the particular setting of confinement strength, indicated in terms of the D/P ratio.

The analysis shows that, in general, the probability of knotting increases with the velocity of pushing. Also, it is shifted towards the occurrence of more complex knots with increasing velocity of pushing. This observation is consistent with previous investigations of knotting in DNA pushed through square channels [34] and also in studies of polymers under compressive force in cylindrical [28,36] and helical channels [36]. The knotting probability also depends on the diameter of the nanochannels. Here, the dependence is similar to the situation of knots compressed in finite channels, where the knotting probability is the largest in the channels with $D/P = 1$.

On the other hand, the distinctive feature of the pushing inside the infinite channels seems to be apparent lower knotting probability in nanochannels with helical geometry as compared to the cylindrical channels. As discussed in Section 3.1, when pushing inside the infinite channels, the resulting compaction is determined by establishing a balance between several forces, i.e., the confinement force, elastic force, and hydrodynamic force intermediated by the pushing force. The resulting difference in compaction observed by means of molecular simulations between finite and infinite channels indicates that opening the channels affects the compaction in channels with helical channels more than in the case of cylindrical channels. For this reason, unlike the previous simulations in finite channels, the helical channels do not enhance the knotting probability.

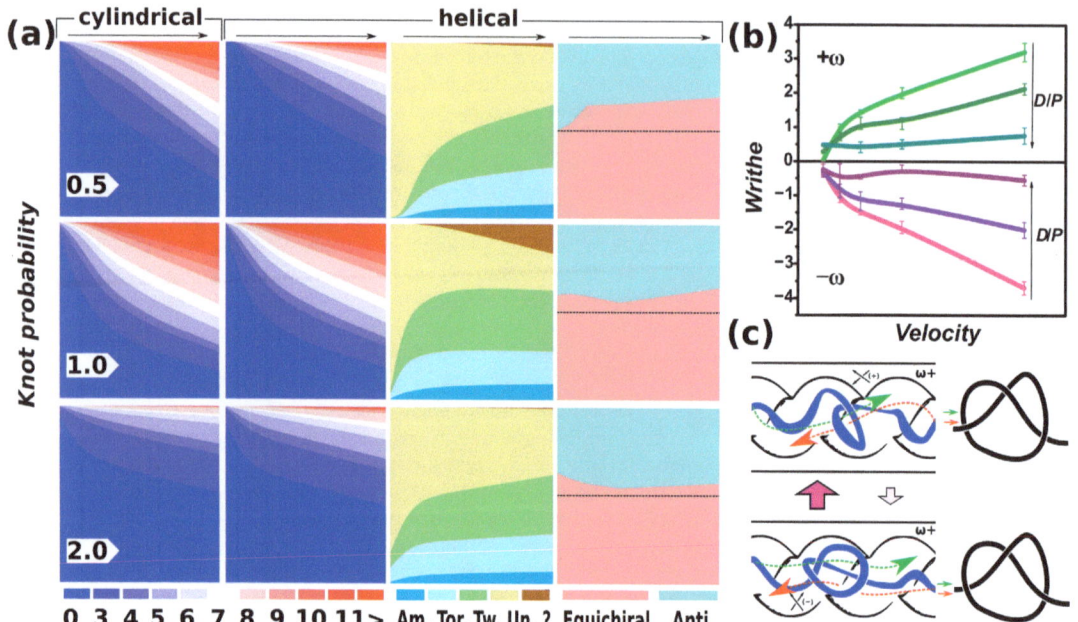

Figure 3. Knotting probabilities. (**a**) The first two columns compare knotting probabilities in terms of crossing numbers, distinguished by a thermometer scale, obtained for DNA polymer pushed inside infinite open channels with cylindrical and helical geometries. The rows correspond to different confinement strength in terms of D/P. In the case of helical channels, the knot types are evaluated in terms of the frequencies of amphichiral (Am.), torus (Tor.), twist knots (Tw.), unknots (Un.), and undefined knots ("?"). The last column shows a comparison of equichiral and antichiral knots, i.e., the knots with the same handedness as or the opposite handedness to that of the chiral helical channel. (**b**) Average writhe of the DNA chains in helical channels as a function of increasing pushing velocities and strength of confinement in terms of the D/P ratio. The $\omega+$ and $\omega-$ indicate handedness of the channels (see Section 2.2). The computed pushing velocities are shown in the inset of Figure 1a. (**c**) A proposed mechanism for how the helical channels control handedness of compression–confinement-induced knotting.

The decreased knotting probability in helical channels can, however, be due to the specific geometric parameters of the helical channels determined by the pitch $k = D/2\pi$, which determines the distance between helical loops or size of the helical turns, $d_H = 2\pi k\sigma$. It is important to note that the current setting of the pitch was chosen based on our previous work where we investigated chiral effects in terms of mobility of localized knots with a given chirality. It is probable that for the current experimental setting of the polymers pushed inside the open channels, the pitch has to be fine-tuned, perhaps towards larger values above the deflection length $\lambda = (D^2/P)^{1/3}$ [71], so that $d_H > 2\pi k\sigma$ for k fixed to $k = D/2\pi$.

In order to gain insight into this behavior, we simulated DNA polymer pushed into a helical channel with the size of helical loops well above persistence length, set to $d_H = 2\,P$, and analyzed the knotting probability, as shown in Supplementary Information, Figure S2.

Figure S2 surprisingly indicates significant compaction associated with high levels of knotting even at strong confinement forces ($D/P = 0.5$) and very small compressive forces ($F\sigma/\varepsilon_0 = 0.1$). Since detailed refinement and thorough exploration of the parameter settings of the channels' geometry are clearly beyond the scope and extent of a single work, we will readdress it in future works, providing more information on polymer behavior for various parameter settings of the helical channels, polymer chain length, and persistence lengths.

The midsection of Figure 3a for helical channels also shows knotting probabilities for different knot groups, represented by amphichiral, torus, twist knots, unknots, and complex knots above 11 crossings for which the notation is not included in the topological software. It has to be noted that the populations are normalized again, but some of the groups overlap, such as the trefoil knot, which is both the twist and torus knot, and the 4_1 knot, which is both the twist knot and the amphichiral knot. The population of the amphichiral knots increases with the pushing velocity and the level of compaction under the compressive force of pushing, which increases the complexity of entanglements.

Note that there are only 20 amphichiral knots out of 801 knots that can be constructed from knotted lines up to 11 crossings [72]. The populations of twist knots seem to be increasing in the case of strong and weak confinement forces, $D/P = 0.5$ and $D/P = 2$. This might be related to the fact that for both of these settings, the knottedness is lower than in the case of $D/P = 1$, and the amounts of existing twist and torus knot types as a function of crossing number do not evolve equally; in other words, with increasing crossing number, there are more twist knots than torus knots. For the cases outside intermediate confinement, $D/P = 0.5$ and 2, we see an abundance of unknots. In the case of the strong confinement, $D/P = 0.5$, the unknots can be related to lower degrees of compaction and prevailing effects of confinement keeping the chain extended. In the case of weak confinement, $D/P = 2$, chain length effects or timescale and velocity effects might be taking place. The polymer at its given length is much more diluted; hence, on one hand, it leads to much more spooling, but also it might have not enough time to explore the geometrical spaces of the larger channels at the fixed rate of pushing. This may lead to the higher extent of writhing indicated in Figure 3b, and perhaps much smaller forces/velocities of pushing should be investigated in the case of weak confinement in the future to obtain a general picture on polymer behavior in channels with geometric modulation. Such regimes of very slow pushing in wide channel will require separate computer experiments, given the heavy computational expenses inevitable for such a computer experiment.

3.4. Handedness of the Channels and Knots

We further investigate whether the effect of the handedness of the helical channels on the knot chirality is preserved to some extent, and if channels with helical geometry and given handedness can be used to control handedness of the knots that are created during the pushing of the DNA through the channels. In the previous study [36], we probed the chirality of the knots by computing writhe of the knotted part additionally to the topological analyses by KymoKnot software (http://kymoknot.sissa.it/) [73].

In the current work, we directly use the information on the chirality of the knots provided by Knoto-ID software [53]. However, the information had to be pretreated before the data could be used to compute statistics of left-handed and right-handed knots. Knoto-ID identifies the knot types by evaluating Jones's polynomial and compares the identified knots with the notation in Rolfsen's table of knots [54]. If the knot has the opposite chiral projection than shown in Rolfsen's table, Knoto-ID identifies such knots as "mirror images" and indicates the chirality by the letter "m" in the name of the knot type in the result. Originally, Rolfsen's table does not distinguish the knots based on their chirality, so knots with positive and negative writhes are mixed. This situation motivated Vázquez et al. to suggest the creation of a new biologically motivated knot table, where the aspect of the knot's handedness in terms of the prevailing knot's writhe would be considered and reflected [55]. In their work, the authors identified the knots with opposite chirality than indicated in the Rolfsen's table. We used this sorted information together with the knot types detected by Knoto-ID software as Rolfsen's analogues or their mirror images in order to identify the left-handed and right-handed knots.

Moreover, mathematically, knots occur only on closed curves, and the algorithms for finding knots often involve some kind of closure method that constructs a connection between the free ends of the linear polymer chain. In order to eliminate the possible bias coming from the closure method, we evaluated only the knots found in conformations with an arbitrarily chosen very short end-to-end distance. It is noteworthy that although we do not know how the end-to-end distance and the bias from the closure method are related quantitatively, one intuitively expects that the number of entanglements introduced by closing the arc grows with the distance spanned by the added closing segments [74]. If we consider, for simplicity, a case where the end-to-end distance is equal to the size of the bond length $\ell = 1\,\sigma$, there would be no need to construct a closure.

Furthermore, evaluating knots at short end-to-end distances can be of practical relevance, as the knots could be chemically embedded in the polymer by closing the polymer ring chemically. The distance was set to $10\,\sigma$ based on the average variation of the end-to-end distance found in consecutive frames in simulated trajectories, and we consider it a ligation distance. After computing numbers of right-handed and left-handed knots, we evaluated their statistics, which are summarized in the last column of Figure 3. Here, the knots with the same handedness as the handedness of the helical channel are summed up as the number of equichiral knots. In the opposite case, the knots with the opposite handedness to that of the channels are summed up as antichiral ones. The graphs in the last column of Figure 3 show the ratio of the populations of the equichiral and antichiral knots. The graphs indicate that even in the regime of DNA compression by pushing inside the nanochannels, the ability of helical channels to give rise to equichiral knots is preserved, as we observed in the case of compression knot factories with finite nanochannels. It should be noted that in the new rigorous table by Vázquez et al. [55], chirality of knots was determined only up to nine crossings, which we used in our analyses and the plots in Figure 3 (others were not included); hence, the resolution of the chiral channels can be even higher but it is beyond current knowledge and is a subject for future investigation.

3.5. On the Mechanism of How Geometry of the Channels Induces Handedness of Knots

In addition to the information on knotting statistics, computer simulations can also help understand the mechanisms by which knotting occurs [28,75]. In the following, we propose a mechanism for how the chirality of the channels may control the chirality of knots and entanglements created on a DNA chain exposed to compression by being pushed inside the nanochannels.

An earlier work investigated this by inspecting positions of emerging knots along the polymer chains [28], and found that the knots were created mainly by backfolding and threading of polymer ends through the loops. Another work also suggested that the knotting mechanism may involve maintaining contacts of the DNA polymer at specific sites, as found on ribosomal surfaces [75,76]. We believe that the process of knotting in chiral channels involves both compression-induced backfolding and threading, as well as the direction of writhing induced in polymer chains that are in contact with the walls of the chiral channels, aligning the polymer with the winding direction of the helical channels.

The occurrence of backfolding is evident from the evolution of the polymer metrics in terms of the chain span, investigated in Section 3.1 and shown in Figure 1. It is also evident in the index-position projections of the monomer distributions, which we investigated in Section 3.2, and that show multiply folded conformations in Figure 2. A multiply folded chain creates loops that preferably turn and twist in the direction controlled by the handedness of the curvature of the chiral channels. This preference in twisting and folding in the direction of the winding of the helical channels is demonstrated by the average value of the writhe whose sign matches the writhe of the channel (Figure 3b). The channel-induced writhe was investigated in the previous work and we show the average writhe computed by evaluating Gauss integral [77] in Figure 3b. Twisting of the loops is also indispensable for creating knots by threading, as illustrated by the drawing in Figure 3c and demonstrated using a physical analog model (Figure S3) for which we used rubber tubes and 3D printed models.

Additionally, we show the evolution of the writhe throughout the simulations in Figure S4. The graphs in Figure S4 indicate that the writhe evolution begins with an abrupt change upon initiation of pushing, sometimes with a arbitrary direction of writhing as the pushing forces come into play. The average value and sign of the writhe further evolve, so that the sign aligns with the chirality of the channels. In the case of cylindrical channels, the sign tends to approach zero after a sufficiently long pushing inside the channels. These evolutions indicate that the handedness of knots is not inherited from a arbitrary initial structure after a long equilibration run, and that the process of controlling handedness benefits from a longer period of pushing inside the channels. Hence, the overall writhe of the chiral spaces plays a pivotal role in determining the chirality of the knots and influences their bias toward a particular chiral form.

4. Conclusions

By means of coarse-grained molecular dynamics simulations, we studied the behavior of polymers in terms of polymer metrics, monomer distributions and topology of the polymer chain pushed inside infinite nanochannels with both cylindrical and helical geometries, using a DNA biopolymer model system. The simulations showed that the polymer undergoes a compaction upon increasing pushing velocity that could be used for controlling knottedness.

When compared to simulations of polymers compressed in finite channels, distinct features emerge. Primarily, the geometry of the channels exerts varying effects on the extent of polymer chain compaction. Consequently, when the polymer is pushed inside open channels, it forms fewer knots compared to when it is compacted by compression within finite channels against an impenetrable wall. The confined environment of the open channels limits the polymer's ability to explore helical loops of the chiral nanochannels, but it still generates equichiral knots and equichiral writhe. The lower degree of knotting observed in helical loops during the pushing inside helical channels may seem inconsistent with Ralf Metzler's argument regarding enhanced knotting resulting from irregularities in nanochannels [78]. However, it is essential to note that the argument did not specify the spacing between these irregularities, whereas in our case, the irregularities are represented by the helical loops. Therefore, we posit that if the spacing between these irregularities is smaller than the DNA's persistence length, their effects become translated into the increased confinement strength, which counteracts polymer folding.

Our findings also prompted additional simulations involving variations in the pitch of the channels and the radius of the helix. Some of these simulations were included in the discussion of the results and provided as Supplementary Information. They unveiled unprecedented backfolding and facilitated control over chirality, setting the stage for further investigation. The simulations demonstrate the feasibility of controlling polymer chirality by pushing them inside helical nanochannels, a feat achievable in experimental settings. This control over chirality can be fine-tuned by adjusting channel geometry, such as the pitch, the pushing regime (velocity, duration), channel length, and chain length. The last aspect, exploring chain lengths, remains a significant focus, especially in the context of wider channels. For a more comprehensive examination of knot spectra, simulation schemes employing Monte Carlo simulations, as developed in existing works, would be more suitable than molecular dynamics simulations. This approach allows for efficient conformational sampling at lower ligating distances, without generating correlated conformations, and minimizing bias induced by the closure method. Investigating knots at the ligating distance holds importance for potential applications such as chemical synthesis and embedding knots onto the polymer chain through methods like photoinitiated polymerization or click chemistry [79,80].

Supplementary Materials: The following supporting information can be downloaded at: https://www.mdpi.com/article/10.3390/polym15204185/s1, Figure S1: Polymer metrics as a function of pushing force in confinements with different strengths; Figure S2: The monomer distributions and knotting probabilities in helical channel with the pitch of the channel set to 2 P; Figure S3: Physical analogue model is used as a demonstrator of how the helical channels induce handedness to knots; Figure S4: Evolution of writhe with the simulation time.

Author Contributions: Conceptualization, D.R.; methodology, R.R. and D.R.; software, R.R. and D.R.; validation, R.R. and D.R.; analysis, R.R.; investigation, D.R. and R.R.; resources, D.R. and R.R.; data curation, D.R. and R.R.; writing—original draft preparation, D.R.; writing—review and editing, R.R. and D.R.; visualization, D.R.; supervision, D.R.; project administration, D.R.; funding acquisition, D.R. All authors have read and agreed to the published version of the manuscript.

Funding: This research was funded by the Grant Agency of the Ministry of Education, Science, Research and Sport of the Slovak Republic, Grant VEGA 2/0102/20, and further support is anticipated by VEGA 2/0038/24. The research was supported also by the Slovak Research and Development Agency SRDA 21-0346 and SRDA SK-AT 20-0011. Support from International Cooperation Project COST 17 139 EUTOPIA (EUropean TOPology Interdisciplinary Action) is also acknowledged. The calculations were performed in the Computing Centre of the Slovak Academy of Sciences using the supercomputing infrastructure acquired in project ITMS 26230120002 and 26210120002 (Slovak infrastructure for high-performance computing) supported by the Research and Development Operational Program funded by the European Regional Development Fund.

Institutional Review Board Statement: Not applicable.

Data Availability Statement: All data are available on request.

Acknowledgments: The authors are very grateful to Andrzej Stasiak (University of Lausanne) and Peter Cifra (Slovak Academy of Sciences) for useful discussions within duration of the project. The authors are grateful to Zdeno Špitálsky (Slovak Academy of Sciences) for helping with preparation of a physical analogue demonstrator and 3D-printed models.

Conflicts of Interest: We do declare no conflicts of interests.

References

1. Hiemenz, P.C.; Lodge, T.P. *Polymer Chemistry*; CRC Press: Boca Raton, FL, USA, 2007; ISBN 1-4200-1827-2.
2. Dietrich-Buchecker, C.; Rapenne, G.; Sauvage, J.-P. Molecular Knots—From Early Attempts to High-Yield Template Syntheses. In *Molecular Catenanes, Rotaxanes and Knots*; John Wiley & Sons, Ltd.: Hoboken, NJ, USA, 1999; pp. 107–142, ISBN 978-3-527-61372-4.
3. Dietrich-Buchecker, C.O.; Sauvage, J.-P. A synthetic molecular trefoil knot. *Angew. Chem. Int. Ed. Engl.* **1989**, *28*, 189–192. [CrossRef]
4. Deibler, R.W.; Mann, J.K.; Sumners, D.W.L.; Zechiedrich, L. Hin-mediated DNA knotting and recombining promote replicon dysfunction and mutation. *BMC Mol. Biol.* **2007**, *8*, 44. [CrossRef] [PubMed]
5. Portugal, J.; Rodríguez-Campos, A. T7 RNA polymerase cannot transcribe through a highly knotted DNA template. *Nucleic Acids Res.* **1996**, *24*, 4890–4894. [CrossRef] [PubMed]
6. Racko, D.; Benedetti, F.; Goundaroulis, D.; Stasiak, A. Chromatin Loop Extrusion and Chromatin Unknotting. *Polymers* **2018**, *10*, 1126. [CrossRef]
7. Rawdon, E.J.; Dorier, J.; Racko, D.; Millett, K.C.; Stasiak, A. How topoisomerase IV can efficiently unknot and decatenate negatively supercoiled DNA molecules without causing their torsional relaxation. *Nucleic Acids Res.* **2016**, *44*, 4528–4538. [CrossRef]
8. Racko, D.; Benedetti, F.; Dorier, J.; Burnier, Y.; Stasiak, A. Generation of supercoils in nicked and gapped DNA drives DNA unknotting and postreplicative decatenation. *Nucleic Acids Res.* **2015**, *43*, 7229–7236. [CrossRef]
9. Dabrowski-Tumanski, P.; Sulkowska, J.I. Topological knots and links in proteins. *Proc. Natl. Acad. Sci. USA* **2017**, *114*, 3415–3420. [CrossRef]
10. Virnau, P.; Mirny, L.A.; Kardar, M. Intricate knots in proteins: Function and evolution. *PLoS Comput. Biol.* **2006**, *2*, e122. [CrossRef]
11. Jackson, S.E.; Suma, A.; Micheletti, C. How to fold intricately: Using theory and experiments to unravel the properties of knotted proteins. *Curr. Opin. Struct. Biol.* **2017**, *42*, 6–14. [CrossRef]
12. Danon, J.J.; Krüger, A.; Leigh, D.A.; Lemonnier, J.-F.; Stephens, A.J.; Vitorica-Yrezabal, I.J.; Woltering, S.L. Braiding a molecular knot with eight crossings. *Science* **2017**, *355*, 159–162. [CrossRef]
13. Sogo, J.; Stasiak, A.; Martínez-Robles, M.; Krimer, D.; Hernd_ndez, P.; Schvartzman, J. Formation of knots in partially replicated DNA molecules. *J. Mol. Biol.* **1999**, *286*, 637–643. [CrossRef] [PubMed]

14. Lim, N.; Jackson, S. Molecular Knots in Biology and Chemistry. *J. Physics. Condens. Matter Inst. Phys. J.* **2015**, *27*, 354101. [CrossRef]
15. Frisch, H.L.; Wasserman, E. Chemical Topology1. *J. Am. Chem. Soc.* **1961**, *83*, 3789–3795. [CrossRef]
16. Dobay, A.; Sottas, P.-E.; Dubochet, J.; Stasiak, A. Predicting optimal lengths of random knots. *Lett. Math. Phys.* **2001**, *55*, 239–247. [CrossRef]
17. Micheletti, C.; Marenduzzo, D.; Orlandini, E.; Summers, D.W. Knotting of random ring polymers in confined spaces. *J. Chem. Phys.* **2006**, *124*, 064903. [CrossRef]
18. Wettermann, S.; Datta, R.; Virnau, P. Influence of ionic conditions on knotting in a coarse-grained model for DNA. *Front. Chem.* **2023**, *10*, 1627. [CrossRef]
19. Binder, K.; de Gennes, P.-G.; Giannelis, E.P.; Grest, G.S.; Hervet, H.; Krishnamoorti, R.; Leger, L.; Manias, E.; Raphael, E.; Wang, S.-Q. *Polymers in Confined Environments*; Springer Science & Business Media: Berlin/Heidelberg, Germany, 1998; Volume 138, ISBN 3-540-64266-8.
20. Kasianowicz, J.J.; Kellermayer, M.; Deamer, D.W. *Structure and Dynamics of Confined Polymers: Proceedings of the NATO Advanced Research Workshop on Biological, Biophysical & Theoretical Aspects of Polymer Structure and Transport Bikal, Hungary 20–25 June 1999*; Springer Science & Business Media: Dordrecht, The Netherlands, 2012; Volume 87, ISBN 94-010-0401-3.
21. Dai, L.; Doyle, P.S. Universal knot spectra for confined polymers. *Macromolecules* **2018**, *51*, 6327–6333. [CrossRef]
22. Frykholm, K.; Müller, V.; KK, S.; Dorfman, K.D.; Westerlund, F. DNA in nanochannels: Theory and applications. *Q. Rev. Biophys.* **2022**, *55*, e12. [CrossRef]
23. Bleha, T.; Cifra, P. Stretching and compression of DNA by external forces under nanochannel confinement. *Soft Matter* **2018**, *14*, 1247–1259. [CrossRef]
24. Bleha, T.; Cifra, P. Compression and stretching of single DNA molecules under channel confinement. *J. Phys. Chem. B* **2020**, *124*, 1691–1702. [CrossRef] [PubMed]
25. Cifra, P.; Bleha, T. Piston Compression of Semiflexible Ring Polymers in Channels. *Macromol. Theory Simul.* **2021**, *30*, 2100027. [CrossRef]
26. Chakrabarti, B.; Liu, Y.; LaGrone, J.; Cortez, R.; Fauci, L.; Du Roure, O.; Saintillan, D.; Lindner, A. Flexible filaments buckle into helicoidal shapes in strong compressional flows. *Nat. Phys.* **2020**, *16*, 689–694. [CrossRef]
27. Chen, W.; Kong, X.; Wei, Q.; Chen, H.; Liu, J.; Jiang, D. Compression and Stretching of Confined Linear and Ring Polymers by Applying Force. *Polymers* **2021**, *13*, 4193. [CrossRef]
28. Micheletto, D.; Orlandini, E.; Turner, M.S.; Micheletti, C. Separation of geometrical and topological entanglement in confined polymers driven out of equilibrium. *ACS Macro Lett.* **2020**, *9*, 1081–1085. [CrossRef] [PubMed]
29. Jung, Y.; Ha, B.-Y. Confinement induces helical organization of chromosome-like polymers. *Sci. Rep.* **2019**, *9*, 869. [CrossRef]
30. Jun, S.; Thirumalai, D.; Ha, B.-Y. Compression and stretching of a self-avoiding chain in cylindrical nanopores. *Phys. Rev. Lett.* **2008**, *101*, 138101. [CrossRef]
31. Hayase, Y.; Sakaue, T.; Nakanishi, H. Compressive response and helix formation of a semiflexible polymer confined in a nanochannel. *Phys. Rev. E* **2017**, *95*, 052502. [CrossRef] [PubMed]
32. Huang, A.; Reisner, W.; Bhattacharya, A. Dynamics of DNA squeezed inside a nanochannel via a sliding gasket. *Polymers* **2016**, *8*, 352. [CrossRef]
33. Bernier, S.; Huang, A.; Reisner, W.; Bhattacharya, A. Evolution of nested folding states in compression of a strongly confined semiflexible chain. *Macromolecules* **2018**, *51*, 4012–4022. [CrossRef]
34. Rothörl, J.; Wettermann, S.; Virnau, P.; Bhattacharya, A. Knot formation of dsDNA pushed inside a nanochannel. *Sci. Rep.* **2022**, *12*, 5342. [CrossRef]
35. Zeng, L.; Reisner, W.W. Organized states arising from compression of single semiflexible polymer chains in nanochannels. *Phys. Rev. E* **2022**, *105*, 064501. [CrossRef]
36. Rusková, R.; Račko, D. Knot Factories with Helical Geometry Enhance Knotting and Induce Handedness to Knots. *Polymers* **2022**, *14*, 4201. [CrossRef] [PubMed]
37. Chelakkot, R.; Winkler, R.G.; Gompper, G. Flow-induced helical coiling of semiflexible polymers in structured microchannels. *Phys. Rev. Lett.* **2012**, *109*, 178101. [CrossRef] [PubMed]
38. Zhou, J.; Wang, Y.; Menard, L.D.; Panyukov, S.; Rubinstein, M.; Ramsey, J.M. Enhanced nanochannel translocation and localization of genomic DNA molecules using three-dimensional nanofunnels. *Nat. Commun.* **2017**, *8*, 807. [CrossRef] [PubMed]
39. Cifra, P.; Bleha, T. Pressure of Linear and Ring Polymers Confined in a Cavity. *J. Phys. Chem. B* **2023**, *127*, 4646–4657. [CrossRef]
40. Micheletti, C.; Marenduzzo, D.; Orlandini, E.; Sumners, D.W. Simulations of knotting in confined circular DNA. *Biophys. J.* **2008**, *95*, 3591–3599. [CrossRef]
41. Khorshid, A.; Zimny, P.; Tétreault-La Roche, D.; Massarelli, G.; Sakaue, T.; Reisner, W. Dynamic compression of single nanochannel confined DNA via a nanodozer assay. *Phys. Rev. Lett.* **2014**, *113*, 268104. [CrossRef]
42. Khorshid, A.; Amin, S.; Zhang, Z.; Sakaue, T.; Reisner, W.W. Nonequilibrium dynamics of nanochannel confined DNA. *Macromolecules* **2016**, *49*, 1933–1940. [CrossRef]
43. Amin, S.; Khorshid, A.; Zeng, L.; Zimny, P.; Reisner, W. A nanofluidic knot factory based on compression of single DNA in nanochannels. *Nat. Commun.* **2018**, *9*, 1506. [CrossRef]
44. Fielden, S.D.P.; Leigh, D.A.; Woltering, S.L. Molecular Knots. *Angew. Chem. Int. Ed.* **2017**, *56*, 11166–11194. [CrossRef]

45. Valdés, A.; Martinez-Garcia, B.; Segura, J.; Dyson, S.; Díaz-Ingelmo, O.; Roca, J. Quantitative disclosure of DNA knot chirality by high-resolution 2D-gel electrophoresis. *Nucleic Acids Res.* **2019**, *47*, e29. [CrossRef]
46. Katsonis, N.; Lancia, F.; Leigh, D.A.; Pirvu, L.; Ryabchun, A.; Schaufelberger, F. Knotting a molecular strand can invert macroscopic effects of chirality. *Nat. Chem.* **2020**, *12*, 939–944. [CrossRef]
47. Shi, X.; Pumm, A.-K.; Isensee, J.; Zhao, W.; Verschueren, D.; Martin-Gonzalez, A.; Golestanian, R.; Dietz, H.; Dekker, C. Sustained unidirectional rotation of a self-organized DNA rotor on a nanopore. *Nat. Phys.* **2022**, *18*, 1105–1111. [CrossRef]
48. Pairault, N.; Niemeyer, J. Chiral mechanically interlocked molecules–applications of rotaxanes, catenanes and molecular knots in stereoselective chemosensing and catalysis. *Synlett* **2018**, *29*, 689–698.
49. Ashbridge, Z.; Fielden, S.D.; Leigh, D.A.; Pirvu, L.; Schaufelberger, F.; Zhang, L. Knotting matters: Orderly molecular entanglements. *Chem. Soc. Rev.* **2022**, *51*, 7779–7809. [CrossRef] [PubMed]
50. Wang, T.; Jiang, L.; Zhang, Y.; Wu, L.; Chen, H.; Li, C. Fabrication of polyimide mixed matrix membranes with asymmetric confined mass transfer channels for improved CO_2 separation. *J. Membr. Sci.* **2021**, *637*, 119653. [CrossRef]
51. Cheng, Q.; Ma, Q.; Pei, H.; Mo, Z. Chiral membranes for enantiomer separation: A comprehensive review. *Sep. Purif. Technol.* **2022**, *292*, 121034. [CrossRef]
52. Zhang, J.; Albelda, M.T.; Liu, Y.; Canary, J.W. Chiral nanotechnology. *Chirality* **2005**, *17*, 404–420. [CrossRef]
53. Dorier, J.; Goundaroulis, D.; Benedetti, F.; Stasiak, A. Knoto-ID: A tool to study the entanglement of open protein chains using the concept of knotoids. *Bioinformatics* **2018**, *34*, 3402–3404. [CrossRef]
54. Rolfsen, D. *Knots and Links*; AMS Chelsea Publishing: Providence, RI, USA, 2003; Volume 346.H, ISBN 978-0-8218-3436-7.
55. Brasher, R.; Scharein, R.; Vázquez, M. New biologically motivated knot table. *Biochem. Soc. Trans.* **2013**, *41*, 606–611. [CrossRef]
56. Di Stefano, M.; Tubiana, L.; Di Ventra, M.; Micheletti, C. Driving knots on DNA with AC/DC electric fields: Topological friction and memory effects. *Soft Matter* **2014**, *10*, 6491–6498. [CrossRef]
57. Lu, Y.; Weers, B.; Stellwagen, N.C. DNA persistence length revisited. *Biopolymers* **2002**, *61*, 261–275. [CrossRef]
58. Chávez, Y.; Chacón-Acosta, G.; Dagdug, L. Unbiased diffusion of Brownian particles in a helical tube. *J. Chem. Phys.* **2018**, *148*, 214106. [CrossRef] [PubMed]
59. Limbach, H.J.; Arnold, A.; Mann, B.A.; Holm, C. ESPResSo—An extensible simulation package for research on soft matter systems. *Comput. Phys. Commun.* **2006**, *174*, 704–727. [CrossRef]
60. Arnold, A.; Lenz, O.; Kesselheim, S.; Weeber, R.; Fahrenberger, F.; Roehm, D.; Košovan, P.; Holm, C. ESPResSo 3.1: Molecular Dynamics Software for Coarse-Grained Models. In *Meshfree Methods for Partial Differential Equations VI*; Griebel, M., Schweitzer, M.A., Eds.; Springer: Berlin/Heidelberg, Germany, 2013; pp. 1–23.
61. Nievergelt, Y. Computing the distance from a point to a helix and solving Kepler's equation. *Nucl. Instrum. Methods Phys. Res. Sect. A Accel. Spectrometers Detect. Assoc. Equip.* **2009**, *598*, 788–794. [CrossRef]
62. Lam, E.T.; Hastie, A.; Lin, C.; Ehrlich, D.; Das, S.K.; Austin, M.D.; Deshpande, P.; Cao, H.; Nagarajan, N.; Xiao, M. Genome mapping on nanochannel arrays for structural variation analysis and sequence assembly. *Nat. Biotechnol.* **2012**, *30*, 771–776. [CrossRef]
63. Rusková, R.; Račko, D. Channels with Helical Modulation Display Stereospecific Sensitivity for Chiral Superstructures. *Polymers* **2021**, *13*, 3726. [CrossRef] [PubMed]
64. Pastorino, C.; Kreer, T.; Müller, M.; Binder, K. Comparison of dissipative particle dynamics and Langevin thermostats for out-of-equilibrium simulations of polymeric systems. *Phys. Rev. E* **2007**, *76*, 026706. [CrossRef]
65. Ripoll, M.; Ernst, M.H.; Español, P. Large scale and mesoscopic hydrodynamics for dissipative particle dynamics. *J. Chem. Phys.* **2001**, *115*, 7271–7284. [CrossRef]
66. Dorfman, K.D.; Gupta, D.; Jain, A.; Muralidhar, A.; Tree, D.R. Hydrodynamics of DNA confined in nanoslits and nanochannels. *Eur. Phys. J. Spec. Top.* **2014**, *223*, 3179–3200. [CrossRef]
67. Hsieh, S.-S.; Wu, F.-H.; Tsai, M.-J. DNA stretching on the wall surfaces in curved microchannels with different radii. *Nanoscale Res. Lett.* **2014**, *9*, 382. [CrossRef] [PubMed]
68. Cannavacciuolo, L.; Winkler, R.G.; Gompper, G. Mesoscale simulations of polymer dynamics in microchannel flows. *Europhys. Lett.* **2008**, *83*, 34007. [CrossRef]
69. Cifra, P.; Teraoka, I. Partitioning of polymer chains in solution with a square channel: Lattice Monte Carlo simulations. *Polymer* **2002**, *43*, 2409–2415. [CrossRef]
70. Fritsche, M.; Heermann, D.W. Confinement driven spatial organization of semiflexible ring polymers: Implications for biopolymer packaging. *Soft Matter* **2011**, *7*, 6906–6913. [CrossRef]
71. Odijk, T. The statistics and dynamics of confined or entangled stiff polymers. *Macromolecules* **1983**, *16*, 1340–1344. [CrossRef]
72. Horie, K.; Kitano, T.; Matsumoto, M.; Suzuki, M. A Partial Order On The Set Of Prime Knots With Up To 11 Crossings. *J. Knot Theory Ramif.* **2011**, *20*, 275–303. [CrossRef]
73. Tubiana, L.; Polles, G.; Orlandini, E.; Micheletti, C. KymoKnot: A web server and software package to identify and locate knots in trajectories of linear or circular polymers. *Eur. Phys. J. E* **2018**, *41*, 72. [CrossRef]
74. Tubiana, L.; Orlandini, E.; Micheletti, C. Probing the entanglement and locating knots in ring polymers: A comparative study of different arc closure schemes. *Prog. Theor. Phys. Suppl.* **2011**, *191*, 192–204. [CrossRef]
75. Dabrowski-Tumanski, P.; Piejko, M.; Niewieczerzal, S.; Stasiak, A.; Sulkowska, J.I. Protein Knotting by Active Threading of Nascent Polypeptide Chain Exiting from the Ribosome Exit Channel. *J. Phys. Chem. B* **2018**, *122*, 11616–11625. [CrossRef]

76. Dabrowski-Tumanski, P.; Stasiak, A. AlphaFold Blindness to Topological Barriers Affects Its Ability to Correctly Predict Proteins' Topology. *Preprints* **2023**, 2023081698. [CrossRef]
77. Katritch, V.; Bednar, J.; Michoud, D.; Scharein, R.G.; Dubochet, J.; Stasiak, A. Geometry and physics of knots. *Nature* **1996**, *384*, 142–145. [CrossRef]
78. Metzler, R.; Reisner, W.; Riehn, R.; Austin, R.; Tegenfeldt, J.O.; Sokolov, I.M. Diffusion mechanisms of localised knots along a polymer. *Europhys. Lett. (EPL)* **2006**, *76*, 696–702. [CrossRef]
79. El-Sagheer, A.H.; Brown, T. Click Nucleic Acid Ligation: Applications in Biology and Nanotechnology. *Acc. Chem. Res.* **2012**, *45*, 1258–1267. [CrossRef] [PubMed]
80. Devaraj, N.K.; Finn, M.G. Introduction: Click Chemistry. *Chem. Rev.* **2021**, *121*, 6697–6698. [CrossRef]

Disclaimer/Publisher's Note: The statements, opinions and data contained in all publications are solely those of the individual author(s) and contributor(s) and not of MDPI and/or the editor(s). MDPI and/or the editor(s) disclaim responsibility for any injury to people or property resulting from any ideas, methods, instructions or products referred to in the content.

Article

Cellulose Nanocrystal Embedded Composite Foam and Its Carbonization for Energy Application

So Yeon Ahn [1], Chengbin Yu [2,*] and Young Seok Song [1,*]

1. Department of Fiber Convergence Materials Engineering, Dankook University, Jukjeon-dong, Yongin 16890, Republic of Korea; soyeon@dankook.ac.kr
2. Research Institute of Advanced Materials (RIAM), Department of Materials Science and Engineering, Seoul National University, Seoul 08826, Republic of Korea
* Correspondence: ycb0107@snu.ac.kr (C.Y.); ysong@dankook.ac.kr (Y.S.S.); Tel.: +82-2-880-8326 (C.Y.); +82-31-8005-3567 (Y.S.S.)

Citation: Ahn, S.Y.; Yu, C.; Song, Y.S. Cellulose Nanocrystal Embedded Composite Foam and Its Carbonization for Energy Application. *Polymers* 2023, 15, 3454. https://doi.org/10.3390/polym15163454

Academic Editors: Hyeonseok Yoon, Alexandre M. Afonso, Luís L. Ferrás and Célio Pinto Fernandes

Received: 23 June 2023
Revised: 15 August 2023
Accepted: 17 August 2023
Published: 18 August 2023

Copyright: © 2023 by the authors. Licensee MDPI, Basel, Switzerland. This article is an open access article distributed under the terms and conditions of the Creative Commons Attribution (CC BY) license (https://creativecommons.org/licenses/by/4.0/).

Abstract: In this study, we fabricated a cellulose nanocrystal (CNC)-embedded aerogel-like chitosan foam and carbonized the 3D foam for electrical energy harvesting. The nanocrystal-supported cellulose foam can demonstrate a high surface area and porosity, homogeneous size ranging from various microscales, and a high quality of absorbing external additives. In order to prepare CNC, microcrystalline cellulose (MCC) was chemically treated with sulfuric acid. The CNC incorporates into chitosan, enhancing mechanical properties, crystallization, and generation of the aerogel-like porous structure. The weight percentage of the CNC was 2 wt% in the chitosan composite. The CNC/chitosan foam is produced using the freeze-drying method, and the CNC-embedded CNC/chitosan foam has been carbonized. We found that the degree of crystallization of carbon structure increased, including the CNCs. Both CNC and chitosan are degradable materials when CNC includes chitosan, which can form a high surface area with some typical surface-related morphology. The electrical cyclic voltammetric result shows that the vertical composite specimen had superior electrochemical properties compared to the horizontal composite specimen. In addition, the BET measurement indicated that the CNC/chitosan foam possessed a high porosity, especially mesopores with layer structures. At the same time, the carbonized CNC led to a significant increase in the portion of micropore.

Keywords: cellulose nanocrystal; chitosan; carbonized structure; voltammetric

1. Introduction

Sustainable development is a top issue in human life, and sustainable and renewable materials are utilized in many application fields [1,2]. The carbon-based materials are selected as fillers to improve the availability of polymer composites [3,4]. Some natural polymers, such as cellulose, chitin, silk, wool, and protein, are primarily water-based and extracted in nature [5,6]. It is a green energy selection for cellulose-based composites [7,8]. For instance, since cellulose has many strong points, such as low cost, good chemical and physical properties, and low environmental load, it has attracted significant attention in material science and engineering [9,10]. In addition, cellulose aerogel can be applied in medical areas, which requires biocompatibility and biodegradability [11,12]. The aerogel produces using cellulose and cellulose derivatives. For example, since nano-fibrillated cellulose (NFC) is a cellulosic nanomaterial with good crystallinity and sizeable specific surface due to the cellulose I structure, it can serve as a structural material for the aerogel [13,14]. Although cellulose can extend its applications, its relatively low mechanical property must be improved significantly [15,16]. On the other hand, much effort has been made to construct a form-stable cellulose compound for energy-related applications by carbonizing cellulosic materials [17,18]. The carbonization of cellulose generally needs to use catalysts or surfactants to promote cellulose transformation at high temperatures and pressure [19,20]. Considering the abundance and cost of cellulose, embedding and carbonizing cellulose can contribute to expanding applications related to carbonized materials [21,22]. Since carbon is one of the most popular materials for electrodes, energy storage materials, and catalysts, the carbonization of natural organic materials such as cellulose has a high potential [23,24]. Furthermore, cellulose particles behave electro-rheologically and can control temporally and spatially using an electric field [25,26]. In the energy harvesting area, carbon is a fascinating material for promoting electron transfer and providing ion storage [27,28]. However, polarized electrical particles disperse irregularly, and that large portion is complete in the non-conducting area due to the absence of electric fields [29,30]. The rigid chain structures formed by the particle arrangement limit the electrical energy harvesting applications [31,32].

Chitosan is widely utilized as a renewable polymer in various applications, such as medical, drug delivery, and tissue engineering [33,34]. It can quickly decompose because of its biodegradable structure, making chitosan a renewable material for various applications [35,36]. Furthermore, chitosan is non-toxic and potentially valuable for drug delivery and green energy development [37]. The chemical structure of chitosan merely includes a primary amine ($-NH_2$) group, and it can adsorb certain kinds of molecules from the aqueous solutions [38,39]. Therefore, chitosan can exhibit intrinsic physical and chemical properties such as non-toxicity, excellent antimicrobial performance, and membrane-associated performance [40,41]. Cellulose aerogel is a porous, lightweight, and flexible material made from cellulose according to the chemical structure (e.g., sugar moieties). It can be a promising agent for targeting some chemical components in bioengineering. Since chitosan has plenty of amine and hydroxyl functional groups, various chemical and physical treatments for it, such as hydrolysis, cross-linking, and polymerization, can be employed for further applications [42,43]. For example, chitosan derivatives can be utilized in electrical energy storage and harvesting applications [44,45] since the chitosan foam is created by adding a foaming agent and evaporating the solvent, which can solidify to an aerogel-like structure [46]. The aerogel-like chitosan foam can break down under external conditions, and this property causes a chance to modify a range of applications [47,48]. The porous chitosan foam is utilized as a power generator due to the continuous skeletons in the porous structure, which performs as an electron carrier to increase the electrical conductivity [49,50]. It is well known that the phase change material (PCM) with a high thermal energy storage (TES) is broadly utilized for thermoelectric energy harvesting due to the Seebeck effect [51,52]. The phase transition process gave rise to the temperature difference at two sides of the thermoelectric power generator (TEG) and induced electron movement in the closed circuit [53,54]. In order to prevent the PCM leakage problem, 3D porous supporting materials such as graphene and silica aerogels are employed to infiltrate pure PCM for fabricating the form-stable PCM composite [54,55]. In this research area, the porous aerogels can maintain the solid state of PCM composite without any leakage during the melting process and increase both thermal and electrical conductivities significantly [56,57]. The pore size distribution seems to control the foam porosity, mechanical strength, and electron.

Based on the typical properties of cellulose and chitosan, the cellulose-incorporated chitosan composites were characterized, and obtained functional structures were evaluated. The cellulose nanocrystal (CNC), glutaraldehyde (Glu), and chitosan ternary composites were fabricated with different CNC/chitosan ratios and characterized using various techniques [58]. The composite showed an emulsion state, and the internal crystal size was affected by pH and temperature variations. This study mentioned that cellulose nanocrystal (CNC)-modified chitosan composite could improve the emulsifying capacity and crystal structures significantly. The composite under the emulsion state was difficult to fabricate as a 3D porous structure due to the immiscible substances in the aqueous solution. There was a need for constructing a porous composite with cellulose nanocrystal (CNC), and the freeze-drying process was utilized to get a sponge-like matrix [59]. It was possible to fabricate cellulose foams with different concentrations. However, different kinds of cellulose-embedded chitosan composite were prepared for an advanced nanocomposite which can reinforce the porosity and cyclic voltammetry.

Since cellulose and chitosan are natural polymers with similar chemical structures, cellulose particles such as nano-fibrillated cellulose (NFC) and cellulose nanocrystal (CNC) were incorporated into the chitosan easily. In this sense, cellulose nanoparticle-embedded chitosan composite can show enhanced physical properties such as mechanical and electrical properties [60,61]. This study embedded CNC into chitosan and composite foam preparation using freeze-drying. After that, the CNC-embedded foam carbonizes at a high temperature. The physicochemical features of the carbonized foam, such as chemical, structural, electrochemical and morphological properties, were analyzed experimentally. The carbonized foam could significantly increase electrical energy harvesting, including CNC nanoparticles.

2. Experimental

2.1. Materials

Microcrystalline cellulose (MCC) was hydrolyzed chemically. The MCC was purchased from Acros Organics. The diameter of the MCC was 50 μm and the Chitosan was provided from Sigma Aldrich. For the hydrolysis, sulfuric acid and filter paper with a pore size of 700 nm were supplied by Ducksan Chemical (Yongin, Republic of Korea) and Hyundai Micro (Seoul, Republic Korea), respectively. A cyanobacterial strain, Synechococcus, was obtained from the Korea Research Institute of Bioscience and Biotechnology (KRIBB) and grown in a BG-11 medium in an incubator with shaking at 25 °C. The microalgae were harvested on day 30. 2,5-dimethyl-1,4-benzoquinone (DMBQ), mediator was purchased from Sigma Aldrich (St. Louis, MI, USA). The platinum mesh was purchased from Ametek Inc. (Berwyn, PA, USA). MEET Co. (Seoul, Republic Korea) supplied carbon felt, and Nafion membrane was purchased from Dupont (Wilmington, DE, USA).

2.2. Preparation of Carbonized CNC/Chitosan Foam

Figure 1 shows a brief schematic of carbonized CNC/chitosan composite foam. Before CNC/chitosan foam fabricates, CNC nanoparticles need to be prepared. First, the MCC was treated with sulfuric acid of 64 wt% at 45 °C for 2 h so that the chemical chain and particle dimension change were made under sulfuric acid. After purification, remove access sulfuric acid and prepare the CNC suspension. Since the size of CNC can affect the generation of foam structure, the treatment time and pH need to be controlled precisely. In order to increase the CNC weight fraction, the suspension rinses increased several times. Distilled water and the suspension were filtered to remove the solvent. After that, the CNC powder is produced by using the freeze-drying method. The sample dries for 48 h. Second, the CNC/chitosan suspension was prepared. The chitosan dissolves in acetic acid of 1 wt% at a 40 °C oil bath. The suspension is stirred for 2 h. Afterward, the CNC powders disperse in the chitosan solution by applying ultrasonication for 30 min, which can enhance the dispersion of the particles. In the current study, the weight fraction of the CNC was set at 2 wt% because we tried to analyze the effect of the existence of the nanoparticle in the carbonized nanoparticle-filled composite foam.

Third, the CNC/chitosan foam structure was generated. In order to do this, the CNC/chitosan solution was poured into the centrifugation tube and cooled down using the liquid nitrogen. All samples should be frozen to ice structures and put into the freeze-dryer (ilShinBioBase Co., Gyeonggi, Republic of Korea). Thereafter, the CNC/chitosan samples were under the freeze-drying process for 48 h to evaporate the solvent. The temperature at the bottom of the sample sets was lower than at the top to employ directional freezing. The temperature gap was less than 5 °C. Finally, using a furnace, the CNC/chitosan composite foam carbonizes in an argon gas environment (GTF0850, GSS, Republic of Korea). The sample was put into the furnace at 150 °C for 30 min to improve the adsorption of chitosan molecules onto the CNC surface. After that, the furnace temperature increased to 1200 °C at 10 C/min, and the specimen was for 2 h places. Finally, this specimen was placed at the room temperature to obtain an aerogel-like CNC/chitosan foam. In this study, the effect of not only the addition of the CNC particles in the foam but also the carbonization of the composite foam is investigated.

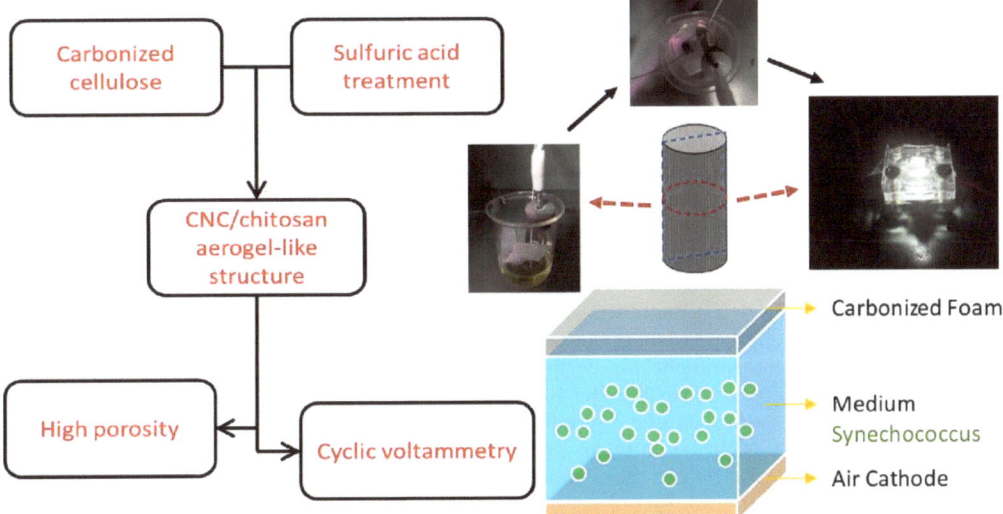

Figure 1. Schematic of fabrication of modified CNC/chitosan composite structure.

2.3. Characterization

An optical image of the sample surface is obtained using a microscope (Olympus SZX7, Olympus, Tokyo, Japan). A field emission scanning electron microscope was used to conduct (FESEM, S-4800, Hitachi, Hitachi, Japan) analysis. For the sample preparation, the samples were fractured and coated with platinum. For analysis, a cryogenic scanning electron microscope was employed (cryo-SEM, Mira-3 FEG, Tescan, Warrendale, PA, USA). After freezing the CNC/chitosan suspension, the suspension, at $-100\,°C$ for 10 min, was sublimed to prevent frozen water recrystallization. The additional morphological analysis uses a transmission electron microscope (TEM, JEM-200CX (JEOL, Tokyo, Japan). A droplet of the particle suspension on a 200 mesh TEM grid is used for the measurement deposits.

Fourier transform infrared spectroscopy (FT-IR) analysis was conducted with the use of an FT-IR spectrometer (Nicolet iS10, Thermo, Waltham, MA, USA) equipped with a smart diamond attenuated total reflection (ATR) accessory. A KBr-pellet method is used in the scan range of 4000–400 cm^{-1}. Raman spectroscopy analysis uses a Raman spectrometer (LabRam Aramis, Horriba Jovin Yvon, Longjumeau, France). An excitation wavelength of 532 nm was employed, and the measurement wavelength was from 1000 to 3500 cm^{-1}. The accumulation time was 100 s. Before the measurement, the Raman spectrometers calibrate to the silicon peak. X-ray diffraction analysis is conducted using a comprehensive angle X-ray scattering system (WAXS, D8 Discover, Bruker, Billerica, MA, USA) at 1000 µA with Cu Kα radiation (wavelength = 0.154 nm) in the 2θ range of 4.5–40° with a step interval of 0.02°.

In order to estimate the possibility of using an electrode, the electrochemical analysis of the sample was conducted using a photo-microbial solar cell (PMSC) system. Before the measurement, a Nafion membrane was treated as below: the membrane was put in a 3 wt% H_2O_2 and rinsed with distilled water for 30 min. After that, it is immersed in 0.5 M sulfuric acid for 1 h at 80 pa. Then, keep the membrane in distilled water. Membrane electrode assembly (MEA) fabricates for a cathode. A carbon felt was coated with platinum using a sputter and hot-pressed with the Nafion membrane at 130 °C for 1 min at a pressure of 5 MPa. The electrochemical tests, including cyclic voltammetry and chronoamperometry, were carried out using a potentiostat (VeraStat 3, Princeton Applied Research, Oak Ridge, TN, USA) [62]. The measurement was conducted in a Faraday cage using a resistance of 500 Ω and a light source.

The porous structure of sample was analyzed using the Brunauer–Emmet–Teller (BET) method. Nitrogen gas adsorption characteristics (Quantachrome NOVAe, 2000) are obtained. The measurement is conducted at a relative vapor pressure of 0.02 to 0.3 at −196 Pa. The average pore size of sample evaluates through the Barrett–Joyner–Halendar (BJH) analysis, which was adopted to determine pore volume and size with adsorption and desorption techniques.

3. Results and Discussion

This study investigated how adding CNCs affects the carbonized structure since a polysaccharide nanoparticle can interact with a natural polymer. Figure 2 presents the morphological analysis results of the samples. After the CNCs were suspended in the chitosan solution, the suspension was freeze-dried (Figure 2a). All of these foam samples were prepared using the freeze-drying process. It shows that the CNCs disperse in the chitosan, generating a foam structure. In order to align the nanoparticle, which applies the directional freezing method, the directional freezing method could align the particles and crystals [63]. The directional freezing method can align the layered structure along the heat transfer direction, i.e., the cooling direction. Figure 2b represents the carbonized CNC/chitosan foam. The color turned black after carbonization, and it was modified sufficiently by the treatment.

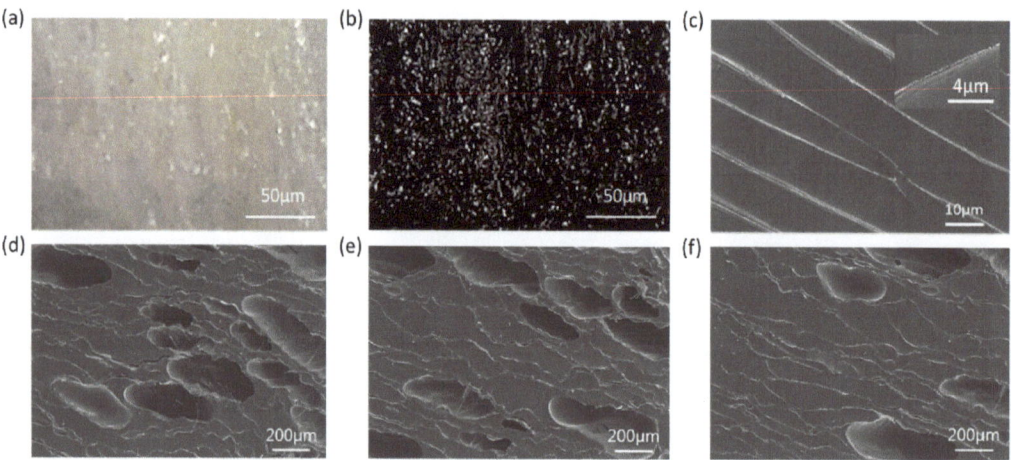

Figure 2. Morphological analysis of the samples: Optical microscopic images of (**a**) the CNC/chitosan foam and (**b**) the carbonized CNC/chitosan foam. (**c**) Cryo-SEM image of (c) the CNC/chitosan suspension and SEM images of (**d**) the carbonized chitosan foam, (**e**) the CNC/chitosan foam, and (**f**) the carbonized CNC/chitosan foam.

Figure 2c shows the cryo-SEM image of the CNC/chitosan suspension. The alignment of ice layers was found, further leading to the directionality of the foam microstructure. The stripe pattern in the image was generates due to the freezing process. For this reason, the freeze-drying method used in this study can cause the anisotropic structure of the foam. Figure 2d shows the SEM image of the carbonized chitosan foam. The sample had an anisotropic porous structure. As a result of the freeze-drying method, the foam structure was formed and maintained even during carbonization. The SEM images of non-carbonized CNC/chitosan foam and carbonized CNC/chitosan foam are presented in Figure 2e,f, respectively. The CNC nanoparticles affected the carbon structure after carbonization. On the other hand, the mechanical behavior of these foams is an important physical property. The CNC/chitosan foam with cysteamine cross-linked graphene aerogel (GCA) has a similar stress–strain effect to that reported in a previous study [64].

Figure 3a illustrates the FT-IR result of both chitosan and carbonized chitosan foams. The stretching vibration of the chitosan sample is related to the peaks between 3500 and 3250 cm^{-1}, indicating the existence of O–H. The adsorption peaks of NH_2 and secondary amides vibration of –NH are associated with peaks between 3500 and 3400 cm^{-1} and between 3300 and 3280 cm^{-1}, respectively. The symmetric and asymmetric C–H vibrations induce the band of 2960–2870 cm^{-1}. The C–O–C vibration indicates the 1160 cm^{-1} peak. After the carbonization, the FTIR spectra changed drastically. The C–C and C=C stretching vibrations lead to the bands at 1200 and at 1650 cm^{-1}, respectively. The FT-IR peaks showed the general difference between the original and carbonized chitosan foam. The characteristic peaks of chitosan were confirmed and the complete carbonization identified. Figure 3b,c show the Raman spectra of the carbonized chitosan and CNC/chitosan foams, and Table 1 presents the corresponding typical peak results. The samples yielded graphitic D and G peaks around 1350 and 1590 cm^{-1}, respectively. Furthermore, the 2D band was around 2680 cm^{-1}. The shifted peak indicated the presence of highly disordered graphite and the formation of aromatic clusters. The ratio of the amount of structured carbon incorporated into the carbonized sample can be estimated using the relative intensity ratio of the D peak, and the G peak was higher than the D peak. However, once the natural nanoparticles were added into the polymer, the D peak of the composite foam showed a relatively high value. This means that the amount of amorphous carbon increases by adding the CNCs. The result of the Raman peaks indicates that the carbonized chitosan and CNC/chitosan foams had different internal structures and could affect the electrical properties.

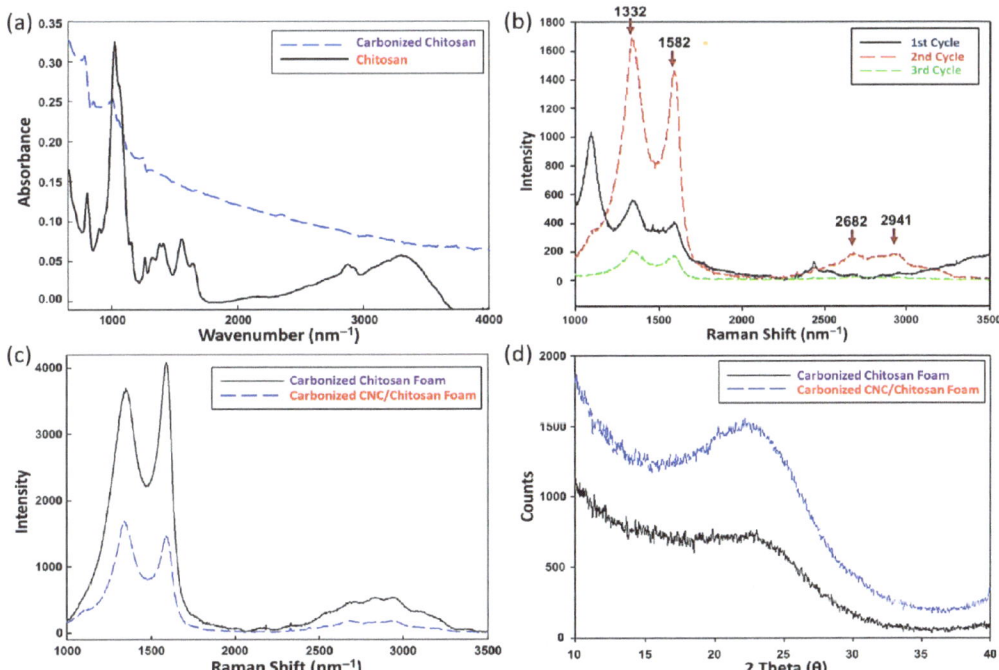

Figure 3. (a) FT-IR result of the raw and carbonized chitosan foam. (b) Scanning results of carbonized CNC/Chitosan foam. (c) Raman spectra of the carbonized chitosan foam and CNC/chitosan foam. (d) WAXS spectra of the carbonized chitosan foam and CNC/chitosan foam.

Table 1. Raman spectrum result for the carbonized chitosan foam and CNC/chitosan foam.

Materials	Peak		I_D/I_G
Carbonized Chitosan	D band	1353.1237	1.1600
	G band	1587.2168	
Carbonized CNC/Chitosan	D band	1338.3350	1.1560
	G band	1588.6597	

Both carbonized chitosan foam and CNC/chitosan foam exhibited graphitic peaks. The carbonized CNC with fewer chemical functional groups had excellent chemical stability to utilize an electron carrier [65].

Figure 3d presents the WAXS result of the samples. The crystal size was obtained by using the Scherrer equation: $D = \frac{K\lambda}{\beta \cos\theta}$, where D is the size of the ordered (crystalline) domains, K is a shape factor, λ is the X-ray wavelength, β is the full width at half maximum (FWHM), and θ is the diffraction angle. The shape factor used in this study was 0.9, and the wavelength is 1.54 Å. The chitosan and CNC-embedded chitosan composite foams showed crystallite sizes of 0.17 and 0.13 nm, respectively. The degree of crystallization of the specimens calculated using the peak height ratio method is expressed as below:

$$\chi_{CR} = \frac{I_{200} - I_{AM}}{I_{200}} \times 100 \tag{1}$$

where χ_{CR} is the crystallinity index of a specimen, I_{200} is the max intensity in 200 plane peaks, and I_{AM} is the min intensity between 200 and 100 plane peaks. From the calculation results in Table 2, the crystallinity was approximately 14.0% and 24.0%, which showed the carbonized chitosan and CNC/chitosan foams, respectively. The pure chitosan specimen showed a larger crystal size but smaller crystallinity than the CNC-filled specimen. This implied that the added nanoparticle could serve as a nucleation site, thus leading to the relatively high crystal portion in the matrix. Figure 4 shows the recyclable voltammetry results of the specimens. The voltammetric behavior's several cycles observe applied voltage and mass (or reactant) transport due to the concentration polarization and overpotential effect. It founds that the carbonized CNC/chitosan foam had a higher peak current than the carbonized chitosan foam. Note that the prepared sample possessed an anisotropic internal structure after freeze-drying, and the carbonized CNC/chitosan foam exhibited excellent recyclable behavior. Therefore, two kinds of samples were prepared for cutting in the vertical or horizontal direction, as shown in the inset. Interestingly, the vertical sample showed more extensive electrochemical characteristics than the horizontal sample. On the other hand, renewable energy harvesting is a comprehensive study to replace fossil fuel energy production. In particular, energy production using living cells, such as microbial fuel cells (MFC) and bio photovoltaic cells (BPV), is an attractive energy harvesting method. In this study, the BPV system using cyanobacteria (i.e., Synechococcus) employs to evaluate the possibility of usage as a carbon electrode. Figure 5 presents the chronoamperometry analysis result of the sample. Like the cyclic voltammetry result, the vertically prepared sample offered a higher current than the horizontally prepared sample. This indicates the existence of an anisotropic structure in the sample [66,67]. In addition, the CNC/chitosan sample possessed higher current values than the only chitosan sample due to the relatively dense internal structure of the CNC-embedded sample.

Table 2. WAXS spectrum result of the carbonized chitosan foam and CNC/chitosan foam.

Materials	Crystal Size (nm)	Crystallinity (%)	Interplanar Distance (nm)
Carbonized Chitosan	0.0166	14.0982	0.1537
Carbonized CNC/Chitosan	0.0125	23.9901	0.0147

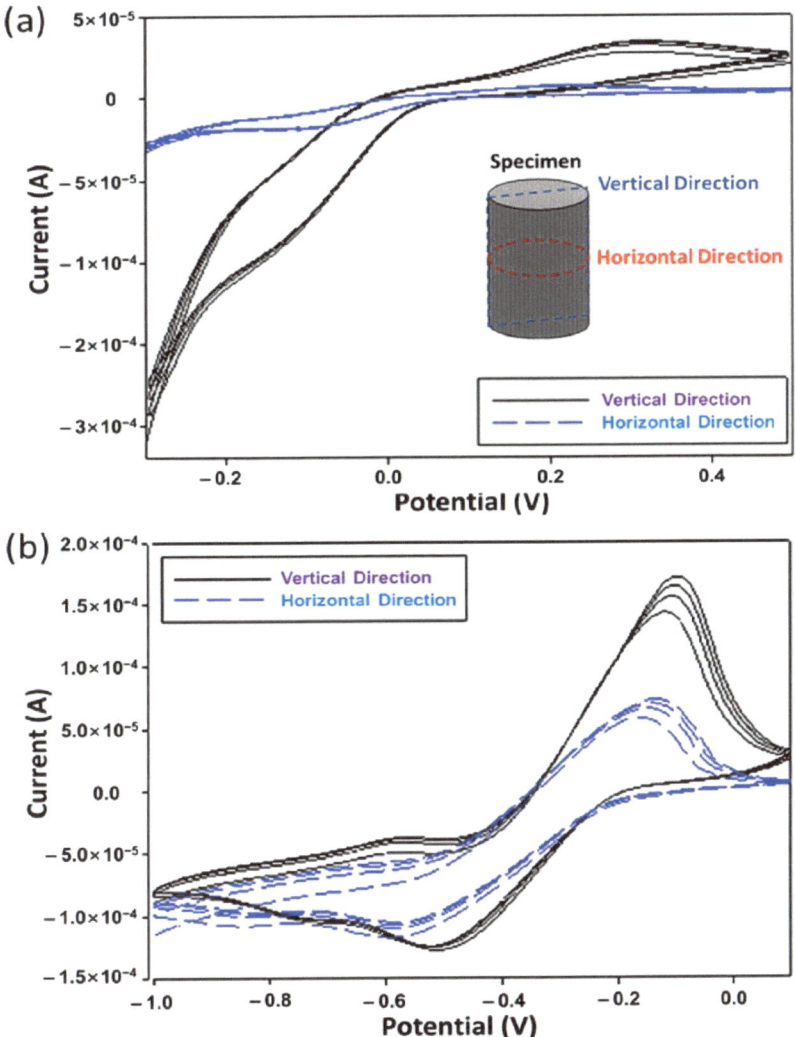

Figure 4. Cyclic voltammetry results of (**a**) the carbonized chitosan foam and (**b**) the carbonized CNC/chitosan foam.

To analyze the porous structure, adsorption and desorption analyses were carried out. BET depends on the assumption that adsorption energy is independent of adsorption sites. The BET equation relates the monolayer capacity as follows:

$$\frac{1}{W\left(\frac{P_0}{P}-1\right)} = \frac{1}{W_m C} + \frac{C-1}{W_m C}\left(\frac{P}{P_0}\right) \quad (2)$$

where W is the mass of gas adsorbed as monolayer at a relative pressure P/P_0, P is the actual vapour pressure of adsorbate, P_0 is the saturated vapour pressure, C is the BET constant, and W_m is the required mass of gas adsorbed in a complete monolayer. Figure 6 shows the result of sample gas physisorption before carbonization, and Table 3 lists related calculations. Gas physisorption is an experimental technique based on the Van der Waals interaction between gas molecules and solid particles. Figure 6a presents the hysteresis loops of the samples. Depending on the particle structure, this obtains the different loops. For instance, a porous material with a solid adsorbent–adsorbate interaction offers Langmuir isotherm (i.e., steep uptake at low pressure), implying a micropore structure. The adsorption and desorption curves of the chitosan and CNC/chitosan foams found characterize reversible isotherms for mesopore/macropores with layer structures. For the BET adsorption isotherm analysis, the linear relationship in the relative pressure range from 0.00 to 0.40 was employed (Figure 6b). The SSAs of the chitosan and CNC/chitosan samples were 372.98 and 360.17 m^2/g, respectively. The samples had pore diameters of 1.21 and 1.19 nm, respectively. Figure 6c,d show that the specimen's structure was analyzed using the micropore (MP) and BJH analyses. The pore size distribution was obtained based on the physisorption equilibrium isotherms. The BJH analysis considers the pore radius of the adsorption layer, meniscus radius, and thickness. In the chitosan foam case, the micropore volume ratio to mesopore volume was 23.5% versus 82.5%. In the CNC/chitosan foam, the ratio was 20.6% versus 78.4%. These results reveal that both the specimens possessed primarily mesopore, and the CNC composite showed a higher portion of micropore than the chitosan foam.

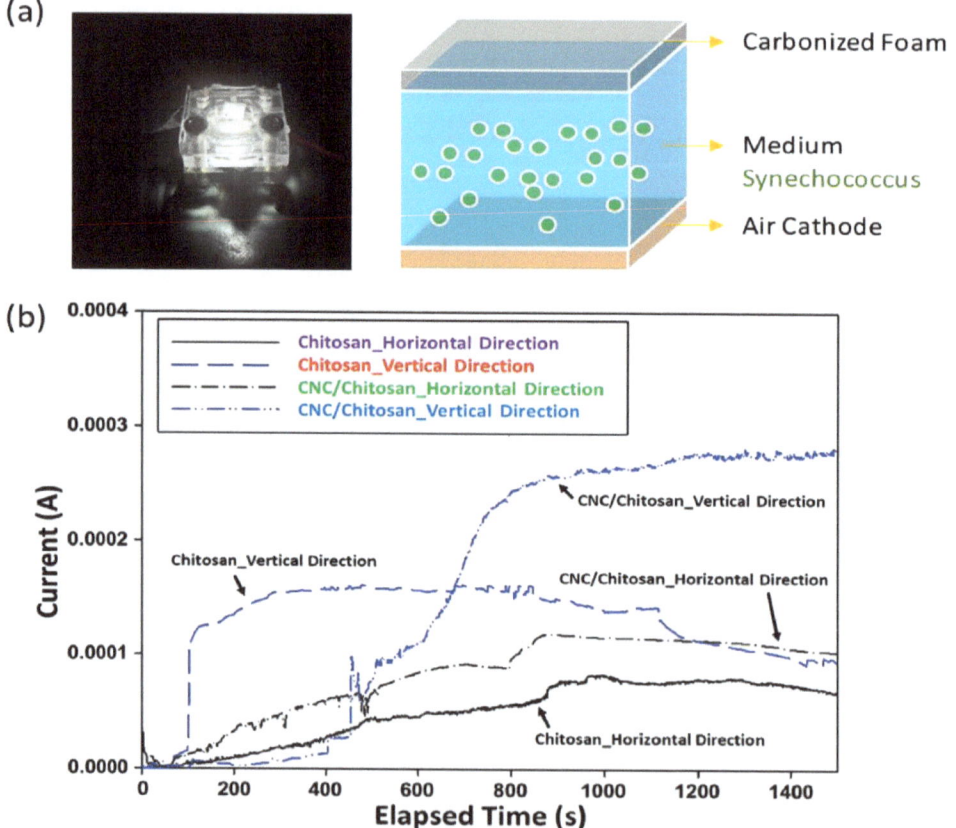

Figure 5. Chronoamperometry analysis of the sample: (**a**) photograph (left) and schematic image (right) of the experimental setup and (**b**) current result to time for the carbonized chitosan foam and CNC/chitosan foam.

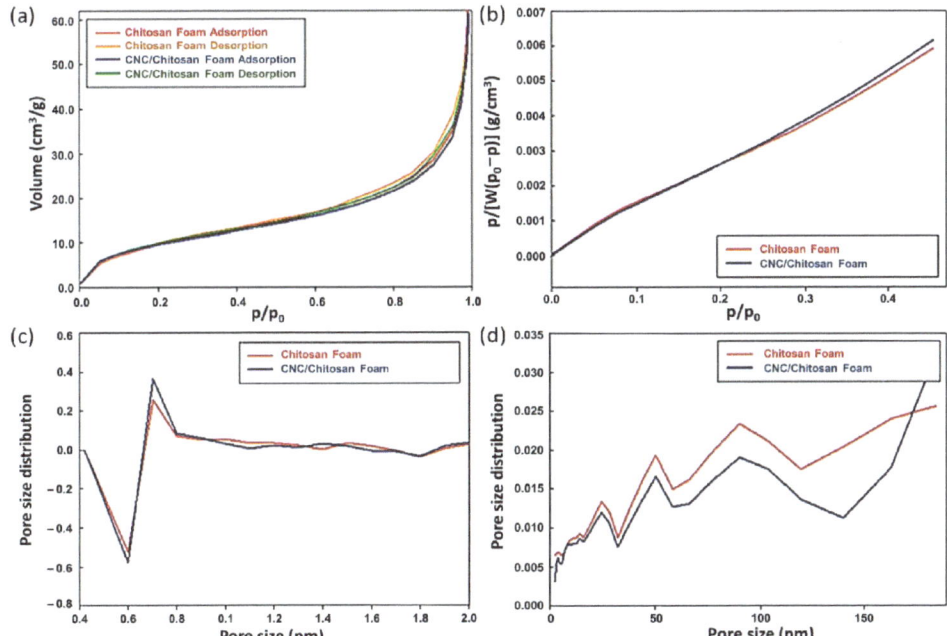

Figure 6. Gas physisorption result of the uncarbonized chitosan foam and CNC/chitosan foam: (a) hysteresis loops; (b) BET analysis of adsorption isotherm and pore size distributions obtained from (c) MP and (d) BJH analyses.

Table 3. Result of BET, MP, and BJH for the uncarbonized chitosan foam and CNC/chitosan foam.

	Materials	Chitosan	CNC/Chitosan
BET	Specific surface area (m^2/g)	372.9814	360.1709
	Total pore volume (cm^3/g)	0.1256	0.1142
	Average pore volume (nm)	1.2050	1.1866
MP	Micropore volume (cm^3/g)	0.2141	0.2021
	Micropore/Total volume	0.4288	0.4107
BJH	Mesopore volume (cm^3/g)	0.0672	0.0453
	Mesopore/Total volume	0.1622	0.1398

Figure 7 presents the results of the microstructural analysis for the carbonized samples. Table 4 shows the difference between carbonization obtained from the specific surface area and total pore volume, indicating a significant increase. The carbonized specimens showed the Langmuir isotherms (Figure 7a,b). This indicates that the samples mainly possess a microporous structure. The carbonized chitosan and CNC/chitosan foams had 891 and 842 m^2/g, respectively. The cellulose nanocrystal (CNC)-modified chitosan composite foam showed a slightly smaller pore size than the carbonized chitosan due to the change of crystal structures. The pore diameters for the samples were 2.03 and 1.933 nm, respectively. The pore diameters can affect the electron movement and cyclic voltammetry according to the electrical analysis. Figure 7c,d present the results obtained through the MP and BJH analyses, respectively. The carbonized chitosan foam showed that the ratio of micropore volume to mesopore volume was 75% versus 25%. In the case of the carbonized CNC/chitosan foam, the ratio was 81% and 19%. The carbonization process was found to increase the portion of the micro-pore significantly. In addition, the addition of CNC increased the micropore content in the foam. Overall, we envision the carbonized CNC/chitosan foam as a porous power generator and functional material for biological energy harvesting applications.

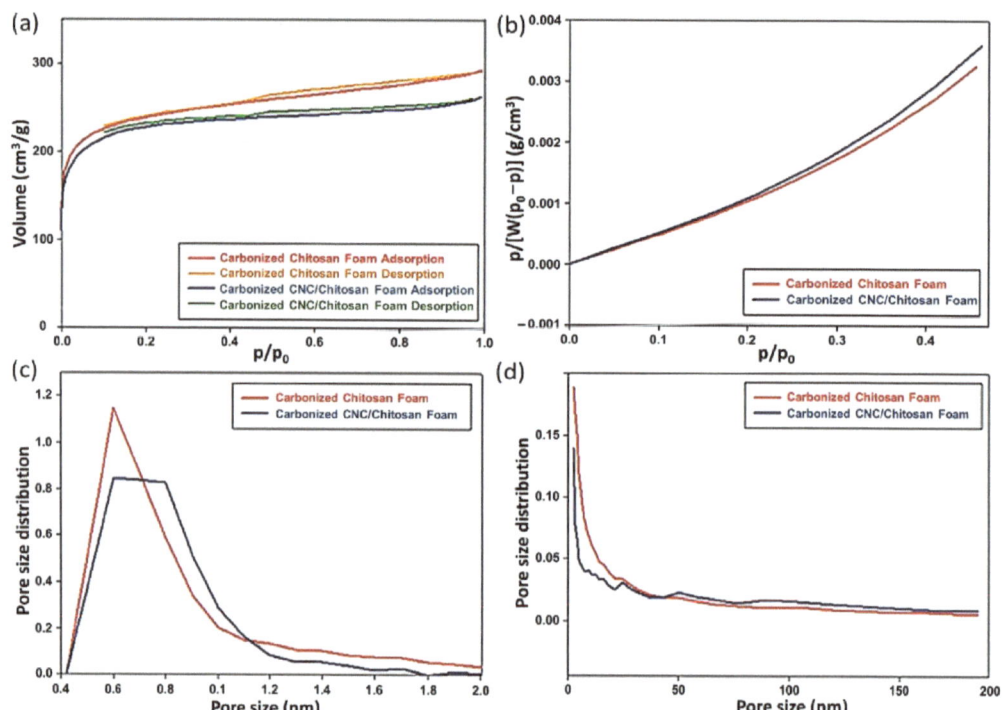

Figure 7. Gas physisorption result of the carbonized chitosan foam and CNC/chitosan foam: (**a**) hysteresis loops; (**b**) BET analysis of adsorption isotherm and pore size distributions obtained from (**c**) MP and (**d**) BJH analyses.

Table 4. Result of BET, MP, and BJH for the carbonized chitosan foam and CNC/chitosan foam.

	Carbonized Materials	Chitosan	CNC/Chitosan
BET	Specific surface area (m^2/g)	890.66	842.37
	Total pore volume (cm^3/g)	0.4531	0.4072
	Average pore volume (nm)	2.0349	1.9337
MP	Micropore volume (cm^3/g)	0.4034	0.3767
	Micropore/Total volume	0.8900	0.9250
BJH	Mesopore volume (cm^3/g)	0.1335	0.0883
	Mesopore/Total volume	0.2950	0.2170

4. Conclusions

In this study, we investigated carbonized CNC-filled chitosan foam. In order to obtain the CNC nanoparticles, MCC was chemically treated using sulfuric acid to prepare the modified chitosan structures. The NCC was suspended in the chitosan solution, and a freeze-drying process formed the foam structure by evaporating the solvent. After the measurements, the content of 2 wt% NCC in the chitosan foam showed excellent mechanical and electrical properties due to the most appropriate crystal structures. The chemical and structural characteristics of carbonized CNC-embedded CNC/chitosan foam were analyzed through FT-IR and WAXS measurements. The degree of crystallization was modified after the CNC treatment and increased the electron movement and cyclic voltammetry. Compared to CNC/chitosan foam, the carbonized composite foam has a relatively larger crystallite size and higher crystallinity, which had a negative effect on the electrical property. The cyclic voltammetric result indicates that the sample with a vertical structure had more electrochemical performance than the sample with a horizontal structure. In addition, the BET result shows that the carbonized CNC/chitosan showed 81% micropore and total pore size, which was a significant increase compared to the non-carbonized sample. The carbonized CNC/chitosan has an improved internal structure and promotes electro mobility, making it a promising electrode for energy harvesting applications. Furthermore, carbonized composite foams have the potential for porous sensors and actuators in biomedical and environmental areas.

Author Contributions: Conceptualization: S.Y.A., C.Y. and Y.S.S.; methodology: C.Y.; validation: C.Y. and Y.S.S.; investigation: S.Y.A. and C.Y.; writing—original draft preparation: C.Y.; supervision: Y.S.S.; funding acquisition: Y.S.S. All authors have read and agreed to the published version of the manuscript.

Funding: National Research Foundation of Korea, Grant/Award Number: 2018R1A5A1024127.

Institutional Review Board Statement: Not applicable.

Informed Consent Statement: Not applicable.

Data Availability Statement: Data will be made available upon reasonable request to the corresponding author.

Acknowledgments: This research was supported by the Basic Science Research Program through the National Research Foundation of Korea (NRF), funded by the Ministry of Education (2018R1A5A1024127). The authors are grateful for the support.

Conflicts of Interest: The authors declare no conflict of interest.

References

1. Del Valle, L.J.; Díaz, A.; Puiggalí, J. Hydrogels for biomedical applications: Cellulose, chitosan, and protein/peptide derivatives. *Gels* **2017**, *3*, 27. [CrossRef]
2. Yu, C.; Song, Y.S. Enhancing energy harvesting efficiency of form stable phase change materials by decreasing surface roughness. *J. Energy Storage* **2023**, *58*, 106360. [CrossRef]
3. Xu, J.; Sun, J.; Zhao, J.; Zhang, W.; Zhou, J.; Xu, L.; Guo, H.; Liu, Y.; Zhang, D. Eco-friendly wood plastic composites with biomass-activated carbon-based form-stable phase change material for building energy conversion. *Ind. Crops Prod.* **2023**, *197*, 116573. [CrossRef]
4. Xie, X.; Yang, Z.; Zhu, W.; Wu, Z.; Hui, J.; Yu, C.; Xiang, M.; Qin, H. Modification of noble metal platinum by constructing stibium heterojunction as a high performance electrocatalyst for hydrogen. *Mater. Lett.* **2023**, *338*, 133984. [CrossRef]
5. Li, Z.; Yang, L.; Cao, H.; Chang, Y.; Tang, K.; Cao, Z.; Chang, J.; Cao, Y.; Wang, W.; Gao, M.; et al. Carbon materials derived from chitosan/cellulose cryogel-supported zeolite imidazole frameworks for potential supercapacitor application. *Carbohydr. Polym.* **2017**, *175*, 223–230. [CrossRef]
6. Chen, X.; Song, Z.; Yuan, B.; Li, X.; Li, S.; Nguyen, T.T.; Guo, M.; Guo, Z. Fluorescent carbon dots crosslinked cellulose nanofibril/chitosan interpenetrating hydrogel system for sensitive detection and efficient adsorption of Cu (II) and Cr (VI). *Chem. Eng. J.* **2022**, *430*, 133154. [CrossRef]
7. Yu, S.; Dong, X.; Zhao, P.; Luo, Z.; Sun, Z.; Yang, X.; Li, Q.; Wang, L.; Zhang, Y.; Zhou, H. Decoupled temperature and pressure hydrothermal synthesis of carbon sub-micron spheres from cellulose. *Nat. Commun.* **2022**, *13*, 3616. [CrossRef] [PubMed]
8. Basarir, F.; Kaschuk, J.J.; Vapaavuori, J. Perspective about Cellulose-Based Pressure and Strain Sensors for Human Motion Detection. *Biosensors* **2022**, *12*, 187. [CrossRef] [PubMed]
9. Aziz, T.; Farid, A.; Haq, F.; Kiran, M.; Ullah, A.; Zhang, K.; Li, C.; Ghazanfar, S.; Sun, H.; Ullah, R.; et al. A review on the modification of cellulose and its applications. *Polymers* **2022**, *14*, 3206. [CrossRef]
10. Yang, Z.; Chen, H.; Xiang, M.; Yu, C.; Hui, J.; Dong, S. Coral reef structured cobalt-doped vanadate oxometalate nanoparticle for a high-performance electrocatalyst in water splitting. *Int. J. Hydrog. Energy* **2022**, *47*, 31566–31574. [CrossRef]
11. Long, L.-Y.; Weng, Y.-X.; Wang, Y.-Z. Cellulose aerogels: Synthesis, applications, and prospects. *Polymers* **2018**, *10*, 623. [CrossRef] [PubMed]

12. Jin, H.; Nishiyama, Y.; Wada, M.; Kuga, S. Nanofibrillar cellulose aerogels. *Colloids Surf. A Physicochem. Eng. Asp.* **2004**, *240*, 63–67. [CrossRef]
13. Li, T.; Chen, C.; Brozena, A.H.; Zhu, J.Y.; Xu, L.; Driemeier, C.; Dai, J.; Rojas, O.J.; Isogai, A.; Wågberg, L.; et al. Developing fibrillated cellulose as a sustainable technological material. *Nature* **2021**, *590*, 47–56. [CrossRef] [PubMed]
14. Banvillet, G.; Grange, C.; Curtil, D.; Putaux, J.L.; Depres, G.; Belgacem, N.; Bras, J. Cellulose nanofibril production by the combined use of four mechanical fibrillation processes with different destructuration effects. *Cellulose* **2023**, *30*, 2123–2146. [CrossRef]
15. Qiu, K.; Wegst, U.G. Excellent Specific Mechanical and Electrical Properties of Anisotropic Freeze-Cast Native and Carbonized Bacterial Cellulose-Alginate Foams. *Adv. Funct. Mater.* **2022**, *32*, 2105635. [CrossRef]
16. Hu, J.; Wu, H.; Liang, S.; Tian, X.; Liu, K.; Jiang, M.; Dominic, C.M.; Zhao, H.; Duan, Y.; Zhang, J. Effects of the surface chemical groups of cellulose nanocrystals on the vulcanization and mechanical properties of natural rubber/cellulose nanocrystals nanocomposites. *Int. J. Biol. Macromol.* **2023**, *230*, 123168. [CrossRef]
17. Shen, Z.; Kwon, S.; Lee, H.L.; Toivakka, M.; Oh, K. Cellulose nanofibril/carbon nanotube composite foam-stabilized paraffin phase change material for thermal energy storage and conversion. *Carbohydr. Polym.* **2021**, *273*, 118585. [CrossRef]
18. Yu, C.; Song, Y.S. Advanced internal porous skeleton supported phase change materials for thermo-electric energy conversion applications. *J. Polym. Res.* **2022**, *29*, 79. [CrossRef]
19. Yang, Z.; Xie, X.; Zhang, Z.; Yang, J.; Yu, C.; Dong, S.; Xiang, M.; Qin, H. NiS2@ V2O5/VS2 ternary heterojunction for a high-performance electrocatalyst in overall water splitting. *Int. J. Hydrog. Energy* **2022**, *47*, 27338–27346. [CrossRef]
20. Zeng, M.; Pan, X. Insights into solid acid catalysts for efficient cellulose hydrolysis to glucose: Progress, challenges, and future opportunities. *Catal. Rev.* **2022**, *64*, 445–490. [CrossRef]
21. Pan, D.; Dong, J.; Yang, G.; Su, F.; Chang, B.; Liu, C.; Zhu, Y.C.; Guo, Z. Ice template method assists in obtaining carbonized cellulose/boron nitride aerogel with 3D spatial network structure to enhance the thermal conductivity and flame retardancy of epoxy-based composites. *Adv. Compos. Hybrid Mater.* **2022**, *5*, 58–70. [CrossRef]
22. Bharti, V.K.; Pathak, A.D.; Sharma, C.S.; Khandelwal, M. Ultra-high-rate lithium-sulfur batteries with high sulfur loading enabled by Mn2O3-carbonized bacterial cellulose composite as a cathode host. *Electrochim. Acta* **2022**, *422*, 140531. [CrossRef]
23. Şentürk, S.B.; Kahraman, D.; Alkan, C.; Gökçe, İ. Biodegradable PEG/cellulose, PEG/agarose and PEG/chitosan blends as shape stabilized phase change materials for latent heat energy storage. *Carbohydr. Polym.* **2011**, *84*, 141–144. [CrossRef]
24. Yu, C.; Song, Y. Modified Supporting Materials to Fabricate form Stable Phase Change Material with High Thermal Energy Storage. *Molecules* **2023**, *28*, 1309. [CrossRef] [PubMed]
25. Yang, Z.; Xiang, M.; Niu, H.; Xie, X.; Yu, C.; Hui, J.; Dong, S. A novel 2D sulfide gallium heterojunction as a high-performance electrocatalyst for overall water splitting. *J. Solid State Chem.* **2022**, *314*, 123365. [CrossRef]
26. Plachy, T.; Kutalkova, E.; Skoda, D.; Holcapkova, P. Transformation of Cellulose via Two-Step Carbonization to Conducting Carbonaceous Particles and Their Outstanding Electrorheological Performance. *Int. J. Mol. Sci.* **2022**, *23*, 5477. [CrossRef] [PubMed]
27. Yu, C.; Song, Y.S. Graphene aerogel supported phase change material for pyroelectric energy harvesting: Structural modification and form stability analysis. *Energy Technol.* **2023**, *11*, 2201108. [CrossRef]
28. Perumal, S.; Kishore, S.C.; Atchudan, R.; Sundramoorthy, A.K.; Alagan, M.; Lee, Y.R. Sustainable synthesis of N/S-doped porous carbon from waste-biomass as electroactive material for energy harvesting. *Catalysts* **2022**, *12*, 436. [CrossRef]
29. Ma, M.; Chu, Q.; Lin, H.; Xu, L.; He, H.; Shi, Y.; Chen, S.; Wang, X. Highly anisotropic thermal conductivity and electrical insulation of nanofibrillated cellulose/Al2O3@ rGO composite films: Effect of the particle size. *Nanotechnology* **2022**, *33*, 135711. [CrossRef]
30. Yang, Z.; Xiang, M.; Zhu, Y.; Hui, J.; Jiang, Y.; Dong, S.; Yu, C.; Ou, J.; Qin, H. Single-atom platinum or ruthenium on C4N as 2D high-performance electrocatalysts for oxygen reduction reaction. *Chem. Eng. J.* **2021**, *426*, 131347. [CrossRef]
31. Jiao, E.; Wu, K.; Liu, Y.; Zhang, H.; Zheng, H.; Xu, C.A.; Shi, J.; Lu, M. Nacre-like robust cellulose nanofibers/MXene films with high thermal conductivity and improved electrical insulation by nanodiamond. *J. Mater. Sci.* **2022**, *57*, 2584–2596. [CrossRef]
32. Yu, C.; Youn, J.R.; Song, Y.S. Encapsulated Phase Change Material Embedded by Graphene Powders for Smart and Flexible Thermal Response. *Fibers Polym.* **2019**, *20*, 545–554. [CrossRef]
33. Khajavian, M.; Vatanpour, V.; Castro-Muñoz, R.; Boczkaj, G. Chitin and derivative chitosan-based structures—Preparation strategies aided by deep eutectic solvents: A review. *Carbohydr. Polym.* **2022**, *275*, 118702. [CrossRef] [PubMed]
34. Hamedi, H.; Moradi, S.; Hudson, S.M.; Tonelli, A.E.; King, M.W. Chitosan based bioadhesives for biomedical applications: A review. *Carbohydr. Polym.* **2022**, *282*, 119100. [CrossRef] [PubMed]
35. Rathod, N.B.; Bangar, S.P.; Šimat, V.; Ozogul, F. Chitosan and gelatine biopolymer-based active/biodegradable packaging for the preservation of fish and fishery products. *Int. J. Food Sci. Technol.* **2023**, *58*, 854–861. [CrossRef]
36. Arafa, E.G.; Sabaa, M.W.; Mohamed, R.R.; Kamel, E.M.; Elzanaty, A.M.; Mahmoud, A.M.; Abdel-Gawad, O.F. Eco-friendly and biodegradable sodium alginate/quaternized chitosan hydrogel for controlled release of urea and its antimicrobial activity. *Carbohydr. Polym.* **2022**, *291*, 119555. [CrossRef]
37. Baharlouei, P.; Rahman, A. Chitin and Chitosan: Prospective Biomedical Applications in Drug Delivery, Cancer Treatment, and Wound Healing. *Mar. Drugs* **2022**, *20*, 460. [CrossRef]
38. Mao, H.; Wei, C.; Gong, Y.; Wang, S.; Ding, W. Mechanical and water-resistant properties of eco-friendly chitosan membrane reinforced with cellulose nanocrystals. *Polymers* **2019**, *11*, 166. [CrossRef]

39. Yu, C.; Youn, J.R.; Song, Y.S. Enhancement of Thermo-Electric Energy Conversion Using Graphene Nano-platelets Embedded Phase Change Material. *Macromol. Res.* **2021**, *29*, 534–542. [CrossRef]
40. Chadha, U.; Bhardwaj, P.; Selvaraj, S.K.; Kumari, K.; Isaac, T.S.; Panjwani, M.; Kulkarni, K.; Mathew, R.M.; Satheesh, A.M.; Pal, A.; et al. Advances in chitosan biopolymer composite materials: From bioengineering, wastewater treatment to agricultural applications. *Mater. Res. Express* **2022**, *9*, 052002. [CrossRef]
41. Islam, M.M.; Shahruzzaman, M.; Biswas, S.; Sakib, M.N.; Rashid, T.U. Chitosan based bioactive materials in tissue engineering applications-A review. *Bioact. Mater.* **2020**, *5*, 164–183. [CrossRef] [PubMed]
42. Woźniak, A.; Biernat, M. Methods for crosslinking and stabilization of chitosan structures for potential medical applications. *J. Bioact. Compat. Polym.* **2022**, *37*, 151–167. [CrossRef]
43. Yu, C.; Youn, J.R.; Song, Y.S. Tunable Electrical Resistivity of Carbon Nanotube Filled Phase Change Material Via Solid-solid Phase Transitions. *Fibers Polym.* **2020**, *21*, 24–32. [CrossRef]
44. Balyan, M.; Nasution, T.I.; Nainggolan, I.; Mohamad, H.; Ahmad, Z.A. Energy harvesting properties of chitosan film in harvesting water vapour into electrical energy. *J. Mater. Sci. Mater. Electron.* **2019**, *30*, 16275–16286. [CrossRef]
45. Yi, H.; Xia, L.; Song, S. Three-dimensional montmorillonite/Ag nanowire aerogel supported stearic acid as composite phase change materials for superior solar-thermal energy harvesting and storage. *Compos. Sci. Technol.* **2022**, *217*, 109121. [CrossRef]
46. Yu, C.; Song, Y.S. Phase Change Material (PCM) Composite Supported by 3D Cross-Linked Porous Graphene Aerogel. *Materials* **2022**, *15*, 4541. [CrossRef] [PubMed]
47. Yu, C.; Park, J.; Youn, J.R.; Song, Y.S. Integration of form-stable phase change material into pyroelectric energy harvesting system. *Appl. Energy* **2022**, *307*, 118212. [CrossRef]
48. Yu, C.; Kim, H.; Youn, J.R.; Song, Y.S. Enhancement of Structural Stability of Graphene Aerogel for Thermal Energy Harvesting. *ACS Appl. Energy Mater.* **2021**, *4*, 11666–11674. [CrossRef]
49. Chadha, N.; Saini, P. Post synthesis foaming of graphene-oxide/chitosan aerogel for efficient microwave absorbers via regulation of multiple reflections. *Mater. Res. Bull.* **2021**, *143*, 111458. [CrossRef]
50. Fu, Y.; Wan, Z.; Zhao, G.; Jia, W.; Zhao, H. Flexible conductive sodium alginate/chitosan foam with good mechanical properties and magnetic sensitivity. *Smart Mater. Struct.* **2021**, *30*, 075027. [CrossRef]
51. Yu, C.; Youn, J.R.; Song, Y.S. Multiple Energy Harvesting Based on Reversed Temperature Difference Between Graphene Aerogel Filled Phase Change Materials. *Macromol. Res.* **2019**, *27*, 606–613. [CrossRef]
52. Yu, C.; Song, Y.S. Analysis of Thermoelectric Energy Harvesting with Graphene Aerogel-Supported Form-Stable Phase Change Materials. *Nanomaterials* **2021**, *11*, 2192. [CrossRef] [PubMed]
53. Yu, C.; Youn, J.R.; Song, Y.S. Enhancement in thermo-electric energy harvesting efficiency by embedding PDMS in form-stable PCM composites. *Polym. Adv. Technol.* **2022**, *33*, 700–709. [CrossRef]
54. Yu, C.; Song, Y.S. Modification of Graphene Aerogel Embedded Form-Stable Phase Change Materials for High Energy Harvesting Efficiency. *Macromol. Res.* **2022**, *30*, 198–204. [CrossRef]
55. Yu, C.; Yang, S.H.; Pak, S.Y.; Youn, J.R.; Song, Y.S. Graphene embedded form stable phase change materials for drawing the thermo-electric energy harvesting. *Energy Convers. Manag.* **2018**, *169*, 88–96. [CrossRef]
56. Yu, C.; Park, J.; Youn, J.R.; Song, Y.S. Sustainable solar energy harvesting using phase change material (PCM) embedded pyroelectric system. *Energy Convers. Manag.* **2022**, *253*, 115145. [CrossRef]
57. Yu, C.; Youn, J.R.; Song, Y.S. Reversible thermo-electric energy harvesting with phase change material (PCM) composites. *J. Polym. Res.* **2021**, *28*, 279. [CrossRef]
58. Li, Z.; Jiang, X.; Yao, Z.; Chen, F.; Zhu, L.; Liu, H.; Ming, L. Chitosan functionalized cellulose nanocrystals for stabilizing Pickering emulsion: Fabrication, characterization and stability evaluation. *Colloids Surf. A Physicochem. Eng. Asp.* **2022**, *632*, 127769. [CrossRef]
59. Ali, A.; Bano, S.; Poojary, S.; Chaudhary, A.; Kumar, D.; Negi, Y.S. Effect of cellulose nanocrystals on chitosan/PVA/nano β-TCP composite scaffold for bone tissue engineering application. Journal of Biomaterials Science. *Polym. Ed.* **2022**, *33*, 1–19.
60. Hao, D.; Fu, B.; Zhou, J.; Liu, J. Efficient particulate matter removal by metal-organic frameworks encapsulated in cellulose/chitosan foams. *Sep. Purif. Technol.* **2022**, *294*, 120927. [CrossRef]
61. Yang, Z.; Xiang, M.; Wu, Z.; Fan, W.; Hui, J.; Yu, C.; Dong, S.; Qin, H. Single-atom lanthanum on 2D N-doped graphene oxide as a novel bifunctional electrocatalyst for rechargeable zinc–air battery. *Mater. Today Chem.* **2022**, *26*, 101147. [CrossRef]
62. Yu, C.; Song, Y.S. Form stable phase change material supported by sensible and thermal controllable thermistor. *Compos. Commun.* **2023**, *40*, 101600. [CrossRef]
63. Zhang, H.; Hussain, I.; Brust, M.; Butler, M.F.; Rannard, S.P.; Cooper, A.I. Aligned two-and three-dimensional structures by directional freezing of polymers and nanoparticles. *Nat. Mater.* **2005**, *4*, 787–793. [CrossRef]
64. Yu, C.; Song, Y.S. Characterization of Phase Change Materials Fabricated with Cross-Linked Graphene Aerogels. *Gels* **2022**, *8*, 572. [CrossRef]
65. Yang, Z.; Yang, J.; Yu, C.; Bai, J.; Xie, X.; Jiang, N.; Chen, B.; Dong, S.; Xiang, M.; Qin, H. Rare-Earth Lanthanum Tailoring Mott–Schottky Heterojunction by Sulfur Vacancy Modification as a Bifunctional Electrocatalyst for Zinc–Air Battery. *Small Struct.* **2023**, *4*, 2200267. [CrossRef]

66. Yang, Z.; Xie, X.; Wei, J.; Zhang, Z.; Yu, C.; Dong, S.; Chen, B.; Wang, Y.; Xiang, M.; Qin, H. Interface engineering Ni/Ni12P5@CNx Mott-Schottky heterojunction tailoring electrocatalytic pathways for zinc-air battery. *J. Colloid Interface Sci.* **2023**, *642*, 439–446. [CrossRef] [PubMed]
67. Yang, Z.; Niu, H.; Xia, L.; Li, L.; Xiang, M.; Yu, C.; Zhang, Z.; Dong, S. Rare-earth europium heterojunction electrocatalyst for hydrogen evolution linking to glycerol oxidation. *Int. J. Hydrog. Energy* **2023**, *in press*. [CrossRef]

Disclaimer/Publisher's Note: The statements, opinions and data contained in all publications are solely those of the individual author(s) and contributor(s) and not of MDPI and/or the editor(s). MDPI and/or the editor(s) disclaim responsibility for any injury to people or property resulting from any ideas, methods, instructions or products referred to in the content.

Article

Investigations of the Laser Ablation Mechanism of PMMA Microchannels Using Single-Pass and Multi-Pass Laser Scans

Xiao Li [1,2,3,*], Rujun Tang [1,3], Ding Li [1,3], Fengping Li [4,5], Leiqing Chen [6,*], Dehua Zhu [6], Guang Feng [4], Kunpeng Zhang [1,3] and Bing Han [7]

1. Zhejiang Provincial Key Laboratory of Laser Processing Robotics, College of Mechanical and Electrical Engineering, Wenzhou University, Wenzhou 325035, China; 22461440079@stu.wzu.edu.cn (R.T.); 23461146041@stu.wzu.edu.cn (D.L.); zkp@wzu.edu.cn (K.Z.)
2. China International Science & Technology Cooperation Base for Laser Processing Robotics, Wenzhou University, Wenzhou 325035, China
3. Institute of Laser and Optoelectronic Intelligent Manufacturing, Wenzhou University, Wenzhou 325035, China
4. Oujiang Laboratory, Zhejiang Laboratory for Regenerative Medicine, Vision and Brain Health, Wenzhou 325035, China; lfp@wzu.edu.cn (F.L.); fengguang_516@126.com (G.F.)
5. Wenzhou Key Laboratory of Ultrafast Laser Precision Manufacturing Technology, Wenzhou 325035, China
6. College of Mechanical and Electrical Engineering, Wenzhou University, Wenzhou 325035, China; zhu_556@163.com
7. School of Electronic and Optical Engineering, Nanjing University of Science and Technology, Nanjing 210094, China; hanbing@njust.edu.cn
* Correspondence: lixiao@wzu.edu.cn (X.L.); 00132009@wzu.edu.cn (L.C.)

Citation: Li, X.; Tang, R.; Li, D.; Li, F.; Chen, L.; Zhu, D.; Feng, G.; Zhang, K.; Han, B. Investigations of the Laser Ablation Mechanism of PMMA Microchannels Using Single-Pass and Multi-Pass Laser Scans. *Polymers* 2024, 16, 2361. https://doi.org/10.3390/polym16162361

Academic Editor: Swee Leong Sing

Received: 16 July 2024
Revised: 9 August 2024
Accepted: 19 August 2024
Published: 21 August 2024

Copyright: © 2024 by the authors. Licensee MDPI, Basel, Switzerland. This article is an open access article distributed under the terms and conditions of the Creative Commons Attribution (CC BY) license (https://creativecommons.org/licenses/by/4.0/).

Abstract: CO_2 laser machining is a cost effective and time saving solution for fabricating microchannels on polymethylmethacrylate (PMMA). Due to the lack of research on the incubation effect and ablation behavior of PMMA under high-power laser irradiation, predictions of the microchannel profile are limited. In this study, the ablation process and mechanism of a continuous CO_2 laser machining process on microchannel production in PMMA in single-pass and multi-pass laser scan modes are investigated. It is found that a higher laser energy density of a single pass causes a lower ablation threshold. The ablated surface can be divided into three regions: the ablation zone, the incubation zone, and the virgin zone. The PMMA ablation process is mainly attributed to the thermal decomposition reactions and the splashing of molten polymer. The depth, width, aspect ratio, volume ablation rate, and mass ablation rate of the channel increase as the laser scanning speed decreases and the number of laser scans increases. The differences in ablation results obtained under the same total laser energy density using different scan modes are attributed to the incubation effect, which is caused by the thermal deposition of laser energy in the polymer. Finally, an optimized simulation model that is used to solve the problem of a channel width greater than spot diameter is proposed. The error percentage between the experimental and simulation results varies from 0.44% to 5.9%.

Keywords: microchannel; PMMA; laser machining; incubation effect; simulation model

1. Introduction

Microfluidic devices have gained important applications in many areas, such as chemical synthesis [1], biology [2], and optofluidic technology [3], because of their advantages, like high sensitivity, rapid analysis, low sample and reagent consumption, and measurement automation, over conventional methods. Polymethylmethacrylate (PMMA) belongs to the class of thermoplastic polymeric materials. The inherent properties of PMMA, such as optical transparency, thermal stability, chemically inertness, biocompatibility, and low cost, have made it one of the most promising polymer materials for microfluidic devices [4].

Up to now, the microfluidic devices manufactured with PMMA generally are used in fuel cells [5], DNA analysis [6], blood detection [7], capillary electrophoresis [8], microreactors [9], and microchannel heat sinks [10]. These applications require 10–500 μm wide microchannels for the transfer, manipulation, and mixing of fluids. To fabricate the microchannels on PMMA, many manufacturing methods have been explored. Conventional methods such as micro-milling [11], hot embossing [12], and injection molding [13] are time-consuming and require many steps to obtain final microfluidics devices. Photolithography [14] and plasma etching [15] produce high-quality microchannels but involve several steps and special facilities, such as a clean room and sophisticated equipment. Moreover, the by-products of these processes are toxic in nature. In comparison to these techniques, the direct laser ablation provides a promising method for fabricating microchannels on polymers. The laser ablation method has various advantages like flexibility, ease in automation, simple steps in machining, less processing time, and lower cost in the fabrication of complicated geometric microchannels [16].

In laser ablation machining, bulk material is removed from the irradiated material surface by melting and vaporization. At present, CO_2 and excimer laser have successfully processed the transparent polymer material because most of the polymers exhibit significant absorptivity at the mid-infrared spectrum and low UV wavelength [17–19]. In addition, femtosecond lasers also have been used to fabricate high-quality microchannels with precise sizes [20,21]. However, the femtosecond laser system requires expensive and highly sophisticated instruments and is associated with a significant cost of fabrication. Compared with UV laser and femtosecond laser systems, CO_2 laser machining is a cost effective and time saving solution for fabricating microchannels on PMMA. CO_2 laser systems provide a higher power output resulting in less processing time and higher ablation efficiency. Meanwhile, PMMA has a low heat capacity and a low heat conduction comparing other polymers, such as polycarbonate (PC) and polypropylene (PP) [22–26]. Therefore, most of the absorbed energy immediately causes the removal of local material without creating thermal cracks. Consequently, CO_2 laser ablation results in clean and clog-free ablation on PMMA with a smaller heat affected zone (HAZ) and better surface finish [27].

CO_2 laser fabricating microchannels on PMMA has been studied by various authors over the years. Chen et al. [28] investigated the influence of laser power, laser scanning speed, and scanning number on surface roughness of PMMA microchannel fabricated by a continuous CO_2 laser. It was found that the scanning number had the greatest impact on the surface roughness. Bilican et al. [29] utilized a CO_2 laser having a pulse duration of microseconds to fabricate microchannels on PMMA and PS substrates. It was observed that the size and profile of the microchannel could be changed based on laser power, scanning speed, pulse frequency, and defocusing distance. In order to investigate the influence of CO_2 laser parameters on the profile of PMMA microchannels, Yuan et al. [30] derived a mathematical model based on the energy balance under a continuous Gaussian laser irradiation. Prakash et al. [31] studied the fabrication of PMMA microchannels using multi-pass CO_2 laser processing based on experiments and numerical simulation. It was found that compared to a single-pass process, a multi-pass laser process produced a smaller channel width and a smaller HAZ. A model based on Yuan's model was proposed, which was able to predict accurately the profile of the microchannel that undergone multiple passes. To investigate pulsed laser machining microchannels, Pazokian [32] used a simulation model of a ns laser ablating polymer that mainly considered the influence of laser spot overlap on channel depth. Meanwhile, other mathematical models [33–35] were utilized to analyze the ablation process of CO_2 laser machining PMMA. These models considered that laser power was balanced by the increasing temperature of an infinitesimal volume of the polymer and the heat losses from decomposition and vaporization. Based on these considerations, the removal depth appeared to be proportional to the incident laser power and the pulse frequency and inversely proportional to the scanning speed. In addition, Anjum et al. [36] proposed a soft computing technique to build the prediction models of the channel depth, width, and surface roughness with multi-pass laser machine. By comparing various approaches, such as random forest, gradient boost, ridge regression, linear regression, support vector regression, and gaussian process regression, it was noticed that the physics-informed machine learning GPR technique could be used effectively in the depth, kerf width, and surface roughness estimations.

Up to now, most mathematical and simulation models mainly focus on predicting the variation in the channel depth with laser parameters, while considering that the channel width is approximately equal to the laser spot diameter. However, when using a high-power laser, the channel width is often greater than the laser spot diameter [34,35]. When the laser interacts with polymers, there is the incubation effect [37–39], which causes a decrease in the ablation threshold and an increase in the channel depth and width by increasing the pulse overlap and scanning number. Due to the lack of research on the incubation effect and ablation behavior of PMMA under high-power laser irradiation, the prediction of the microchannel profile is limited.

This paper investigates the ablation threshold of CO_2 laser machining microchannels on PMMA using single-pass and multi-pass laser scan modes and analyzes the thermal decomposition behavior and incubation effect. Meanwhile, the influences of laser parameters on channel width, depth, volume ablation rate, and mass ablation rate are studied. Finally, a mathematical model is proposed to predict the profile of the microchannel based on the experimental results.

2. Experimental Set-Up and Procedure

A continuous CO_2 laser system working at a constant output power of 7.8 W with a Gaussian profile was used to perform the experiments. The experimental set-up is schematically represented in Figure 1. The laser beam is concentrated into a minimum spot radius of 70 μm using a telecentric f-theta lens. A set of galvanometer systems is used to deflect the laser beam on the scanning surface. These movements are controlled using 2D CAD software, which allows the machining path, the scanning speed, and the scanning number to be set. Since channels are produced by moving the laser beam along a single direction, the f-theta lens allows the beam to ablate the PMMA surface perpendicularly. This enables the same spot dimensions on the whole working area and a uniform removal rate.

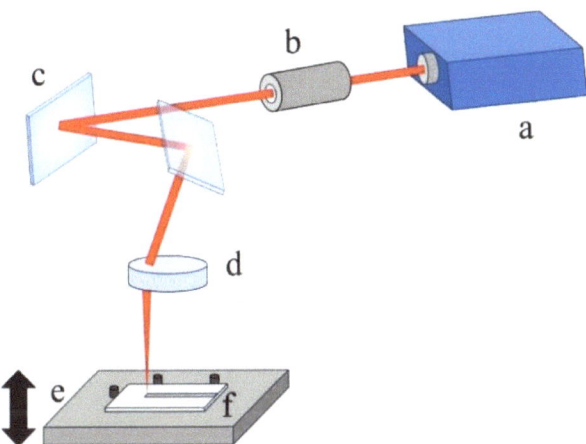

Figure 1. Schematic diagram of the experimental set-up: (**a**) CO_2 laser source; (**b**) beam expander; (**c**) galvanometer systems; (**d**) telecentric f-theta lens; (**e**) adjustable platform; (**f**) PMMA sample.

The size of the PMMA samples used in the experiments is 60 mm × 60 mm × 2 mm. The cleaned PMMA sample is placed on an adjustable platform to allow the setting of the focal point onto the PMMA surface. The ablation behavior of the laser machining straight channels on PMMA has been studied based on various laser fluences, scanning speeds and scanning numbers. The profiles and morphologies of the channels were obtained using laser scanning confocal microscopy (LSCM) and scanning electron microscopy (SEM). The depth, width, and cross-sectional area of the channel are measured using LSCM, and the measurement method is shown in Figure 2. The measurement tests are conducted 5–7 times at different length positions in each channel. To obtain an accurate cross-sectional profile of the microchannel, the samples containing channels are cut across the channel length using a ps pulsed laser. The cross-sectioned samples are polished using a variety of mesh sizes of emery papers and velvet cloth for a good surface finish. Then, accurate microscopic images of the cross-sectioned profile of the channel are captured using LSCM. A less than 5% error is found between the results obtained from direct measurements using LSCM and the results from cutting and polishing.

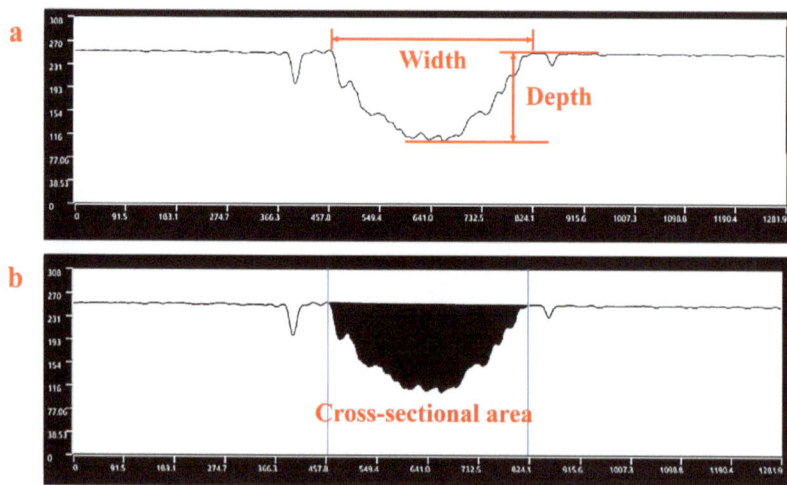

Figure 2. Measurement method of the channel profile using LSCM: (**a**) width and depth; (**b**) cross-sectional area.

The ablation mass of PMMA is measured using a high-precision electronic balance with an accuracy of 0.1 mg. The measurement steps include the following. First, the sample is subjected to ultrasonic cleaning using purified water. Then, the weight of the sample after drying is measured using the balance. Some parallel microchannels are fabricated on the sample based on the same laser parameters, and the weight of the sample after machining is measured. Then, the ablation mass under the laser parameters is calculated by dividing the weight difference of the sample by the number of the channels on the sample. Finally, the above process is repeated to gain the influence of different laser parameters on the ablation mass. The measurements and calculations of the ablation mass under the same laser parameters are repeated 3 times.

3. Experimental Results and Discussion

3.1. Laser Ablation Threshold

In the experiments, the laser power is a constant value. Considering that the beam spot size remains unchanged on the surface of the PMMA sample, the laser power density irradiated on the PMMA is a constant value. The total laser energy acted on the PMMA is related to the total laser irradiation time. Here, the laser power density I, the total laser energy density F_{tot}, the total laser energy E_{tot}, and the total laser irradiation time t_{tot} are represented using the following equations [30,31,40]:

$$I = \frac{P}{\pi R_0^2} \tag{1}$$

$$F_{tot} = \frac{nP}{2vR_0} \tag{2}$$

$$E_{tot} = Pt_{tot} \tag{3}$$

$$t_{tot} = \frac{nL}{v} \tag{4}$$

where the laser power $P = 7.8$ W, the spot radius $R_0 = 70$ µm, and the channel length fabricated by laser ablation $L = 20$ mm. The laser scanning speed v and scanning number n vary in the experiments. For the case of a single-pass laser process, $n = 1$ and F_{tot} can be represented as F_{sin}.

Figure 3 shows the microscope images of the laser ablating PMMA surface with a single-pass laser scan at a scanning speed from 1800 mm/s to 2700 mm/s. According to Equation (2), the laser energy density F_{sin} increases from 2.06 J/cm^2 to 3.10 J/cm^2 with the scanning speed decreasing from 2700 mm/s to 1800 mm/s. When the laser energy density is 2.06 J/cm^2, small pores appear at the laser irradiation center on the surface of PMMA as shown in Figure 3a. The irradiated surface is generally flat with a slight outward expansion. When the laser energy density is 2.14 J/cm^2, obvious ablation marks appear at the laser irradiation center as shown in Figure 3b. The ablated area is no longer flat, and some small bumps appear. When the laser energy density is 3.10 J/cm^2, the ablated area is widened, and a large number of pores of various sizes and depths appear in the ablated area as shown in Figure 3c. In the ablated area, the pores with black spots have a greater depth than those with white spots. Meanwhile, partial surface retreat phenomenon occurs in the ablated area, and bulges appear at the edges of the ablated area.

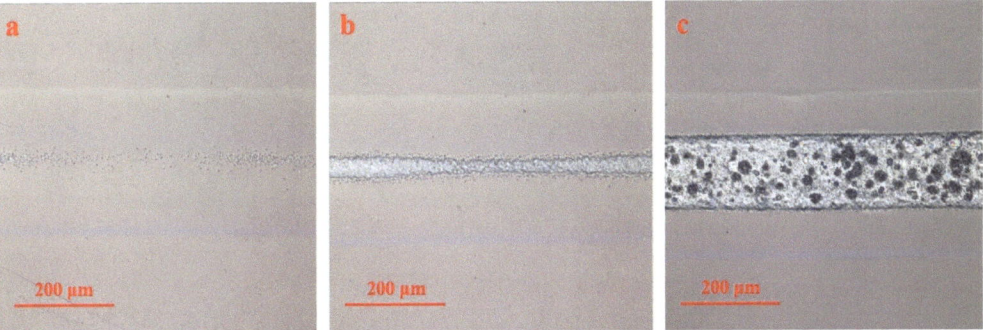

Figure 3. Ablation morphology on the surface of PMMA obtained with a single-pass laser scan with the following scanning speeds and laser energy densities: (**a**) 2700 mm/s, 2.06 J/cm^2; (**b**) 2600 mm/s, 2.14 J/cm^2; (**c**) 1800 mm/s, 3.10 J/cm^2.

Figure 4 shows the microscope images of the laser ablating PMMA surface obtained with a multi-pass laser scan at scanning speeds of 3000 mm/s and 6000 mm/s. In the case of 3000 mm/s, the laser energy density F_{tot} increases from 1.86 J/cm^2 to 5.57 J/cm^2 as the scanning number increases from 1 pass to 3 passes. In the case of 6000 mm/s, the laser energy density increases from 4.64 J/cm^2 to 6.50 J/cm^2 as the scanning number increases from 5 passes to 7 passes. It can be found that the irradiated surface undergoes a process of no ablation, the occurrence of ablation, and surface retreat as the total laser energy density F_{tot} increases. The ablation morphology using a multi-pass laser scanning system is similar to that obtained with a single-pass laser scan. However, the ablation threshold and the surface retreat threshold vary with the changes in the scanning speed and the scanning number.

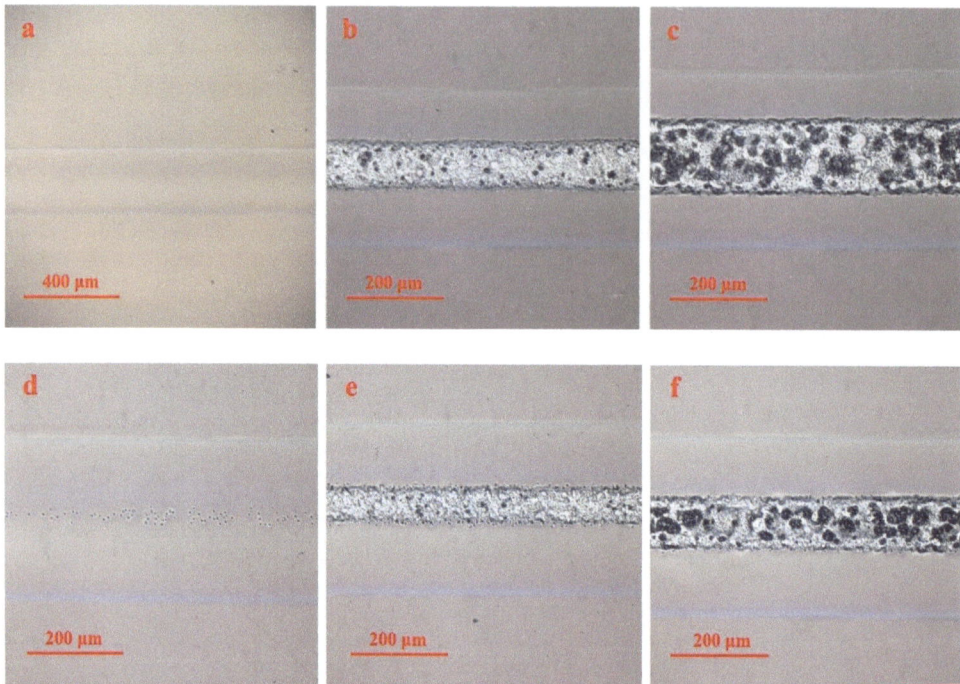

Figure 4. Ablation morphology on the surface of PMMA obtained with a multi-pass laser scanning system with the following scanning speeds, scanning numbers, and total laser energy density: (**a**) 3000 mm/s, 1 pass, 1.86 J/cm^2; (**b**) 3000 mm/s, 2 pass, 3.71 J/cm^2; (**c**) 3000 mm/s, 3 pass, 5.57 J/cm^2; (**d**) 6000 mm/s, 5 pass, 4.64 J/cm^2; (**e**) 6000 mm/s, 6 pass, 5.57 J/cm^2; (**f**) 6000 mm/s, 7 pass, 6.50 J/cm^2.

The ablation threshold and the surface retreat threshold with different scan modes are shown in Table 1. Table 1 shows that the single-pass mode has a lower ablation threshold and a lower surface retreat threshold compared to the multi-pass mode. For the multi-pass mode, a faster laser scanning speed, or a lower laser energy density of a single-pass F_{sin}, causes a higher ablation threshold and a higher surface retreat threshold. Meanwhile, by increasing the scanning number, the multi-pass mode with a low F_{sin} can achieve the same ablation effect as the laser scan mode with a high F_{sin} [39]. These phenomena are attributed to the incubation effect of laser ablating PMMA.

Table 1. The ablation threshold and the surface retreat threshold of different laser scan modes.

Laser Scan Mode	Ablation Threshold (J/cm^2)	Surface Retreat Threshold (J/cm^2)
Single pass	2.14	3.10
Multi-pass at 3000 mm/s	3.71	5.57
Multi-pass at 6000 mm/s	5.57	6.50

The authors believe that regardless of whether it is a continuous CO_2 laser or a ns pulsed CO_2 laser, the machining channels on PMMA mainly utilize the photothermal effect of the interaction mechanism between the laser and PMMA. The laser energy absorbed by PMMA is converted into material thermal energy. When sufficient thermal energy is deposited onto the material, a series of reactions such as pyrolysis, ablation, and gasification occur, ultimately leading to the material removal from the channel. Meanwhile, there are other mechanisms such as heat conduction, surface convective cooling, and environmental radiation that lead to a decrease in material temperature during the ablation process. Therefore, decreasing the laser scanning speed and increasing the pulse overlap rate can promote the deposition of material thermal energy in a short period of time until complete material ablation is achieved, thereby improving the ablation efficiency of the laser energy. On the other hand, under the same total laser energy density F_{tot}, increasing the laser scanning speed and the scanning number and decreasing the pulse overlap rate may lead to more heat loss and dissipation by thermal conduction, surface convective cooling, and environmental radiation, thereby increasing the ablation threshold and surface retreat threshold.

3.2. Thermal Decomposition Behavior of PMMA

Figure 5a,d,g show that the surface of PMMA after laser ablation can be divided into three regions, which are called the ablation zone, incubation zone, and virgin zone in this paper. In the ablation zone, there are dense pores and microcavity, while the entire surface retreats. The virgin zone is far away from the laser irradiation area. There is no visible change between the virgin zone and the unirradiated sample. The incubation zone is located between the ablation zone and the virgin zone, and there are clear boundaries between them. In the incubation zone, the material surface slightly expands outward. Meanwhile, some small pores appear at the boundary between the ablation zone and the incubation zone. It is indicated that the polymer in the virgin zone does not undergo the thermal decomposition reaction. The polymer in the incubation zone undergoes the early stages of the thermal decomposition reaction. Moreover, a part of polymer in the ablation zone undergoes the complete thermal decomposition reaction and is ultimately removed.

The bulge formation noted at the boundary between the ablation zone and the incubation zone is probably due to two possible reasons [34]. One possible reason is the softening and expansion of the polymer after laser irradiation. A second possible reason is that the molten polymer is ejected from the center to the edges of the ablation zone by gaseous products from the decomposition reaction.

PMMA melts at 130 °C, and it almost does not undergo decomposition reactions up to 200 °C. When the temperature is between 220 °C and 300 °C, PMMA mainly undergoes a depolymerization reaction, which decomposes PMMA into methyl methacrylate (MMA) monomers. Within this temperature range, the depolymerization reaction accounts for 95% of the mass decrease in PMMA. Meanwhile, due to the boiling point of MMA being 100 °C, the MMA generated from the depolymerization reaction is maintained as a volatile gas during the entire decomposition reaction. When the temperature is between 360 °C and 400 °C, the MMA monomer further decomposes to yield small molecular gaseous products, such as methane (CH_4), methanol (CH_4O), formaldehyde (CH_2O), etc. Finally, the gaseous products undergo combustion to yield the final products, such as CO_2, CO, H_2O, and energy [41].

It can be found that the density and size of pores inside the channel increase as the number of laser scans increases from 2 to 12 as shown in Figure 5b,c,e,f,h,i. In Figure 5c, a sheet-like PMMA is noted that has melted and resolidified inside and outside the channel at a scanning speed of 3000 mm/s with 2 passes of the laser scan. It is indicated that the low temperature of PMMA leads to a low depolymerization reaction level due to a low total laser energy density F_{tot}. Hence, a portion of the polymer in the ablation zone melts and resolidifies without undergoing the depolymerization reaction. A part of the molten polymer leaves the channel due to splashing and then resolidifies in the incubation zone. In Figure 5f,i, small pores appear inside the large pores. The extent of depolymerization reaction is high with a high number of laser scans. Only a small amount of resolidified sheet-like PMMA appears in the channel, with most concentrating at the boundary between the ablation zone and the incubation zone as shown in Figure 5e,h. It is indicated that the PMMA in the ablation zone center undergoes a thorough depolymerization reaction as the total laser energy density F_{tot} increases. A large amount of polymer directly transitions from solid to gas due to incubation effect from the previous laser scans. Therefore, it is found that the channel depth and the material porosity of the ablation zone increase as the total laser energy density F_{tot} increases. Meanwhile, it reveals that an increase in the scanning number improves the decomposition reactions of the PMMA under the same laser energy density of a single-pass F_{sin}.

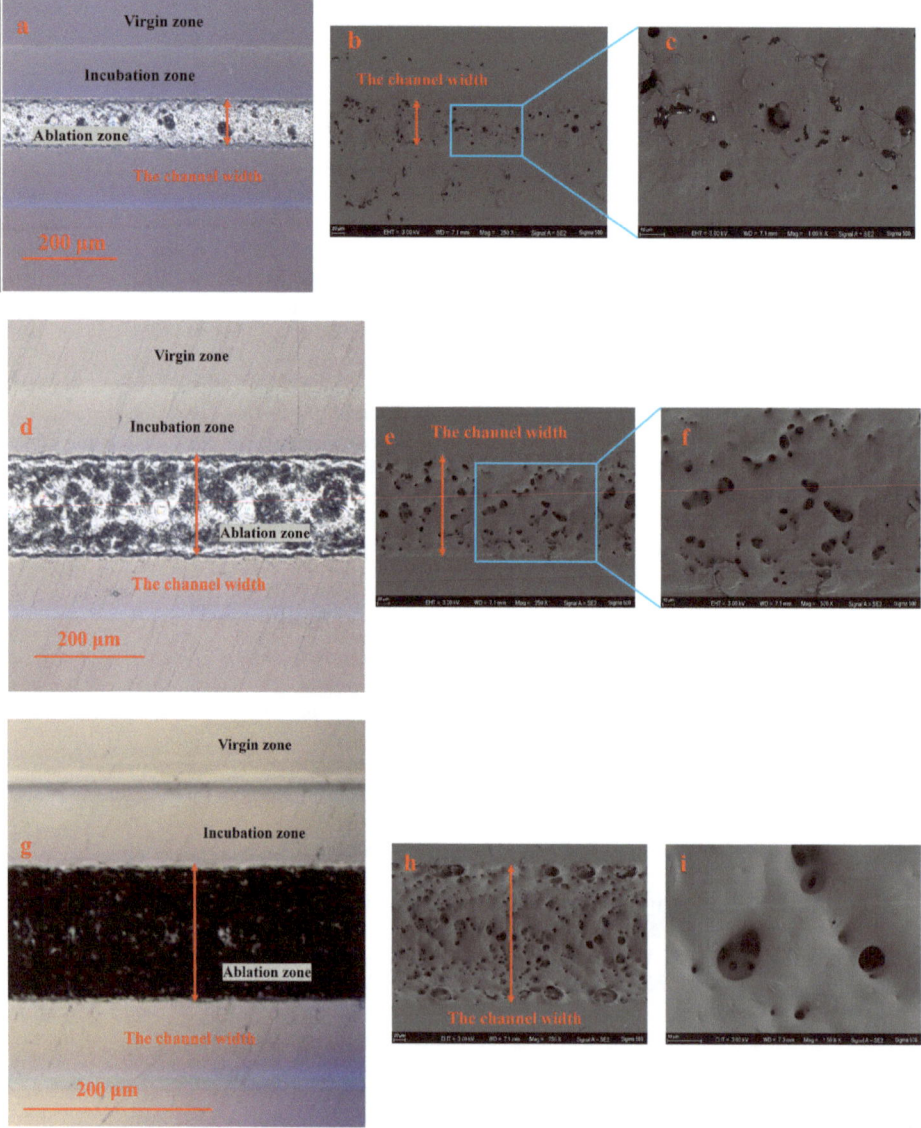

Figure 5. Ablation morphology images obtained using LSCM and SEM: (1) Scanning speed of 3000 mm/s for 2 passes: (**a**) LSCM, 20×; (**b**) SEM, 250X; (**c**) SEM, 1000×; (2) Scanning speed of 3000 mm/s for 4 passes: (**d**) LSCM, 20×; (**e**) SEM, 250×; (**f**) SEM, 500×; (3) Scanning speed of 3000 mm/s for 12 passes: (**g**) LSCM, 20×; (**h**) SEM, 250×; (**i**) SEM, 1500×.

3.3. The Profile, Width, and Depth of the Channel

Figure 6 shows that the cross-sectional profile of the channel is approximately V-shaped at a scanning speed of 250 mm/s to 600 mm/s in single-pass mode. The cross-sectional profile of the channel becomes shallow and gentle as the scanning speed increases. Then, the cross-sectional profile of the channel become approximately trapezoidal at a scanning speed of 700 mm/s to 1000 mm/s. It can be found that the depth, width, and aspect ratio of the channel increase as the laser scanning speed decreases. Figure 7 shows that the cross-sectional profile of the channel gradually changes from trapezoidal to V-shaped as the scanning number increases from 5 to 16 at a scanning speed of 3000 mm/s. It can be found that the depth, width, and aspect ratio of the channel increase as the number of laser scans increases. The result indicates that the higher the total laser energy density F_{tot} is, the greater the aspect ratio of the channel, and the cross-section profile of the channel is closer to V-shaped. The smaller the total laser energy density F_{tot} is, the smaller the aspect ratio of the channel, and the cross-section profile of the channel is closer to trapezoidal.

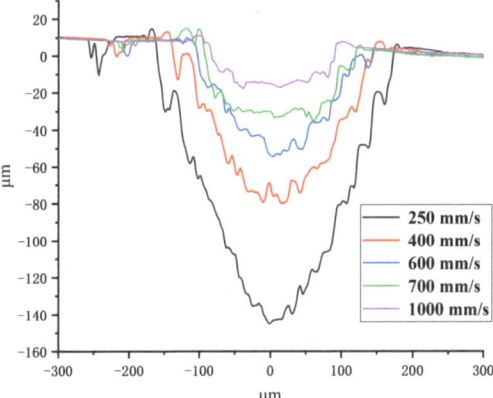

Figure 6. The cross-sectional profile of the channel obtained in single-pass mode at a scanning speed of 250 mm/s to 1000 mm/s.

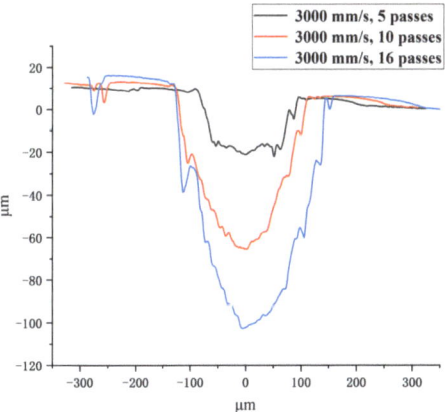

Figure 7. The cross-sectional profile of the channel obtained in multi-pass mode at a scanning speed of 3000 mm/s from 5 passes to 16 passes.

The influence of the total laser energy density F_{tot} on channel width and depth are shown in Figures 8 and 9, respectively. They show that the width and depth of the channel ablated by different single-pass and multi-pass laser scan modes increase as the total laser energy density F_{tot} increases. The width and the depth of the channel obtained in single-pass mode are much larger than those obtained in multi-pass mode under the same total laser energy density F_{tot}. Meanwhile, the width and depth of channel obtained in a multi-pass mode of 3000 mm/s are larger than those obtained at 6000 mm/s under the same total laser energy density F_{tot}. As mentioned earlier, according to the theory of laser ablation and thermal deposition, under the same total laser energy density F_{tot}, the higher the laser energy density of single pass F_{sin} is, the higher the utilization efficiency of laser energy for material ablation, and the less heat dissipated during the channel manufacturing process. Therefore, the channel width and depth of single-pass mode are greater than those of multi-pass mode. The channel width and depth obtained at 3000 mm/s are greater than those obtained at 6000 mm/s under the same total laser energy density.

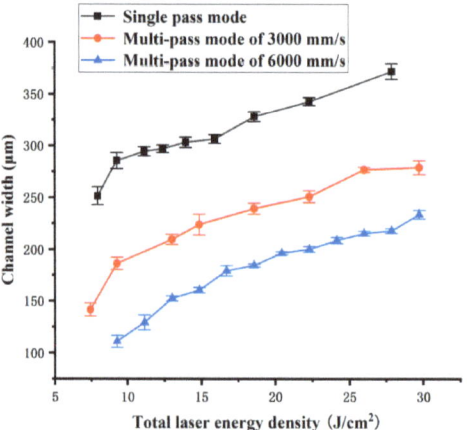

Figure 8. Channel width dependence on the total laser energy density of different laser scan modes.

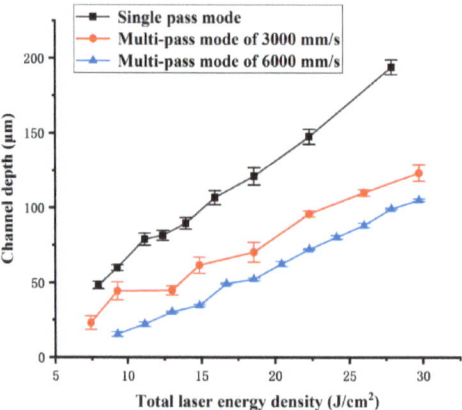

Figure 9. Channel depth dependence on the total laser energy density of different laser scan modes.

In addition, it can be found that the channel width under many laser scan modes is much larger than the laser spot diameter. The authors believe that in the final stage of the PMMA thermal decomposition reaction, the combustion exothermic reaction of small molecule gas products provides additional energy. This allows the surface material outside the laser spot to melt and thermally decompose, thereby expanding the channel width outward.

3.4. Volume Ablation Rate and Mass Ablation Rate

The volume of the channel V can be expressed as follows:

$$V = SL \qquad (5)$$

where S is cross-sectional area of the channel. The volume ablation rate can be expressed as follows:

$$\eta_V = \frac{V}{E_{tot}} \qquad (6)$$

Substituting Equations (3)–(5) into Equation (6), the volume ablation rate η_V can be written as follows:

$$\eta_V = \frac{Sv}{Pn} \qquad (7)$$

The mass ablation rate η_m of the channel can be calculated as follows:

$$\eta_m = \frac{\Delta m}{E_{tot}} \qquad (8)$$

where Δm is the mass loss caused by laser ablation.

Substituting Equations (3) and (4) into Equation (8), the mass ablation rate η_m can be written as follows:

$$\eta_m = \frac{\Delta m v}{PnL} \qquad (9)$$

Figure 10 shows the influence of the laser scanning speed on the cross-sectional area S of the channel obtained in single-pass mode. As the laser scanning speed increases, the cross-sectional area S decreases. According to Equation (2), as the laser scanning speed increases from 100 mm/s to 300 mm/s, the laser energy density F_{sin} acting on the material surface decreases by 37.14 J/cm². However, as the laser scanning speed increases from 300 mm/s to 1600 mm/s, the laser energy density F_{sin} only decreases by 15.09 J/cm². Therefore, the curve in Figure 10 shows a sharp change between 100 mm/s and 300 mm/s, following a slow change between 300 mm/s and 1600 mm/s. The volume ablation rate is calculated using Equation (7), and the influence of the laser scanning speed on the volume ablation rate in single-pass mode is shown in Figure 11. The curve in Figure 11 is approximately linear. In Figure 10, it can be found that there is an approximate functional relationship between the scanning speed and the cross-sectional area, which is noted as follows:

$$S = \frac{\chi}{v^2} \qquad (10)$$

where χ is a constant. Substituting Equation (10) into Equation (7), the volume ablation rate of single pass mode can be written as follows:

$$\eta_V = \frac{\chi}{Pv} \qquad (11)$$

Due to the laser power P in the experiment being a constant, the volume ablation rate is inversely proportional to the scanning speed as shown in Figure 11. It indicates that the volume ablation rate increases as the laser scanning speed decreases. The utilization efficiency of the laser energy used for material ablation is also inversely proportional to the scanning speed.

According to incubation effect, the higher the pulse overlap rate and the lower the laser scanning speed are, the lower the laser ablation threshold of polymer material, which makes it more susceptible to ablation. The incubation effect is consistent given the theory of laser energy thermal deposition. It reveals that the higher the heat flux density from a higher F_{sin} is, the faster the thermal energy deposition inside the material, and the faster the material is ablated. It leads to a lower laser ablation threshold due to rapid ablation reducing thermal dissipation by reducing the interaction time between the laser and the irradiated material.

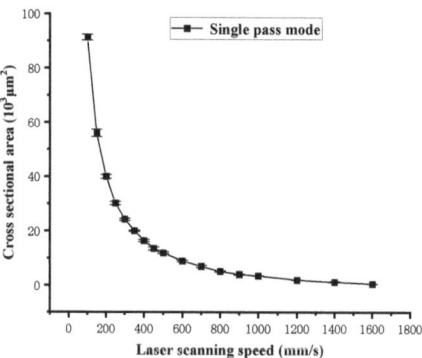

Figure 10. Cross-sectional area dependence on the laser scanning speed in single-pass mode.

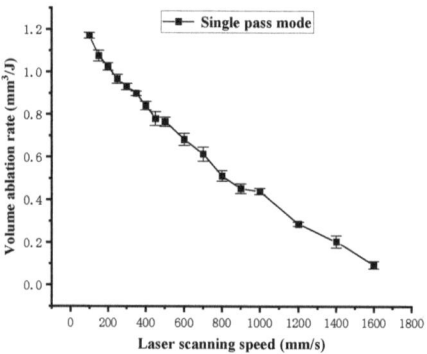

Figure 11. Volume ablation rate dependence on the laser scanning speed in single-pass mode.

Figure 12 shows the influence of the number of laser scans on the cross-sectional area S of the channel obtained in multi-pass mode. As the number of laser scans increases, the cross-sectional area increases. No strict linear relationship is noted between the scanning number and the cross-sectional area of the channel. The cross-sectional area of 3000 mm/s is larger than that of 6000 mm/s under the same laser scan passes. Although the laser energy density F_{tot} of 3000 mm/s is twice that obtained at 6000 mm/s under the same scanning number, the cross-sectional area of 3000 mm/s is more than 5.4 times that obtained at 6000 mm/s. It is attributed to the incubation effect. A higher F_{sin} causes a lower ablation threshold, leading to a faster channel volume increase.

Figure 13 shows the influence of the number of laser scans on the volume ablation rate of the channel for the multi-pass mode. It is found that the overall volume ablation rate increases as the number of laser scans increases, but the curve obtained at 3000 mm/s also shows some downward trends at some specific points. The volume ablation rate obtained at 3000 mm/s is more than 2.6 times that obtained at 6000 mm/s. The curve obtained at 3000 mm/s rapidly increases in the initial stage, indicating a rapid increase in the volume ablation rate due to the incubation effect caused by the early laser scan passes. Compared with the virgin material, the polymer, which has undergone an early depolymerization reaction after some passes of laser scanning, is more susceptible to ablation. Therefore, the early laser scan passes produce a small ablation zone and a large incubation zone, showing a low volume ablation rate. As the number of laser scans increases, the polymer in the incubation zone is ablated rapidly, leading to a rapid increase in the volume ablation rate.

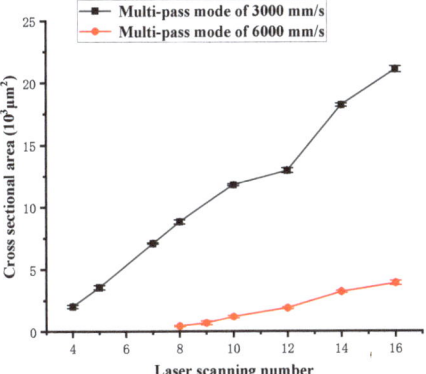

Figure 12. Cross-sectional area dependence on the number of laser scans in multi-pass mode.

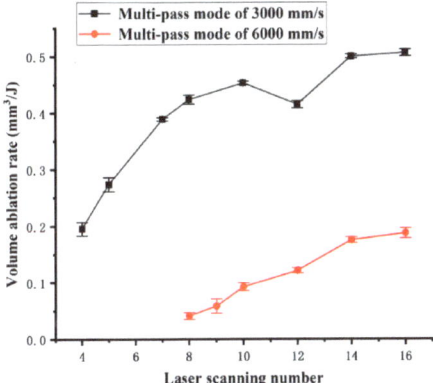

Figure 13. Volume ablation rate dependence on the number of laser scans in multi-pass mode.

Figure 14 shows the influence of the laser scanning speed on the mass loss obtained in single-pass mode. It can be observed that the mass loss decreases as the laser scanning speed increases. Similar to the curve in Figure 10, the curve in Figure 14 shows a sharp change between 50 mm/s and 300 mm/s, following a slow change between 300 mm/s and 1800 mm/s. It is also attributed to the change in the laser energy density F_{sin} in different speed ranges. The mass ablation rate is calculated using Equation (9), and the influence of the laser scanning speed on the mass ablation rate is shown in Figure 15. It shows that the overall mass ablation rate decreases as the laser scanning speed increases, and this curve is similar to the curve in Figure 13. However, the curve in Figure 15 has an upward trend or a plateau at some specific points.

Figure 16 shows the influence of the number of laser scans on the mass loss in multi-pass mode. As the number of laser scans increases, the mass loss increases. A linear relationship is noted between the scanning number and the mass loss of the channel. Although the laser energy density F_{tot} of 3000 mm/s is twice that of 6000 mm/s under the same number of scans, the mass ablation rate of 3000 mm/s far exceeds twice that of 6000 mm/s, even reaching 3.8 times when the number of scans is greater than 5. It is also attributed to the incubation effect, which is similar to the relationship observed in Figure 12.

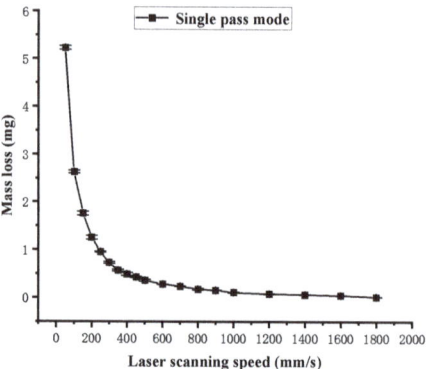

Figure 14. Mass loss dependence on the laser scanning speed in single-pass mode.

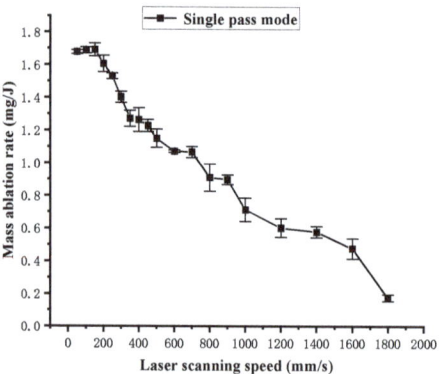

Figure 15. Mass ablation rate dependence on the laser scanning speed in single-pass mode.

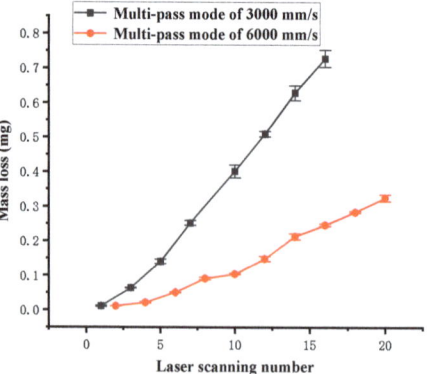

Figure 16. Mass loss dependence on the number of scans in multi-pass mode.

Figure 17 shows the influence of the laser scanning number on the mass ablation rate in the multi-pass mode. The curve obtained at 3000 mm/s rises rapidly at first and then rises slowly as the scanning number increases. However, the curve obtained at 6000 mm/s generally rises as the scanning number increases and exhibits a downward trend at some special points.

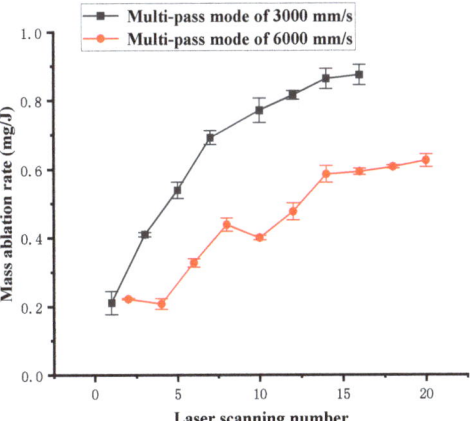

Figure 17. Mass ablation rate dependence on the number of scans in multi-pass mode.

Figures 10–17 indicate that the influence of the scanning speed on the volume ablation rate and the mass ablation rate is greater than that noted for the number of scans. Based on the calculations, the volume ablation rate and the mass ablation rate obtained in single-pass mode are larger than those obtained in multi-pass mode under the same total laser energy density. Due to the fact that the thermal decomposition reaction of PMMA is divided into several stages, there is no linear relationship between the reduction of material volume and mass and the absorbed laser energy. In some cases, the polymer has absorbed an amount of laser energy and accumulated a large of thermal energy; however, the changes in volume and mass may not be significant because the material has not undergone a sufficient depolymerization reaction. It leads to a plateau, a slower change or a reverse in the curves of the volume ablation rate and mass ablation rate in Figures 11, 13, 15 and 17. When the laser scanning number increases and the scanning speed decreases, the heat input of laser irradiation increases, allowing the partially molten and incompletely depolymerized polymer to obtain sufficient energy for further decomposition and gasification. It causes a rapid decrease in the volume and mass of the irradiated material, leading to a rapid change in the volume ablation rate and mass ablation rate curves.

4. Simulation Calculation

The effectiveness of the simulation models proposed by Yuan [30] and Prakash [31] et al. has been validated in the low-power laser manufacturing of PMMA microchannels. This paper combines existing simulation models to derive the profile expression of the channel cross-section after a single-pass laser scan. The laser power density at different positions on the laser focal plane can be expressed as follows:

$$I_s = \frac{\beta P}{\pi R_0^2} e^{-\frac{\beta(x^2+y^2)}{R_0^2}} \tag{12}$$

where β is the laser Gaussian shape coefficient. Here, x and y represent the position coordinates along the channel length and the channel width directions, respectively. The channel depth z_{max} and the depth z at different positions of the channel cross-section can be expressed respectively as follows:

$$z_{max} = \frac{\alpha}{\rho(C_p \Delta T + H_L)} \sqrt{\frac{\beta}{\pi R_0^2} \frac{P}{v}} \tag{13}$$

$$z = z_{max} e^{-\frac{\beta y^2}{R_0^2}} \tag{14}$$

where the laser absorption coefficient α is 0.95, the PMMA density ρ is 1070 kg/m³, the specific heat capacity C_p is 1466 J/(kg K), the temperature difference ΔT from room temperature (20 °C) to completion of the depolymerization reaction (393 °C) is 373 K, and the gasification latent heat H_L is 1800 kJ/kg [31,42]. According to Equation (13), a great consistency in channel depth is achieved between the experimental and simulation results as the laser scanning speed varies from 150 mm/s to 600 mm/s when β is equal to 2.444. The experimental and simulation results of the channel depth are shown in Figure 18. The error percentages between the experimental and simulation results are shown in Table 2. The error percentage of the channel depth varies from 0.44% to 5.9%.

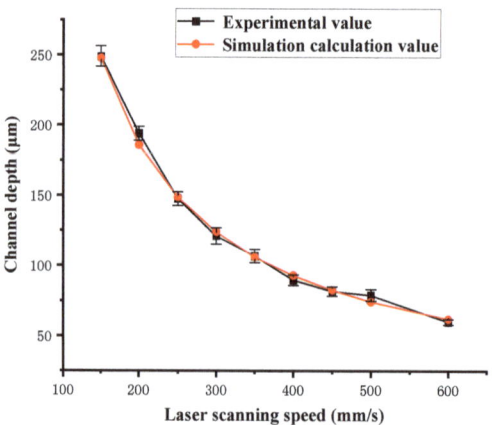

Figure 18. Dependence of channel depth of the experimental values and the simulation calculation values on laser scanning speed in single-pass mode.

Table 2. The error percentages of the channel depth and width between the experimental and simulation results.

Laser Scanning Speed (mm/s)	Error Percentage of Depth (%)	Error Percentage of Width (%)
150	0.44	4.5
200	4.2	5.7
250	0.77	1.5
300	2.4	1.1
350	0.52	2.4
400	3.8	2.7
450	1.3	3.0
500	5.9	1.7
600	3.3	2.9

However, this calculation model ignores the effects of heat conduction, surface convective cooling, environmental radiation, and heat dissipation, so it is not suitable for single-pass laser ablation with a low energy density F_{sin}, and multi-pass laser ablation is severely affected by heat dissipation. Meanwhile, the experimental channel depth is greater than the calculated channel depth when the laser energy density of single-pass F_{sin} is too high, such as 55.7 J/cm² at the scanning speed of 100 mm/s. It is hypothesized that the gasification latent heat used in the calculation model does not include the contribution of exothermic oxidation of the small molecule organic gas in the last stage of the decomposition reaction.

When the laser scanning speed varies from 150 mm/s to 600 mm/s, the channel width is larger than the laser spot diameter. An approximate reciprocal relationship exists between the channel width and the logarithm of the scanning speed. The relationship can be written as follows:

$$w[\mu m] = \frac{c}{\ln(v[s/mm])} \tag{15}$$

where w is the channel width, and c is the correlation coefficient. According to Equation (15), a great consistency in channel width is achieved between the experimental and simulation results when c is equal to 1862.37. The experimental and simulation results of the channel width are shown in Figure 19. The error percentages between the experimental and simulation results are shown in Table 2. The error percentage of the channel width varies from 1.1% to 5.7%. When a high-power laser is used to fabricate channels on the surface of PMMA, the mechanism of channel width variation involves complex heat transfer and thermal decomposition reaction processes. Thus, the channel width cannot be calculated using the simplified models of Yuan [30] and Prakash et al. [31].

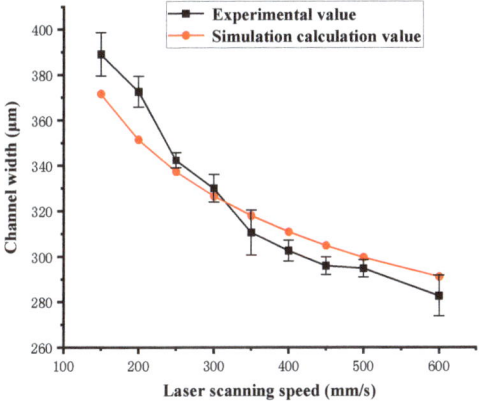

Figure 19. Dependence of channel width of the experimental values and the simulation calculation values on laser scanning speed in single-pass mode.

Combining Equations (13)–(15), a corrected expression describing the cross-sectional profile of the channel obtained in single-pass mode can be obtain as follows:

$$z = \frac{\alpha}{\rho(C_p \Delta T + H_L)} \sqrt{\frac{\beta}{\pi R_0^2} \frac{P}{v}} e^{-\frac{\beta y^2}{w^2}} \tag{16}$$

The cross-sectional profile of the channel is calculated using Equation (16), and the experimental and simulation results are shown in Figure 20. A great consistency in channel profile is achieved between the experimental and simulation results. Meanwhile, the effectiveness of the model proposed in this paper has been validated when the laser scanning speed varies from 150 mm/s to 600 mm/s.

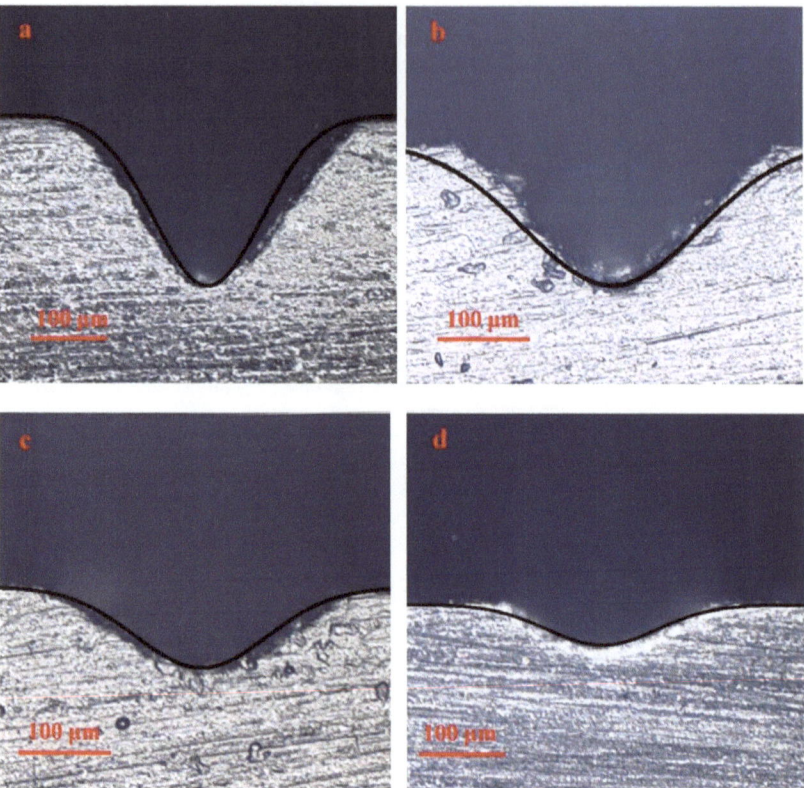

Figure 20. The cross-sectional profile of the channels obtained by using different laser scanning speeds: (**a**) 150 mm/s; (**b**) 250 mm/s; (**c**) 350 mm/s; (**d**) 600 mm/s.

5. Conclusions

The ablation threshold and the ablation morphology of PMMA are investigated under single-pass laser scanning and multi-pass laser scanning, separately. It is found that a higher laser energy density of a single-pass laser scan causes a lower ablation threshold. The ablated surface can be divided into three regions: the ablation zone, incubation zone, and virgin zone. The ablation process of PMMA is mainly attributed to the thermal decomposition reactions, as well as the splashing of molten polymer.

The influences of laser scanning speed, scanning number, and energy density on the channel profile are investigated. The depth, width, aspect ratio, volume ablation rate, and mass ablation rate of the channel increase as the laser scanning speed decreases and the laser scanning number increases. Under the same total laser energy density, a higher laser energy density of a single-pass scan causes a larger channel width, depth, volume ablation rate, and mass ablation rate.

The change in the ablation threshold and the differences in the ablation results under the same total laser energy density are attributed to the incubation effect, which is caused by the thermal deposition of laser energy in the polymer. When the laser energy density of a single-pass scan is high enough, less heat dissipation causes more material ablation, leading to a lower ablation threshold and a higher ablation efficiency of the microchannel with laser machining.

Finally, an optimized simulation model is proposed that is used to solve the problem of obtained a channel width greater than spot diameter with single-pass laser scanning. A great consistency in the depth, width, and cross-sectional profile of the channel is achieved between the experimental and simulation results. The error percentage between the experimental and simulation results varies from 0.44% to 5.9%.

Author Contributions: Conceptualization, D.Z. and F.L.; methodology, B.H. and L.C.; software, D.L.; validation, K.Z. and G.F.; formal analysis, X.L.; investigation, X.L. and R.T.; resources, F.L.; data curation, R.T.; writing—original draft preparation, X.L.; writing—review and editing, X.L.; visualization, R.T. and D.L.; supervision, L.C.; project administration, X.L.; funding acquisition, F.L. All authors have read and agreed to the published version of the manuscript.

Funding: This research was funded by the Major Industrial Technology Projects of Wenzhou (Grant No. ZR2022001), the Key Research and Development Program of Zhejiang Province (Grant No. 2022C01058), the Major Scientific and Technological Innovation Projects of Wenzhou (No. ZG2022009), and the National Natural Science Foundation of China (Grant No. 52261135544).

Institutional Review Board Statement: Not applicable.

Data Availability Statement: The original contributions presented in the study are included in the article, and further inquiries can be directed to the corresponding authors.

Conflicts of Interest: Author Xiao Li was employed by the China International Science & Technology Cooperation Base for Laser Processing Robotics. The remaining authors declare that the research was conducted in the absence of any commercial or financial relationships that could be construed as a potential conflict of interest.

References

1. Abendroth, J.M.; Bushuyev, O.S.; Weiss, P.S.; Barrett, C.J. Controlling motion at the nanoscale: Rise of the molecular machines. *ACS Nano* **2015**, *9*, 7746–7768.
2. Xia, Y.; Si, J.; Li, Z. Fabrication techniques for microfluidic paper-based analytical devices and their applications for biological testing: A review. *Biosens. Bioelectron.* **2016**, *77*, 774–789.
3. Kam, W.; Ong, Y.; Lim, W.; Zakaria, R. Laser ablation and waveguide fabrication using CR39 polymer. *Opt. Lasers Eng.* **2014**, *55*, 1–4.
4. Ma, X.; Li, R.; Jin, Z.; Fan, Y.; Zhou, X.; Zhang, Y. Injection molding and characterization of PMMA-based microfluidic devices. *Microsyst. Technol.* **2020**, *26*, 1317–1324.
5. Tanveer, M.; Ambreen, T.; Khan, H.; Kim, G.M.; Park, C.W. Paper-based microfluidic fuel cells and their applications: A prospective review. *Energy Convers. Manag.* **2022**, *264*, 115732.
6. Bruijns, B.; Van Asten, A.; Tiggelaar, R.; Gardeniers, H. Microfluidic devices for forensic DNA analysis: A review. *Biosensors* **2016**, *6*, 41.
7. Al-Aqbi, Z.T.; Yap, Y.C.; Li, F.; Breadmore, M.C. Integrated microfluidic devices fabricated in poly (methyl methacrylate)(PMMA) for on-site therapeutic drug monitoring of aminoglycosides in whole blood. *Biosensors* **2019**, *9*, 19.
8. Goya, K.; Yamachoshi, Y.; Fuchiwaki, Y.; Tanaka, M.; Ooie, T.; Abe, K.; Kataoka, M. Femtosecond laser direct fabrication of micro-grooved textures on a capillary flow immunoassay microchip for spatially selected antibody immobilization. *Sens. Actuators B Chem.* **2017**, *239*, 1275–1281.
9. Su, Y.; Song, Y.; Xiang, L. Continuous-flow microreactors for polymer synthesis: Engineering principles and applications. In *Accounts on Sustainable Flow Chemistry*; Springer: Cham, Switzerland, 2020; pp. 147–190.
10. Van Erp, R.; Kampitsis, G.; Matioli, E. A manifold microchannel heat sink for ultra-high power density liquid-cooled converters. In Proceedings of the 2019 IEEE Applied Power Electronics Conference and Exposition (APEC), Anaheim, CA, USA, 17–21 March 2019; pp. 1383–1389.
11. Chen, P.-C.; Pan, C.-W.; Lee, W.-C.; Li, K.-M. An experimental study of micromilling parameters to manufacture microchannels on a PMMA substrate. *Int. J. Adv. Manuf. Technol.* **2014**, *71*, 1623–1630.
12. Kim, M.; Moon, B.-U.; Hidrovo, C.H. Enhancement of the thermo-mechanical properties of PDMS molds for the hot embossing of PMMA microfluidic devices. *J. Micromech. Microeng.* **2013**, *23*, 095024.
13. Marson, S.; Attia, U.M.; Lucchetta, G.; Wilson, A.; Alcock, J.R.; Allen, D.M. Flatness optimization of micro-injection moulded parts: The case of a PMMA microfluidic component. *J. Micromech. Microeng.* **2011**, *21*, 115024.

14. Kotz, F.; Arnold, K.; Wagner, S.; Bauer, W.; Keller, N.; Nargang, T.M.; Helmer, D.; Rapp, B.E. Liquid PMMA: A high resolution polymethylmethacrylate negative photoresist as enabling material for direct printing of microfluidic chips. *Adv. Eng. Mater.* **2018**, *20*, 1700699.
15. Pan, C.; Chen, K.; Liu, B.; Ren, L.; Wang, J.; Hu, Q.; Liang, L.; Zhou, J.; Jiang, L. Fabrication of micro-texture channel on glass by laser-induced plasma-assisted ablation and chemical corrosion for microfluidic devices. *J. Mater. Process. Technol.* **2017**, *240*, 314–323.
16. Rodríguez, C.F.; Guzmán-Sastoque, P.; Gantiva-Diaz, M.; Gómez, S.C.; Quezada, V.; Muñoz-Camargo, C.; Osma, J.F.; Reyes, L.H.; Cruz, J.C. Low-cost inertial microfluidic device for microparticle separation: A laser-Ablated PMMA lab-on-a-chip approach without a cleanroom. *HardwareX* **2023**, *16*, e00493. [PubMed]
17. Dudala, S.; Rao, L.T.; Dubey, S.K.; Javed, A.; Goel, S. Experimental characterization to fabricate CO_2 laser ablated PMMA microchannel with homogeneous surface. *Mater. Today Proc.* **2020**, *28*, 804–807.
18. Hu, Z.; Chen, X.; Ren, Y. A study on the surface qualities of four polymer substrate microchannels using CO_2 laser for microfluidic chip. *Surf. Rev. Letters.* **2019**, *26*, 1850160.
19. Rakebrandt, J.-H.; Zheng, Y.; Besser, H.; Scharnweber, T.; Seifert, H.J.; Pfleging, W. Laser-assisted surface processing for functionalization of polymers on micro-and nano-scale. *Microsyst. Technol.* **2020**, *26*, 1085–1091.
20. Go'mez, D.; Goenaga, I.; Lizuain, I.; Ozaita, M. Femtosecond laser ablation for microfluidics. *Opt. Eng.* **2005**, *44*, 051105.
21. Singh, S.S.; Samuel, G. Near-infrared femtosecond laser direct writing of microchannel and controlled surface wettability. *Opt. Laser Technol.* **2024**, *170*, 110214.
22. Agarwal, S.; Saxena, N.S.; Kumar, V. Study on effective thermal conductivity of zinc sulphide/poly (methyl methacrylate) nanocomposites. *Appl. Nanosci.* **2015**, *5*, 697–702.
23. Patti, A.; Acierno, D. Thermal conductivity of polypropylene-based materials. In *Polypropylene—Polymerization and Characterization of Mechanical and Thermal Properties*; IntechOpen: Rijeka, Croatia, 2020.
24. Gunel, E.; Basaran, C. Damage characterization in non-isothermal stretching of acrylics. Part II: Experimental validation. *Mech. Mater.* **2011**, *43*, 992–1012.
25. Sojiphan, K. Finite Element Modeling of Residual Stress Formation in Polycarbonate Welds. Ph.D. Thesis, Ohio State University, Columbus, OH, USA, 2008.
26. Passaglia, E.; Kevorkian, H.K. Specific heat of atactic and isotactic polypropylene and the entropy of the glass. *J. Appl. Phys.* **1963**, *34*, 90–97.
27. Moghadasi, K.; Tamrin, K.F.; Sheikh, N.A.; Jawaid, M. A numerical failure analysis of laser micromachining in various thermoplastics. *Int. J. Adv. Manuf. Technol.* **2021**, *117*, 523–538.
28. Chen, X.; Li, T.; Zhai, K.; Hu, Z.; Zhou, M. Using orthogonal experimental method optimizing surface quality of CO_2 laser cutting process for PMMA microchannels. *Int. J. Adv. Manuf. Technol.* **2017**, *88*, 2727–2733.
29. Bilican, I.; Tahsin Guler, M. Assessment of PMMA and polystyrene based microfluidic chips fabricated using CO_2 laser machining. *Appl. Surf. Science.* **2020**, *534*, 147642.
30. Yuan, D.; Das, S. Experimental and theoretical analysis of direct-write laser micromachining of polymethyl methacrylate by CO_2 laser ablation. *J. Appl. Phys.* **2007**, *101*. [CrossRef]
31. Prakash, S.; Kumar, S. Profile and depth prediction in single-pass and two-pass CO_2 laser microchanneling processes. *J. Micromech. Microeng.* **2015**, *25*, 035010.
32. Hedieh, P. Theoretical and experimental investigations of the influence of overlap between the laser beam tracks on channel profile and morphology in pulsed laser machining of polymers. *Optik.* **2018**, *171*, 431–436.
33. Snakenborg, D.; Klank, H.; Kutter, J.P. Microstructure fabrication with a CO_2 laser system. *J. Micromech. Microeng.* **2003**, *14*, 182.
34. Romoli, L.; Tantussi, G.; Dini, G. Experimental approach to the laser machining of PMMA substrates for the fabrication of microfluidic devices. *Opt. Lasers Eng.* **2011**, *49*, 419–427.
35. Anjum, A.; Shaikh, A. Experimental and analytical modeling for channel profile using CO_2 laser considering gaussian beam distribution. *J. Eng. Res.* **2023**, *11*, 100035.
36. Anjum, A.; Shaikh, A.; Tiwari, N. Experimental investigations and modeling for multi-pass laser micro-milling by soft computing-physics informed machine learning on PMMA sheet using CO_2 laser. *Opt. Laser Technol.* **2023**, *158*, 108922.
37. Gómez, D.; Goenaga, I. On the incubation effect on two thermoplastics when irradiated with ultrashort laser pulses: Broadening effects when machining microchannels. *Appl. Surf. Sci.* **2006**, *253*, 2230–2236.
38. Xiao, S.; Gurevich, E.L.; Ostendorf, A. Incubation effect and its influence on laser patterning of ITO thin film. *Appl. Phys. A* **2012**, *107*, 333–338.
39. Singh, S.S.; Khare, A.; Joshi, S.N. Fabrication of microchannel on polycarbonate below the laser ablation threshold by repeated scan via the second harmonic of Q-switched Nd: YAG laser. *J. Manuf. Process.* **2020**, *55*, 359–372.
40. Romoli, L.; Fischer, F.; Kling, R. A study on UV laser drilling of PEEK reinforced with carbon fibers. *Opt. Lasers Eng.* **2012**, *50*, 449–457.

41. Ali, U.; Karim, K.J.B.A.; Buang, N.A. A review of the properties and applications of poly (methyl methacrylate) (PMMA). *Polym. Rev.* **2015**, *55*, 678–705.
42. Xiang, H.; Fu, J.; Chen, Z. 3D finite element modeling of laser machining PMMA. In Proceedings of the 2006 1st IEEE International Conference on Nano/Micro Engineered and Molecular Systems, Zhuhai, China, 18–21 January 2006; pp. 942–946.

Disclaimer/Publisher's Note: The statements, opinions and data contained in all publications are solely those of the individual author(s) and contributor(s) and not of MDPI and/or the editor(s). MDPI and/or the editor(s) disclaim responsibility for any injury to people or property resulting from any ideas, methods, instructions or products referred to in the content.

Article

Analysis of the Dynamic Cushioning Property of Expanded Polyethylene Based on the Stress–Energy Method

Yueqing Xing *, Deqiang Sun and Guoliang Chen

College of Bioresources Chemical and Materials Engineering (College of Flexible Electronics), Shaanxi University of Science & Technology, Xi'an 710021, China
* Correspondence: xingyueqing@sust.edu.cn; Tel.: +86-18700022396

Citation: Xing, Y.; Sun, D.; Chen, G. Analysis of the Dynamic Cushioning Property of Expanded Polyethylene Based on the Stress–Energy Method. *Polymers* **2023**, *15*, 3603. https://doi.org/10.3390/polym15173603

Academic Editors: Célio Pinto Fernandes, Luís L. Ferrás and Alexandre M. Afonso

Received: 20 July 2023
Revised: 27 August 2023
Accepted: 29 August 2023
Published: 30 August 2023

Copyright: © 2023 by the authors. Licensee MDPI, Basel, Switzerland. This article is an open access article distributed under the terms and conditions of the Creative Commons Attribution (CC BY) license (https://creativecommons.org/licenses/by/4.0/).

Abstract: This paper aimed to experimentally clarify the dynamic crushing performance of expanded polyethylene (EPE) and analyze the influence of thickness and dropping height on its mechanical behavior based on the stress–energy method. Hence, a series of impact tests are carried out on EPE foams with different thicknesses and dropping heights. The maximum acceleration, static stress, dynamic stress and dynamic energy of EPE specimens are obtained through a dynamic impact test. Then, according to the principle of the stress–energy method, the functional relationship between dynamic stress and dynamic energy is obtained through exponential fitting and polynomial fitting, and the cushion material constants a, b and c are determined. The maximum acceleration-static stress curves of any thickness and dropping height can be further fitted. By the equipartition energy domain method, the range of static stress can be expanded, which is very fast and convenient. When analyzing the influence of thickness and dropping height on the dynamic cushioning performance curves of EPE, it is found that at the same drop height, with the increase of thickness, the opening of the curve gradually becomes larger. The minimum point on the maximum acceleration-static stress curve also decreases with the increase of the thickness. When the dropping height is 400 mm, compared to foam with a thickness of 60 mm, the tested maximum acceleration value of the lowest point of the specimen with a thickness of 40 mm increased by 45.3%, and the static stress is both 5.5 kPa. When the thickness of the specimen is 50 mm, compared to the dropping height of 300 mm, the tested maximum acceleration value of the lowest point of the specimen with a dropping height of 600 mm increased by 93.3%. Therefore, the dynamic cushioning performance curve of EPE foams can be quickly obtained by the stress–energy method when the precision requirement is not high, which provides a theoretical basis for the design of cushion packaging.

Keywords: expanded polyethylene (EPE); thickness; dropping height; dynamic cushioning performance curve; stress–energy method

1. Introduction

The product is vulnerable to vibration, shock and other influences during the circulation process, thus being damaged. The most effective way to reduce or avoid product damage due to shock, vibration and other mechanical loads is to use cushion packaging materials to protect the product. Cushion packaging materials can convert the dynamic energy generated by external shocks or vibrations into other forms of energy and absorb some of the energy so that the external force or energy acting on the packaged product is reduced to a certain degree to protect the packaged product [1]. In packaging practice, different types of cushion packaging materials are generally used to meet different packaging requirements. Expanded polyethylene (EPE) foam used for cushion packages has the advantages of moderate price, good cushioning performance and easy processing. EPE is mostly used in large and medium-sized electronic products. Expanded polyethylene (EPE) is a low-density, semi-rigid, closed-cell structure of plastic foam with a lightweight, soft, surface and resistance to multiple impacts. It is an excellent packaging material with good cushioning performance, widely used in construction, electronics and other fields [2–5].

The dynamic impact performance can reflect the protection ability of the packaging cushioning materials to withstand external impact and can provide a variety of options for the cushioning packaging design. Dynamic impact curve is an important reference for product cushioning packaging design, which is also the main method to effectively evaluate the cushioning performance of packaging cushioning materials by the density, thickness and dropping height of the three conditions determined. As a necessary reference for packaging design, it needs to obtain multiple dynamic impact curves for comparison and selection [6]. There are many methods to determine the dynamic impact curve, and the most common are ASTM D1596 [7] and GB 8167-2008 [8], but these two methods require a large number of test data, time-consuming and laborious stare. For this reason, cushion material manufacturers rarely update the dynamic impact curves, and some curves are even outdated, which cannot be used to carry out a good cushion packaging design [9]. For this reason, researchers continue to explore and develop some simplified methods to determine the dynamic impact curve, including stress–energy method [1].

Luo lan [10] conducted dynamic compression tests on commonly used buffer materials to study their dynamic compression characteristics. Wu Lijuan and Jiang Shuai [11] studied the dynamic compression characteristics of three commonly used buffer materials, including expanded polystyrene (EPS), EPE and honeycomb paperboard. Yang Shuai [12] also studied the buffering performance of EPE. Deng Zhen [13] studied and analyzed the performance of pulp-molded products under dynamic experimental conditions. Yan Lirong and Xie Yong [14] studied the dynamic buffering performance of honeycomb paperboard and EPE composite material. EPE dynamic impact curve is a curve that studies the relationship between static stress and maximum impact acceleration value, and the curve is related to materials density [15–17], thickness and drop height. In addition, there is a certain relationship between temperature and humidity [18,19].

DAUM [20] proposed the application theory of stress–energy method. Zhang Hui [21] studied the content of stress–energy method. The buffer coefficient curve of foam was determined by linear method. Based on the energy method, Ding Yi [22] took EPE of a certain density as the experimental material and converted the experimental results through Excel software to the maximum acceleration-static stress curve, and on this basis, the dynamic impact curves of EPE with any thickness at a certain height could be obtained. Chen Manru [23] studied the impact results of EPS under different density gradients by stress–energy method. The relationship between the two parameters a and b of the stress–energy method and the density is discussed. A quadratic function relationship of the density and a, b are obtained by mathematical fitting. Wang Jinmei [24] mathematically fitted the experimental results of the stress–energy method. Through the analysis of the experimental curve, the fitting degree of the exponential function and polynomial function is mainly compared, and the fitting degree of the polynomial is finally obtained.

DAUM [25] described the dynamic impact curve by stress–energy method and obtained it through 25 sets of experiments. The dynamic impact curves under different thickness foams and drop height greatly obtained by this method reduce the number of experiments. According to the stress–energy method, Zhang Botao [26] has done a lot of verification and reached the following conclusions: the maximum acceleration and static stress curve are calculated based on the tested data and according to scientific theory, so the horizontal coordinate range of the curve can be from 0 to infinity, including all stress points; The maximum acceleration-static stress curve of the cushion material at any dropping height and thickness can be obtained every time. Yueqing Xing [27] studied the static crushing responses of expanded polypropylene (EPP) foam and found that EPP foam density has a significantly greater influence on static compressive performance than foam thickness. Yueqing Xing [28] studied the dynamic crushing behavior of Ethylene Vinyl Acetate Copolymer Foam (EVA) based on the energy method.

The above studies are either comparative analyses of different cushion materials or studies of the influence of external conditions on cushion materials. In order to make full use of EPE foam, the thickness of EPE should be selected rationally, and the dynamic cushioning property should be understood. The aim of this paper is to analyze the influence of thickness and dropping height on the dynamic cushioning performance of EPE by testing dynamic cushioning curves and fitted dynamic cushioning curves based on stress–energy method. In this paper, the dynamic impact curves of EPE with a density of 18 kg/m^3 under different thicknesses and dropping heights are obtained by exponential fitting and polynomial fitting based on stress–energy method and equipartition dynamic energy domain, which provides an idea and method for other similar cushion materials to obtain the fitted dynamic impact curves quickly.

2. Materials and Methods

2.1. Materials

The EPE foams hereby are acquired from Suzhou Shunsheng Packaging Material Co., Ltd., Suzhou, China, with a density of 18 kg/m^3 and a mean pore size of 75 μm. Specimens have a cross-section of 150 mm × 150 mm and three different thicknesses of 40 mm, 50 mm and 60 mm, respectively. Specimen labels include a letter and 2 sets of numbers: for example, D represents the uniform code of the test specimen, the following two-digit number is the thickness, and the last three-digit number is the dropping height. Thus, specimen D40-400 is a specimen with a thickness of 40 mm and a dropping height of 400 mm.

2.2. Impact Test Equipment and Method

The test equipment is an XG-HC impact testing machine made by Xian Guangbo Testing Equipment Co., Ltd., Xian, China. The experimental method is in strict accordance with GB/T 8167-2008 [8]. The impact test specimens are prepared for more than 24 h in an environment of 27 °C and 72% relative humidity. In all tests, the specimens are placed centered on the center point of the lower pressure plate of the testing machine. The maximum dropping height of the XG-HC testing machine is 1200 mm, and the minimum mass and the maximum mass of the impact sliding table are 7 kg and 50 kg, respectively. Meanwhile, according to GB/T 8167-2008, the thickness of the test specimen should not be less than 25 mm. Under each static stress value, 5 specimens are taken, and each specimen is impacted for 5 times in succession. The first impact acceleration value is discarded, and the average value of the last 4 times is the maximum impact acceleration value of the specimen. In order to fit the curve more accurately, 7~8 points of static stress are taken [13]. As shown in Figure 1, it is the XG-HC impact testing machine system, which is composed of the XG-HC impact testing machine, data acquisition and processing system, charge-amplifier, data collector and tester controller.

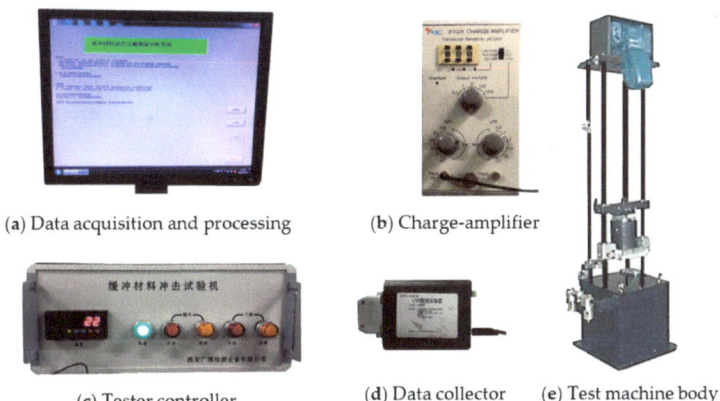

(a) Data acquisition and processing (b) Charge-amplifier

(c) Tester controller (d) Data collector (e) Test machine body

Figure 1. XG-HC impact testing machine system.

2.3. Impact Compression Test Equipment

Figure 2 shows the dynamic impact process of a heavy hammer on EPE foam. The specimen is placed in the middle of the rigid support platform of the testing machine. The heavy hammer is fixed on the sliding table through the fixing device. The weight of the heavy hammer and the dropping height of the sliding table can be set and changed according to the requirements of the test. The heavy hammer and sliding table could drop freely along the smooth guide column of the testing machine, and the dropping height h can be set according to the test requirements. The coordinate system is shown in Figure 2. The data of the acceleration–time curve, stress–strain curve and maximum acceleration-static stress curve can be obtained by the impact testing machine.

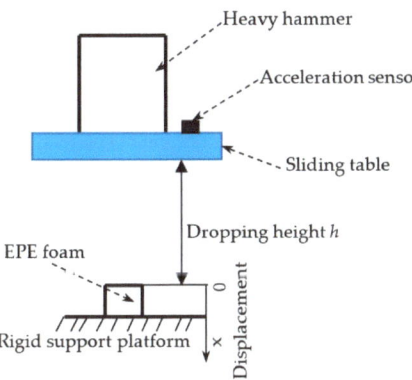

Figure 2. Diagram of drop impact test of heavy hammer [28].

2.4. Impact Characteristic Criteria

The performance of the cushion material is mainly expressed through the dynamic cushion curve [29], and the cushion curve can be used to effectively carry out the cushioning packaging design [30]. The stress–energy method uses the dynamic impact test data to analyze the functional relationship between dynamic stress and dynamic energy in the impact process.

In the process of dynamic impact test, assuming that all the gravitational potential energy of the weight is converted into the dynamic energy of the weight, and after impact, all the dynamic energy of the weight is converted into the deformation energy of the EPE foams, and there is no energy loss in the middle [24], then the deformation energy per unit volume E can be expressed as Equation (1):

$$E = \frac{mgh}{At} = \frac{\sigma_{st} h}{t} \qquad (1)$$

where E is the dynamic energy, kN/m^2; m is the impactor mass, kg; g is the acceleration of gravity, m/s^2; h is the dropping height, m; A is the force area, m^2; σ_{st} is static stress, kPa; t is the thickness, m.

The functional relationship between dynamic stress and static stress can be expressed as Equation (2):

$$\sigma = G \times \sigma_{st} \qquad (2)$$

where σ is the dynamic stress, kPa; G is the acceleration value.

Dynamic stress and dynamic energy can be expressed as an exponential function as Equation (3) [20]:

$$\sigma = f(E) = ae^{bE} \qquad (3)$$

Dynamic stress and dynamic energy can also be expressed as a polynomial function as Equation (4) [24]:

$$\sigma = f(E) = aE^2 + bE + c \qquad (4)$$

where a, b, and c are the material constants, which are parameters determined by the properties of the cushioning material, e is the natural constant, $e = 2.71828$.

The exponential maximum acceleration expression and polynomial maximum acceleration expression of the dynamic impact curve of the specimen can be obtained as Equations (5) and (6):

$$G_m = \frac{\sigma}{\sigma_{st}} = \frac{f(E)}{\sigma_{st}} = \frac{ae^{b\frac{\sigma_{st} h}{t}}}{\sigma_{st}} \qquad (5)$$

$$G_m = \frac{\sigma}{\sigma_{st}} = \frac{f(E)}{\sigma_{st}} = \frac{ah^2}{t^2}\sigma_{st} + \frac{bh}{t} + c\sigma_{st}^{-1} \qquad (6)$$

where G_m is the maximum acceleration.

In the actual dynamic impact process, the impact energy cannot be fully converted into the absorbed energy of the cushioning material, and some of the energy is taken away by the impact rebound of the heavy hammer. The relationship between the gravitational potential energy of the heavy hammer, the deformation energy of the cushioning material and the energy taken away by the impact rebound of the heavy hammer is as follows:

$$mgh = E + \frac{1}{2}mv^2 \qquad (7)$$

where v is the initial velocity at which the heavy hammer begin to bounce, m/s [31].

3. Results

The dynamic performance of the EPE foam is usually expressed by the curve of maximum acceleration-static stress. Maximum acceleration refers to the maximum acceleration transmitted to the protected product under a certain impact. The smaller the acceleration is, the lower the chance of damage to the protected product is, and the better the protection effect of the cushioning material is. All the tests in this paper are established within the elastic range, and due to the limitations of impact mass and dropping height of the dynamic compression testing machine, the relationship between maximum acceleration and static stress of the EPE foam is only explained, which does not represent the maximum impact acceleration and dropping height that the EPE foam can withstand.

The steps of using the stress–energy method are as follows: (1) According to GB/T 8167-2008, the maximum acceleration G is recorded (acceleration G can be obtained directly from the test). The dropping height h, the EPE specimen thickness t, the surface area of the cushion material A, and the weight mass m are known. 5 groups of specimens are prepared for impact test. In order to be precise, 5 specimens with the same m, A, h and t are selected as a group, and each group of single specimens is subjected to 5 times impact tests to obtain the average value of the maximum acceleration. The maximum acceleration-static stress tested curves are obtained. The above process is repeated, and 5 groups of such tests are carried out. According to Equations (1) and (2), the dynamic energy and dynamic stress are calculated. (2) According to the tested dynamic stress and dynamic energy data distribution, exponential fitting and polynomial fitting are carried out according to Equations (3) and (4) to determine the values of EPE material constants a and b. (3) According to Equations (5) and (6), the maximum acceleration-static stress curves are obtained.

3.1. Traditional Dynamic Cushioning Curve Test

The traditional dynamic cushioning curve is tested by the impact tester. The test requires 5 groups of specimens, each group of 5 pieces, and each specimen is impacted at least 5 times. To obtain a maximum acceleration-static stress curve, at least 125 dropping tests need to be performed, which is time-consuming and laborious, and the range of the tested curve is limited [26].

When the dropping height is 400 mm and the thickness of EPE is 40 mm, 50 mm and 60 mm, respectively, the tested maximum acceleration-static stress curves are shown in Figure 3a. The dynamic impact curve of the EPE foam is a "U" shape. The reason is that as the weight of the impactor increases, the static stress increases. During the impact process, the deformation degree of EPE foam increases, the absorbed impact energy increases, and the corresponding maximum acceleration decreases. When the deformation of EPE foam reaches the maximum value, the impact energy absorbed by the material reaches the maximum value, and the maximum acceleration decreases to the minimum. When the weight of the impactor continues to increase, the deformation space of the EPE foam can only absorb a certain amount of impact energy, and the remaining energy will be transferred to the impactor, resulting in an increase in the maximum acceleration and showing an upward trend [32].

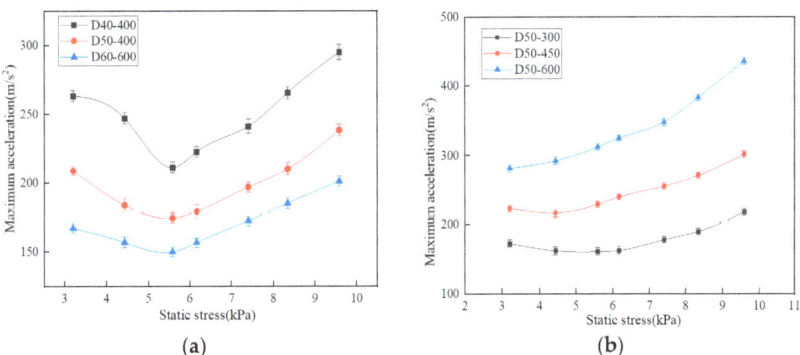

Figure 3. Maximum acceleration-static stress curves (a) specimens with different thicknesses (b) specimens with different dropping heights.

At the same drop height, with the increase of EPE thickness, the overall curve shifts downward. When the specimen thickness is 40 mm, 50 mm and 60 mm, the maximum acceleration corresponding to the lowest point of the maximum acceleration-static stress curve is 211.5 m/s^2, 174.6 m/s^2 and 150.3 m/s^2, respectively, and the static stress of this three kinds of thicknesses is almost the same, 5.63 kPa. It is concluded that the thicker the EPE foam is, the more absorbed energy under the same static stress is, and the better the specimen cushioning performance is [32].

When the specimen thickness is 50 mm, and the dropping height is 300 mm, 450 mm and 600 mm, respectively, the measured maximum acceleration-static stress curves are shown in Figure 3b. For EPE specimens with dropping heights of 300 mm and 450 mm, the trend of the maximum acceleration-static stress curve is to decline first and then rise. The greater the dropping height is, the greater the lowest point corresponding to the maximum acceleration-static stress curve. With the increase of dropping height, the lowest point corresponding to the maximum acceleration-static stress curve moves upward, and the cushioning performance of EPE specimens becomes worse. For the EPE specimen with a dropping height of 600 mm, the trend of the maximum acceleration-static stress curve is gradually increasing, which is different from the other two curves. This is because the concave valley of the curve moves upward to the left with the increase of the dropping height, and the value of the static stress point of the test is limited by the impact testing machine, and a smaller static stress point cannot be obtained.

The curves of maximum acceleration-static stress of specimen with the same thickness shift downward and to the right direction as the drop height decreases. When the drop height increases, the maximum acceleration corresponding to the lowest point of the maximum acceleration-static stress curve gradually increases, and the static stress gradually decreases. When the dropping height of the impactor is 300 mm, 450 mm and 600 mm, the maximum acceleration corresponding to the lowest point of the curve is 160.8 m/s^2, 216.9 m/s and 281.5 m/s^2, respectively, and the static stress is 5.58 kPa, 4.43 kPa, and 3.2 kPa, respectively. It shows that when the specimen thickness and static stress are the same, with the increase of dropping height, the cushioning efficiency and the absorbed energy increase. This is the same trend as that of EPP foams [33].

3.2. Stress–Energy Method

The dynamic stress and dynamic energy test data of specimens with dropping height of 400mm and thickness of 40 mm, 50 mm and 60 mm, respectively, can be obtained from Equations (1) and (2), as shown in Table 1. The dynamic stress and dynamic energy test data of specimens with thickness of 50 mm and dropping height of 300 mm, 450 mm and 600 mm, respectively, is shown in Table 2.

Table 1. Dynamic stress-dynamic energy test data of specimens with different thicknesses.

Specimens	Dynamic Energy (kN/m^2)	Dynamic Stress (kPa)
D40-400	32	843
	44	1095
	56	1180
	62	1372
	74	1786
	83	2215
	96	2826
D50-400	26	669
	35	816
	45	974
	49	1106
	59	1459
	67	1755
	77	2286
D60-400	21	535
	30	696
	37	839
	41	967
	49	1277
	56	1546
	64	1928

Table 2. Dynamic stress-dynamic energy test data of specimens with different dropping heights.

Specimens	Dynamic Energy (kN/m²)	Dynamic Stress (kPa)
D50-300	19	553
	27	718
	33	897
	37	1000
	44	1316
	50	1580
	57	2087
D50-450	29	716
	40	961
	50	1279
	55	1477
	67	1888
	75	2263
	86	2891
D50-600	38	535
	53	696
	67	839
	74	967
	89	1277
	100	1546
	115	1928

Figure 4a shows the dynamic stress-dynamic energy data points for the specimens with different thicknesses. Figure 4b shows the dynamic stress-dynamic energy data points for specimens with different dropping heights. As shown in Figure 4a,b, the dynamic stress-dynamic energy data points are presented by each thickness present curve distribution. In order to explore the function expression of the curve corresponding to the data points, two models of exponential and polynomial are used to fit the data points in Figure 4a,b, and then the fit degree of the two fitted curves was compared.

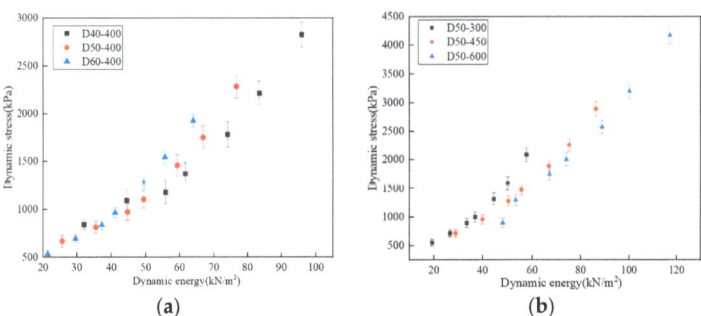

Figure 4. Dynamic stress-dynamic energy data points (a) specimens with different thicknesses (b) specimens with different dropping heights.

The dynamic stress σ_m and dynamic energy E_D of specimens with thicknesses of 40, 50 and 60 mm are exponentially fitted. The resulting functional relationship is as Equation (8):

$$\sigma_m = f(E_D) = 437.69 e^{0.02 E_D} \tag{8}$$

The exponential fitting curve is shown in Figure 5a, in which the material constants a and b are 437.69 and 0.02, respectively.

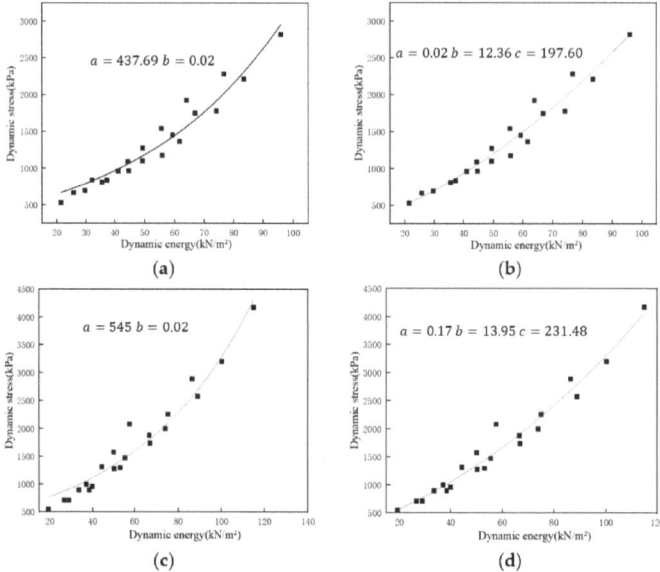

Figure 5. Dynamic stress-dynamic energy fitted curves (**a**) exponential fitted curve for specimens with different thicknesses; (**b**) polynomial fitted curve for specimens with different thicknesses; (**c**) exponential fitted curve for specimens with different dropping heights; (**d**) polynomial fitted curve for specimens with different dropping height.

The dynamic stress σ_m and dynamic energy E_D of specimens with a thickness of 40, 50 and 60 mm, respectively, are fitted by polynomial. The resulting functional relationship is as Equation (9):

$$\sigma_m = f(E_D) = 0.02E_D^2 + 12.36E_D + 197.60 \tag{9}$$

The polynomial fitting curve is shown in Figure 5b, in which the material constants a, b and c are 0.02, 12.36 and 197.60, respectively.

The dynamic stress σ_m and dynamic energy E_D of specimens with dropping heights of 300 mm, 450 mm and 600 mm are exponentially fitted. The resulting functional relationship is as Equation (10):

$$\sigma_m = f(E_D) = 545e^{0.02E_D} \tag{10}$$

The exponential fitting curve is shown in Figure 5c, in which the material constants a and b are 545 and 0.02, respectively.

The dynamic stress σ_m and dynamic energy E_D of specimens with dropping height of 300 mm, 450 mm, and 600 mm, respectively, are fitted by the polynomial. The resulting functional relationship is as Equation (11):

$$\sigma_m = f(E_D) = 0.17E_D^2 + 13.95E_D + 231.48 \tag{11}$$

The polynomial fitting curve is shown in Figure 5d, in which the material constants a, b and c are 0.17, 13.95 and 231.48, respectively.

As can be obtained from Equations (1) and (2), the exponential functional relationship between maximum acceleration and static stress of EPE specimens at any dropping height and thickness is as Equations (12) and (13):

$$G_m = \frac{\sigma_m}{\sigma_{st}} = \frac{f(E_D)}{\sigma_{st}} = \frac{437.69e^{0.02E_D}}{\sigma_{st}} \tag{12}$$

$$G_m = \frac{437.69e^{0.02\frac{\sigma_{st}h}{t}}}{\sigma_{st}} \tag{13}$$

According to Equation (13), the maximum acceleration-static stress exponential fitting curve of any four thicknesses for the specimen at the dropping height of 400 mm can be obtained, as shown in Figure 6a.

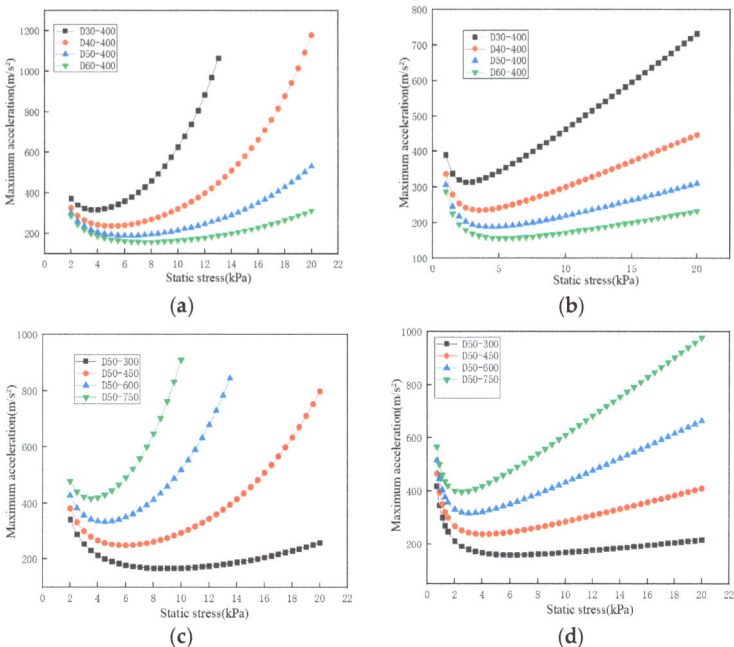

Figure 6. Maximum acceleration-static stress fitted curves (**a**) exponential fitted curve for specimens with different thicknesses; (**b**) polynomial fitted curve for specimens with different thicknesses; (**c**) exponential fitted curve for specimens with different dropping heights; (**d**) polynomial fitted curve for specimens with different dropping height.

As can be obtained from Equations (2) and (9), the polynomial functional relationship between maximum acceleration and static stress of EPE specimens at any dropping height and thickness is as Equations (14) and (15):

$$G_m = \frac{f(E_D)}{\sigma_{st}} = \frac{0.02E_D^2 + 12.36E_D + 197.60}{\sigma_{st}} \tag{14}$$

$$G_m = \frac{0.02h^2}{t^2}\sigma_{st} + \frac{12.36h}{t} + 197.60\sigma_{st}^{-1} \tag{15}$$

According to Equation (16), the polynomial fitting curves of any four thicknesses for the specimen at the drop height of 400 mm can be obtained, as shown in Figure 6b.

As can be obtained from Equations (2) and (10), the exponential functional relationship between maximum acceleration and static stress of EPE specimens at any dropping height and thickness is as Equations (16) and (17):

$$G_m = \frac{f(E_D)}{\sigma_{st}} = \frac{545e^{0.02E_D}}{\sigma_{st}} \tag{16}$$

$$G_m = \frac{545e^{0.02\frac{\sigma_{st}h}{t}}}{\sigma_{st}} \quad (17)$$

According to Equation (17), the exponential fitting curve of any four dropping height for the specimen with the thickness of 50 mm can be obtained, as shown in Figure 6c.

As can be obtained from Equations (2) and (11), the polynomial functional relationship between maximum acceleration and static stress of EPE specimens at any dropping height and thickness is as Equations (18) and (19):

$$G_m = \frac{f(E_D)}{\sigma_{st}} = \frac{0.17E_D^2 + 13.95E_D + 231.48}{\sigma_{st}} \quad (18)$$

$$G_m = \frac{0.17h^2}{t^2}\sigma_{st} + \frac{13.95h}{t} + 231.48\sigma_{st}^{-1} \quad (19)$$

According to Equation (19), the polynomial fitting curves of any four dropping height for the specimen with the thickness of 50 mm can be obtained, as shown in Figure 6d.

In Figure 7, the maximum acceleration-static stress curves obtained by the test and two kinds of fitting methods, respectively, are compared for specimens with thicknesses of 50 mm and dropping heights of 400 mm. The maximum acceleration and static stress data corresponding to the lowest point of the maximum acceleration-static stress curves and fitting parameters R^2 in Figure 7 are compared, as shown in Table 3.

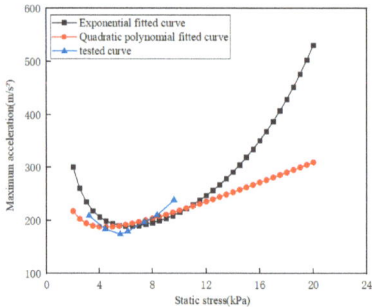

Figure 7. Comparison of fitted curves with tested curves for EPE specimen.

Table 3. Comparison of fitted curves with tested curves for EPE specimen.

Curve Types	Maximum Acceleration (m/s²)	Static Stress (kPa)	R^2
Exponential fitted curve	187.7	6	0.9751
Polynomial fitted curve	188.9	4.5	0.9685
Tested curve	184.6	5.7	

As can be shown from Figure 7 and Table 3, compared with a polynomial fitted curve, the exponential fitted curve is closer to the tested curve. However, when the static stress exceeds 13 kPa, the two kinds of fitted curves are quite different; the reason is that the original dynamic stress-dynamic energy data points of the fitted curve are mainly concentrated in the range of 0–10 kPa, so the fitting effect becomes worse when the static stress exceeds this range.

3.3. Equipartition Dynamic Energy Domain

The dynamic energy value above is scattered and disorderly, and the effect is better in the fitting of specific thickness and height. However, the fitting of EPE specimen curves with arbitrary thickness and drop height is not very good. Because the numerical points of dynamic energy used for fitting are concentrated in part of the interval, once the interval is exceeded, the fitting effect will greatly decrease. Therefore, by dividing the dynamic energy domain equally, the stress–energy method has a better fitting effect at any drop height and thickness.

In this method, 7 groups of dynamic energy values are set in the dynamic energy range of 25–175 kN/m^2, which is 25, 50, 75, 100, 125, 150 and 175 kN/m^2, respectively. For each dynamic energy value, three different sets of s, h, and t are set, which makes the result more accurate. Table 4 shows the test scheme.

Table 4. Testing scheme.

Dynamic Energy (kN·m^{-2})	Static Stress (kPa)	Thickness (mm)	Dropping Height (mm)	Specimen (mm)
25	3.2	40	313	200 × 200
25	3.2	50	391	200 × 200
25	3.2	60	469	200 × 200
50	5.69	40	351	150 × 150
50	5.69	50	439	150 × 150
50	5.69	60	527	150 × 150
75	5.69	40	527	150 × 150
75	5.69	50	659	150 × 150
75	5.69	60	791	150 × 150
100	12.8	40	313	100 × 100
100	12.8	50	391	100 × 100
100	12.8	60	469	100 × 100
125	17	40	294	100 × 100
125	17	50	368	100 × 100
125	17	60	441	100 × 100
150	17	40	353	100 × 100
150	17	50	441	100 × 100
150	17	60	529	100 × 100
175	17	40	412	100 × 100

The impact test is completed according to GBT8168-2008 [8]. The dynamic stress is obtained according to Equation (2), as shown in Table 5.

Table 5. Dynamic energy and dynamic stress.

Dynamic Energy (kN·m^{-2})	Mean Dynamic Stress (kPa)
25	62
50	110
75	158
100	235
125	379
150	488
175	605

Since there is energy loss in the actual impact process, the kinetic energy value is calculated according to Equation (8), and the actual dynamic stress-dynamic energy data are shown in Table 6.

Table 6. Actual dynamic energy and dynamic stress.

Dynamic Energy (kN·m^{-2})	Dynamic Stress (kPa)
20	62
40	110
59	158
78	235
90	379
109	488
127	605

Exponential fitting and polynomial fitting is performed by using the data of dynamic stress and dynamic energy in Table 6 and the functional relationship of maximum acceleration and static stress is obtained as shown in Equations (21) and (23).

$$f(E_D) = 432.47e^{0.02E_D} \tag{20}$$

$$G_m = \frac{f(E_D)}{\sigma_{st}} = \frac{432.47e^{0.02E_D}}{\sigma_{st}} = \frac{432.4714e^{0.02\frac{\sigma_{st}h}{t}}}{\sigma_{st}} \tag{21}$$

$$f(E_D) = 0.39E_D^2 - 13.64E_D + 853.11 \tag{22}$$

$$G_m = \frac{f(E_D)}{\sigma_{st}} = \frac{0.39E_D^2 - 13.63E_D + 853.11}{\sigma_{st}} = \frac{0.39h^2}{t^2}\sigma_{st} - \frac{13.64h}{t} + 853.11\sigma_{st}^{-1} \tag{23}$$

The actually tested curves of EPE specimens with a thickness of 40, 50 and 60 mm are compared with the curves of exponential fitted and polynomial fitted at a dropping height of 400 mm, and the results are shown in Figure 8a,b.

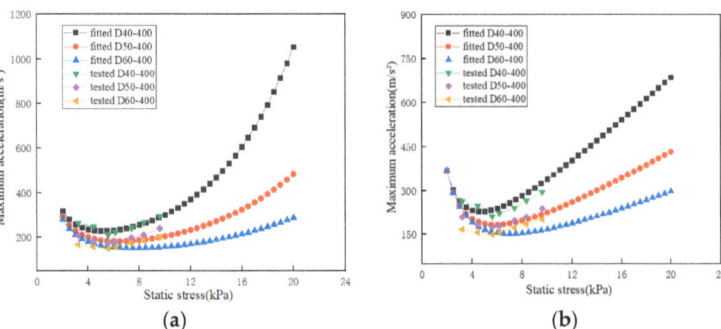

Figure 8. Cont.

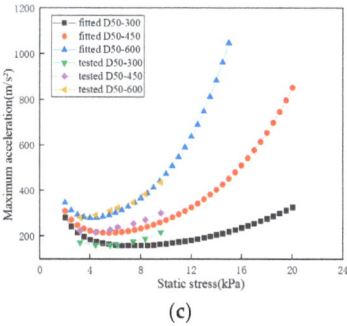
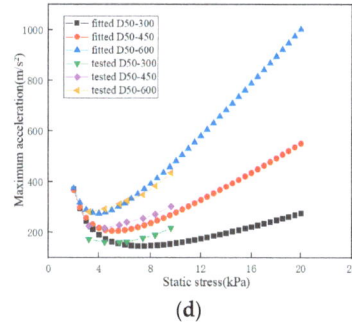

Figure 8. Comparison of fitted curves and tested curves (**a**) exponential fitted curves for EPE specimens of different thicknesses (**b**) polynomial fitted curves for EPE specimens of different thicknesses (**c**) exponential fitted curves for EPE specimens at different dropping height (**d**) polynomial fitted curves for EPE specimens at different dropping height.

The exponential fitting curve is shown in Figure 8a, in which the material constants a and b are 432.47 and 0.02, respectively. The polynomial fitting curve is shown in Figure 8b, in which the material constants a, b and c are 0.39, 13.64 and 853.11, respectively.

The actual tested curves of EPE specimens with a thickness of 50 mm are compared with the fitted curves of exponential fitting and polynomial fitting at the dropping height of 300, 450 and 600 mm. The results are shown in Figure 8c,d. The exponential fitting curve is shown in Figure 8c, in which the material constants a and b are 432.47 and 0.02, respectively. The polynomial fitting curve is shown in Figure 8d, in which the material constants a, b and c are 0.39, 13.64 and 853.11, respectively.

4. Discussion

The fitting coefficient R^2 of the exponential fitted curve is 0.9928, while the fitting coefficient R^2 of the polynomial fitted curve is 0.9880, so the effect of the exponential fitting curve is better than that of the polynomial fitted curve. From the comparison of Figure 8a–d, it can also be seen that the exponential fitted curve is closer to the tested curve than the polynomial fitted curve.

As can be seen from Figure 8a–d, the top curve has the best fitting effect, and the bottom curve has the worst fitting effect. It can be seen from the changing trend of the curves that under the same thickness, with the increase of the dropping height, the opening of the curve also becomes smaller, the bottom of the curve also moves upward to the left, the maximum acceleration corresponding to the lowest point of the maximum acceleration-static stress curves also increases, and the cushion property of the specimens becomes worse.

5. Conclusions

Based on the principle of stress–energy method, the maximum acceleration and static stress curves of EPE specimens with any thicknesses and dropping heights are obtained by exponential fitting and polynomial fitting. The following conclusions are drawn:

(1) Using a dynamic stress-dynamic energy curve to obtain the cushioning characteristics of EPE under any thickness and dropping heights will reduce the number of tests, avoid the system error caused by the test environment and equipment, and obtain stable data. By the equipartition dynamic energy domain method, it can be seen that the fitted curve and the tested curve coincide within a certain range of static stress. Therefore, the maximum acceleration-static stress curve of an EPE specimen of a certain density with any thickness and dropping height can be obtained by exponential fitting and polynomial fitting.

(2) Due to the diversified processing methods of EPE for packaging, the density of EPE is unstable, resulting in different actual densities with different thickness specifications

of the same theoretic density batch. Therefore, there is still a little difference between the fitted maximum acceleration and the actual maximum acceleration in the stress–energy method. In addition, because the test environment and equipment debugging for obtaining the actual acceleration are not suitable for control, too high dropping height, too light and too heavy static load will cause the actual acceleration to deviate from the fitted acceleration phenomenon; therefore, this part of the research needs to be further tested and discussed. The influence of EPE density and other factors on dynamic cushioning performance will be studied later.

Author Contributions: Conceptualization, Y.X.; methodology, Y.X.; validation, Y.X. and D.S.; formal analysis, Y.X. and D.S.; investigation, Y.X. and D.S.; resources, Y.X.; data curation, Y.X. and G.C.; writing—original draft preparation, Y.X.; writing—review and editing, Y.X.; supervision, D.S.; project administration, Y.X. and D.S. All authors have read and agreed to the published version of the manuscript.

Funding: This research was funded by the National Natural Science Foundation of China (51575327).

Institutional Review Board Statement: Not applicable for studies not involving humans or animals.

Informed Consent Statement: Not applicable for studies not involving humans.

Acknowledgments: This work is supported by the Program for Innovative Talents promotion of Shaanxi Province (2017KCT–02), the National First-class Specialty Construction Project (Packaging Engineering, 2022), sub-project of Specialty Comprehensive Reform of Shaanxi Provincial Education Department (Innovative Trial Area with Cultivation modes of Inter-disciplinary and Practical Talents of Packaging Engineering with Strong Ability, 2014), Course Construction Project of Ideological and Political Education of SUST (Fundamentals of Packaging Technology (Bilingual), 2022), Cultivation Project of National (Provincial) High-quality Online Open Course of SUST (Packaging Materials, 2019).

Conflicts of Interest: The authors declare no conflict of interest.

References

1. Yan, S.H.I.; Guang, L.I. Research on Stress–energy Method for Determining Cushioning Curve. *Packag. J.* **2014**, *6*, 35–40.
2. Yang, F.; Li, Z.; Liu, Z.; Zhuang, Z. Shock Loading Mitigation Performance and Mechanism of the PE/Wood/PU/Foam Structures. *Int. J. Impact Eng.* **2021**, *155*, 103904. [CrossRef]
3. Feng, Y. Comparison and Analysis for Double Direction Performances of Cushioning Materials. *China Packag. Ind.* **2007**, *10*, 47–48.
4. Zhang, Y.; Rodrigue, D.; Ait-Kadi, A. High density polyethylene foams. II. Elastic modulus. *J. Appl. Polym. Sci.* **2003**, *90*, 2120–2129. [CrossRef]
5. Ramirez-Arreola, D.E.; Rosa, S.D.L.; Haro-Mares, N.B.; Ramírez-Morán, J.A.; Pérez-Fonseca, A.A.; Robledo-Ortíz, J.R. Compressive strength study of cement mortars lightened with foamed HDPE nanocomposites. *Mater. Des.* **2015**, *74*, 119–124. [CrossRef]
6. Lu, B. Study on static compression buffer curve of foamed polyethylene buffer material. *Packag. Eng.* **2007**, *28*, 42–44.
7. American Society of Testing Materials. ASTMD1596 Standard Test Method for Shock Absorbing Properties of Package Cushioning Materials. In *Annual Book of ASTM Standards*; ASTM International: Conshohocken, PA, USA, 2003.
8. General Administration of Quality Supervision. *Inspection and Quarantine of the Peoples Republic of China, Standardization Administration of the Peoples' Republic of China*; GB 8167-2008 Testing Method of Dynamic Compression for Packaging Cushioning Materials; Standards Press of China: Beijing, China, 2009; pp. 1–3.
9. Navarro-Javierre, P.; Garcia-Romeu-Martinez, M.-A.; Cloquell-Ballester, V.-A.; de-la-Cruz-Navarro, E. Evaluation of Two Simplified Methods for Determining Cushion Curves of Closed Cell Foams. *Packag. Technol. Sci.* **2012**, *25*, 217–231. [CrossRef]
10. Luo, L. Experimental Research on Dynamic Cushioning Performance of the Buffer Material. *Packag. Eng.* **1996**, *17*, 10–13.
11. Wu, L.; Jiang, S. Experimental Research on Dynamic Cushioning Performance of the Three Kinds of Cushioning Materials. *China Water Transp.* **2006**, *11*, 62–64.
12. Yang, S. Research on Cushioning Properties of EPE. Ph.D. Thesis, Tianjin University of Science and Technology, Tianjin, China, 2015.
13. Deng, Z. Research on Performances of Molded Pulp Products under Dynamic Test. Ph.D. Thesis, Tianjin University of Science and Technology, Tianjin, China, 2011.
14. Yan, L.; Xie, Y. Dynamic Cushioning Properties of Combination of Honeycomb Paperboard and EPE. *Packag. Eng.* **2010**, *31*, 13–16.
15. Jin, Q.; Sun, D.; Li, G.; Xing, Y. Effect of density on dynamic impact properties of polypropylene foams. *Packag. Eng.* **2018**, *39*, 88–92.

16. Koohbor, B.; Kidane, A. Design optimization of continuously and discretely graded foam materials for efficient energy absorption. *Mater. Des.* **2016**, *102*, 151–161. [CrossRef]
17. E, Y.P.; Li, Y. Effect of density gradient change on mechanical properties of multilayer foamed polyethylene. *Packag. J.* **2016**, *8*, 30–35.
18. Mcgee, S.D.; Batt, G.S.; Gibert, J.M.; Darby, D. Predicting the Effect of Temperature on the Shock Absorption Properties of Polyethylene Foam. *Packag. Technol. Sci.* **2017**, *30*, 477–494. [CrossRef]
19. Kaewunruen, S.; Ngamkhanong, C.; Papaelias, M.; Roberts, C. Wet/dry influence on behaviors of closed-cell polymeric cross-linked foams under static, dynamic and impact loads. *Constr. Build Mater.* **2018**, *187*, 1092–1102. [CrossRef]
20. Daum, M. *A Simplified Process for Determining Cushion Curves: The Stress–Energy Method*; Hewlett-Packard Company: Palo Alto, CA, USA, 1999.
21. Zhang, H.; Xie, Y.; Wang, J. Determination of Cushioning Performance of Foam Using Improved Stress–Energy Method. *Packag. Eng.* **2011**, *32*, 36–38.
22. Ding, Y.; Chen, L.; Su, J. Discussion on measuring buffer curve of packaging materials based on energy method. *Packag. Food Mach.* **2011**, *29*, 66–69.
23. Chen, M.; Liu, L. Study on Properties of buffer packaging materials based on stress–Energy method. *Packag. Eng.* **2018**, *39*, 44–47.
24. Wang, J.; Liu, C. Study on function model for obtaining foam buffer curve by stress–Energy method. *Packag. Eng.* **2014**, *35*, 79–82.
25. Daum, M. A New Method for Cushion Curves: The Stress–Energy Method. In Proceedings of the International Conference on Transport Packaging, San Antonio, TX, USA, 11–14 December 2006.
26. Zhang, B. Application of stress–energy method in measurement of foam buffer curve. *Packag. Eng.* **2008**, *29*, 59–60, 65.
27. Xing, Y.; Sun, D.; Zhang, M.; Shu, G. Crushing responses of expanded polypropylene foam. *Polymers* **2023**, *15*, 2059. [CrossRef] [PubMed]
28. Xing, Y.; Guo, X.; Shu, G.; He, X. Dynamic Crushing Behavior of Ethylene Vinyl Acetate Copolymer Foam Based on Energy Method. *Polymers* **2023**, *15*, 3016. [CrossRef]
29. Peng, G. *Transportation Packaging Design*; Printing Industry Press: Beijing, China, 1999.
30. Chen, S. The Research Progress of Foaming Material Buffer Performance Characterization Methods. *Print. Qual. Stand.* **2013**, *1*, 25–27.
31. Xu, W. *Packaging Test Technology*, 1st ed.; Printing Industry Press: Beijing, China, 1994.
32. Yin, X.; Chen, Z.; Cui, J.; Zheng, Q. Dynamic Buffering Performance of EPE/Corrugated Board Used in Express Packaging. *Packag. Eng.* **2022**, *43*, 52–57.
33. Sun, D.; Jin, Q.; Li, G. Analysis on Dynamic Cushioning Property of Expanded Polypropylene Materials. *Packag. Eng.* **2019**, *40*, 114–119.

Disclaimer/Publisher's Note: The statements, opinions and data contained in all publications are solely those of the individual author(s) and contributor(s) and not of MDPI and/or the editor(s). MDPI and/or the editor(s) disclaim responsibility for any injury to people or property resulting from any ideas, methods, instructions or products referred to in the content.

MDPI AG
Grosspeteranlage 5
4052 Basel
Switzerland
Tel.: +41 61 683 77 34

Polymers Editorial Office
E-mail: polymers@mdpi.com
www.mdpi.com/journal/polymers

Disclaimer/Publisher's Note: The title and front matter of this reprint are at the discretion of the Guest Editors. The publisher is not responsible for their content or any associated concerns. The statements, opinions and data contained in all individual articles are solely those of the individual Editors and contributors and not of MDPI. MDPI disclaims responsibility for any injury to people or property resulting from any ideas, methods, instructions or products referred to in the content.

www.ingramcontent.com/pod-product-compliance
Lightning Source LLC
LaVergne TN
LVHW072317090526
838202LV00019B/2300